Metalworking Fluids

Second Edition

MANUFACTURING ENGINEERING AND MATERIALS PROCESSING

A Series of Reference Books and Textbooks

SERIES EDITOR

Geoffrey Boothroyd

Boothroyd Dewhurst, Inc.

Wakefield, Rhode Island

1. Computers in Manufacturing, *U. Rembold, M. Seth, and J. S. Weinstein*
2. Cold Rolling of Steel, *William L. Roberts*
3. Strengthening of Ceramics: Treatments, Tests, and Design Applications, *Harry P. Kirchner*
4. Metal Forming: The Application of Limit Analysis, *Betzalel Avitzur*
5. Improving Productivity by Classification, Coding, and Data Base Standardization: The Key to Maximizing CAD/CAM and Group Technology, *William F. Hyde*
6. Automatic Assembly, *Geoffrey Boothroyd, Corrado Poli, and Laurence E. Murch*
7. Manufacturing Engineering Processes, *Leo Alting*
8. Modern Ceramic Engineering: Properties, Processing, and Use in Design, *David W. Richerson*
9. Interface Technology for Computer-Controlled Manufacturing Processes, *Ulrich Rembold, Karl Armbruster, and Wolfgang Ülzmann*
10. Hot Rolling of Steel, *William L. Roberts*
11. Adhesives in Manufacturing, *edited by Gerald L. Schneberger*
12. Understanding the Manufacturing Process: Key to Successful CAD/CAM Implementation, *Joseph Harrington, Jr.*
13. Industrial Materials Science and Engineering, *edited by Lawrence E. Murr*
14. Lubricants and Lubrication in Metalworking Operations, *Elliot S. Nachtman and Serope Kalpakjian*
15. Manufacturing Engineering: An Introduction to the Basic Functions, *John P. Tanner*
16. Computer-Integrated Manufacturing Technology and Systems, *Ulrich Rembold, Christian Blume, and Ruediger Dillman*
17. Connections in Electronic Assemblies, *Anthony J. Bilotta*
18. Automation for Press Feed Operations: Applications and Economics, *Edward Walker*
19. Nontraditional Manufacturing Processes, *Gary F. Benedict*
20. Programmable Controllers for Factory Automation, *David G. Johnson*
21. Printed Circuit Assembly Manufacturing, *Fred W. Kear*
22. Manufacturing High Technology Handbook, *edited by Donatas Tijunelis and Keith E. McKee*

23. Factory Information Systems: Design and Implementation for CIM Management and Control, *John Gaylord*
24. Flat Processing of Steel, *William L. Roberts*
25. Soldering for Electronic Assemblies, *Leo P. Lambert*
26. Flexible Manufacturing Systems in Practice: Applications, Design, and Simulation, *Joseph Talavage and Roger G. Hannam*
27. Flexible Manufacturing Systems: Benefits for the Low Inventory Factory, *John E. Lenz*
28. Fundamentals of Machining and Machine Tools: Second Edition, *Geoffrey Boothroyd and Winston A. Knight*
29. Computer-Automated Process Planning for World-Class Manufacturing, *James Nolen*
30. Steel-Rolling Technology: Theory and Practice, *Vladimir B. Ginzburg*
31. Computer Integrated Electronics Manufacturing and Testing, *Jack Arabian*
32. In-Process Measurement and Control, *Stephan D. Murphy*
33. Assembly Line Design: Methodology and Applications, *We-Min Chow*
34. Robot Technology and Applications, *edited by Ulrich Rembold*
35. Mechanical Deburring and Surface Finishing Technology, *Alfred F. Scheider*
36. Manufacturing Engineering: An Introduction to the Basic Functions, Second Edition, Revised and Expanded, *John P. Tanner*
37. Assembly Automation and Product Design, *Geoffrey Boothroyd*
38. Hybrid Assemblies and Multichip Modules, *Fred W. Kear*
39. High-Quality Steel Rolling: Theory and Practice, *Vladimir B. Ginzburg*
40. Manufacturing Engineering Processes: Second Edition, Revised and Expanded, *Leo Alting*
41. Metalworking Fluids, *edited by Jerry P. Byers*
42. Coordinate Measuring Machines and Systems, *edited by John A. Bosch*
43. Arc Welding Automation, *Howard B. Cary*
44. Facilities Planning and Materials Handling: Methods and Requirements, *Vijay S. Sheth*
45. Continuous Flow Manufacturing: Quality in Design and Processes, *Pierre C. Guerindon*
46. Laser Materials Processing, *edited by Leonard Migliore*
47. Re-Engineering the Manufacturing System: Applying the Theory of Constraints, *Robert E. Stein*
48. Handbook of Manufacturing Engineering, *edited by Jack M. Walker*
49. Metal Cutting Theory and Practice, *David A. Stephenson and John S. Agapiou*
50. Manufacturing Process Design and Optimization, *Robert F. Rhyder*
51. Statistical Process Control in Manufacturing Practice, *Fred W. Kear*
52. Measurement of Geometric Tolerances in Manufacturing, *James D. Meadows*
53. Machining of Ceramics and Composites, *edited by Said Jahanmir, M. Ramulu, and Philip Koshy*
54. Introduction to Manufacturing Processes and Materials, *Robert C. Creese*

55. Computer-Aided Fixture Design, *Yiming (Kevin) Rong and Yaoxiang (Stephens) Zhu*
56. Understanding and Applying Machine Vision: Second Edition, Revised and Expanded, *Nello Zuech*
57. Flat Rolling Fundamentals, *Vladimir B. Ginzburg and Robert Ballas*
58. Product Design for Manufacture and Assembly: Second Edition, Revised and Expanded, *Geoffrey Boothroyd, Peter Dewhurst, and Winston A. Knight*
59. Process Modeling in Composites Manufacturing, *edited by Suresh G. Advani and E. Murat Sozer*
60. Integrated Product Design and Manufacturing Using Geometric Dimensioning and Tolerancing, *Robert Campbell*
61. Handbook of Induction Heating, *edited by Valery I. Rudnev, Don Loveless, Raymond Cook and Micah Black*
62. Re-Engineering the Manufacturing System: Applying the Theory of Constraints, Second Edition, *Robert Stein*
63. Manufacturing: Design, Production, Automation, and Integration, *Beno Benhabib*
64. Rod and Bar Rolling: Theory and Applications, *Youngseog Lee*
65. Metallurgical Design of Flat Rolled Steels, *Vladimir B. Ginzburg*
66. Assembly Automation and Product Design: Second Edition, *Geoffrey Boothroyd*
67. Roll Forming Handbook, *edited by George T. Halmos*
68. Metal Cutting Theory and Practice: Second Edition, *David A. Stephenson and John S. Agapiou*
69. Fundamentals of Machining and Machine Tools: Third Edition, *Geoffrey Boothroyd and Winston A. Knight*
70. Manufacturing Optimization Through Intelligent Techniques, *R. Saravanan*
71. Metalworking Fluids: Second Edition, *Jerry P. Byers*

Metalworking Fluids

Second Edition

edited by

Jerry P. Byers

A CRC title, part of the Taylor & Francis imprint, a member of the Taylor & Francis Group, the academic division of T&F Informa plc.

Published in 2006 by
CRC Press
Taylor & Francis Group
6000 Broken Sound Parkway NW, Suite 300
Boca Raton, FL 33487-2742

Society of Tribologists and Lubrication Engineers
840 Busse Highway
Park Ridge, IL 60068-2376

Printed in the United States of America on acid-free paper
10 9 8 7 6 5 4 3 2 1

International Standard Book Number-10: 1-57444-689-4 (Hardcover)
International Standard Book Number-13: 978-1-57444-689-0 (Hardcover)
Library of Congress Card Number 2005028713

Library of Congress Cataloging-in-Publication Data

Metalworking fluids / edited by Jerry P. Byers.-- 2nd ed.
p. cm. -- (Manufacturing engineering and materials processing ; 71)
Includes bibliographical references and index.
ISBN-13: 978-1-57444-689-0 (alk. paper)
ISBN-10: 1-57444-689-4 (alk. paper)
1. Metal-working lubricants. I. Byers, Jerry P., 1948- II. Series.

TJ1077.M457 2006
671--dc22 2005028713

Visit the Taylor & Francis Web site at
http://www.taylorandfrancis.com

and the CRC Press Web site at
http://www.crcpress.com

Foreword

The first edition of *Metalworking Fluids* was written more than ten years ago and has been one of those key industry references often referred to as "the Bible," in this case for the metalworking industry. One of the reasons the book acquired this status is that it has been one of the few that cover the broad range of topics important to the industry: metalworking fluid technology, application, maintenance, testing methods, health and safety, governmental regulations, recycling, and waste treatment. Indeed, the Society of Tribologists and Lubrication Engineers' (STLE) Metalworking Certification Committee cited this book in its Body of Knowledge as a key reference for preparation for the Certified Metalworking Fluid Specialist® certification examination and, being a peer-reviewed document, as a source for verification of examination questions.

Considerable technical progress has occurred in some areas since the publication of the first edition, and the second edition reflects this progress. For example, more is understood of the microbiology of metalworking fluids and its impact on performance and employee health and safety. Additionally, the waste treatment section has been thoroughly updated, and, as would be expected, the chapter on government regulation, which had become outdated, has now been totally rewritten. Thus, the second edition is very much on target for today's metalworking industry and replaces the first edition in the STLE Certified Metalworking Fluids Specialist® certification program's Body of Knowledge.

Today's metalworking fluids are in fact sophisticated materials. They have to provide lubrication, cooling, and corrosion control in order to machine or form parts at the highest rate of speed, with maximum tool life, minimum downtime, and fewest possible reject parts, while maintaining dimensional accuracy and finish requirements. This well-written book does an impressive job of putting it all into perspective, especially as it impacts value-in-use for a given operation, which further includes the cost of mist control, regulatory compliance, employee health and safety, and waste treatment.

Metalworking Fluids, Second Edition is on the highly recommended list for anyone interested in the metalworking industry. It is written in a style that is easy to read and readily understandable by people with a basic technical background. Thus, machine operators, plant managers, foremen, engineers, chemists, biologists, and hygienists will find the book appealing and informative. Especially important are the many references at the end of each chapter and the extensive glossary of more than 300 terms at the back of the book.

The Society of Tribologists and Lubrication Engineers is proud to team with Taylor & Francis to co-publish this important book.

Dr. Robert Gresham
Director of Professional Development
Society of Tribologists and Lubrication Engineers
Park Ridge, Illinois

Preface

For as long as people have been cutting metal, they have used a fluid to aid in the process. Water may have been the first fluid used, followed by animal fats, vegetable oils, mineral oil, oil-in-water emulsions, and in recent years by clear synthetic chemical solutions. Today, a broad range of coolants and lubricants for metalworking continue to be the key components of the manufacturing process around the world.

Early in my career, I became acutely aware of the importance of metalworking fluids. While walking down the main aisle of a manufacturing plant, I passed a lathe that was dry-machining cast iron. A large, smoking-hot metal chip flew out from the machine and lodged between the inside of my shoe and my foot, burning a good sized hole in my foot. While my foot healed completely after several weeks, the impression upon my memory was permanent. I wondered how the operator, standing much closer to the machine, avoided being seriously injured! Although conventional wisdom held that lubricating fluids were not needed for cast iron machining due to its graphite content, application of a fluid would have prevented this injury, cleaned the metal dust particles out of the air, and prevented flash rusting of the parts.

There are some today who suggest that all metal cutting should be done dry. However, anyone who has spent time visiting manufacturing plants has experienced numerous situations in which a machine was having problems producing quality parts, and either the proper choice of fluids or better application of the fluid made the critical difference. Further, most operations that can be run dry will be greatly improved through the application of the correct fluid when considering the quality of the finished part and productivity (number of parts in the pan at the end of the shift). Experience has shown that it is less expensive to make a part with a fluid than without.

Popular use of the terms "oils" or "cutting oils" to broadly refer to all fluids used for metal cutting is grossly inaccurate. Many are not "oils" at all! The term "metalworking fluid" (MWF) is used in this book. This term includes both metal removal fluids (MRF) designed for cutting and grinding applications, and metal forming fluids used to bend, shape, and stretch metal. By some estimates, about 340 million gallons of metal removal fluids and 185 million gallons of metal forming fluids are in use globally (a total of 210 to 225 million gallons in the U.S.).

This book was written to serve the current needs of industry by presenting a review of the state of the art in metalworking fluid technology, application, maintenance, testing methods, health and safety, governmental regulations, recycling, and waste minimization. First published in 1994, *Metalworking Fluids* was widely acclaimed and considered to be an authoritative resource. More than ten years have passed since that first publication, and it is time to update many subjects and address new issues that have surfaced in recent years. Other texts on the market have tended to ignore or give light treatment to important aspects of the use of fluids for metalworking. It is hoped that this second edition will fill those gaps and cover new ground. The contributors are people well known and respected in the field: formulators, physicians, college professors, fluid users, industry consultants, and suppliers of both chemicals and equipment.

This revised and expanded second edition of the book contains 19 chapters that summarize the latest thinking on various technologies relating to metalworking fluid development, evaluation, and application. Most of the chapters have been updated, and some new ones have been added, and there are several new contributors.

Chapter 1 traces the historical development of the use of lubrication in metalworking. Since metalworking fluids are used to shape various metal alloys, Chapter 2 covers important aspects of the metallurgy of common ferrous and nonferrous metals. Chapters 3, 4, and 5 describe fluid application in metal cutting, grinding, and forming, respectively. Chapter 3 on metal cutting is completely new. Chapter 6 explains the chemistries of straight oil, soluble oil, semisynthetic, and synthetic (nonoil-containing) fluids. Chapter 7 familiarizes the reader with methods for evaluating fluid performance, and includes much new material.

Two aspects of metalworking performance and evaluation are so important and complex that separate chapters have been devoted to them: corrosion control and microbial control (Chapters 8 and 9, respectively). The chapter on microbial control has been completely rewritten to address the latest issues within the industry. Handling aspects of the fluid within a manufacturing facility are covered in Chapters 10, 11, and 12 on the subjects of filtration, management and troubleshooting, and recycling. Disposal of the fluid after a long, useful life is covered in Chapter 13 on waste treatment processes. This newly revised chapter covers some emerging and exciting new technology that is able to eliminate certain organic contaminants from water that were previously quite expensive to remove.

Personal concerns of the machine operator are addressed in the chapters on dermatitis (Chapter 14) and health and safety (Chapter 15). While the basics of skin protection have not changed, Chapter 15 was rewritten to address new health and safety issues. Chapter 16, a new addition from an end-user, investigates air quality and fluid mist in the workplace. Chapter 17 leads the reader through the tangled maze of U.S. government regulations affecting both the manufacture and use of metalworking fluids, explaining the impact of these laws on industry. Chapter 18 is another new addition written by a fluid user, covering the costs and benefits of using metalworking fluids, and addressing some wildly inaccurate fluid cost information being supplied from certain segments of industry. Finally, Chapter 19 offers a comprehensive glossary defining more than 420 terms common to industry and to related disciplines. These terms and definitions were supplied by the contributors of the chapters to which they pertain.

The information provided herein will appeal to a broad readership including machine operators, plant managers, foremen, engineers, chemists, biologists, governmental and industrial hygienists, as well as instructors of manufacturing and industrial disciplines and their students. I hope that this second edition will help modern industry to meet the worldwide competitive demands for improved productivity, improved part quality, reduced manufacturing costs, and a cleaner environment.

Editor

Jerry P. Byers is the Manager of Cimcool® Product Research & Development at Milacron, Inc., in Cincinnati, Ohio. He oversees the laboratory development of synthetic, semisynthetic, soluble oil, and straight oil products for use in the processing of metals, glass, plastics, ceramics, and other materials. He initially joined the company in the area of customer laboratory services, and then became supervisor of the stamping and drawing product development group, before attaining his current position.

Jerry received his bachelor's degree in chemistry from Ball State University, and a master's degree in chemistry from the University of Cincinnati. He is an active member of the Society of Tribologists and Lubrication Engineers (STLE) and an STLE Certified Metalworking Fluids Specialist. He has served on the Board of Directors for that organization, and has held several offices in the Cincinnati section of STLE, including chairman. He has also served as an associate editor for STLE publications and as an instructor in the STLE Metalworking Fluids education course. He holds a patent for a chlorine-free metalworking lubricant package, authored several journal articles, and published the first edition of *Metalworking Fluids* (Marcel Dekker, 1994).

Contributors

Carolina C. Ang
General Motors R&D Center
Warren, Michigan

Giles J.P. Becket
Milacron Global Industrial Fluids
Cincinnati, Ohio

Robert H. Brandt
Brandt & Associates, Inc.
Pemberville, Ohio

John M. Burke
Houghton International
Valley Forge, Pennsylvania

Jean C. Childers
Consultant
Naperville, Illinois

James B. D'Arcy
General Motors R&D Center
Warren, Michigan

Jean M. Dasch
General Motors R&D Center
Warren, Michigan

James E. Denton (Ret.)
Cummins Engine Company
Columbus, Indiana

Raymond M. Dick
Milacron Global Industrial Fluids
Cincinnati, Ohio

Gregory J. Foltz
Milacron Global Industrial Fluids
Cincinnati, Ohio

William A. Gaines
Ford Motor Company
Dearborn, Michigan

John K. Howell
D.A. Stuart Company
Warrenville, Illinois

Lloyd J. Lazarus
Honeywell FM&T, LLC
Kansas City, Missouri

William E. Lucke
Compoundings (ILMA)
Cincinnati, Ohio

C.G. Toby Mathias
Group Health Associates
Cincinnati, Ohio

Jeanie S. McCoy
Consultant
Lombard, Illinois

Frederick J. Passman
Biodeterioration Control Associates, Inc.
Princeton, New Jersey

Stuart C. Salmon
Advanced Manufacturing Science & Technology
Rossford, Ohio

Cornelis A. Smits
Tech Solve
Cincinnati, Ohio

Kevin H. Tucker
Oak International division of Milacron, Inc.
Cincinnati, Ohio

Eugene M. White
Milacron Global Industrial Fluids
Cincinnati, Ohio

Table of Contents

1 Introduction: Tracing the Historical Development of Metalworking Fluids

Jeanie S. McCoy

CONTENTS

I. WHAT ARE THEY?

Metalworking fluids are best defined by what they do. Metalworking fluids are engineering materials that optimize the metalworking process. Metalworking is commonly seen as two basic processes, metal deformation and metal removal or cutting. Comparatively recently, metal cutting has also been considered a plastic deformation process — albeit on a sub-micro scale and occurring just before chip fracture.

In the manufacturing and engineering communities, metalworking fluids used for metal removal are known as cutting and grinding fluids. Fluids used for the drawing, rolling, and stamping processes of metal deformation are known as metal forming fluids. However, the outcome of the two processes differs. The processes by which the machines make the products, the mechanics of the operations, and the requirements for the fluids used in each process, are different.

The mechanics of metalworking govern the requirements demanded of the metalworking fluid. As all tool engineers, metalworking fluid process engineers, and machinists know, the fluid must provide a layer of lubricant to act as a cushion between the workpiece and the tool in order to reduce friction. Fluids must also function as a coolant to reduce the heat produced during machining or forming. Otherwise, distortion of the workpiece and changed dimensions could result. Further, the fluid must prevent metal pick-up on both the tool and the workpiece by flushing away the chips as they are produced. All of these attributes function to prevent wear on the tools and reduce energy requirements. In addition, the metalworking fluid is expected to produce the desired finish on an accurate piece-part. Any discussion of metalworking fluid requirements must include the fact that the manufacturing impetus since the days of the industrial revolution is to machine or form parts at the highest rate of speed with maximum tool life, minimum downtime, and the fewest possible part rejects (scrap), all while maintaining accuracy and finish requirements.

II. CURRENT USAGE IN THE U.S.

The number of gallons of metalworking fluids produced and sold in the U.S. represents a significant slice of the gross national product, as indicated in the 1990 report by the Independent Lubricant Manufacturer's Association. Of the 632 million gallons of lubricants produced by the independent manufacturers, 92 million gallons were metalworking fluids and 32 million gallons were greases, some of which are used in the metal deformation processes.[1]

The National Petroleum Refiners Association, in their annual survey on U.S. lubricating oil sales, reported 2472 million gallons of automotive and industrial lubricants and 56 million gallons of grease sold in 1990; of that, 42% were industrial lubricants. Of the total industrial oil sales, 16% were industrial process oils and 11% were classed as metalworking oils.[2]

These statistics indicate the importance and wide usage of metalworking fluids in the manufacturing world. How they are compounded, used, managed, and how they impact health, safety, and environmental considerations, will be described in subsequent chapters. This chapter will take the reader through the history of the evolution of metalworking fluids, one of the most important and least understood tools of the manufacturing process.

It is surprising that it is not possible to find listings for metalworking fluids in the available databases. The National Technology Information Service, Dialog Information Service, the well-known Science Index, the Encyclopedia of Science and Technology Index, and the *Materials Science Encyclopedia* all lack relevant citations. The real story appears to be buried in technical magazines written by engineers and various specialists for other engineers and specialists, and is obscured in books on related topics. Clearly, this is an indication that this information needs to be collected and published.

III. HISTORY OF LUBRICANTS: EVIDENCE FOR EARLY USAGE OF METALWORKING FLUIDS

The histories of Herodotus and Pliny, and even the Scriptures, indicate that humankind has used oils and greases for many applications. These include lubrication uses such as hubs on wheels, axles, and bearings, as well as for nonlubrication uses such as embalming fluids, illumination, waterproofing of ships, setting of tiles, unguents, and medicines.[3] However, records documenting the use of lubricants as metalworking fluids are not readily available. Histories commonly report that man first fashioned weapons, ornaments, and jewelry by cold working the metal, then as the ancient art of the blacksmith developed, by hot working the metal. Records show that animal and vegetable oils were used by early civilizations in various lubrication applications. Unfortunately, the use of lubricants as metalworking fluids in the metalworking crafts is not described in those early historical writings.[4]

Reviewing the artifacts and weaponry of the early civilizations of Mesopotamia, Egypt, and later the Greek and Roman eras on through the Middle Ages, it is obvious that forging and then wire drawing were the oldest of metalworking processes.[5] Lubricants must have been used to ease the wire drawing process. Since the metalworking fluid is, and always has been, an important part of the process, it may not be unreasonable to presume that the fluids used then were those that were readily available. These included animal oils and fats (primarily whale, tallow, and lard), as well as vegetable oils from various sources such as olive, palm, castor, and other seed oils.[6] Even today, these are used in certain metalworking fluid formulations. Some of the most effective known lubricants have been provided by Nature. Only by inference, since records of their early use has not been found, can we speculate that these lubricants must have been used as metalworking fluids in the earliest metalworking processes.

IV. HISTORY OF TECHNOLOGY

A. Greek and Roman Era

The explanation for the lack of early historical documentation might be found by examining the writings of the ancient Greek and Roman philosophers on natural science. It is readily seen that there was little interest among the "intelligentsia" for the scientific foundations of the technology of the era.

As Singer points out in his *History of Technology*, the craftsman of that era was relegated to a position of social inferiority because knowledge of the technology involved in the craft process was scorned as unscientific. It was neither studied nor documented, evidently not considered as being worthy of preservation.[7] Consequently, the skills and experience of the craftsman became valuable personal possessions to be protected by secrecy; the only surviving knowledge was handed down through the generations.[8]

B. The Renaissance (1450 to 1600)

During the Renaissance, plain bearings of iron, steel, brass, and bronze were used increasingly, especially da Vinci's roller disc bearings in clock and milling machinery as early as 1494; Agricola confirmed the wide use of conventional roller bearings in these applications.[9] Although machines were developed to make these parts, there is no record that any type of metalworking lubricant was used in the bearing, gear, screw, and shaft manufacture. It is possible that those parts which were made of soft metals such as copper and brass did not require much, if any, lubrication in the manufacturing process, but it would seem logical that the finish requirements of iron and steel parts would demand the use of some type of metalworking fluid.

John Schey, in his book *Metal Deformation Processes*, points out that metalworking is probably humankind's first technical endeavor and, considering the importance of lubricants used in the process, he was amazed to find no record of their use until fairly recent times.[10]

C. Toward the Industrial Revolution (1600 to 1750)

It was shortly after the turn of the 17th century that scientific inquiry into the mechanics of friction and wear became the seed that promoted an appreciation for the value of lubrication for moving parts and metalworking processes. The first scant references to lubrication were in the descriptions of power driven machinery (animal, wind, and water) by early experimenters on the nature of friction.

In China, Sung Ying-Hsing (1637) wrote of the advantage of oil in cart axles. Hooke (1685) cautioned on the need for adequate lubrication for carriage bearings, and Amontons (1699) elucidated laws of friction in machines through experimentation. In the same year (1699), De la Hire described the practice of using lard oil in machinery. Desaugliers (1734) suggested that the role of the lubricant was to fill up the imperfections on surfaces and act as tiny rollers, and Leupold (1735) recommended that tallow or vegetable oil should be used for lubricating rough surfaces.[11]

It is interesting to note that although Amontons' endeavors are often considered to be experiments in dry friction, his notes carefully recorded the use of pork fat to coat the sliding surfaces of each experiment. As Dowson points out, Amontons was really studying frictional characteristics of lubricated surfaces under conditions now depicted as boundary lubrication,[12] the mechanism operating most frequently in metalworking operations.[13] These concepts were basic to the development of theories of friction and wear that occurred during the 18th century, culminating in the profound works of Coulomb, who theorized that both adhesion and surface roughness caused friction.

In the 19th century, the means to mitigate friction and wear through lubrication were investigated, leading to the Reynolds' theory of fluid film lubrication. In the early part of the 20th century, Hardy with Doubleday introduced the concept of boundary lubrication, which to this day is still a cornerstone of our current foundation of knowledge on the theory of lubrication.[4] It should be noted that William Hardy's works on colloidal chemistry paved the way for the development of water "soluble" cutting fluids.

However, it was not the development of scientific theory that ultimately led to the explosion of research in this area, and especially on the mechanics of metalworking and metalworking fluids in the 20th century. Rather, it was the wealth of mechanical inventions and evolving technologies that created the need for understanding the nature of friction and wear, and how these effects can be mitigated by proper lubrication.

Interest in craft technologies soared during this period with the founding of the Royal Society of England in 1663 by a group identifying themselves as the "class of new men," interested in the application of science to technology.[8] Their most significant contribution was the sponsorship of *Histories of Nature, Art or Works*, which for the first time contained scientific descriptions of the craft technologies as practiced in the 17th century for popular use. Although the *Histories* published surveys on a wealth of subjects, and long lists of inventions as described by Thomas Sprat, the only reference to a metalworking operation was in the treatise on "An Instrument for Making Screws with Great Dispatch." No mention was made of metalworking fluid usage.[14]

The lack of early information on machining fluids can only be attributed to a reluctance on the part of the craftsman, seen even today on the part of manufacturers, to disclose certain aspects regarding the compounding of the fluids. The revelation of "trade secrets" which might yield a competitive advantage, is not done unless the publicity for market value is seen to outweigh the consequence of competitors learning "how to do it."

Some information on lubrication in metal deformation processes, however, has been documented. K.B. Lewis relates that, in the 17th century, wire drawing was accomplished with grease or oil, but only if a soft, best quality iron was used. High friction probably caused steel wire to break.[15] Around 1650, Johann Gerdes accidentally discovered a method of surface preparation that permitted easy drawing of steel wire. It was a process called "sull-coating" whereby iron was steeped in urine until a soft coating developed. This procedure remained in practice for the next 150 years; later, diluted, sour beer was found to work as effectively. By about 1850 it was discovered that water worked just as well.[16] Although the process of rolling was applied to soft metals as early as the 15th century — and in the 18th century, wire rod was regularly rolled — lubricants were not, and still are not, used for rolling rounds and sections.[17]

Since research into the history of lubrication and the history of technology has not yielded documentation on the early use of metalworking fluids, consideration of the elements involved in the metalworking process led to a search through the history of machine tool evolution for answers. A few surprising facts came to light.

V. EVOLUTION OF MACHINE TOOLS AND METALWORKING FLUIDS

L.T.C. Rolt, writing on the history of machine tools, states unequivocally that through all the ages, the rate of man's progress has been determined by his tools. Indeed, the pace of the industrial revolution was governed by the development of machine tools.[18] This statement is echoed by R.S. Woodbury who points out that historians traditionally have described the political, social, and economic aspects of human endeavor; including the inventions concerned with power transmission, new materials (steel), transportation, and the textile industry. Most have overlooked the technological development of the machine tool "without which the steam engines and other machinery could not have been built, and steels would have little significance."[19]

This same observation could be further extended to include the significance of the technological development of metalworking fluids, without which the machine tool industry could not have progressed to where it is today. The development of metalworking fluids was the catalyst permitting the development of energy-efficient machine tools having the high speed and feed capacities required for today's production needs for extremely fast metal forming and metal cutting operations.

In general, machine tool historians seem to believe that the bow drill was the first mechanized tool as seen in bas-relief and carvings in Egypt in approximately 2500 B.C.[20] The lathe, probably developed from the mechanics of the potter's wheel, can be seen in paintings and woodcuts as early as 1200 B.C.[21] In the Greek and Roman era (first century B.C. and the first century A.D.) the writings of three authors on technical processes describing various mechanisms have survived:

> Hero of Alexandria (50 to 120 A.D.) whose works include mechanical subjects.
> Frontinus (Sextus Julius, 35 B.C. to 37 A.D.) who concentrated on water engineering mechanisms.
> Vitruvius, whose ten books, *De architectura* (31 B.C.) were the only "work of its kind to survive from the Roman world." Book VIII, devoted to water supplies and water engineering, refers to the use of a metalworking fluid. Vitruvius describes a water pump with a bronze piston and cylinders that were machined on a lathe with *oleo subtracti*, indicating the use of olive oil to precision turn the castings.[22]

The first record of a mechanized grinding operation that was accomplished by use of a grinding wheel for sharpening and polishing is evidenced in the Utrecht Psalter of 850 A.D., which depicted a grinding wheel operated by manpower turning a crank mechanism.[23] The first grinding fluid probably was water, used as the basic metal removal process in the familiar act of sharpening a knife on a whetstone, as is still done today.

A. Early Use of Metalworking Fluids in Machine Tools

Undoubtedly, water was used as the cutting fluid as grinding machines became more prevalent. Evidence for this presumption is seen in a 1575 copper engraving by Johannes Stradanus, which is a grinding mill similar to drawings by Leonardo da Vinci. The engraving depicts a shop set up to grind and polish armor where "the only addition appears to be chutes to supply water to some of the wheels."[24]

It was common practice in Leonardo's day to use tallow on grinding wheels. An indication that oil was also used as a metalworking fluid is illustrated in Leonardo's design for an internal grinding machine (the first hint of a precision machine tool) which had grooves cut into the face of the grinding wheel to permit a mixture of oil and emery to reach the whole grinding surface.[25]

The development of machine tools was slow during the following 200 years. In this period, the manufacturing of textiles flourished in England with the invention of Hargreaves' spinning jenny and Awkwright's weaving machinery. Carton Ironworks was founded in 1760, no doubt resulting in improvement of iron smelting and steel making. These inventions, plus the introduction of cast iron shafts in machinery, all gave impetus to design machine tools in order to produce these kinds of new machine parts. Still, by 1775 the available machine tools for industry had barely advanced beyond those that were used in the Middle Ages.[26]

The troubles between England and the colonies that began in 1718, resulted in a series of events that in time actually promoted machine tool development and the use of metalworking fluids. At that time, American colonial pig iron was exported to England. This alarmed the British iron-masters because they considered the colonies a good market for their iron production. They were successful in getting a ban on the importation of American manufactured iron. In addition, in 1750, the government of England prohibited the erection of steel furnaces, plating forges, and rolling mills in the colonies. In 1785, Britain passed laws that prohibited the export of tools, machines, engines, or persons connected with the iron industry or the trades evolving from it to the newly formed U.S.[27] The rationale for this edict was to impact the economy of the colonies by hindering the developing American manufacturing industries and forcing the colonies to purchase English manufactured items. Rather than impeding this American technical development, the British ban stimulated the ingenuity of the American manufacturing pioneers to develop tools, machines, and superior manufacturing skills.

These events encouraged the development of the American textile industry. It was quickened by the inventions of Eli Whitney, first with his cotton gin permitting the use of very "seedy" domestic cotton, followed by his unique system of rifle manufacturing. The munitions industry began to flourish in America. Whitney developed the system of "interchangeable parts," made possible by more precise machining of castings by which parts of duplicate dimension were effected through measurement with standard gauges. Whitney has been called the father of mass production in that he dedicated each machine to a specific machining operation, and then assembled rifles from baskets of parts holding the product of each machine.[28] This system of manufacture was quickly adopted by other American and European manufacturers. Whitney continued to be a forerunner of machine tool invention in order to keep pace with the new manufacturing demand. He is credited with the invention of the first milling machine, a multipoint tool of great value.[29] However, there is no mention of any metalworking fluid used in any of the machining process — probably known only to the machinist as one of the skills of his trade.

B. Growth of Metalworking Fluid Usage

The practice of using metalworking fluids was concomitant with machine tool development both in the U.S. and in England. R.S. Woodbury relates further evidence for the use of water as a metalworking fluid. In 1838, James Whitelaw developed a cylindrical grinding machine for grinding the surface of pulleys wherein "a cover was provided to keep in the splash of water."[30] James H. Nasmyth, in his 1830 autobiography, describes the need for a small tank to supply water, or soap and water, to the cutter to keep it cool. This consisted of a simple arrangement of a can to hold the coolant supply and an adjustable pipe to permit the coolant to drip directly on the cutter.[31]

Woodbury relates that the more common practice of applying cutting fluid during wet grinding (using grinding lathes), was holding a wet sponge against the workpiece. That practice was soon abandoned. A December 1866 drawing shows that a supply of water was provided through a nozzle, and an 1867 drawing shows a guard installed on the slideways of that same lathe to prevent the water and emery from corroding and pitting the slideways.[30] In retrospect, after reviewing the developments in machine tools and machine shop practice, it is obvious that the majority of modern machine tools had been invented by 1850.[26]

C. After the Industrial Revolution (1850 to 1900)

The next 50 years saw rapid growth in the machine tool industry and concurrently in the use of metalworking fluids. This came about as a result of the new inventions of this period, which in response to the great needs for transportation, saw the development of nationwide railways. The next century saw the development of the automobile and aircraft. In order to build these machines, machine tools capable of producing large heavy steel parts were rapidly designed (Figure 1.1).

In this period there was growing awareness of the value of metalworking fluids as a solution to many of the machining problems emerging from the new demands upon the machine tools. However, there were four significant happenings that altogether made conditions ripe for rapid progress in the development of compounded metalworking fluids, which paralleled the sophistication of machine tools.

1. Discovery of Petroleum in the U.S.

One of the most important factors was the discovery of huge quantities of petroleum in the U.S. in 1859, which eventually had a profound influence on the compounding of metalworking fluids. Petroleum at that time was largely refined for the production of kerosene used for illumination and

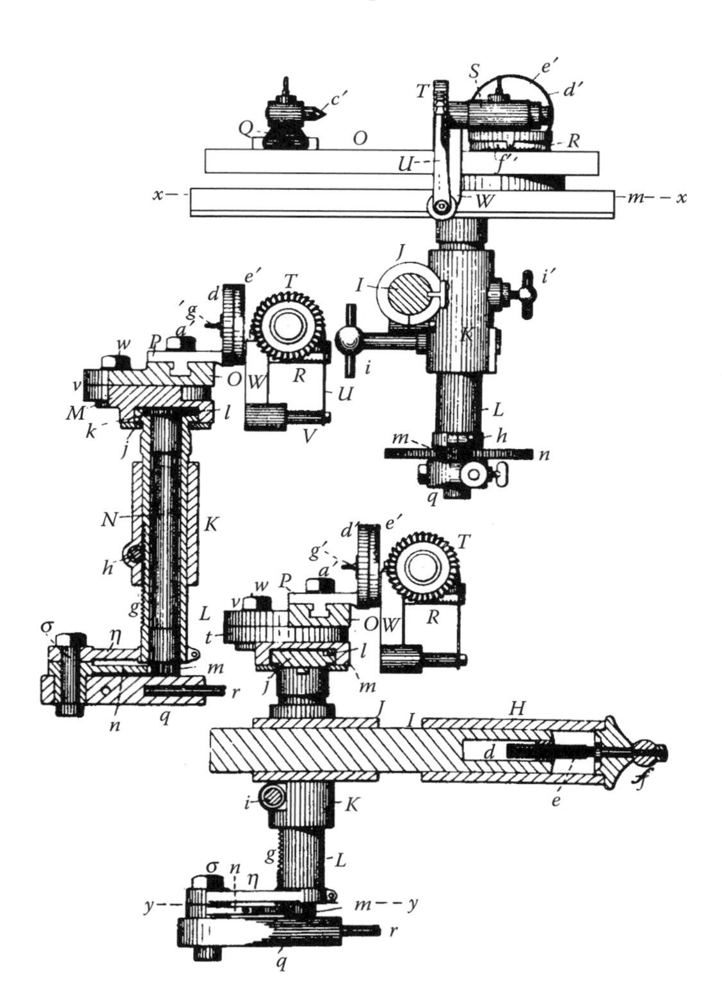

FIGURE 1.1 Diagram of cutter grinder developed by F. Holz, U.S. Patent # 439154 (1890).

fuel. The aftermath of the Civil War with its depressed economic climate led refiners to find a use for oil, which was considered a by-product and had been discarded as useless. This caused an environmental problem for the city of Cleveland. The refiners, forced to find a solution to the oil "problem," induced industry to use oil for lubricant applications, with the result that mineral oils then began to replace some of the popular animal and vegetable oil-based lubricants.[32] During this period, some of today's famous independent lubricant manufacturing companies came into existence, offering a variety of compounded lubricants and cutting oils to improve the machining process and permit greater machine output. Some of these original specialty lubricant manufacturers have since been absorbed into the prevailing industrial conglomerates.[33]

2. Introduction of Better Alloy Steels

The second factor influencing the development of metalworking fluids was the development of alloy steels for making tools. David Mushet, a Scotch metallurgist, developed methods of alloying iron to make superior irons. One of his sons, Robert Forest Mushet, also a metallurgist, founded a method of making Bessemer's pneumatic furnace produce acceptable steels. Some writers claim that Bessemer's furnace was predated by 7 years with the "air-boiling" steels produced by the American inventor, William Kelly.[34]

R.F. Mushet made many contributions to the steel industry with his various patents for making special steels. Perhaps his most important legacy is his discovery that certain additions of vanadium and chromium to steel would cause it to self-harden and produce a superior steel for tool making. In the U.S., Taylor and White experimented with different alloying elements and also produced famous grades of tool steels. The significance is that these tough tool steels permitted tools to be run at faster speeds, enabling increased machine output.[35]

3. Growth of Industrial Chemistry

The third development that had great impact was the budding petrochemical industry. Chemistry had long been involved in the soap, candle, and textile industries. Chemists' endeavors turned to opportunities that the petroleum industry offered, resulting in the creation of a variety of new compounds; many were used in the "new" lubricants needed for growing industrial and manufacturing applications.

4. Use of Electricity as a Power Source

The fourth factor was the development of electric power stations that permitted the use of the electric motor as a power source. Before the use of electric motors to drive machines, power was transmitted by a series of belts to permit variable gearing, and then replaced by the clutch. The electric motor permitted connection directly to machine drive shafts. This eliminated some of the machining problems caused by restricted and inconsistently delivered power, which had resulted in problems such as "chatter." The introduction of steam turbines to drive Edison's dynamos for the generation of electric power in the 1890s[36] was a boon to machine tool designers. Increased sophistication of design and heavier duty capability in machine tools were required in order to produce the machinery needed for the petroleum and electrical power industries, and to make the steam engines and railroad cars for the growing railway transportation ventures. Electric power made the design of more powerful machine tools possible, but the stresses between the tool and the workpiece were increasing in heavy duty machining operations. The need to mitigate these conditions brought about the natural evolution of sophisticated metalworking fluids.

This period also heralded the beginnings of the investigation into the scientific phenomena operating in the metal removal process and the effectiveness of metalworking fluids in aiding the process. Physicists, chemists, mechanical engineers, and metallurgists all contributed to unravel the mystique of what happens during metalworking, and the effect the metalworking fluid has upon the process.

D. Early Experimentation with Metalworking Fluids

It appears that the first known publication on actual cutting fluid applications was in 1868, in *A Treatise on Lathes and Turning* by Northcott. He reported that lathe productivity could be materially increased by using cutting fluids.[37] However, the use of metalworking fluids, especially in metal removal operations, was widespread in both England and the U.S., as evidenced in a report on the Machine Tool Exhibition of 1873 held in Vienna. Mr. J. Anderson, superintendent of the Arsenal at Woolrich, England, wrote that in his opinion the machine tools made in continental Europe were not up to the standards of those in England and America, in that there was a conspicuous absence of any device to supply coolant to the edge of the cutting tool.[38] This observation was confirmed by a drawing of the first universal grinding machine, which was patented by Joseph R. Brown in 1868[39] and appeared in a Brown and Sharpe catalogue of 1875. It included a device for carrying off the water or other fluids used in grinding operations.[40] Obviously, the use of metalworking fluids had become standard machine shop practice.

Curiosity regarding the lubrication effect of metalworking fluids in machining had its beginnings in the publication of the *Royal Society of London Proceedings* in 1882. In that publication, Mallock wondered about the mystery of how lubricants appear to mitigate the effects of friction by going between "the face of the tool and the shaving," noting that it was impossible to see how the lubricant got there.[41] In that same time frame, evidence for the use of various types of oil in metal cutting operations appeared in Robert H. Thurston's *Treatise on Friction and Lost Work in Machining and Mill Work*, which described various formulas for metalworking. For example, he stated that the lubricants used in bolt cutting must have the same qualities as those required for "other causes of lubrication." He cautioned that the choice of lubricant will be determined by the oil giving the smoothest cut and finest finish with "minimum expenditure of power … whatever the market price." His advice was that the best lard oil should be commonly used for this purpose, although he agreed with current practice that mineral oil could be used. Thurston also advised in opposition to "earlier opinions, that in using oil on fast running machinery, the best method is to provide a supply as freely as possible, recovering and reapplying after thorough filtration."[42]

Thurston was an engineer who chaired the Department of Mechanical Engineering at the Stevens Institute of Technology in 1870. His important contributions were in the areas of manufacturing processes, winning him "fame on both sides of the Atlantic."[43] His well-known lubricant testing machines enabled him to provide advice to machinists. Typically, his studies found that sperm oil was superior to lard oil when cutting steel. In cutting cast iron, he recommended a mixture of plumbago (black lead oxide) and grease, claiming a lower coefficient of friction.[44]

It was during this period that chemical mixtures with oils came into usage as metalworking fluids. Most notable was the advent of the sulfurized cutting oils dating back to 1882. The proper addition of sulfur to mineral oil, mineral–lard oil, and mineral–whale oil mixtures, was found to ease the machining of difficult metals by providing better cooling and lubricating qualities and prevented chips from welding onto the cutting tools. Sulfur has the ability to creep into tiny crevices to aid lubrication.[45]

Around this same time, another famous engineer was engaged in an endeavor that forever changed the way machining was carried out and how machine shops were managed. Thuston's contemporary, Fredrick W. Taylor, was a tool engineer in the employ of the Midvale Steel Company, Philadelphia. As foreman of the machine shop, he aspired to discover a method to manage the cutting of metals so that by optimizing machine speeds and work feed rates, production rates could be significantly increased. In 1883, his various experiments in cutting metal proved that directing a constant heavy stream of water at the point of chip removal so increased the cutting speed that the output of the experimental machine rose by 30 to 40%.[46] This was a discovery of prime importance when it is considered that it contradicted Mushet, who insisted that as standard practice his "self hardening" tools must be run "dry."[47] Taylor's experiments revealed that the two most important elements of the machining processes were left untouched by experimenters, even those in academia.[48] Those two elements were the effect of cooling the tool with a rapid cooling fluid, and the contour of the tool.

Taylor published his findings in an epochal treatise *On the Art of Cutting Metal* in 1907, based on the results of 50,000 tests in cutting 800,000 pounds of metal. He reported that the heavy stream of water, which cooled the cutting tool by flooding at the cutting edge, was saturated with carbonate of soda to prevent rusting. The cutting fluid was termed "suds." This practice was incorporated onto every machine tool in the new machine shop built by the Midvale Steel Works in 1884. At Taylor's direction, each machine was set in a cast iron pan to collect the suds, which were drained by piping into a central well below the floor. The suds were then pumped up to an overhead tank from which the coolant was returned to each machine by a network of piping.[49] This was the first central coolant circulation system, the forerunner of those huge 100,000 plus gallons central coolant systems in use today for supplying cutting fluids to automated machine transfer lines in machining centers.

No secret was made of Taylor's coolant system, and by 1900 the idea of a circulating coolant system was copied in a machine designed by Charles H. Norton. It had a built-in suds tank and a pump capable of circulating fifty gallons of coolant per minute, evidence that Norton appreciated the need to avoid heat deformation at high cutting rates.[50]

E. Status of Metalworking Fluids (1900 to 1950)

As a result of engineers seeking more productive machining methods in upgrading the design of machine tools, and metallurgists producing stronger and tougher alloy steels, the compounding of metalworking fluids likewise improved. At the turn of the century, the metalworking fluids industry provided machinists with a choice of several metalworking fluids: straight mineral oils, combinations of mineral oils and vegetable oils, animal fats (lard and tallow), marine oils (sperm, whale, and fish), mixes of free sulfur and mineral oil used as cutting oils, and of course "suds."

The lubricant manufacturers of this era were well versed in the art of grease making, having learned the value of additives as early as 1869 with E.E. Hendrick's patented "Plumboleum," a mixture of lead oxide and mineral oil. Grease, in many cases, was the media of choice used for metal deformation. They were simple compounds, mixtures of metallic soaps, mineral or other oils and fats, and sometimes fibers.[51]

World War I had a significant effect on the course of metalworking fluid development. In the early stages of the European involvement, white oil could no longer be imported from Russia. An American entrepreneur, Henry Sonneborn, who made petroleum jelly and white oil for the pharmaceutical industry since 1903, found his white mineral oil and related products in great demand by lubricant manufacturers.[52] Chemists entered the endeavor by using a chemical process, the acidification of neutral oil with sulfuric acid, which resulted in a reaction product, a mixture of white oil and petroleum sulfonate. The white oil was extracted with alcohol. The sulfonate was discarded until it was discovered to be most useful as a lubricating oil additive and also in compounding metalworking fluids. Sulfonates eventually were found to combine with fatty oils and free fatty acids to make emulsions.[53]

1. Development of Compounded Cutting Oils

As tougher alloy steels became more common, and as machine tool and cutting tool speeds increased, the stresses incurred in the machining process tended to overwork the cutting oils. These were mostly combinations of mineral oils and lard oil, or mixtures of free sulfur and mineral oil.

Overworking caused a chemical breakdown resulting in objectionable odors, rancidity, and very often dermatitis.

The disadvantages of those cutting oils had to be addressed. In 1918, no doubt spurred on by the demands of the munitions industries and the need for greater precision in machining, serious research into better compounding of sulfurized cutting oils began and continued into the late 1920s. The problems to overcome were to extend the limits of sulfur combined with mineral oil by effecting a means of chemically reacting sulfur with the hydrocarbon molecules. This inhibited the natural corrosiveness of sulfur, yet gained the maximum benefit of sulfur for the machining process. In 1924, a special sulfo-chlorinated oil was patented by one of the oldest lubricant compounding companies in the U.S. and marketed as Thread-Kut 99. It is still used today for such heavy duty machining operations as thread cutting and broaching on steels.[54]

However, these chemically compounded oils did not solve all cutting difficulties. The new, highly sulfo-chlorinated cutting oils could not be used for machining brass or copper since sulfur additives stained those metals black and contributed to eventual corrosion.[55]

2. Development of "Soluble Oils"

The worth of Taylor's experience was not lost on the engineering and manufacturing community. His demonstration of the profound effect that an aqueous chemical fluid had on machine productivity began the search for water/oil/chemical-based formulas for metalworking fluids. W.H. Oldacre has written that, although "water-mixed oil" emulsions were used extensively in the first quarter of the 20th century, and the wide range of formulations made a very important contribution to machine shop practice, it is not clear when the first crude emulsions were made by mixing "suds" (soda water) with fatty lubricants. History has neglected the commercial development of soluble oils.[56]

Around 1905, when chemists began to look at colloidal systems, the scientific basics of metalworking fluid formulation began to unfold. Industrial chemists focused their attention upon emulsions, colloidal systems in which both the dispersed and continuous phases are liquids. Two types of emulsions were recognized: a dispersion of oil or hydrocarbon in aqueous material, such as milk and mayonnaise; and dispersions of water in oil such as butter, margarine, and oil field emulsions. Theories of emulsification began with the Surface Tension Theory, the Adsorption Film Theory, the Hydration Theory, and the Orientation Theory put forth by Harkins and Langmuir. These theories explained the behavior of emulsifying agents, which eventually found a direct application in the formulation of cutting fluids.

It has been reported that an English chemist, H.W. Hutton, discovered a way to emulsify oil in water in 1915. What it comprised and how it was made is not described.[57] However, in the U.S. in 1915, an early brochure (Technical Bulletin 16, still available from the Sun Refining Co., Tulsa, Oklahoma) by one of the oldest oil companies claimed the innovation of the first "all petroleum based (naphthenic) soluble oil." This was first marketed under the name of Sun Seco during World War I.

The growing body of knowledge on colloid and surfactant chemistry led to the compounding of various "soluble oils" using natural fatty oils. H.W. Hutton was granted a patent for the process of producing water-soluble oils by compounding sulfonated and washed castor oil with any sulfonated unsaponified fatty oil (other than castor oil), and then saponifying the sulfonated oils with caustic alkali.[58]

After World War I, new developments in lubrication science through the work of Hardy and Doubleday (1919 to 1933) elucidated the mechanism of boundary lubrication.[59] The petrochemical industry began to flourish while applications for new synthetic chemicals, such as detergents and surfactants, found many commercial and industrial uses. The automobile industry recovered. The effort to speed up mass production of cars required stronger machine tools capable of faster cutting speeds. Oil in water emulsions were the preferred fluids, except in heavy duty machining operations such as broaching, gear hobbing, and the thread cutting of tough alloy steels where chemically compounded oils were used.

The need for stable emulsions in the food, cosmetics, and soap making industries, as well as by the metalworking fluid manufacturers, maintained high interest in oil/water emulsions. The research of B.R. Harris, expanding upon the Orientation Theory of emulsions, focused on the synthesis of many new compounds relating their chemical structure to various types of surface modifying activity. Reporting in *Oil and Soap* magazine, Harris established that all fatty interface modifiers have two essential components: a hydrophilic part which makes the compound water-soluble and a lipophilic part which makes the compound fat-soluble. These must be in balance to effect a good emulsion.[60] As research in this area continued, many emulsifying agents were developed for the previously mentioned industries. Some, the amine-soaps, wetting agents, and other special function molecules, were compounded with mineral and/or vegetable oils by metalworking fluid compounders to effect stable "soluble oils."[61]

3. Influence of World War II

With the growth of the aircraft industry, exotic alloys of steel and nonferrous metals were introduced, creating the need for even more powerful machine tools having greater precision capability. Better metalworking fluids to effectively machine these new tough metals were also needed. The circumstances of World War II, which demanded aircraft, tanks, vehicles, and other war equipment, began a production race of unknown precedent. Factories ran 24 h daily, never closing in the race to produce war goods. The effort centered on new machine tool design to shape the new materials and to make production parts as fast as possible. The cover of the February 24, 1941, edition of *Newsweek* magazine featured a huge milling machine carrying the title "The Heart of America's Defense: Machine Tools." In fact, metalworking fluids along with machine tools are at the heart of the cutting process. The demand for more effective war production translated into faster machining speeds. Higher feed rates using the available fluids led to problems such as poor finishes, excessive tool wear, and part distortion. The need to satisfy the war production demand mandated inquiry into the mechanics of the machining process in both Europe and the U.S.

4. Mechanisms of Cutting Fluid Action

In 1938 in Germany, Schallbroch, Schaumann, and Wallichs tested machinability by measuring cutting temperature and tool wear, and in so doing derived an empirical relationship between tool life and cutting tool temperature.[62] In the U.S. in about the same period, H. Ernst, M.E. Merchant, and M.C. Shaw studied the mechanics of the cutting process. Ernst studied the physics of metal cutting and determined that a rough and torn surface is caused by chip particles adhering to the tool causing a built-up edge (BUE) on the nose of the cutting tool due to high chip friction. Application of a cutting fluid lowered the chip friction and reduced or eliminated the BUE.[63] This confirmed Rowe's opinion that the BUE was the most important consideration to be addressed in the machining process.[64] Many studies by many engineers and scientists were made, but the researchers who made the most important discoveries affecting the course of metalworking fluid development were employed by one of the largest machine tool builders in the U.S.

Ernst and Merchant, seeking to quantify the frictional forces operating in metal cutting, developed an equation for calculating static shear strength values.[65] Merchant, in another study, was able to measure temperatures at the chip–tool interface. He found that in this area heat evolves from two sources, the energy used up in deforming the metal and the energy used up in overcoming friction between the chip and the tool. Roughly two thirds of the power required to drive the cutting tool is consumed by deforming the metal, and the remaining third is consumed in overcoming chip friction. Merchant found that the right type of cutting fluid could greatly reduce the frictional resistance in both metal deformation and in chip formation, as well as reduce the heat produced in overcoming friction.[66]

Ernst and Merchant began a 3-year study to scientifically quantify the friction between the cutting tool and the chip it produced. They found temperatures at the tool–chip interface ranging between 1000 and 2000°F (530 to 1093°C) and the pressure at the point was frequently higher than 200,000 psi (1,380,000 kPa).[67] Bisshopp, Lype, and Raynor also investigated the role of the cutting fluid in machining experiments to determine whether or not a continuous film existed in the chip–tool interface. They admitted that in some experiments the cutting fluid did appear to penetrate, as indicated by examination of the tool and the workpiece under ultraviolet light. They concluded that a continuous film, as required for hydrodynamic lubrication, could not exist in the case where a continuous chip was formed. Neither was it possible for fluid to reach the areas where there was a chip–tool contact in the irregularity of the surfaces.[68] Other researchers, A.O. Schmidt, W.W. Gilbert, and O.W. Boston, investigated radial rake angles in face milling and the coefficient of friction with drilling torque and thrust for different cutting fluids.[69] Schmidt and Sirotkin investigated the effects of cutting fluids when milling at high cutting speeds. Depending upon which of the various cutting fluids were used, tool life increased approximately 35 to 150%.[70]

Ernst and Merchant studied further into the relationship of friction, chip formation, and high quality machined surfaces. Their research belied the conclusions of Bisshopp et al.[68] They found that cutting fluid present in the capillary spaces between the tool and the workpiece was able to lower friction by chemical action.[71] Shaw continued this study of the chemical and physical reactions occurring in the cutting fluid and found that even the fluid's vapors have constituents that are highly reactive with the newly formed chip surfaces. The high temperatures and pressures at the contact point of the tool and chip effect a chemical reaction between the fluid and the tool–chip interface, resulting in the deposition of a solid film on the two surfaces which becomes the friction reducing agent.[72,73]

Using machine tool cutting tests on iron, copper, and aluminum with pure cutting fluids, Merchant demonstrated that this reaction product, which "plated out" as a chemical film of low shear strength, was indeed the friction reducer at the tool–chip interface. He stated that materials such as free fatty acids react with metals to form metallic soaps, and that the sulfurized and sulfo-chlorinated additives in turn form the corresponding sulfides and chlorides acting as the agents that reduce friction. However, he quickly cautioned that as cutting speeds increase, temperature increases rapidly and good cooling ability from the fluid is essential. At speeds of over 50 feet per minute (254 mm/sec), the superior cutting fluid must have the dual ability to provide cooling as well as friction reduction capacity.[74]

Having learned which chemical additives are effective as friction reducers, Ernst, Merchant, and Shaw theorized that if they could combine these chemicals with water in the form of a stable chemical emulsion, a new cutting fluid having both friction reducing and cooling attributes could be created. In 1945, as a result of this research, their company compounded a new type of "synthetic" cutting fluid.[75] The new product, described as a water-soluble cutting emulsion with the name of CIMCOOL®, appeared as a news item in a technical journal in October 1945.[76] Two years later, the first semisynthetic metalworking fluid was introduced by this same company at the 1947 National Machine Tool Builders Show. It was a preformed emulsion very similar to a soluble oil but with better rust control and chip washing action.[77] This research was one of the important developments in metalworking fluid formulation, in that it provided the impetus for a whole new class of metalworking fluids, facilitating the new high-speed machining and metal deformation processes developed in the next quarter century.

5. Metalworking Fluids and the Deformation Process

During the same period of investigation into cutting fluid effects upon the metal removal process, many papers appeared in technical journals on the ameliorating effects of lubricants and "coolants," as the aqueous based fluid came to be termed. In the next decade, much research appeared in the technical literature on the theories of metal forming and how the lubricants used affected the metal deformation processes of extrusion, rolling, stamping, forging, drawing, and spinning. Notable among them is the often quoted work by Bowden and Tabor on the friction and lubrication of solids,[78] Nadai's theory of flow and fracture of solids,[79] Bastian's works on metalworking lubricants discussing their theoretical and practical aspects,[80] theories of plasticity by Hill in 1956,[81] and by Hoffman and Sacks,[82] followed by Leug and Treptow's discussion of lubricant carriers used in the drawing of steel wire.[83] Also notable are the investigations of Billingmann and Fichtl on the properties and performance of the new cold rolling emulsions,[84] and Schey's investigations of the lubrication process in the cold rolling of aluminum and aluminum alloys.[85]

Metalworking deformation processes involve tremendous pressures on the metal being worked. Consequently, very high temperatures are produced demanding a medium to effect friction reduction and cooling. If these stresses are not mitigated, there is the imminent danger of wear and metal pick-up on the dies, producing scarred work surface finishes.[86] To prevent these maleffects of metal forming, a suitable material must be used to lubricate, cool, and cushion both the die and the workpiece. In general, metal deformation processes rely upon the load carrying capacity and the frictional behavior of metalworking lubricants as their most important property. In some cases, however, friction reduction is critical, as in rolling operations. Insufficient friction would permit the metal to slide edgewise in the mill and cause the rolls to slip on the entering edge of the sheet or strip. Lack of friction also causes a problem in forging, a condition known as "flash," which prevents sufficient metal from filling the die cavities.[87]

VI. METALWORKING FLUIDS TODAY

At mid-century, metalworking fluids had acquired sufficient sophistication and proved to be the necessary adjunct in high speed machining and in the machining of difficult material: the exotic steels and specially alloyed nonferrous metals. They began to be regarded as the "corrector" of many machining problems and sometimes, by the uninitiated or inexperienced, were expected to be a cure-all for most machining problems. In the next decade, many cutting fluid companies sprang into existence offering a multitude of metalworking formulations to ameliorate machining problems and increase rates of production. Listings of metalworking fluids are to be found in a great number of publications of technical papers and handbook publications of various societies that cater to the lubrication engineering, tool making, and metallurgical communities.

Considering the many processes and the myriad of products available, there was, and is, confusion and controversy as to the best choice of fluid in any given situation. In the 1960s, the literature published by various technical organizations on the subject of how and what to use in metalworking processes was profuse. It was recognized by the metalworking community that direction was desirable, but there seems to be an isolation of those involved in the metalworking process from those involved in metalworking lubrication. As Schey has pointed out, the province of the metalworking process has traditionally been within the sphere of mechanical engineers and metallurgists; while the area of metalworking lubrication was within the expertise of chemists, physicists, and manufacturing process engineers. The National Academy of Science, observing this division, realized the need for communication among these specialists to integrate current knowledge and further the expansion of metalworking fluid technology and metalworking processes. They directed their Materials Advisory Board to institute the "Metalworking Processes and Equipment Program," a joint effort of the Army, Navy, Air Force, and NASA. One of the outcomes of this program was a comprehensive monograph containing the interdisciplinary knowledge of metalworking processes and metalworking lubricants to serve both as a text and a reference book.[88]

This brief history of the evolution of metalworking fluids shows that the dynamics of metalworking fluid technology are dependent upon the dynamics of metalworking processes as created by the parameters of machine tool design. These dynamics are mutually dependent parts of the total process and can only be investigated jointly. The body of knowledge evolving from metalworking fluid technology developed by these "cross culture"[89] engineers and scientists contributed significantly to the growing body of science and technology in the area of friction, lubrication, and wear. In the late 1960s, this technology blossomed into a new science named "tribology." A "veritable explosion of information" has followed since 1970.[90]

Today's metalworking fluid compounders find themselves having to harken to and abide by the edicts of new government regulations regarding the impact of formulation chemicals on the environment, as well as machine operator health and safety. Inattention to these edicts can well lead to cases of product liability with dire ramifications. It has been pointed out that "societal concerns about jobs often clash with society's demand for a risk-free environment" in which to live.[91] Regulatory issues, as well as health and safety aspects, will be covered thoroughly in later chapters.

During the mid-1980s, the automotive industries realized that metalworking fluids, as an integral part of the metalworking process, were fully as important as the metals used in the manufacture of assembly parts. Industrial management wanted guarantees that the metalworking fluids would be maintained in such a condition as to enable production of certain quantities of parts without interruption. Negotiations in this area produced a new form of metalworking fluid management[92] in which the supplier, working with a committee of factory personnel, supplied the fluids and technical expertise, and guaranteed the fluid performance in terms of the number of parts produced. More recently, there has been a trend toward "independent" chemical managers that are not lubricant manufacturers. This "Tier One" manager buys the fluid from the "Tier Two" fluid supplier, and then maintains the fluid on-site within the end-user's plant. In 2005, the Society of Tribologists and Lubrication Engineers (STLE) began offering courses and certification for in-plant fluid managers.

New methods of metalworking are constantly being developed. For example, water-jets and lasers[93] are being used to cut both metal and nonmetal parts. Some machining is being carried out either using dry or near-dry methods.[94] Just as in the days of R.F. Mushet, dry machining is often touted by tooling suppliers.[95] Cooled, compressed air has been used to cool the metal cutting process and move the chips out of the way, while vacuum systems remove the chips to a collection bin. In other cases, a small amount of vegetable oil is introduced into the air stream to help lubricate without using enough fluid to require collection or recirculation.[96] However, the demand for high speed production machining will continue, and metalworking fluids continue to be the "enabler."[97,98]

REFERENCES

1. Cleves, E., *Report on the Volume of Lubricants Manufactured in the United States by Independent Lubricant Manufacturers in 1990*, Annual Meeting Independent Lubricant Manufacturers Association, Alexandria, VA, pp. 1–5, 1991.
2. National Petroleum Refiners Association. *1990 Report on U.S. Lubricating Oil Sales,* Washington, DC, p. 5, 1991.
3. Wills, J. G., *Lubrication Fundamentals*, Marcel Dekker, New York, pp. 1–2, 1980.
4. Schey, J. A., *Metal Deformation Processes: Friction and Lubrication*, Marcel Dekker, New York, p. 1, 1970.
5. Bastian, E. L. H., *Lubr. Eng.*, 25(7), 278, 1968.
6. Dowson, D., *History of Tribology*, Longmans Green, New York, p. 253, 1979.
7. Singer, C. et al. *A History of Technology, Vol. III*, Oxford University Press, London, p. 668, 1957.
8. Singer, C. et al. *A History of Technology, Vol. III*, Oxford University Press, London, p. 663, 1957.
9. Dowson, C., *History of Tribology*, Longmans Green, New York, p. 126, 1979.

10. Schey, J. A., *Metal Deformation Processes: Friction and Lubrication*, Marcel Dekker, New York, pp. 1–2, 1970.
11. Dowson, D., *History of Tribology*, Longmans Green, New York, pp. 177–178, 1979.
12. Dowson, D., *History of Tribology*, Longmans Green, New York, p. 154, 1979.
13. Bastian, E. L. H., *Metal Working Lubricants*, McGraw-Hill, New York, p. 3, 1951.
14. Singer, C. et al. *A History of Technology, Vol. III*, Oxford University Press, London, p. 669, 1957.
15. Lewis, K. B., *Wire Ind.*, 1, 4–8, 1936.
16. Lewis, K. B., *Wire Ind.*, 2, 49–55, 1936.
17. Schey, J. A., *Metal Deformation Processes: Friction and Lubrication*, Marcel Dekker, New York, p. 5, 1970.
18. Rolt, L. T. C., *A Short History of Machine Tools*, MIT Press, Cambridge, MA, p. 11, 1965.
19. Woodbury, R. S., *Studies in the History of Machine Tools*, MIT Press, Cambridge, MA, p. 1, 1972.
20. Dowson, D., *History of Tribology*, Longmans Green, New York, pp. 21–23, 1979.
21. Woodbury, R. S., *History of the Lathe to 1850*, MIT Press, Cambridge, p. 23, 1961.
22. Landels, J. G., *Engineering in the Ancient Worlds*, University of California Press, Berkley, CA, p. 77 and pp. 199–215, 1978.
23. Woodbury, R. S., *History of the Grinding Machine*, MIT Press, Cambridge, MA, p. 13, 1959.
24. Woodbury, R. S., *History of the Grinding Machine*, MIT Press, Cambridge, MA, p. 23, 1959.
25. Woodbury, R. S., *History of the Grinding Machine*, MIT Press, Cambridge, MA, p. 21, 1959.
26. Gilbert, K. B. et al. *The Industrial Revolution 1750–1850, History of Technology, Vol. IV*, Oxford University Press, London, p. 417, 1958.
27. Rolt, L. T. C., *A Short History of Machine Tools*, MIT Press, Cambridge, MA, p. 138, 1965.
28. Handlin, O., *Eli Whitney and the Birth of Modern Technology*, Little Brown Company, Boston, MA, pp. 119–143, 1956.
29. Handlin, O., *Eli Whitney and the Birth of Modern Technology*, Little Brown Company, Boston, MA, p. 170, 1956.
30. Woodbury, R. S., *History of the Grinding Machine*, MIT Press, Massachusetts Institute of Technology, Cambridge, MA, p. 41, 1959.
31. Naysmyth, J., *James Naysmyth, an Autobiography*, Stiles, S., Ed., Harper and Brothers, London, pp. 437, 1883.
32. Dowson, D., *History of Tribology*, Longmans Green, New York, pp. 286–287, 1979.
33. Million Dollar Directory. *America's Leading Public and Private Companies*, Dunn and Bradstreet Corporation, Parsippany, NJ, 1991.
34. Osborn, F. M., *The Story of the Mushets*, Thomas Nelson and Sons, London, pp. 38–52, 1952.
35. Osborn, F. M., *The Story of the Mushets*, Thomas Nelson and Sons, London, pp. 89–95, 1952.
36. Morris, R. B., Ed., *Encyclopedia of American History*, Harper and Rowe, New York, p. 562–565, 1965.
37. Northcott, W. H., *A Treatise on Lathes and Turning*, Longmans Green and Company, London, 1868.
38. Steed, W. H., *History of Machine Tools 1700–1910*, Oxford University Press, London, p. 91, 1969.
39. Woodbury, R. S., *History of the Grinding Machine*, MIT Press, Massachusetts Institute of Technology, Cambridge, MA, p. 167, 1959.
40. Woodbury, R. S., *History of the Grinding Machine*, MIT Press, Massachusetts Institute of Technology, Cambridge, MA, p. 65, 1959.
41. Mallock, A., Action of cutting tools, *Royal Society of London Proceedings*, 33, 127, 1881.
42. Thurston, R. H., *A Treatise on Friction and Lost Work in Machining and Millwork*, Wiley, New York, p. 141, 1885.
43. Dowson, D., *History of Tribology*, Longmans Green, New York, p. 551, 1979.
44. Thurston, R. H., *A Treatise on Friction and Lost Work in Machining and Millwork*, Wiley, New York, p. 284, 1885.
45. *Lubrication*, The Texas Oil Company, New York, Vol. 39, p. 10, 1944.
46. Taylor, F. W., *On the Art of Cutting Metals*, Society of Mechanical Engineers, New York, pp. 138–143, 1907.
47. Rolt, L. T. C., *A Short History of Machine Tools*, MIT Press, Massachusetts Institute of Technology, Cambridge, p. 198, 1965.

48. Taylor, F. W., *On the Art of Cutting Metals*, Society of Mechanical Engineers, New York, p. 14 and pp. 137–138, 1907.
49. Taylor, F. W., *On the Art of Cutting Metals*, Society of Mechanical Engineers, New York, p. 9, 1907.
50. Rolt, L. T. C., *A Short History of Machine Tools*, MIT Press, Massachusetts Institute of Technology, Cambridge, p. 213, 1965.
51. *Lubrication*, The Texas Oil Company, New York, Vol. 39, p. 73, 1944.
52. *The Oil Daily*, August 6, pp. B7–B8, 1960.
53. Hutton, H. W., Improvements in or Relating to the Acid Refining of Mineral Oil, British Patent No. 13, 888/20, 1920.
54. Oldacre, W. H., Cutting Fluid and Process of Making the Same, U.S. Patent No. 1,604,068, 1923.
55. *Lubrication*, The Texas Oil Company, New York, Vol. 39, pp. 9–16, 1944.
56. Oldacre, W. H., *Lubr. Eng.*, 1(8), 162, 1944.
57. Kelly, R., *Carbide Tool J.*, 117(3), 28, 1985.
58. Hutton, H. W., Process of Producing Water-Soluble Oil, British Patent No. 13,999, 1923.
59. Dowson, D., *History of Tribology*, Longmans Green, New York, p. 351, 1979.
60. Harris, B. R., Epstein, S., and Cahn, R., *J. Am. Oil Chem. Soc.*, 18(9), 179–182, 1941.
61. *Emulsions*, 7th ed., Carbide and Carbon Chemicals, pp. 28–39, 1946.
62. Schallbock, H., Schaumann, H., and Wallichs, R., *Testing for Machinability by Measuring Cutting Temperature and Tool Wear, Vortrage der Hauptversammlung, Der Deutsche Gesellschaft fur Metalkunde*, V.D.I. Verlag, Dusseldorf, pp. 34–38, 1938.
63. Ernst, H., *Physics of Metal Cutting, Machining of Metals*, American Society of Metals, Cleveland, OH, p. 1–34, 1938.
64. Rowe, G. W., *Introduction of Principles of Metalworking*, St. Martins Press, New York, pp. 265–269, 1965.
65. Ernst, H. and Merchant, M. E., *Surface Friction of Clean Metals, Proceedings of Massachusetts Institute of Technology Summer Conference on Friction and Surface Finish*, pp. 76–101, 1940.
66. Merchant, M. E., *J. Appl. Phy.*, 16, 267–275, 1945.
67. Kelly, R., *Carbide Tool J.*, 17(3), 26, 1985.
68. Bisshopp, K. E., Lype, E. F., and Raynor, S., *Lubr. Eng.*, 6(2), 70–74, 1950.
69. Schmidt, A. O., Gilbert, W. W., and Boston, O. W., *Trans. ASME*, 164(7), 703–709, 1942.
70. Schmidt, A. O. and Sirotkin, G. B. V., *Lubr. Eng.*, 4(12), 251–256, 1948.
71. Ernst, H. and Merchant, M. E., *Surface Treatment of Metals*, American Society of Metals, Cleveland, OH, pp. 299–337, 1948.
72. Shaw, M. C., *Met. Prog.*, 42(7), 85–89, 1942.
73. Shaw, M. C., *J. Appl. Mech.*, 113(3), 37–44, 1948.
74. Merchant, M. E., *Lubr. Eng.*, 6(8), 167–181, 1950.
75. Schwartz, J., *1884–Cincinnati Milacron–1984: Finding Better Ways*, The Hennegan Co., Florence, KY, p. 88, 1984.
76. Jennings, B. H., New products and equipment: water soluble emulsion, *Lubr. Eng.*, 1, 79, 1945.
77. Kelly, R., *Carbide Tool J.*, 17(3), 29, 1985.
78. Bowden, F. P. and Tabor, F., *Friction and Lubrication of Solids*, Clarendon Press, Oxford, 1950.
79. Nadai, A., *Theory of Flow and Fracture of Solids*, McGraw-Hill, New York, 1950.
80. Bastian, E. L. H., *Metalworking Lubricants*, McGraw-Hill, New York, 1951.
81. Hill, R., *The Mathematical Theory of Plasticity*, Oxford University Press, London, 1956.
82. Hoffman, O. and Sachs, G., *Theory of Plasticity*, McGraw-Hill, New York, 1957.
83. Leug, W. and Treptow, K. H., *Stahl u. Eisen*, 76, 1107–1116, 1956.
84. Billigman, J. and Fichtl, W., *Stahl u. Eisen*, 78, 344–357, 1958.
85. Schey, J., *J. Inst. Met.*, 89, I-6, 1960.
86. Rowe, G. W., *Wear*, 7, 204–216, 1964.
87. Sargent, L. B., *Lubr. Eng.*, 21(7), 286, 1965.
88. Schey, J. A., *Metal Deformation Processes: Friction and Lubrication*, Marcel Dekker, New York, 1970.
89. Ludema, K. C., *Lubr. Eng.*, 44(6), 447–452, 1988.
90. Schey, J., *Tribology in Metalworking*, American Society for Metals, Metals Park, OH, p. V, 1983.

91. Nachtman, E. and Kalpakjian, S., *Lubricants and Lubrication in Metalworking Operations*, Marcel Dekker, New York, p. 215, 1985.
92. Sullivan, T., Older wiser, CMS providers take stock, *Lubes'N'Greases*, 24–29, 2003.
93. Gotz, D., Waterjet gives steel ribbon maker a productivity lift, *Mod. Appl. News*, 78–81, 2000.
94. Canter, N., The possibilities and limitations of dry machining, *Tribol. Lubr. Technol.*, 30–35, 2003.
95. Anon, Dry steel turning pays off with lower part costs, longer tool life, *Mod. Appl. News*, 44–47, 2002.
96. Anon, Making the case for near-dry machining, *Mod. Appl. News*, 54–55, 2001.
97. Krueger, M. et al. New technology in metalworking fluids and grinding wheels achieves 130-fold improvement in grinding performance, *Abrasives Mag.*, 8–15, 2000.
98. Ezugwu, E. et al. The effect of argon enriched environment in high speed machining of titanium alloy, *Tribol. Trans.*, 48, 18–23, 2005.

2 Metallurgy for the Nonmetallurgist with an Introduction to Surface Finish Measurement

James E. Denton

CONTENTS

I. INTRODUCTION

In one sense, it is unfortunate that metallurgy is such a mature science. Metals have been in use for so long and have been so effective in performing their role that they tend to be taken for granted. In the modern high-tech world of materials, it is ceramics, engineered polymers, and fiber reinforced composites that monopolize the scientific literature. Meanwhile, metals continue to quietly go about their business of supporting the infrastructure of our society.

Metalworking fluids, as the name implies, exist solely to facilitate the shaping of a useful metal object during a cutting, grinding, stamping, or drawing process. Before selecting the proper fluid for an operation, it is necessary to understand something about the metal that will be encountered. It is the intent of this chapter to give the reader an appreciation for the technology of metals and an understanding of the fundamentals of metallurgy, with special focus on the ways in which these fundamentals influence the behavior of metals during the various forming and fabrication processes they undergo on their way to becoming useful products. We will begin the chapter with a discussion of general topics regarding structure and properties that are common to all metals. Later in the chapter, we will concentrate on the more specific topics of composition and thermal treatments. A short section on the topic of surface texture measurement is also included.

II. THE STRUCTURE OF METALS

As is all matter in our universe, metals are composed of atoms. Further, metals are also crystalline in structure. By crystalline we understand that the individual atoms in metals are arranged in a regular and predictable three-dimensional array. This array is called the crystal lattice. The metallic bonding between the neighboring atoms in the crystal lattice is somewhat different than chemical bonding and gives rise to the inherent strength and malleability of the metals, properties not exhibited by other chemical compounds. It is also a special property of this metallic bond that the outer electrons of the atoms are generally shared by all the atoms within the structure and the electrons are free to circulate throughout the whole of the metal. This feature gives rise to the property of electrical conductivity, one of the properties that distinguish metals from nonmetals. Further, the atoms vibrate about their nominal position within the crystal lattice giving rise to the thermal properties of conductivity, thermal expansion, specific heat, and, ultimately, the melting point of the metal.

While there are a large number of possible crystalline arrangements, there are four that are particularly important in understanding most of the metals we come into contact with in our daily work. These are termed body centered cubic, face centered cubic, body centered tetragonal, and close packed hexagonal. When these structures are depicted, the arrangement of only the smallest number of atoms that completely describe their spatial relationship, the so-called unit cell, need be shown. Repetition of these unit cells in all three dimensions builds up the total structure. The unit cell should be thought of as the basic building block. Figure 2.1 shows the basic unit cells of the four important crystal structures.

In Figure 2.1, the atom sites are represented by small points, which might be considered as representing the nucleus of the atom. In reality, the atoms, including their orbiting electrons, are more nearly spheres and the outer electron shells of neighboring atoms actually touch each other. As might be deduced, the body centered cubic structure consists of a cube with atoms located at each corner and one atom at the very center of the cube. The face centered cubic structure consists of a cube with atoms located at each corner and at the center of each face, like dice with a five showing on every face. The body centered tetragonal cell is like the body centered cubic cell except that it is stretched in one direction. The top and bottom faces are squares while the four-side faces are rectangles. The hexagonal cell is most easily recognized as a hexagonal prism with atoms located at every corner and the center of the top and bottom ends. Since the hexagon is symmetrical about its center, the simplest unit cell is one whose end faces are parallelograms, arrived at by slicing the hexagon into three pieces. This is the simple hexagonal structure. There is an additional

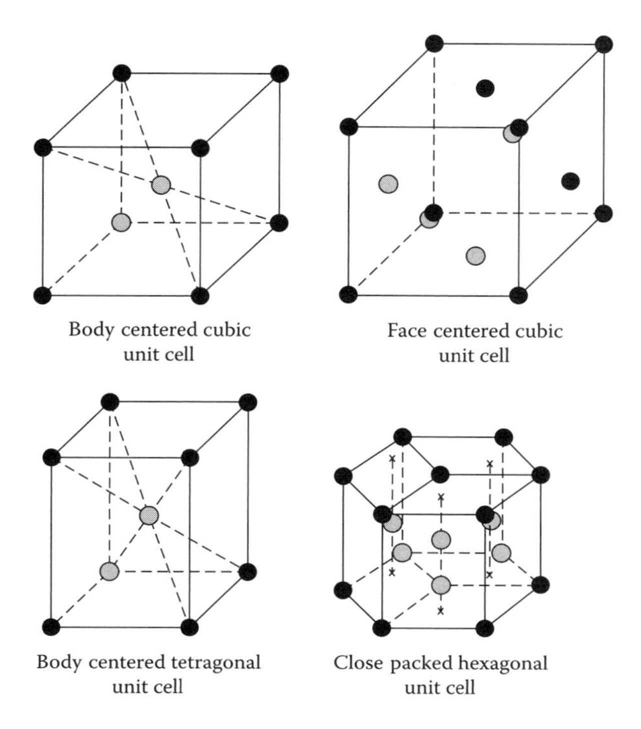

FIGURE 2.1 Crystal structures in metals.

potential atom site in each of the three unit cells making up the hexagon. If this site is occupied, the unit cell is termed close packed hexagonal. The crystal structures are frequently referred to by their acronyms: body centered cubic is abbreviated BCC, face centered cubic is abbreviated FCC, body centered tetragonal is abbreviated BCT, and close packed hexagonal is abbreviated HCP.

Visualizing these four bodies is, at first, difficult but there are very good reasons for understanding the structures. Malleability, or the ability of a metal to undergo deformation without breaking into fragments, is explained by the slipping of the various planes of atoms past one another. Slip tends to occur most readily on planes that have the highest atomic density. For example, in the FCC structure, a diagonal plane that intersects three corners, as shown in Figure 2.2, has the densest atom population of any of the possible planes in the FCC system. It is called the 111 plane, accounting for most of the slip that occurs when an FCC structured metal is deformed. The fact that the slip is confined to certain discrete crystallographic planes explains the development of strain lines and texture on the surface of formed sheet metal components. It also accounts for "earing," which is the concentration of excess material at specific points on the edges of drawn or ironed metal components.[1] Table 2.1 shows some common metals and their normal, room temperature, crystal structures.

It is necessary to specify conditions when identifying the crystal structure of a metal because many metals can exist in more than one crystal structure at different temperatures. This phenomenon is called allotrophism. Iron, for example, has three equilibrium crystal structures. Up to 1670°F (910°C) iron is BCC, a structure called alpha ferrite. From 1670 to 2552°F (910 to 1400°C), iron is FCC, a structure called austenite. From 2552°F (1400°C) to the melting point it is BCC again, a structure called delta ferrite. The transition between crystal structures is accompanied by a change in volume. This allotropic change in crystal structure is the basis for heat treatment strengthening of metals, which will be discussed in some detail later in this chapter.

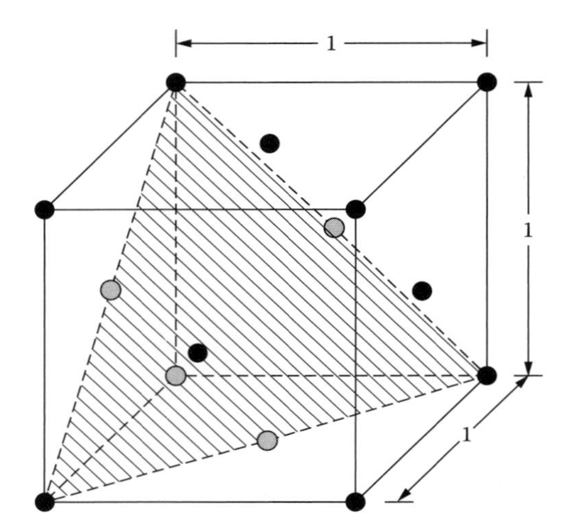

FIGURE 2.2 [111] Primary slip plane in face centered cubic unit cell.

All is not perfect in the world of crystal structure. Several types of defects in the normal atomic arrangement are frequently encountered. There may be atoms missing from a normally occupied site and this type of defect is called a vacancy. Occasionally, an extra plane of atoms may cause a disruption in the otherwise normal lattice and this type of defect is called an edge dislocation. There may also be foreign or alloy atoms substituting in a normal atom site in the parent structure. Foreign atoms are likely to have a different diameter than the matrix atoms, giving rise to a local distortion in the lattice. A special type of foreign atom defect is the interstitial atom. Elements such as boron, oxygen, carbon, and nitrogen have very small atomic diameters and typically have the correct size to fit into interstitial lattice sites or spaces between adjacent atoms. While these defects may be thought of as imperfections in the crystal structure, they have many beneficial effects and therefore are not necessarily detrimental. Figure 2.3 illustrates some of the common crystal defects.

Up until this point, we have discussed crystal structure within the context of a single crystal. To be sure, there are examples in nature, as well as intentionally prepared mechanical components, that exist as single crystals. Very large, naturally occurring quartz crystals are quite common. Some jet engine turbine blades, for example, are directionally solidified to result in the entire component being composed of a single crystal. The silicon crystals used to construct transistors are single crystals in which extreme care has been taken in their preparation to create as perfect a crystal structure

TABLE 2.1
Normal Crystal Structure of Some Common Metallic Elements

Metal	Crystal Structure at Room Temperature
Aluminum	FCC
Chromium	BCC
Copper	FCC
Iron	BCC
Nickel	FCC
Tin	T (simple tetragonal)
Titanium	HCP
Zinc	HCP

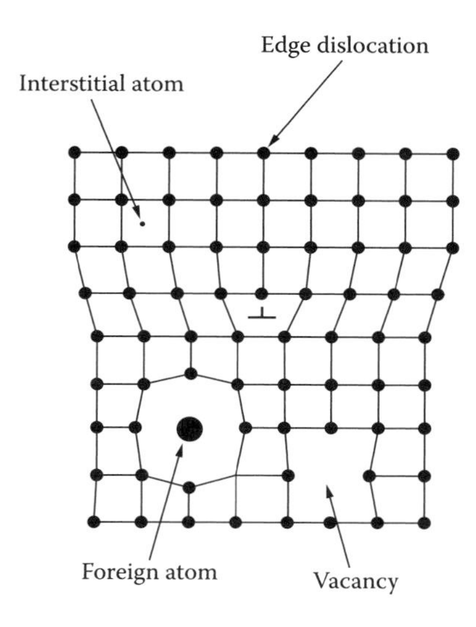

FIGURE 2.3 Types of crystal defects.

as possible. However, most metallic bodies are polycrystalline or composed of many crystals. Within each individual crystal, the arrangement of the atoms is near perfect as we have previously described them. Adjacent crystals, however, may have completely different orientations as shown in Figure 2.4, so that where two adjacent crystals meet, their atom planes do not line up exactly.

The individual crystals in a polycrystalline metal are called grains, and the contact regions between adjacent grains are called grain boundaries. In a properly manufactured and processed metal, the grain boundaries are stronger than the grains themselves; and when the metal is broken, failure occurs transgranular or through the grain. Under some adverse conditions such as overheating or oxidation, the grain boundaries may develop problems, which diminish their strength. When such a metal is fractured, the failure is intergranular or within the grain boundaries. The strength and ductility of the metal are significantly reduced in this type of failure. The fracture in this case may appear grainy and faceted giving rise to the old folklore conclusion that

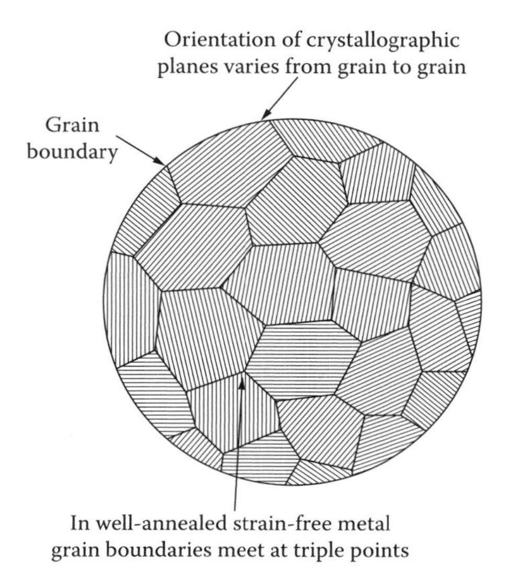

FIGURE 2.4 Polycrystalline grain structure.

"the metal crystallized" and caused the failure. As we now know, the metal was always crystalline but the intergranular fracture made the crystallinity more visually apparent. The randomness in grain orientation in most finished metal products results in approximately equal mechanical properties in all directions, a property called isotrophism. In fabricated forms where extreme forming deformation has been confined to one specific direction, for example as in wire drawing or cold drawn bar stock, the strength and ductility in the longitudinal direction of the wire or bar will be dramatically different than in the transverse direction.

III. THE PROPERTIES OF METALS

A. Physical and Mechanical Properties

The properties of metals are divided into two categories: physical and mechanical. Physical properties depend on the electronic configuration of the metal's atoms and include those inherent characteristics that remain essentially unchanged as the metal is processed. Such characteristics as density, electrical conductivity, thermal expansion, and modulus of elasticity are some of the more typical examples of physical properties.

B. Mechanical Properties

Mechanical properties are those that can be drastically changed by processing and thermal treatment. Such characteristics as hardness, strength, ductility, and toughness are the properties that govern the performance of the metal in use. A common and easily determined mechanical property of a metal is hardness. One of the earliest hardness tests adapted from the science of geology was a scratch test based on the Moh's scale. This test involves a scale of 1 to 10 with talc as 1 and diamond as 10. The test is based on a determination of which material will scratch the other but is somewhat subjective and not particularly quantitative. Most contemporary hardness testing is based on indentation where an indenter of a specified geometry is pressed into the sample under a specified load. The projected area of the indentation or depth of penetration is determined and the hardness is expressed in the dimensions of pressure; for example, kilograms per square millimeter. In the United States, Brinell and Rockwell tests are the most common. In Europe, the Vickers test is more common. Obviously, in indentation hardness testing the indenter must be harder than the material being tested to prevent deformation of the indenter itself. Typically, hardness testing indenters are made of hardened steel, tungsten carbide, or diamond.

The Brinell test uses a 10 mm diameter ball with a load of 3000 kg for ferrous metals or 500 kg for softer nonferrous metals. The Rockwell test uses either a conical diamond indenter, a 1/16 in. diameter or a 1/8 in. diameter hardened steel ball with loads of 60, 100, or 150 kg. The combinations of three different loads with three different indenters provides nine possible Rockwell scales that are useful for metals ranging from very soft to very hard. There is also a special Rockwell test for thin or fragile samples called the Superficial tester that uses lighter loads of 15, 30, or 45 kg. The Vickers test is similar to the Rockwell test but uses a pyramidal shaped diamond indenter with equal diagonals. There are numerous other hardness tests that have been devised for special purposes, but the ones described above are the most frequently encountered. In specifying a hardness value, it is necessary not only to give the numerical value but also to indicate the scale or type of test used. Recognized abbreviations appended to the hardness number are of a format that begins with H for hardness followed by additional letters and numbers indicating the specific type of test. HV indicates the Vickers test. HV and DPH (for diamond pyramid hardness), are used interchangeably. HBN stands for the Brinell hardness test. If a particular test is conducted with a variety of loads, the load may be subscripted after the abbreviation, e.g., HBN_{500}. HRC is Rockwell hardness on the C scale, HRB is Rockwell hardness on the B scale, etc. Table 2.2 shows a summary of these various tests and the hardness ranges over which they are used.[2]

TABLE 2.2
Approximate Hardness Conversions[a]

Brinell		Vickers	Rockwell			Rockwell Superficial						Applications
500 kg	3000 kg		A 60	B 100	C 150	15N	30N brale	45N	15T	30T 1/16 in. ball	45T	
—	—	1076	87	—	70	—	86	—	—	—	—	
—	757	860	84	—	66	92	83	70	—	—	—	
—	682	737	82	—	62	91	79	68	—	—	—	
—	560	605	79	—	56	88	74	62	—	—	—	Carbide
—	496	528	77	—	51	86	69	56	—	—	—	
—	429	455	73	—	45	83	64	49	—	—	—	Hard Steel
—	372	393	70	—	40	80	60	43	—	—	—	
—	332	353	68	—	36	78	56	38	—	—	—	
—	290	309	66	—	31	75	51	32	—	—	—	
—	265	278	64	—	27	73	48	28	—	—	—	
195	234	247	61	99	21	70	42	21	93	83	72	
179	216	222	59	96	16	—	—	—	92	80	69	
163	195	202	56	92		—	—	—	90	78	65	Soft
151	176	182	54	88		—	—	—	89	75	61	Steel
135	156	163	51	82		—	—	—	87	71	55	Bronze
118	135	140	46	74		—	—	—	85	66	47	Hard Brass
110	125	131	44	70		—	—	—	83	63	43	Aluminum
104	117	113	42	68		—	—	—	82	60	39	Copper
95	107	107	40	60		—	—	—	80	56	33	
85	—	100	—	52		—	—	—	76	51	25	
75	—	93	—	41		—	—	—	74	48	14	

[a]Approximate conversions for a variety of metals based on ASTM E-140.

There is a proportional relationship that exists between hardness and tensile properties of steel that is useful in estimating strength. If the Brinell hardness number is multiplied by 500, the result is approximately equal to the tensile strength in pounds per square inch (lb/in.^{2}). If the Brinell hardness number is multiplied by 400, the result is approximately equal to the yield strength in lb/in.^{2} This relationship is reasonably valid over a wide range of hardness including annealed and hardened steels. The proportionality, however, is not so good for extremely soft or extremely hard steels.

The tensile test is another common test used to describe the properties of a metal and illustrates several important characteristics. There are a large variety of tensile test specimens, but all share a common design, which has a controlled cross-section gauge length and enlarged ends for gripping. During testing, the specimen is subjected to an increasing pull load or stress at a carefully controlled rate of increase called the strain rate. Most metals are sensitive to strain rate and exhibit different properties when tested at different strain rates. Figure 2.5 shows the various points depicted by a tensile test.

First, it may be noted that this is a plot of load versus extension, or in engineering terms, stress versus strain. It is an expected characteristic of all metals that they behave elastically. That is, as stress is applied, the metal elongates in a linearly proportional response. The first part of the tensile curve, then, is a straight line. The slope of this straight line portion, stress divided by strain, is called the elastic modulus or Young's modulus of elasticity. In the elastic portion of the curve, the elongation is completely recoverable. The metal behaves as a rubber band. If the load is removed the metal returns exactly to its original, unstressed size. If, however, the load is continually increased, a point is reached where the metal is no longer capable of stretching elastically and permanent plastic deformation is produced. This point is called the proportional limit, above which stress is no longer linearly proportional to strain. If the proportional limit is exceeded, the metal undergoes permanent plastic deformation. If the load is removed, the sample will be found to be longer than its original length. The proportional limit is difficult to precisely discern on the tensile

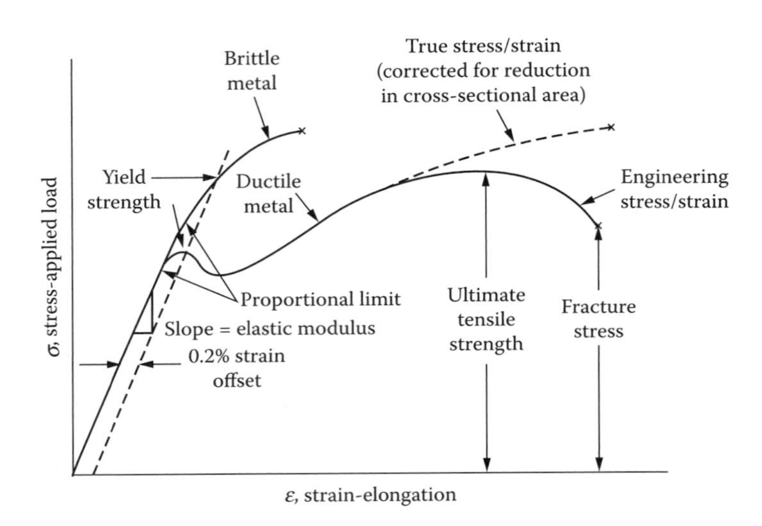

FIGURE 2.5 Stress/strain behavior of a ductile and brittle metal in the tensile test.

curve, so an arbitrary point is defined and called the yield strength. To determine the yield strength, a line offset 0.2% from the origin is drawn parallel to the straight-line portion of the tensile curve. Where it intersects the curve is called the yield point and the stress corresponding to this point is called the yield strength. There are other methods of defining the yield point but the 0.2% offset method is, by far, the most common. The appearance of the stress/strain curve above the yield point varies dramatically as a function of the thermal and mechanical history of the metal, as well as the composition and inherent characteristics of the metal itself.

For soft ductile metals, the curve may take a small dip after yield before continuing upward to the maximum load, which is called the ultimate tensile strength. While the load is being applied, the sample is responding by elongating and becoming smaller in diameter. Local "necking-down" eventually occurs, so that the sample assumes an hourglass shape. Because of the reduction in diameter, the stress/strain curve shows a drop-off in load. When the stress/strain curve is corrected for the reduction in diameter, it may be seen that the work hardening effect persists up to final failure. The corrected stress/strain curve is called the true stress/true strain curve. This is the normal behavior of ductile metals that have a capacity to work harden.

In very hard or brittle metals, the curve bends only slightly at yield and ultimate failure occurs soon after. The point of ultimate failure is called the fracture strength. A metal that exhibits a lot of plastic deformation after the yield point is called a ductile metal, while a metal that fails with very little deformation is called a brittle metal. The area under the stress/strain curve is a product of a force times a distance and may be thought of as a work function whose physical equivalent is toughness. High toughness, then, is exhibited by a material that has the combination of high strength with the ability to deform in order to maximize the area under the stress/strain curve.

A more common procedure for determining toughness is the impact test. This procedure rigidly supports the test specimen, which is then struck by a falling pendulum, thus producing a relatively high strain rate representative of many collision-like events that a component may experience during its service life. In the Charpy impact test, the specimen is supported at both ends as a simple beam and struck in the center by the pendulum. In the Izod impact test, the specimen is clamped at one end in a vise as a cantilever beam and struck at the unsupported end by the pendulum. In both tests the specimens are sometimes notched to concentrate the impact energy from the falling pendulum in a smaller volume of material. Samples may be tested over a range of temperatures to determine if they are embrittled at low temperature. BCC metals typically exhibit a substantial reduction in toughness

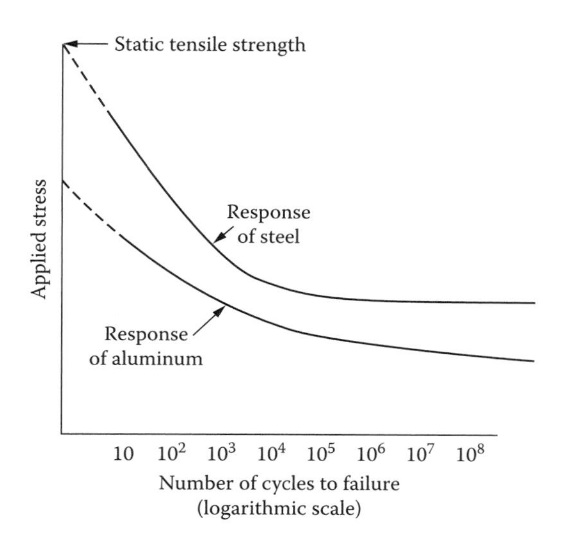

FIGURE 2.6 Fatigue test endurance limit curve.

at low temperatures, and this phenomenon is called the ductile-to-brittle transition temperature. Results of impact testing typically are expressed in units of pound-feet or Joules.

Metals are susceptible to a peculiar and insidious failure mode called fatigue, where fracture occurs at a stress level substantially below the yield strength of the material. This type of failure requires cyclical stress or repeated loading and unloading of the component. Cracks are initiated and grow until the remaining unfractured area is insufficient to support the applied load, at which point failure occurs catastrophically. There is no observable deformation or bending before failure, so there is little or no warning that failure is imminent. Inspection of the fractured surface of a fatigue failure shows a series of so-called "beach marks" or approximately concentric lines whose focus point coincides with the crack initiation site. The beach marks or lines are actually crack arrest fronts. The crack grows in stages during successive load applications and stops between cycles, leaving a track of the characteristic beach marks. Since the effects of fatigue can be so serious, a measure of fatigue strength is an important property of a metal. A graph plotting number of loading cycles to failure versus applied load yields a curve of the shape shown in Figure 2.6.

At lower stress values, the slope of the curve levels out and approaches a flat line for some metals, like steel. Other metals, such as aluminum, never really approach horizontal and do not have a well-defined fatigue limit. This threshold level, below which failure does not occur, is called the endurance limit. Fatigue tests are normally terminated at 10 million load cycles if failure does not occur and samples which survive this long are assumed to have infinite life for engineering purposes. Fatigue life is drastically affected by numerous factors including internal cleanliness of the metal, surface finish, geometry, residual stresses, and environmental conditions.[3]

IV. PURE METALS VS. ALLOYS

Of the approximately 100 stable elements that comprise the periodic table, 18 are considered to be nonmetals, seven are considered metalloids (having some metallic characteristics and some nonmetallic characteristics), and all the remaining elements are metals. It is somewhat astonishing to realize that so many of the basic elements are actually metals. Many of these basic metallic elements are quite useful in their pure form. Copper and aluminum both have their greatest thermal and electrical conductivity at highest purity. Pure zinc is applied to steel in a process called galvanizing to provide sacrificial corrosion protection. Pure, annealed iron has its highest magnetic saturation and least magnetic hysterisis. Many similar examples of the utility of pure metals exist.

Discovery of naturally occurring deposits of gold, silver, copper, and other elements have been noted throughout history. The brilliant luster and malleability of these pure metals were no doubt mystical to prehistoric cultures. However, the softness of these metals made them unsuitable for anything more than jewelry and decorative use. Distinct from a pure metal is an alloy or a combination of two or more pure metals. The serendipitous discovery of the alloy of copper and tin that we call bronze ushered in the Bronze Age, a new technological era in our history. Combining two metals into an alloy produces higher strength and hardness than either pure metal possessed alone, thereby extending their application to many more useful tools.

Metals may combine to make an alloy in a variety of ways.[4] If the combining metals have the same crystal structure, the same chemical valence and very nearly the same atomic diameter, they may form a substitutional solid solution. In this type of alloy. the metals are completely miscible in the liquid and solid states and the two different atoms simply replace each other interchangeably at the normal crystal lattice sites. The copper–nickel system is an example of this type of alloying.

A second type of alloy, called a eutectic, is formed by two metals that have different crystal structures and widely disparate atomic diameters. Obviously, these atoms cannot be substituted for each other in a mixture. They are miscible in the liquid state but immiscible in the solid state. The most defining characteristic of such an alloy is that the melting point of the alloy is lower than either of the two pure metals making it up. Further, since the metals are immiscible in the solid state, the microstructure of the solid has two distinctly different phases. The lead–tin system is an example of this type of alloying. In this system, the eutectic composition, the alloy having the lowest melting point, occurs at 38% lead plus 62% tin, making it very useful as a solder in the electronics industry.

A third possible combining tendency is shown by two elements where one has a plus chemical valence and the other has a negative valence. These two elements would have a strong chemical affinity and tend to form an intermetallic compound having a specific atomic ratio of composition. The magnesium–silicon system forms an intermetallic compound, Mg_2Si, since magnesium has a valence of $+2$ and silicon has a valence of -4. This type of compound can be manipulated in a heat treat process called precipitation hardening and can produce a dramatic increase in hardness and strength in the alloy.

These example reactions are only a small view of the possibilities in alloying metals. Most alloys are composed of more than just two metals and it can be seen that the complexities increase exponentially with three or more metals and even nonmetals in combination.

V. FERROUS METALS

The use of iron and steel as structural materials is so dominant that all metals are classified into two broad categories: ferrous and nonferrous. Ferrous metals include all the alloys whose major alloying element is iron. This broad category includes cast iron, carbon steel, alloy steel, stainless steel, and tool steel. Nonferrous materials include, basically, everything else. The more common materials in the nonferrous group are aluminum, magnesium, titanium, zinc, copper, and nickel-based alloys.

A. CARBON AND ALLOY STEELS

In the simplest form, steel is an alloy of iron and carbon. The melting and reduction of iron ore was traditionally carried out using charcoal or coal as a fuel. As the molten iron was in contact with the carbon-rich fuel, it absorbed excess carbon into its structure resulting in an inherently brittle product called pig iron. To develop the desired ductile and tough properties of steel it was necessary to reduce the carbon content. A technological breakthrough in the production of low carbon steel from high carbon pig iron occurred in 1856 when Henry Bessemer developed the Bessemer converter. In this process. compressed air was blown through the molten iron to convert the excess

TABLE 2.3
Alloy Content of Common Steel Grades

Alloy Series	Nominal Alloy Content
10XX	Carbon and manganese (up to 0.90%)
11XX	Carbon, manganese, and sulfur
12XX	Carbon, manganese, sulfur, and phosphorus
13XX	Carbon and manganese (1.75%)
15XX	Carbon and manganese (1.25%)
40XX	Carbon, manganese, and molybdenum
41XX	Carbon, manganese, chromium, and molybdenum
43XX	Carbon, manganese, nickel (1.80%), chromium, and molybdenum
48XX	Carbon, manganese, nickel (3.50%), and molybdenum
51XX	Carbon, manganese, and chromium (0.90%)
52XX	Carbon, manganese, and chromium (1.45%)
61XX	Carbon, manganese, chromium, and vanadium
86XX	Carbon, manganese, nickel, chromium, and molybdenum

carbon to carbon dioxide. The basic differentiation between cast iron and steel is the carbon content, ranging up to approximately 1.5% in steel and up to 4% in cast iron.

Steel comes in a very wide range of alloys and carbon contents. The most common specifying bodies for steel composition are the American Iron and Steel Institute (AISI), The Society of Automotive Engineers (SAE), and the American Society for Testing Materials (ASTM). AISI does not actually write standards but, acting as the voice for the steel industry, determines which grades are manufactured and sold in such quantities as to be considered standard grades. In the format of AISI/SAE standard alloy steels, the composition is represented by a four digit number. The first two numbers designate the major metallic alloys present and the last two numbers designate the carbon content in hundredths of a percent. As an example, a frequently used alloy steel grade is 4140. According to this designation system it is possible to tell that the material contains chromium and molybdenum as the principle alloying elements along with 0.40% carbon. Table 2.3 shows the alloying elements associated with the more popular standard grades.[5]

The alloying elements used in steel have specific functions and their concentration ranges have been developed for the various grades to produce specific properties. Carbon is the essential element that determines the ultimate hardness steel is capable of achieving. Steels with low ranges of up to 0.20% carbon do not respond well to heat treatment and can achieve maximum hardness up to approximately 35 HRC. Steels with medium carbon, up to about 0.50%, may be hardened fully to as high as 60+ HRC. This is about the limit of martensitic hardness in medium alloyed steels. Carbon in excess of 0.50% has little additional effect on hardness. The excess carbon above 0.70% forms carbide particles that, while they do not increase the hardness, do have a beneficial effect on wear resistance and compressive strength. The high carbon steels, notably 52100 and 1095, find application in bearings and cutting edge use.

B. Hardness and Hardenability

Hardness and hardenability are two separate terms and have distinctly different meanings. Simply defined, hardenability governs the rate at which a particular grade of steel must be cooled from the hardening temperature to achieve full hardness. Steels with low hardenability must be cooled rapidly, usually requiring a water quench. Steels with high hardenability can be cooled more slowly and may be quenched in oil or air. The final hardness produced may be the same in low and high hardenability steel grades, but different cooling rates would have been required to achieve that hardness. Since cooling rate is also a function of mass, steels with high hardenability may through harden in very thick section sizes. Steels with low hardenability will harden only at the surface or not at all in thick section sizes.

Manganese is the most effective metallic alloy at increasing the hardenability of steel. In resulfurized steels, some of the manganese combines with sulfur to form manganese sulfide inclusions. Vanadium, chromium, and molybdenum are also very effective at increasing hardenability, but unlike manganese, are also strong carbide formers. These elements promote increased wear resistance and also resist softening upon exposure to elevated temperatures. Nickel generally improves hardenability but its primary effect is as an austenite stabilizer. Austenite is the allotropic form of iron having the FCC crystal structure that exists at the elevated temperatures used for heat treating. Steels that contain appreciable amounts of nickel can typically be hardened from lower heat treating temperatures. Nickel also promotes toughness and is frequently used for applications requiring high impact resistance.

Sulfur and phosphorus are usually thought of as contaminants in steel but when intentionally added, as in the 11XX and 12XX series, they promote machinability. They are essentially insoluble in iron and form nonmetallic stringer type inclusions having relatively low melting points. During machining operations these stringers lubricate the cutting tool and also act as chip breakers. Both sulfur and phosphorus have a negative effect on strength and toughness of the steel and would not be used for critically stressed applications requiring high reliability, such as aircraft components. Lead is another alloy that falls into the category of a free machining additive but due to its toxicity is being used less and less. In an attempt to mitigate the negative effects of sulfur on strength, the so-called shape controlled steels were developed. Calcium and tellurium are being added to some proprietary steels to reshape the stringer type sulfides into more globular inclusions that have a less detrimental effect on mechanical properties. The alloy steels contain a maximum of about 5% alloying elements. Steels for more demanding applications such as the tool steels require much higher alloy additions.

C. Tool Steels

The obvious definition of this class of steels hardly fits. In addition to tools, these steels are used for molds, bearings, wear parts, and a wide variety of structural components. They are likewise difficult to categorize. Tool steels evolved in a highly proprietary market where a diverse population of specialty steel mills developed materials for unique applications. Each manufacturer put their own special twist in the chemical composition to provide some real or perceived commercial market advantage. Even today, more than a few brand names still persist and are widely recognized and specified by name. Attempts to organize the extensive array of tool steels eventually settled on a system of classification by function. In the AISI designation system, tool steels are identified by a letter followed by one or two numbers. The letters classify the grades by function or application while the numbers were assigned chronologically within the grade.[6]

1. High Speed Steels

In the late 1800s and early 1900s, all metal cutting was carried out with straight high carbon steels. This grade is capable of developing high hardness but tends to soften rapidly when exposed to elevated temperature. For this reason cutting speeds had to be restrictively low to prevent the tool from overheating. In the late 1800s, it was discovered that the addition of tungsten and chromium to cutting steel made it much more resistant to softening when exposed to elevated temperature and, therefore, made it possible to increase the cutting speed to a remarkable degree. These steels came to be known as high speed steel. A classical composition of 18% tungsten, 4% chromium, and 1% vanadium evolved and became the basis of the tungsten type high speeds now designated as type T1. The tungsten-type high speed steels dominated the metal cutting industry until World War II when, due to a shortage of tungsten, molybdenum was substituted for most of the tungsten. The lower cost and domestic availability of molybdenum (there are substantial deposits of molybdenum ore near Climax, CO) led to the growing popularity of the M-type high speed steels. Type M2 now accounts for the greatest proportion of high speed cutting tools.

2. Hot Work Die Steels

This group of tool steels, designated the H series, was developed basically for die casting of zinc, magnesium, and aluminum, or for such high temperature forming operations as extrusion and forging dies. It has three subgroups: the H1 through H19 chromium type, the H20 through H39 tungsten type, and the H40 through H59 molybdenum type. H13 is probably the most popular grade in this series.

3. Cold Work Steels

This group of steels is restricted to forming operations that do not exceed 500°F (260°C) owing to the lack of refractory alloys that would resist softening at elevated temperature. There are three subgroups: the O-type oil hardening grades, the A type medium alloy air hardening grades, and the D-type high carbon high chromium series.

4. Shock Resisting Steels

This group of steels, designated the S-type, are used for chisels, punches, and other applications requiring extreme toughness at high hardness levels. S7 is the most popular grade in this category.

5. Mold Steels

This group of steels, designated the P series, is used primarily for plastic molds. Types P2 to P6 are carburizing steels that have very low hardness in the annealed condition, permitting the mold cavity to be generated by hubbing. Hubbing is a cold work process where a tool having the geometry of the desired cavity is pressed into the mold blank rather than creating the cavity by conventional machining operations. The mold is subsequently carburized and hardened for long-term durability. Types P20 and P21 are normally supplied in the preheat treated condition in the 30 HRC hardness range. Following final machining, the molds made of type P21 steel are ready for service without further heat treatment.

6. Water Hardening Steels

This group of steels, designated the W series, are low alloy high carbon grades that have very little resistance to softening at elevated temperature. When used for cutting tools they are restricted to woodworking tools and slow cutting speeds. When heat treated in moderate section thicknesses they must be water quenched and develop full hardness only at the surface, the core remaining somewhat softer and tougher. This grade is also useful for springs.

7. Special Purpose Steels

This group of steels, designated the L series, are low in alloy and carbon content compared with the other tool steels. The characteristics of good hardness, wear resistance, and high toughness make this grade useful for machinery parts such as collets, cams, and arbors.

Since tool steels normally contain large amounts of expensive alloying elements and are required to withstand very severe service conditions, particular care is devoted to their production. Double melting practice is common, wherein the steel is produced by electric furnace or vacuum induction melting and then subjected to a secondary refining process, such as vacuum arc or electroslag remelting. An alternate manufacturing procedure uses high purity powders to produce moderate size billets by the powder metallurgy process. The PM billets are then consolidated by hot isostatic pressing and rolling to final size. An added advantage of this process is that it prevents segregation of the alloying elements that naturally occurs during solidification from the molten stage and also produces very fine carbide particle size. These factors make tool steels a premium priced commodity. A highly alloyed grade in a double melted or PM form can cost well over $10 per pound.[5,6]

D. Stainless Steels

In the AISI numbering system the primary stainless steels are designated by a three digit number; 2XX, 3XX, 4XX, and 5XX. In addition, there is one other class called the precipitation hardening grades, which are designated by their chromium and nickel contents. For example, 17-7 PH is a common precipitation hardening grade containing 17% chromium and 7% nickel, along with minor amounts of other elements that promote the precipitation hardening behavior. As one can imagine, there is also a never ending variety of proprietary and special purpose grades beyond the scope of this discussion. Table 2.4 shows a generalized format of these compositional classifications.

Aside from the compositionally based classification, stainless steels are also categorized by their crystallographic structures: ferritic, martensitic, and austenitic. Both the ferritic and martensitic stainless steels are magnetic, while the austenitic grade is nonmagnetic. The most corrosion resistant grades are to be found in the austenitic series, followed by the ferritic series and the precipitation hardening grades. The martensitic grades, which are hardenable by a quench and temper heat treatment, generally have the poorest corrosion resistance.

The primary mechanism by which stainless steels gain their corrosion resistance is through the development of a stable, protective surface oxide film. It is generally accepted that a minimum chromium content of 12% is necessary to form the protective oxide film. To a great extent, stainless steels form this film naturally by reaction with oxygen in the atmosphere. However, a denser, more stable oxide may be forced through a process called passivation. This process subjects the material to a strong oxidizing acid, usually concentrated nitric acid, at an elevated temperature. During passivation, minute particles of nonstainless metals which may have become embedded in the surface during machining, and contact with nonstainless forming equipment, are dissolved while the oxide film is being generated. This treatment leaves the metal in its most corrosion resistant condition.

Stainless steel is subject to a serious reduction in corrosion resistance through a mechanism called sensitization. Chromium has a very strong affinity for carbon and tends to form very stable carbides. Moreover, a small amount of carbon can combine with a large amount of chromium, thereby effectively negating the effect of the chromium in promoting corrosion resistance. When heated in the range of 1000 to 1200°F (540 to 650°C) a precipitation reaction between carbon and chromium occurs at the grain boundaries, resulting in the sensitization. Since a significant portion

TABLE 2.4
AISI/SAE Stainless Steel Designation System

Alloy Series	Nominal Alloy Content
2XX	Chromium, manganese plus nickel
3XX	Chromium plus nickel
4XX	Chromium (12–20%)
5XX	Chromium (5%)
PH	Chromium, nickel, molybdenum plus aluminum or copper

of the chromium has been tied up, corrosion can readily proceed at the grain boundaries in a form of corrosion called intergranular attack. One common fabrication process that can induce sensitization is welding. At some point along the edges of the weld, a temperature in the sensitization range will occur, rendering the stainless steel subject to intergranular attack in the heat affected zone of the weld. Sensitization can be reversed by a heat treating process, which consists of a high temperature cycle that redissolves the precipitated chromium carbides, followed by rapid cooling through the sensitization range. While this corrective treatment is effective, it can be expensive and result in unacceptable distortion of welded fabrications. Specific grades for welding have been developed that minimize sensitization by controlling the carbon content to very low limits, as in types 304L and 316L, or by incorporating elements that have a stronger affinity for carbon than chromium, such as in type 347.[7,8]

E. Cast Iron

Cast iron is a very important class of engineering materials. The relatively low melting point and fluidity of cast iron makes it readily cast into complex shapes. It has good mechanical properties and is easily machined due to its unique microstructure. In its broadest description, cast irons are alloys of iron, carbon, and silicon. The carbon is usually present in the range of 2.0 to 4.0% and the silicon in the range of 1.0 to 3.0%. This composition results in a microstructure that has excess carbon present as a second phase. The form taken by the excess carbon is the basis of the three major subdivisions of cast irons: gray iron, white iron, and ductile iron.

The most common variety of cast iron has the excess carbon present in the form of graphite flakes and is called gray iron. This terminology derives from the appearance of a freshly fractured surface which has a dull gray texture owing to the presence of the graphite flakes. During fracture, cracking propagates along the graphite flakes, since graphite has practically no strength.

In irons with the carbon and silicon content minimized, and where a very rapid solidification rate was attained, the excess carbon is present as a carbide and there is no free graphite. This type is called white iron because a freshly fractured surface has a smooth, white appearance. White iron is very hard and brittle in the as-cast condition. It is frequently used in applications where extreme wear and abrasion resistance is required. White iron can be converted to so-called malleable iron by a heat treatment which causes the carbide to decompose into compact clumps of graphite called temper carbon. The compact clump form of free graphite interrupts the crack path more effectively than the flake form found in gray iron, thereby providing some measure of ductility.

Another method of producing a ductile form of cast iron is by a nodulizing inoculation. If magnesium or rare earth metals are added to the molten iron, the excess carbon forms spheroidal nodules of graphite, rather than the flake form of carbon found in gray iron. The nodular graphite structure results in a substantial increase in strength and ductility. Ductile iron castings can compete with steel castings or forgings in many near net shape applications.

Commercial iron castings are seldom melted to strict chemical composition in the way steel is produced. It is more common that the required mechanical properties are specified and the foundry selects the composition that will meet those specifications. The mechanical properties of cast iron are drastically affected by cooling rate during solidification, which is a function of metal section thickness. In order to meet the specified mechanical property requirements, the foundry must have the latitude to adjust the chemical composition to suit the weight and section thickness of the particular casting. The parameter that is most often used for controlling mechanical properties is the carbon equivalent. This parameter is the sum of the total carbon plus one third of the silicon content plus one third of the phosphorus content. These three elements in the given ratios affect the rejection of carbon from the melt during solidification, and determine the resulting graphite size and distribution.

Gray cast iron, therefore, is commercially designated by its tensile strength. Class 25 iron has a tensile strength of 25,000 lb/in.^{2} (172,000 kPa); class 30 has 30,000 lb/in.^{2} (206,000 kPa), etc. Since gray cast iron has essentially no ductility, there is no measurable yield strength. Ductile iron, by contrast, is designated by three numbers; the tensile strength, yield strength, and percentage elongation from the tensile test. Common varieties of ductile iron are 60-40-18, 80-55-06, 100-70-03, and 120-80-02. The minimum mechanical properties of type 60-40-18, for example, would be 60,000 lb/in.^{2} (415,000 kPa) ultimate tensile strength, 40,000 lb/in.^{2} (275,000 kPa) yield strength, and 18% elongation.[9]

VI. STRENGTHENING MECHANISMS AND THE MICROSTRUCTURES OF IRON AND STEEL

There are several strengthening mechanisms in the metallurgy of iron and steel. Steel exhibits a substantial increase in hardness and strength as the result of cold work or plastic deformation. Distortion of the crystal lattice and creation of extensive dislocations that occurs during plastic deformation results in strengthening the metal. Addition of alloying elements can also cause strengthening through two separate mechanisms. Alloys that form substitutional solid solutions strengthen through inhibiting crystallographic slip. Atoms of solid solution alloying elements replace the base metal at some atom sites. Since they have a slightly different atomic diameter, they cause protuberances or depressions in the atomic planes they occupy and interfere with slip, thereby strengthening the metal. Another mode of alloy strengthening is through the formation of a second phase. The iron/carbon system is a striking example of this mechanism. Carbon is essentially insoluble in iron at room temperature. In steel, carbon forms a second phase called cementite having the composition Fe_3C. This was previously discussed under alloying as the formation of an intermetallic compound. From the composition it may be noted that one carbon atom unites with three iron atoms and so a small addition of carbon forms a lot of the second phase, cementite.

A. Microstructure

The normal microstructure of iron that has no carbon is the single phase body centered cubic form of iron, called ferrite. As carbon is added, grains of a second phase involving cementite appear in the ferrite. The typical form that cementite takes in steel is a microconstituent called pearlite. Pearlite is a mixture of ferrite and layers of cementite arranged in a lamellar or plywood-like sandwich. Within the grains of the pearlite phase, the equilibrium carbon content is 0.8 wt%. As additional carbon is added to the steel, the amount of the pearlite phase increases. The ferrite phase is essentially free of carbon. Therefore, a medium carbon steel such as 1040 would have a microstructure of approximately 50% ferrite phase and 50% pearlite. At an overall carbon content in the steel of 0.8%, the entire microstructure is composed of pearlite. As the carbon content is increased above 0.8%, discrete particles of cementite appear in the microstructure. Figure 2.7 illustrates the effect of increasing carbon content on the pearlite content.

B. Heat Treatment of Steel

Heat treatment offers the most profound option for strengthening and hardening this material. Steel of the appropriate composition may increase its tensile strength by a factor of eight times through heat treatment. The heat treat process consists of three distinct steps: austenitizing, quenching, and tempering.

If steel is heated above its critical temperature, the BCC crystal structure of ferrite changes to the FCC form called austenite. This process is called austenitizing, and is the first step in the heat treatment process for steel. The Fe_3C cementite that was insoluble in the BCC ferrite is readily soluble in the FCC austenite and quickly forms a single phase solution of high carbon austenite when heated above the critical temperature. Figure 2.8 shows a modified view of the crystal structure previously shown in Figure 2.1. From this figure, which shows the atoms as spheres rather

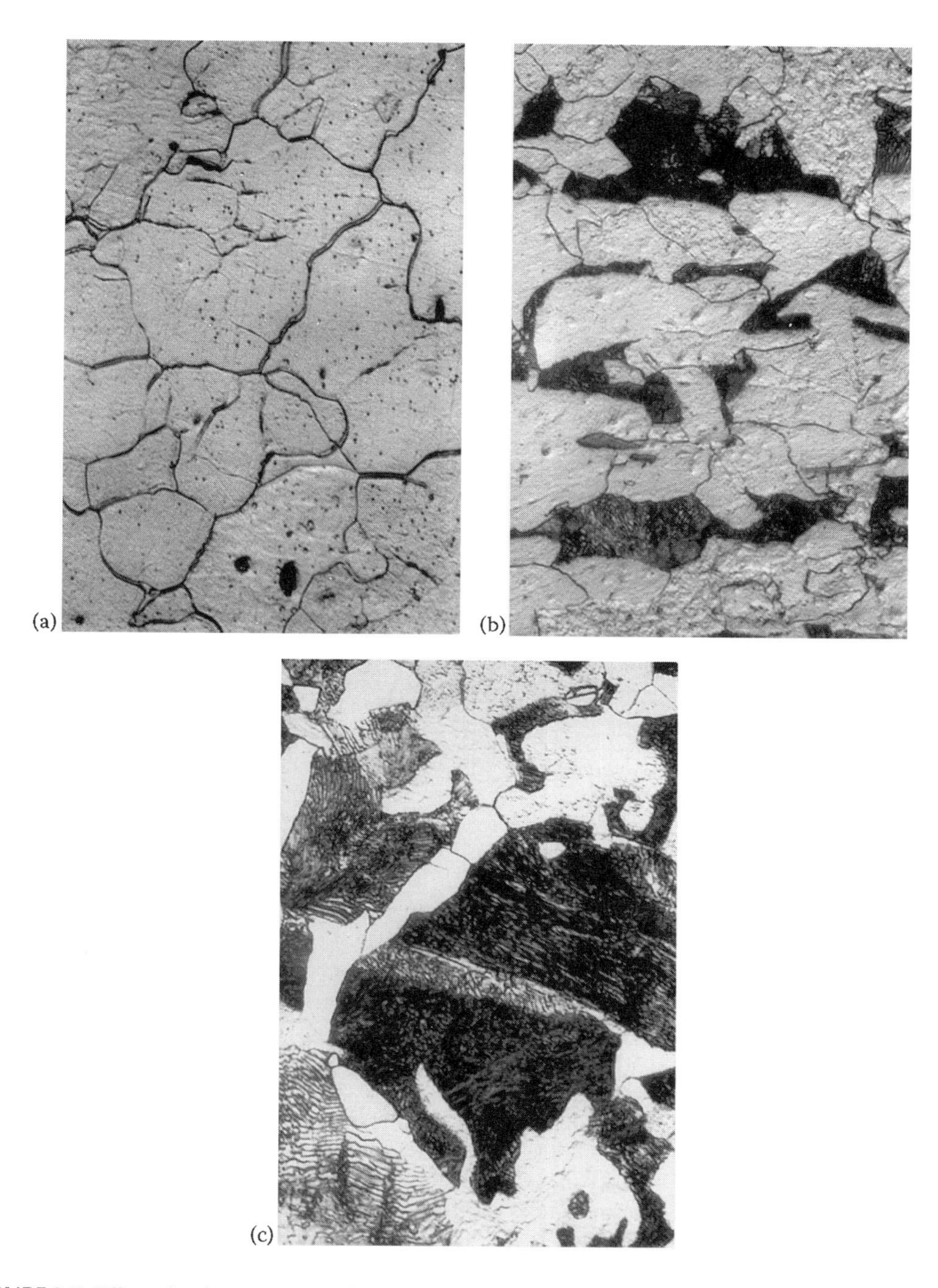

FIGURE 2.7 Effect of carbon content on the microstructure of steel. Original magnification of all micrographs was 500×, 2% nital etchant. (a) 1005 steel with 0.05% carbon. Microstructure is completely ferritic. (b) 1045 steel with 0.45% carbon. Microstructure is a mixture of ferrite (white grains) and pearlite (dark etching phase). (c) 1075 steel with 0.75% carbon. Microstructure is predominantly pearlite (dark etching phase) with lesser amounts of ferrite (white grains).

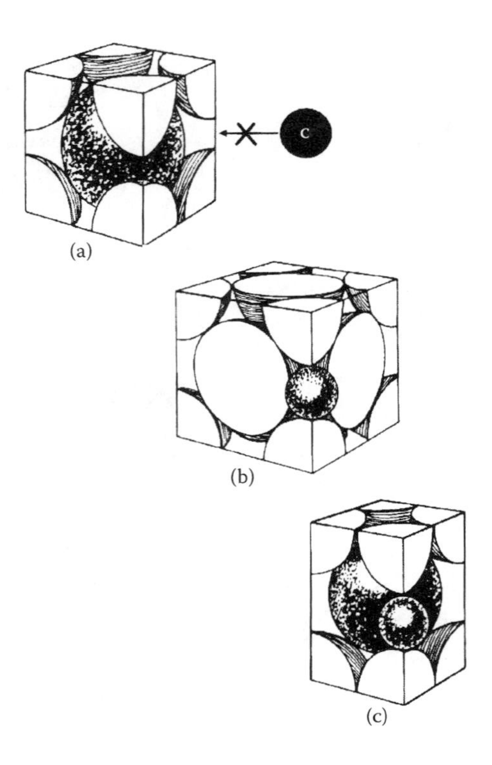

FIGURE 2.8 The solubility of carbon in heat treatment and its effect on the microstructure of steel. (a) BCC ferrite has very low solubility for carbon due to interstitial site too small for the carbon atom. (b) FCC austenite has optimum size interstitial site for carbon atom and high solubility for carbon. (c) BCT martensite has carbon atom trapped in an interstitial site that would be too small and is distorted.

than points, it is possible to visualize why the carbon is soluble in FCC austenite and not in BCC ferrite.

Note that the interstitial space between the iron atoms in the BCC structure is just barely too small to accommodate the carbon atom. The slightly larger interstitial space between the atoms of the FCC structure is perfectly sized for the carbon atom.

Once austenitized with the carbon in solution, if the steel were to be cooled slowly back to room temperature, the carbon would be rejected from solution and the structure would revert to its original microstructure of the ferrite and cementite mixture. However, if the steel is rapidly cooled, the dissolved carbon is trapped in solution in the austenite and transforms to the metastable BCT structure called martensite. This second step in the heat treat process is called quenching.

The BCT martensite crystal structure is the hard, strong form in steel but tends to be very brittle in the as-quenched condition. This necessitates the third step in the heat treat process called tempering. Freshly transformed martensite has a highly distorted and stressed crystal lattice. Although it is very hard, it is also very brittle. A degree of toughness can be restored with only a slight sacrifice in hardness by performing a tempering operation. Tempering is carried out by heating in the range of 300 to 1000°F (150 to 540°C), which is below the critical temperature where the steel would again transform to austenite. Increasing the tempering temperature lowers the resulting hardness. At 300°F (150°C) the reduction in hardness is only one or two points on the Rockwell C scale. At 1000°F (540°C) the reduction in hardness may be as much as 30 Rockwell C points. Hardened parts are never used in the as quenched condition without tempering. Figure 2.9 shows the response of a typical alloy, 4140 steel, to thermal tempering.[10]

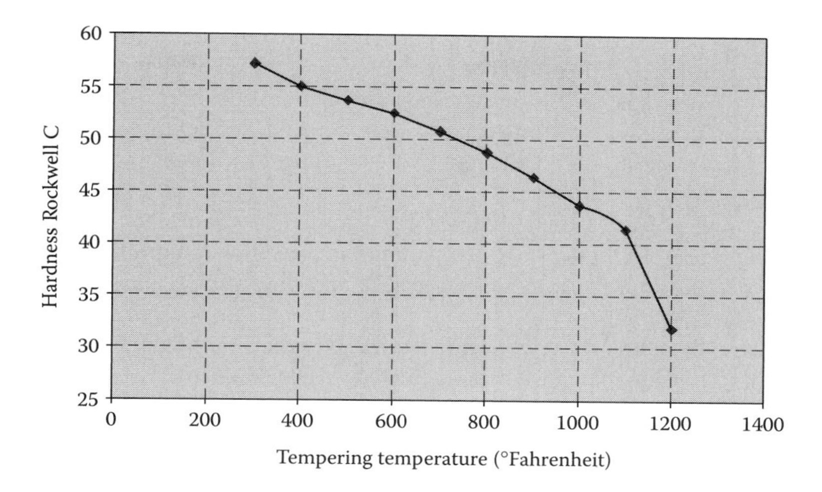

FIGURE 2.9 Temper response of 4140 alloy steel.

In summary, there are three distinct steps in the heat treatment of steel. Initially, of course, the steel must have adequate carbon and alloy composition for the intended application. Step one is austenitizing or heating to the critical temperature where it transforms to the FCC structure and dissolves the carbon. Step two, it must be quenched or cooled at a rate that is sufficiently fast to prevent reversion to ferrite, trapping the crystal structure in the hard martensite form. Finally, in step three, the steel must be tempered to restore some degree of toughness and to achieve the specified hardness.

Manipulation of these criteria (carbon plus heating above the critical temperature, followed by rapid cooling, followed by tempering) forms the basis for all of the various heat treatments commonly used for martensitic hardening of steel. The same principles apply to cast iron, the matrix of which, exclusive of the graphite particles, may be thought of as high carbon steel. This most straightforward hardening technique is called neutral or through hardening, and subjects the entire part being hardened to all of the criteria as described above. This process produces uniform hardness throughout the part.

Frequently it is desired to selectively harden only a small area of the part, the end of a shaft or the teeth of a gear, for example. In this case it is possible to heat only that portion of the component to be hardened. When quenched, only the heated area will be transformed to martensite. Alternately, it would be possible to heat the entire shaft or gear but only quench the areas to be hardened. The remainder of the shaft would be allowed to cool slowly, thus reverting to the soft ferrite plus cementite form.

Another method, called carburizing, begins with a low carbon steel that has insufficient carbon to harden. When austenitized or heated above its critical temperature in the presence of an atmosphere containing a source of carbon, such as methane, the surface of the part absorbs carbon which then diffuses inward into the steel. The depth to which the extra carbon diffuses is a function of the time and temperature during exposure to the carbon-rich atmosphere. When quenched, only the high carbon surface case transforms to hard martensite, while the center or core of the part, where carbon did not reach, will remain softer and tougher. A similar type of case hardening may be accomplished using an atmosphere of ammonia instead of methane in a process called nitriding. A hybrid process using both methane and ammonia is called carbonitriding. The fact that the chemical composition of the surface of the steel is modified in these processes distinguishes them from the neutral or through hardening process.

There are other reasons for heat treatment besides hardening. Normalizing is a conditioning heat treatment applied to steel bars, plate, castings, and forgings. This process is usually performed during fabrication at the steel mill, forge shop, or foundry. Its purpose is to homogenize and soften to prepare the steel for machining or further processing. It consists of heating the metal above the critical temperature where austenite is formed and allowing it to cool normally in the open air under ambient conditions. Since the cooling rate is not controlled, the results are not uniformly predictable but it is the most energy efficient and economical means of producing a machinable hardness.

Another softening process is called annealing. This process is carried out like normalizing except the cooling rate from the elevated temperature is slower and controlled, usually inside the annealing furnace. The results are more predictable and the material may be put in its softest condition by this process.

Another process called stress relieving is applied to precision manufactured components to remove residual stresses that may have been induced by the manufacturing processes. The purpose is usually to render the part dimensionally stable. Although normalizing and annealing are also effective at relieving stresses, the resulting change in hardness and microstructure may not always be desirable. For this reason, stress relieving is usually carried out below the critical temperature so there is no transformation in crystal structure and distortion is minimized.

The heat treating processes described above are generally applicable to steel, cast iron, and the martensitic grades of stainless steel. Exceptions are that cast iron is not a candidate for carburizing since it already contains excess levels of carbon, and stainless steel is not usually carburized because of its adverse effects on corrosion resistance. The precipitation hardening grades of stainless steel are hardened by a completely different mechanism, one that is analogous to the heat treatment applied to aluminum.

VII. NONFERROUS METALS

It is implicit in the term nonferrous that this category of materials includes all metals where iron is not the major element present, and includes a great number of different metals. Although titanium, nickel, and cobalt-based alloys are widely used in the aerospace industry, for the sake of brevity, we will concentrate on the two most common industrial nonferrous alloy systems, aluminum and copper.

A. Aluminum

The most striking property of the aluminum alloys is density, which is only about one third that of steel. Other important properties of aluminum are high thermal and electrical conductivity, good corrosion resistance, and ease of fabrication. This unique collection of properties makes aluminum well suited to a variety of commercial applications. Aluminum is available in practically all wrought product forms such as plate, sheet, foil, bar, rod, wire, tubing, forgings, and complex cross-section extrusions. In addition, its low melting point and fluidity make aluminum ideal for casting. Sand, plaster, permanent mold, and pressure die castings are readily available.

There are two main classification systems for aluminum alloys; one system for wrought products and one for castings. The major specifying body is the Aluminum Association. The designation system for wrought products is based on a four digit system, while the cast form is designated by a three digit number. Both product forms typically carry a suffix of one letter and one to four numbers that describe the temper or strengthening process applied to the product. Table 2.5 shows the Aluminum Association designation system for wrought aluminum products, while Table 2.6 shows a similar designation system for cast aluminum products.

Pure aluminum is a chemically reactive metal and will form many chemical compounds. The good corrosion resistance of aluminum is due to the natural tendency of aluminum to form a tenacious oxide, Al_2O_3, on the surface when exposed to air. Even if scratched the oxide will quickly renew itself. In addition to this natural oxide forming tendency, a thicker and more protective oxide layer can be induced by an electrochemical process called anodizing. In this process the aluminum

TABLE 2.5
Aluminum Association Designation System for Wrought Aluminum Products

Alloy Series	Nominal Alloy Content
1XXX	99.00% purity, <1% alloy
2XXX	Copper plus minor additions of manganese and magnesium
3XXX	Manganese and magnesium
4XXX	Silicon
5XXX	Magnesium plus minor additions of manganese and chromium
6XXX	Silicon and magnesium plus minor additions of copper, manganese, or chromium
7XXX	Zinc and magnesium plus minor additions of copper and chromium

part is made the anode in an electrolyte of a strong oxidizing acid such as sulfuric or chromic acid. The cathodes are inert lead bars. Passing a current through this system develops a heavy, porous aluminum oxide coating. After rinsing in cold water, the still porous coating may be dyed a variety of colors by immersion in a dye bath. When the desired color has been achieved, the anodized coating is sealed by immersing the part in hot water which hydrates the oxide, thus sealing the porosity. If the acid anodizing bath is refrigerated during the anodizing process, exceptionally thick and ceramic-like oxide coatings can be developed. This process is called hard anodizing and produces very wear-resistant coatings. Aluminum alloys containing appreciable quantities of silicon are difficult to anodize, especially for decorative purposes.[11]

B. Heat Treatment of Aluminum Alloys

Certain wrought alloys in the 2XXX, 6XXX, and 7XXX series, and the 2XX, 3XX, and 7XX series casting alloys are heat treatable by a process called precipitation hardening. The alloys 2024, 6061, and 7075 are common high strength wrought alloys, while 319, 355, and 356 are common casting alloys that are subject to hardening through a precipitation or age hardening heat treatment. The mechanism as described here also applies to the precipitation hardening grades of stainless, albeit the temperatures required are substantially higher. In addition to strengthening, precipitation hardening provides the most machinable condition in most aluminum alloys. Copper, magnesium, and zinc form intermetallic compounds or secondary phases with the aluminum microstructure, as shown in Figure 2.10.

TABLE 2.6
Aluminum Association Designation System for Cast Aluminum Products

Alloy Series	Major Alloy Addition
1XX	99.0% purity, <1.0% alloy
2XX	Copper
3XX	Silicon with minor additions of copper and/or magnesium
4XX	Silicon
5XX	Magnesium
6XX	Unused series
7XX	Zinc
8XX	Tin

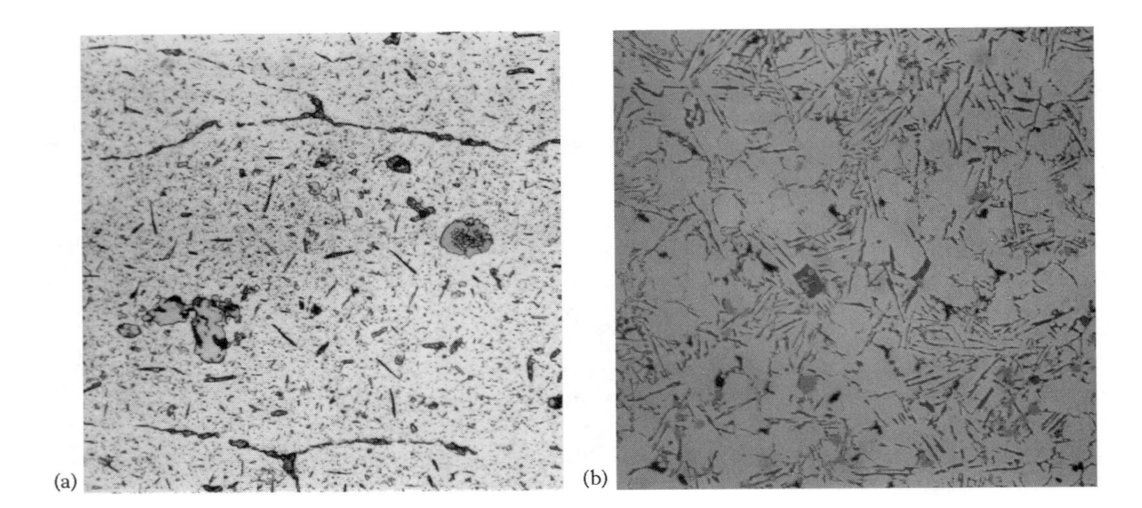

FIGURE 2.10 Comparison of wrought and cast microstructure of aluminum. Original magnification of all micrographs was 500×, 2% HF etchant. (a) Wrought alloy 3003. Microstructure is particles of $MnAl_6$ in a solid solution matrix. Particles are elongated and oriented in the rolling direction (left to right). (b) Die cast alloy 308. Microstructure is particles of $CuAl_2$ and silicon eutectic in a solid solution matrix. There is no preferred orientation as in the wrought alloy.

In the equilibrium or slowly cooled condition, these intermetallic compounds form and grow to relatively large microconstituents within the metal. As such, they have a relatively minor effect on hardening or strengthening the alloy. If the alloy is heated to a point where the intermetallic compounds are redissolved into the matrix, and then rapidly cooled at a rate faster than would allow them to reform, the alloy is said to be solution annealed. A low temperature heating then encourages the precipitation of the intermetallics in very fine particles that are practically undetectable by conventional optical microscopy. An alloy processed in this manner is said to be precipitation or age hardened. The extensive distribution of the very fine precipitates inhibits slip on the critical crystallographic slip planes and results in the strengthening effect. Typical solution annealing is carried out by heating in the range of 1000°F (540°C) followed by water quenching. This leaves the alloy in its softest and most formable condition. A few alloys, 6061 for example, will allow the precipitate to form at room temperature and these are said to be naturally aging grades. Other alloys, such as 2024, must be heated to the range of 350°F (180°C) to induce the precipitation to occur, and these are said to be the artificial aging grades. The heat-treated condition

TABLE 2.7
Commonly Specified Temper Designations for Aluminum Products

Temper Designation	Process Description
F	As fabricated, no additional thermal processing
O	Annealed to lowest strength for maximum ductility and dimensional stability
H	Strain hardened (applies to wrought products only). May be followed by one or more numbers that further describe the strain hardening process
W	Solution heat treatment, applies only to alloys that naturally age harden
T	Heat treated to produce a stable temper. Always followed by one or more numbers that define the specifics of the process used

TABLE 2.8
Unified Numbering System for Copper Alloy Designation

Alloy Group	UNS Designation	Principle Alloy Element	Solid Solubility (at. %)
Copper and high copper alloys	C10000	None	
Brasses	C20000, C30000, C40000, C66400–C69800	Zinc	37
Phosphor bronze	C50000	Tin	9
Aluminum bronze	C60600–C64200	Aluminum	19
Silicon bronze	C64700–C66100	Silicon	8
Cupronickel	C70000	Nickel	100

is described by a system of suffixes, called the temper designation, which is appended to the alloy number. Table 2.7 shows the common temper designations used for aluminum products.[7]

C. Copper and Copper Alloys

Copper is one of the few metals found in its metallic form in nature. The metal, along with its principle alloys, is an important group of engineering materials. Among the outstanding characteristics of copper and copper alloys are excellent thermal and electrical conductivity, formability, castability, and corrosion resistance. Certain bronze and brass alloys have excellent tribological compatibility with steel and are frequently used in bearing and bushing applications. Copper alloys protect themselves from corrosion by forming a tenacious oxide which is the familiar green patina seen on exposed architectural elements and marine fittings. The oxide coating on copper is electrically conductive and as a result, copper wiring does not have the problem with electrical contact resistance that aluminum wiring does. Copper alloys are susceptible to stress corrosion cracking, especially in the presence of ammonia. The Copper Development Association (CDA) carried out much of the classification of copper alloys. The CDA numbers were adapted into the Unified Numbering System (UNS), which is now the most widely recognized designation system. Table 2.8 shows the UNS designation system for copper alloys.

The significance of the solid solubility limits shown in Table 2.8 is that copper alloys having less than the indicated limit of alloying element will have a single phase microstructure. In these systems the alloying element is present in the form of a substitutional solid solution. Such alloys exhibit substantially increased strength in combination with good ductility and formability. When the alloy content exceeds the limit of solid solubility, a second phase appears in the microstructures, and even higher strengthening results. Two phase alloys lose some of their ductility, and the capability for rolling thin sheets may be diminished or completely eliminated. Lead, tellurium, and selenium are elements added to copper alloys to promote machinability. These elements are insoluble in the matrix, produce a separate phase in the microstructure, and behave much as sulfur does as a free machining additive to steel. Lead also imparts a self-lubricating characteristic to bearing alloys.

Additional hardening and strengthening above that produced by alloying can be obtained by cold working. The hardness of cold worked wrought copper and brass is usually expressed as a fraction related to the degree of cross-sectional area reduction accomplished during the rolling process. Table 2.9 shows the temper designations commonly produced in wrought copper alloys.[11]

VIII. MEASUREMENT OF SURFACE FINISH

Specifying surface finish is basically a process of describing the topography and texture of the boundary surface of a solid body in quantifiable terms. Surface finish is an important parameter of

TABLE 2.9
Temper Designations for Rolled Copper Alloys

Temper Designation	Reduction in Thickness/Area (%)
1/4 hard	10.9
1/2 hard	20.7
3/4 hard	29.4
Hard	37.1
Extra hard	50.1
Spring	60.5

a component part. It determines how the part will respond to sliding friction, how well it will retain a lubricant, the wear rate that will be experienced, and how well it will retain a coating, such as electroplating or painting to name just a few. Surface finish measurement is also closely linked to dimensional tolerancing. It would be irrational to reference a very precise dimension from a very rough surface.

A. Terminology

Although there are various noncontact methods of measuring surface texture, such as electrical capacitance and laser interferometry, the most common and widespread method currently in use is by the contact stylus method. In this technique, a diamond or gemstone tipped stylus having a tip radius typically 0.0004 in. (10 μm), is dragged across the surface being measured; the up and down motion of the stylus is tracked electronically. Much of the terminology in this discussion is appropriate to either contact or noncontact measuring methods. The language of surface finish measurement contains a number of unique terms, and before going much further it would be well to provide definitions for these terms. Figure 2.11 serves to illustrate the physical significance of these terms.

Nominal surface is a hypothetical surface that defines the shape of the body, such as would be depicted by an engineering drawing. The nominal surface is smooth and serves only as a reference for dimensioning and assigning the allowable tolerance for deviations from the surface.

The surface topography (also called texture) parameter is the composite of all the deviations from the nominal surface. The irregularities comprising the texture are several, including the following:

Form: These are deviations from the specified surface geometry such as taper, concavity, convexity, twist, etc.

Waviness: These are relatively long range periodic deviations that may have resulted from such sources as cutter chatter, machine vibrations, or an out of balance grinding wheel, which alter the path of the cutter from that actually intended.

Roughness: Relatively closely spaced deviations resulting from the interaction of the tool and workpiece such as tears, gouges, cutter marks, and built up edge sloughing. The actual cutter path need not vary from the path intended.

Flaws: Defects not necessarily related to the cutting process. Scratches or dents occurring after the surface was cut, and cracks or porosity in the material are examples of flaws. These defects are random in orientation and spacing with respect to other surface texture features.

Lay: Surfaces generated by a cutting process such as lathe turning, milling, planing, or grinding have an obvious directional quality. Processes such as flame cutting and welding produce directionality on a very gross scale. This directional orientation is called "lay,"

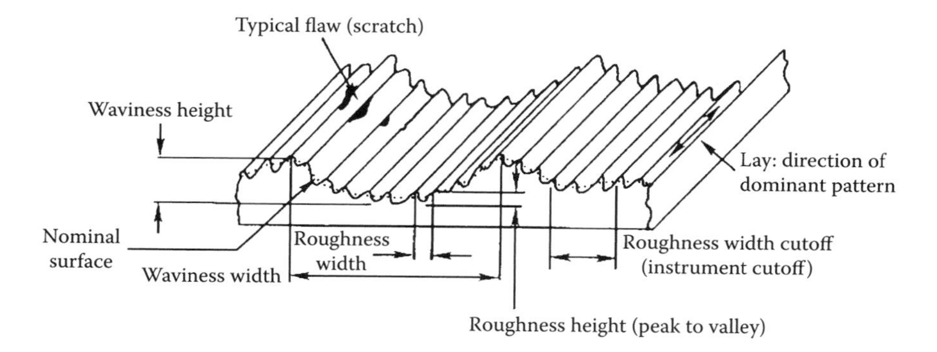

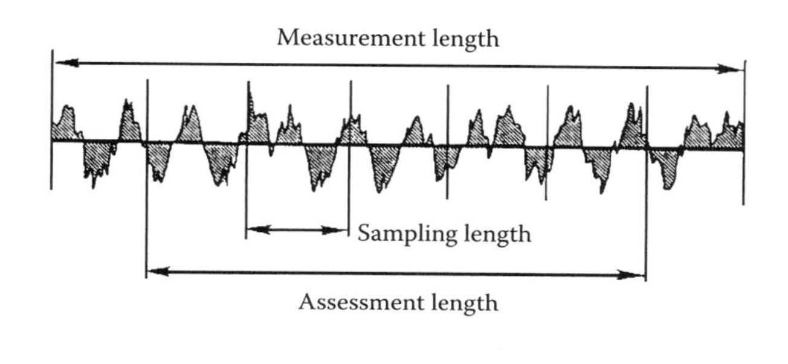

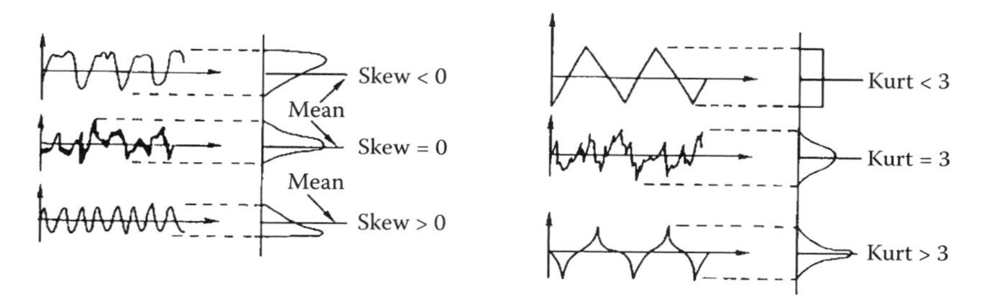

FIGURE 2.11 Surface finish terminology. (Reprinted with permission from AS291E © 2001 SAE International.)

and the lay direction is parallel to the major lines defining the lay. Some processes such as casting, shot blasting, or electrical discharge machining produce surfaces with no discernible directional characteristics.

Nominal profile is the hypothetical line created by the intersection of a plane at right angle with the nominal surface. The nominal profile serves as the reference base line for superposition of the measured profile. The nominal profile through a flat surface will be a straight line, through a cylindrical surface will be a circle, etc.

Measured profile is the wavy, zigzag line that is defined by the intersection of a plane at right angle to the measured surface and at right angle to the lay of the finish. It is generally this profile that the stylus of the surface measurement device tries to trace.

Peaks and valleys: Peaks are points or ridges that protrude above the plane of the surface, while valleys are holes or troughs that lie beneath the plane of the surface.

Mean line is a straight constructed line that divides the measured profile line, such that the area enclosed by the peaks is equal to the area enclosed by the valleys.

Sampling length is also called cutoff length. This is a preselected distance over which the measurements are taken and parameters computed. By judicious selection of the cutoff length, longer scale effects of waviness and form errors can be filtered out. A commonly used cutoff length is 0.030 in. (0.76 mm).

Assessment length is the stylus travel distance over which the average profile is determined. The stylus stroke should encompass at least five cutoffs plus some amount of overtravel on each end of the assessment length.

B. Concepts and Parameters

The reader has likely seen traces from surface finish measuring equipment and noted that the profile is very sharp with the slope of the peaks and valleys being very steep. The deviations of actual surfaces on real bodies are much more gently undulating hills, bumps, and valleys. Indeed, if actual bodies had such sharp sloped peaks and valleys, friction would be infinite and sliding of two mating surfaces past each other would likely be impossible. The reason that the surface finish trace looks different from the actual surface is that the vertical scale of the trace must be greatly magnified to resolve the minute fluctuations of the surface. If the profile were equally magnified in both the vertical and horizontal directions, the profile trace would be a proportional magnification of the surface but the trace would require yards and yards of chart paper. Typical magnifications are 200× in the horizontal direction and 5000× in the vertical direction, and this makes the profile look much sharper and steeper than it actually is.

Once the profile trace has been acquired, a number of mathematical manipulations of the data are made to generate the following roughness parameters:

R_a is the average surface roughness computed as the arithmetic mean of the absolute value of the distance between the baseline to the maximum peak or valley height. The average roughness is easier to visualize if the bottom half of the trace (the valley part) is flipped up onto the top half (the peak part) of the trace as shown in Figure 2.11(b). The line defining the mean height of this flipped trace is the R_a roughness. R_a is the most universally used surface roughness parameter. The units of surface roughness are μin. in the English system and μm in the metric system.

R_q is the equivalent of R_a except the root mean square method is used as the averaging technique instead of arithmetic averaging.

R_{sk} denotes skewness, which refers to the distribution of peaks and valleys about the mean line as shown in Figure 2.11(c). Surfaces that have peaks and valleys of equal height and depth have zero skew. If the valleys are deeper than the peaks are high the surface has negative skew. If the peaks are higher than the valleys are deep the surface has positive skew.

R_{ku} denotes kurtosis, which is a parameter that describes the sharpness of the profile as shown in Figure 2.11(d). A typical surface has a kurtosis of approximately 3. If the points of the peaks and valleys are more obtuse or flatter than average, the kurtosis is less than three. If the points on the peaks and valleys are very acute or sharper than average, the kurtosis is greater than 3.

t_p is the percent bearing ratio. It is determined by drawing a construction line parallel to the mean line at a specified height above the mean line thus cutting off the peaks. It simulates the effect of wearing away the peaks and projects the resulting bearing area that would be in contact with a perfectly flat mating surface.

Various methods of creating surfaces produce their own characteristic surface textures. Figure 2.12 shows the range of surface roughness that can be expected for a number of common commercial production processes. Under ideal conditions, the actual surface roughness may be controlled to higher or lower values, but this chart gives a good approximation of the surface finish that can be achieved by the various manufacturing methods.[12–14]

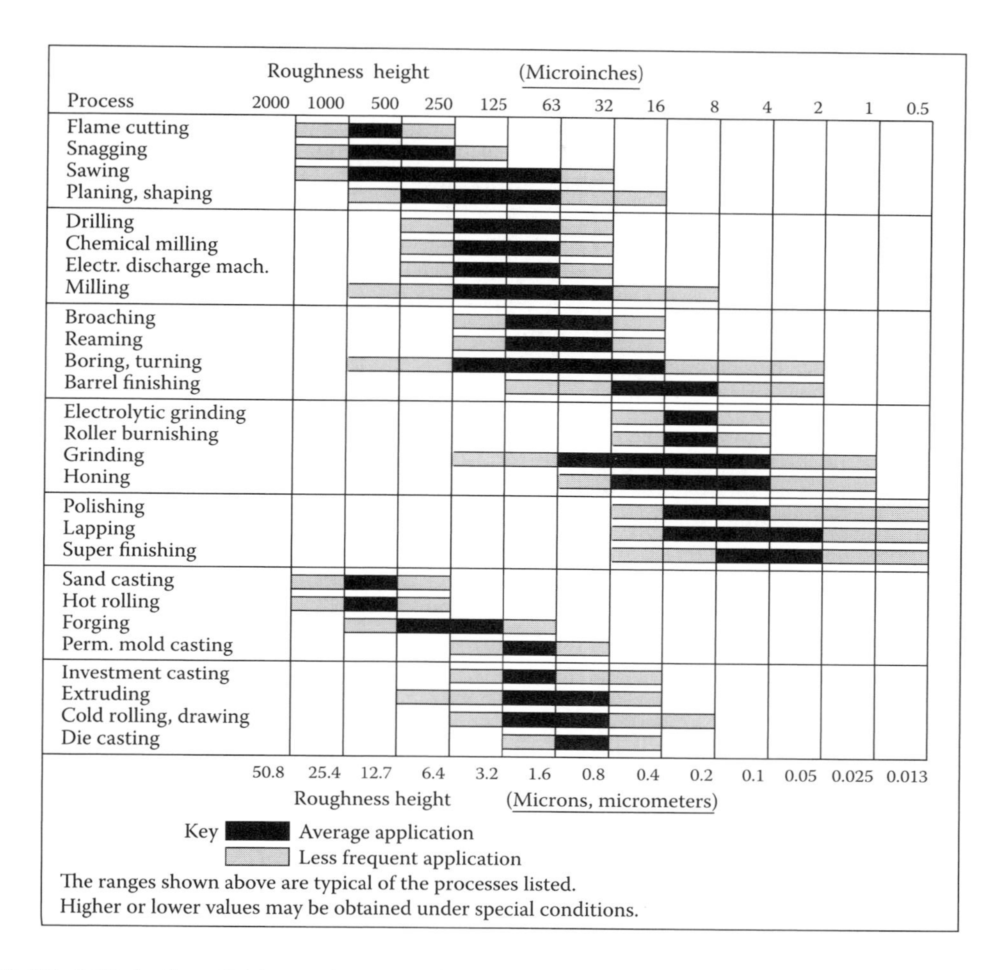

FIGURE 2.12 Surface finish roughness produced by common production methods. (Reprinted with permission from AS291E © 2001 SAE International.)

C. Interpretation of Engineering Symbols

Engineering drawing symbols are used to convey the designer's intentions to the machinist or manufacturer who must create the actual component. As shown in Figure 2.13, surface finish

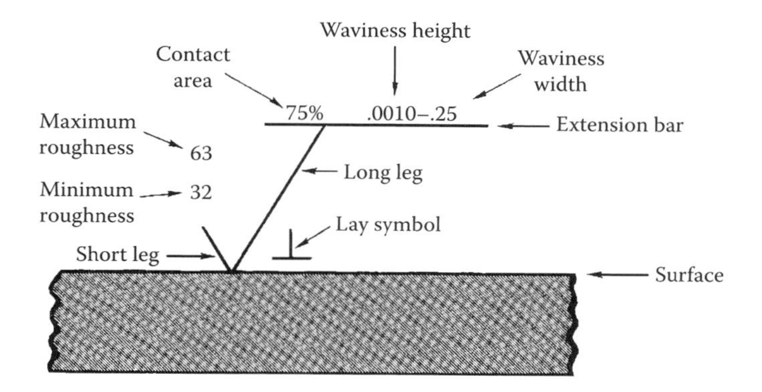

FIGURE 2.13 Engineering symbols for surface texture.

requirements are indicated by a check-like mark, which may sometimes have an extension bar extending from the top of the long leg of the check.

The point of the symbol touches the surface to which it applies, or the dimension extension line from that surface. The required surface roughness appears as a number just above the short leg of the check. If only one number appears here it indicates maximum allowable roughness. If the roughness must be controlled between a maximum and minimum value, two numbers appear. The units of the numbers, English or metric, are consistent with the other dimensions of the drawing. The maximum allowable waviness height is shown on the extension bar to the right of the intersection of the long leg of the check. Waviness width, when required, is placed to the right of the waviness height separated by a dash. Desired contact area with the mating surface is expressed as a percentage, and appears directly above the intersection of the extension bar and long leg of the check. The lay direction is shown to the right of the point of the check and may specify perpendicular, parallel, angular, multi directional, circular, or radial lay direction. All of the above symbols may not appear on an engineering drawing unless it is intended that those feature tolerances must apply to the manufactured component.

The precision and functional performance of any manufactured component is strongly dependant on its surface texture. Understanding the fundamentals of specifying surface texture parameters will help insure that this finished component will perform as intended.

REFERENCES

1. Guy, A. G., *Elements of Physical Metallurgy*, Addison-Wesley, Reading, MA, pp. 72–78, 1959.
2. Lysaght, V. E. and DeBellis, A., *Hardness Testing Handbook*, Wilson Instrument Division, American Chain and Cable Co., Reading, PA, 1969.
3. Keyser, C. A., *Basic Engineering Metallurgy*, Prentice Hall, Englewood Cliffs, NJ, pp. 59–78, 1959.
4. Dalton, W. K., *The Technology of Metallurgy*, Macmillan, New York, pp. 79–106, 1993.
5. American Society for Metals. Properties and selection: stainless steels, tool materials and special-purpose metals, In *Metals Handbook*, 9th ed., Vol. 3, American Society for Metals, Metals Park, OH, pp. 3–56; see also pp. 114–143, 1978.
6. Roberts, G. A. and Cary, R. A., *Tool Steels*, 4th ed., American Society for Metals, Metals Park, OH, pp. 227–229, 1985.
7. American Society for Metals. Properties and selection: stainless steels, tool materials and special-purpose metals, In *Metals Handbook*, 9th ed., Vol. 3, American Society for Metals, Metals Park, OH, pp. 3–6 and 421–433, 1980.
8. Peckner, D. and Bernstein, I. M., *Handbook of Stainless Steels*, McGraw-Hill, New York, pp. 1.1–1.10, 1977.
9. Walton, C. F. and Opar, T. J., *Iron Casting Handbook*, American Foundrymen's Society, Cast Metals Institute, Des Plains, IL, pp. 207; see also pp. 225, 326, 446, 1981.
10. ASM International, *Heat Treater's Guide, Practices and Procedures for Irons and Steels*, 2nd ed., ASM International, Metals Park, OH, pp. 319–325, 1995.
11. American Society for Metals. Properties and selection: nonferrous alloys and pure metals, In *Metals Handbook*, 9th ed., Vol. 2, American Society for Metals, Metals Park, OH, pp. 3–43; see also pp. 140–143, 239–251, 1979.
12. Amstutz, H., *Surface Texture: The Parameters*, Sheffield Measurements, Fond du Lac, WI, 1978.
13. *Aerospace Standard AS 291E*, Society of Automotive Engineers, Warrendale, PA.
14. The American Society of Mechanical Engineers. *Surface Texture Standard ANSI B46.1*, The American Society of Mechanical Engineers, New York, 1978.

3 Metal Cutting Processes

Stuart C. Salmon

CONTENTS

I. BACKGROUND

Metal cutting may take the form of a number of production and manufacturing processes. Metal may be cut by sawing, shearing, and blanking; sliced with slitting saws and grinding wheels; cut by lasers, sonic, electro-chemical processes, and water jets; milled, drilled, planed, broached, turned, and ground. Indeed, industry cuts metal with a variety of techniques and technologies. However, in this chapter, we will be concentrating on the machining aspect of metal cutting. In particular, we will be looking at the influence of the cutting fluid on the machining process, with respect to "large" chip making process such as turning, milling, planning, drilling, and broaching. The "small" chip making processes of grinding and abrasive machining are given to another chapter.

Metal cutting operations have been performed as far back as the Greek and Roman era, and perhaps even before. The skills of the metal worker during those times were kept secret and held tightly to the chest. Job security was felt to be important back then too. Trial and error was the mode of operation. It was not until the industrial revolution had taken place, roughly between the years 1760 and 1850, that metal cutting was academically researched and studied. Machine tools had to be invented. The concepts and designs for metal cutting machines were founded in the apparatus and mechanical schemes of early wood cutting lathes. The earliest depiction of a lathe comes from a Ptolemaic Period tomb painting, around the third century B.C.[1] So was the progression made from woodworking to metalworking. Just as today, our research and development programs are still driven by materials technology, as we move from metals to composites and ceramics, with an even keener eye on higher levels of precision and surface integrity.

Though industry today still uses the trial and error approach, and the experience of the skilled operator to shorten the time to a workable solution, there has been a surge in the academic community to move away from the pure sciences and toward the applied sciences in order to understand, predict, and model the complexities of metal cutting with the myriad of variables associated with manufacturing processes.

Iain Finnie[2] published a work in 1956, which reviewed the history of metal cutting theories and cited the work of Cocquilhat, in 1851, as the earliest academic researcher who focused initially on drilling. However, it was Frederick Winslow Taylor (1856 to 1915) who first applied a scientific method to developing a process model for metal cutting, in particular that relating to the prediction of tool life.[3]

Taylor's basic tool life equation establishes a relationship between tool life and cutting speed. His equation is:

$$VT^n = C$$

where T is the tool life in minutes, V is the cutting speed, and both C and n are constants depending on the cutting conditions (the depth of cut, cutting fluid, material, etc.). Notice that for a tool life of 1 min, $V = C$.

Taylor's equation has been modified only slightly since 1907, to take into account depth of cut and feed rate. With some minor modifications, his model holds, even today through carbide, ceramic, and superabrasive tools. The modified equation is:

$$VT^n f^m a^p = K$$

where T is the tool life in minutes, V is the cutting speed, f is feed rate, a is depth of cut, and n, m, p, and K are constants. These constants will vary with tool properties, the most influential being tool coatings. Coatings have dramatically changed the life and performance of cutting tools in recent times. They provide not only wear resistance, but also oxidation resistance, heat resistance, and even lubrication. Nevertheless, Taylor's fundamental equation and relationship still applies and has stood the test of time.

Taylor had modeled and understood the tool wear relationship with respect to feeds and speeds, but there was still a lack of understanding as to how machining really worked. No one was sure by what means a chip was formed or what influenced the chip formation mechanism. Time in 1870, and Hartig, Tresca and Von Mises in 1873 made the first attempts at trying to explain chip formation, but it was Mallock,[5] in 1881, who brought about the theory of shearing and how friction plays a part in the cutting mechanism. He was the first to observe the effects of lubricants and tool sharpness and their affects on vibration and chatter. Franz Reuleaux, around 1900, who was famed more for his geometry and kinematic studies,[6] suggested that a crack forms ahead of the shearing action and that metal cutting was very much akin to splitting wood. His comment was not taken

FIGURE 3.1 Single point machining of 4340 steel where a crack can be seen extending out in front of the cutting tool and moving up along the shear plane.

favorably! It was as though a backward step had been taken by associating cutting metal with wood. However, to some extent he was right, as seen in Figure 3.1.

In 1937, the Fin, Piispanen,[7] first modeled the shear zone and theorized chip formation using shear stress diagrams based on tool geometry and the conservation of energy principle. Later in 1944, Ernst and Merchant[8] published their work, which became the basis for further refinements by Lee and Shaffer[9] in 1951. A series of researchers working with Oxley, from 1959 to 1977,[10] then added their interpretations and refinements. The answer to chip formation in the shear zone had been found. But it was Prof. Ramalingham, at the University of Buffalo, who looked beyond the shear zone. Using polarized light, Ramalingham took pictures to show that machining stresses not only occur in the shear zone, but also penetrate deep into the surface of the material beneath the cutting tool, and perhaps more surprisingly, ahead of the cutting tool action (see Figure 3.2). At relatively low speeds, the stresses within the surface of the workpiece are very high, so there is a pressure wave ahead of the tool which causes the workpiece material to spring back once the tool has passed, causing the surface to rub on the flank face of the tool. The flank face is the face or surface of the tool that supports and trails the cutting edge (see Figure 3.3). The cutting edge is the line boundary between the flank face and the rake face. It is the machined surface that rubs across the flank face and the chip being removed that rubs against the rake face. It is therefore important to have a large flank face clearance angle, or relief, when machining at slow speeds, as well as more of a lubricating type of fluid to assist in minimizing the friction from the rubbing of the tool against the machined surface and preventing any build-up of material on the tool tip. As the machining speed increases, the pressure wave in front of the tool decreases, so there is less spring back, less rubbing,

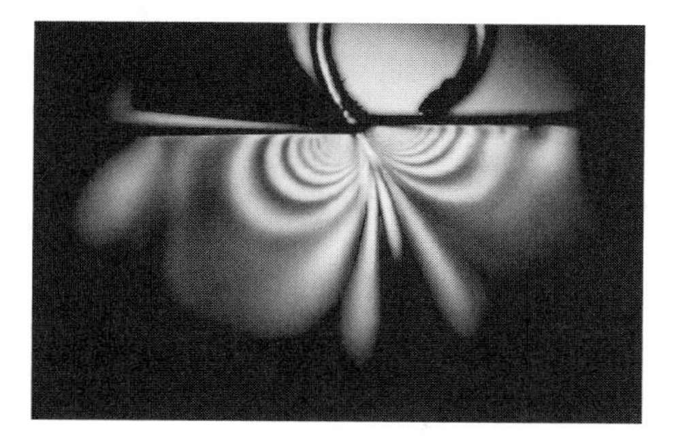

FIGURE 3.2 Researchers concentrated on the shear stresses which occur in chip formation; however, from this polarized light picture, it can been seen that machining stresses penetrate deep into the surface of the material and even ahead of the cutting action.

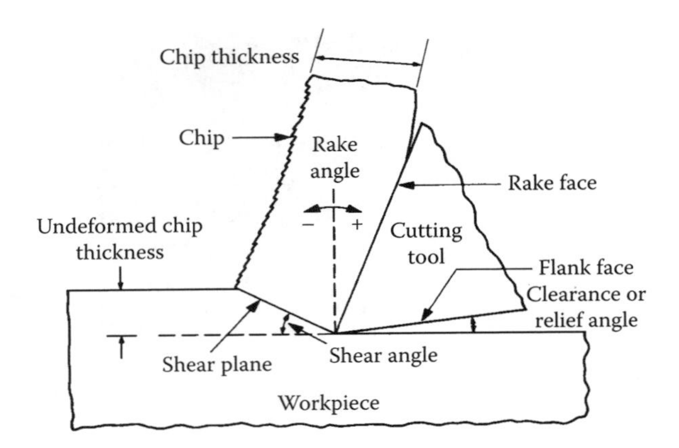

FIGURE 3.3 The geometry and terminology of the cutting action as a section through a single point orthogonal cut.

and any built-up-edge virtually disappears. A cutting fluid that is more of a coolant is, therefore, preferred for higher speed machining, over a fluid which has a great deal of lubricity.

It will be worthwhile, at this juncture, to become familiar with the tool geometry nomenclature, as well as the basic principles of machining for four of the major processes: turning, milling, drilling, and broaching. Though the chip formation at the tool point is virtually identical in each of these processes, with respect to shear and fracture and wear, each process imposes quite different demands on the cutting fluid and its application.

A. Drilling

Drilling is a process whereby a spiral fluted tool with two symmetrical cutting edges, in the form of tapered blades, called the drill point, remove material in a circular motion and transport the chips up and out of the hole by way of the spiral flutes. Drill geometry is quite complex, as shown in Figure 3.4, and it is well to realize that the cutting speed is a maximum at the drill periphery and drops to zero at the center of the drill point. The angle of the drill point is key to successful drilling. The angle is approximately 118 to 135° for general purpose drilling of steels, 90 to 140° for

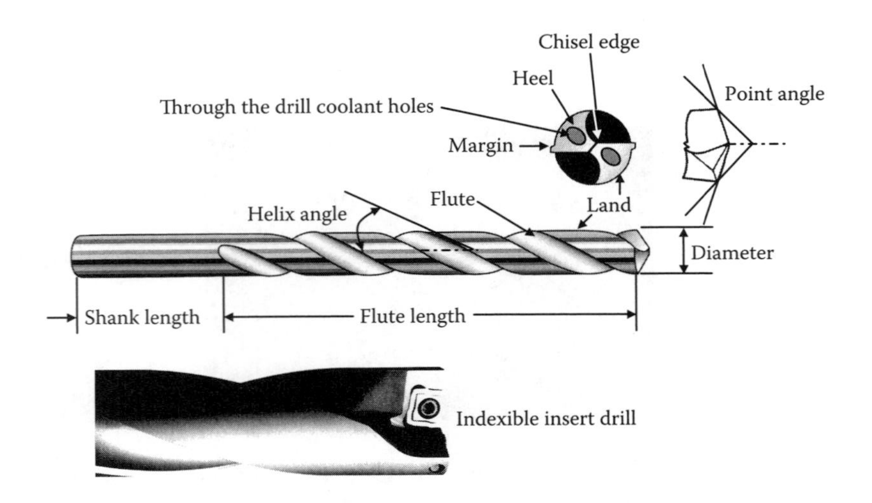

FIGURE 3.4 The basic geometry and nomenclature for a twist drill and an indexible insert drill.

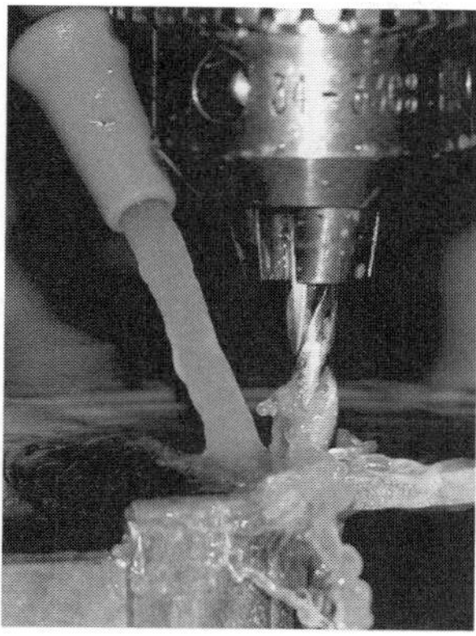

FIGURE 3.5 A twist drill about to drill a hole. The cutting fluid is flooding the drill tip, whereas the center picture shows the drill buried deep into the workpiece with the likelihood of virtually no fluid getting to the drill point. The third picture shows a through the drill fluid application which will take and maintain a flow of fluid to the drill point and evacuate the chips up the flutes.

aluminum and soft alloys and 80° for plastics and rubber. Figure 3.4 shows typical twist drill geometry, as well as an indexible insert drill that adheres to the same geometry, yet is significantly more robust and provides a better means of through the drill fluid application.

The proper application of the cutting fluid, when drilling, is particularly difficult. Figure 3.5 shows how the drill is exposed to the flood of fluid prior to entering the hole but once drilling begins, the tool is buried into the workpiece, cutting with an action that evacuates material from the point of the drill up and out of the hole. If the chips are being evacuated, then so too is any fluid. Some twist drills have through holes that allow the cutting fluid to be pumped down the body of the drill exiting at the heel, close to the drill point. This not only helps to cool and lubricate the drill, but assists in the evacuation of the chips, improving the surface of the walls of the hole.

Gun-drilling, as shown in Figure 3.6, is a variant of conventional drilling in that the hole is made using a tool configuration that cuts to one side. In this case the cutting point is not symmetrical. Unlike the twist drill and indexible insert drill, it uses the wall of the hole being drilled as its support for cutting and maintaining straightness. Gun-drilling therefore requires a cutting fluid that has a high degree of lubricity and low foam; lubrication to minimize the friction between the

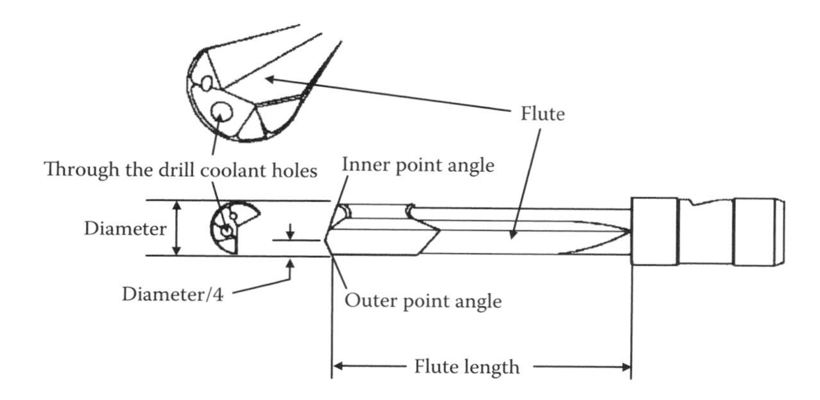

FIGURE 3.6 The basic geometry and nomenclature for a gun-drill.

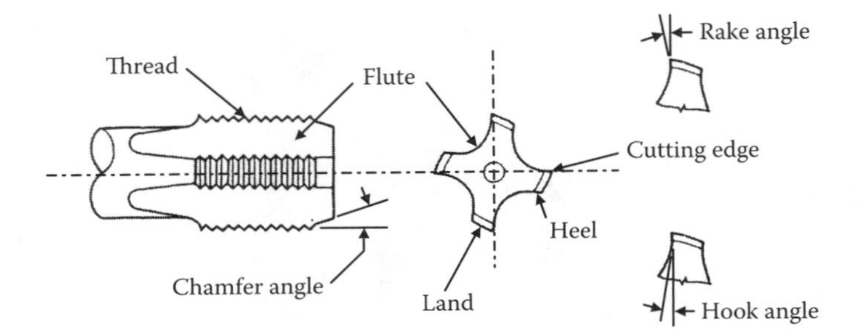

FIGURE 3.7 The basic geometry and nomenclature of a tap.

rubbing support surfaces, and low foam due to the high pressure and high flow rate of the through the tool application of the cutting fluid.

B. Tapping

A further variant of drilling is tapping. Tapping cuts or forms a thread into a workpiece by removing the material from the walls of a previously drilled hole, using a tap (see Figure 3.7). Tapping is performed with a slower cutting speed than drilling, and careful attention must be paid to the axial feed so that it precisely follows the lead of the thread. Chip clearance is provided by the flutes of the tap, but the chip is generally broken by sporadic reversal and withdrawal of the tap, providing an opportunity to clear the flutes and apply fluid to the tap and hole. There are two general types of tapping operations: cut and form tapping. Cut taps make a thread by cutting the metal, generating chips in the process. Form taps make threads by pushing the metal aside, making no chips. A more lubricious fluid is necessary when form tapping, due to the higher amount of rubbing that occurs.

C. Turning

Turning is a process whereby a stationary tool is moved axially along a rotating workpiece (see Figure 3.8). Such an action may produce a straight cylindrical shaft, or by offsetting the tool path or by interpolating in two axes, a tapered shaft may be produced. Straight and tapered cylinders, or with

FIGURE 3.8 The turning operation from the left of the tool and from the right.

FIGURE 3.9 This figure shows that just as with a drill in Figure 3.6, a turning tool may also benefit from a high-pressure jet of fluid, more as a liquid chip breaker than an evacuator of the chips.

more complex axis interpolation, spheres and threads, in fact almost any shape may be turned around the surface of the rotating workpiece with respect to the turning tool shape, geometry, and tool path. Turning may be performed on the outside of a shaft, as well as on the inside of a tubular shaft.

When turning, the chip formation is continuous as the tool is in continual contact with the workpiece. The chips may therefore become long and stringy and difficult to handle. A chip-breaker, which is usually an angled piece of carbide, may be placed behind the point of cut so that the chip, instead of moving along the entire rake face, is curled tightly by the chip-breaker causing it to break the chip into small curls. This not only makes chip handling easier, it may also lead to cooler cutting as the amount of frictional rubbing that occurs across the rake face is minimized. Some of these geometries may be cast into indexible carbide and ceramic inserts. Another method for controlling the chip formation in turning is the use of a high pressure jet of cutting fluid onto the back of the chip as it is being sheared (see Figure 3.9). A jet of fluid in the order of 0.5 l/sec (8 gpm) flow and 100 bar (1500 psi) pressure blast at the back of the chip and into the nip between the chip and the rake face provides cooling and an efficient and nonwearing chip-breaking action.

The lathe or turning machine may also be used to bore holes into workpieces, first by drilling into the face of the workpiece with a drill mounted in line with the rotational axis of the part. Once drilled, the hole may be opened to a larger diameter using a boring tool. The length to diameter ratio of the boring bar, which holds the cutting tool, is critical in that the system has to be vibrationally stable. In extreme cases the boring bar can whirl as well as vibrate torsionally, resulting in poor surface integrity and a great deal of noise.

The turning tool is in contact with the workpiece continuously, therefore, thermal fatigue is rarely an issue unless there is an intermittent cut — for example, if the shaft being turned were to have a slot down its length. Flood cooling and high pressure generally works well in turning. The tool and workpiece are bathed in fluid keeping them both cool and well lubricated. The flood of fluid also helps to flush away the chips and maintain thermal stability of the machine tool.

D. Milling

Milling is a process whereby a cutter of multiple teeth is rotated and moved across a workpiece using the face of the cutter to produce a flat surface (see Figure 3.10), or using the periphery of the cutter to produce a form or a slot. Figure 3.11 shows the comparative tool nomenclature for a turning tool and a milling cutter.

FIGURE 3.10 A face milling operation. The cutter is shown in the top right photo; top left shows the cutter at speed; bottom, the cutting action is slowed by a strobe so that the cutter appears stationary, but the chip formation can be seen as the insert is leaving the cut.

Although there are common angles to the cutters, there is a fundamental difference between turning and milling: the milling process is an intermittent cut. Each tooth is in contact with the workpiece for only a short span of time. Milling cutters are, therefore, most susceptible to thermal fatigue resulting in chipping and poor tool life. Minimum quantity lubrication (MQL) works extremely well for milling since the quenching effect of flood application and high-pressure application methods is eliminated. Figure 3.10 shows the indexible insert cutter and two shots of a face milling process, one at speed and the other using a strobe to stop the cutting action in order to observe the chip formation at the point of cut. The direction of the cutter is inconsequential when face milling, but when slot milling or form cutting, there is a major difference. The terms used are up-cutting (or conventional milling) and down-cutting (or climb milling.) Figure 3.12 shows the cutter rotation for each of the methods. Up-cutting is also termed "conventional milling" since, in the days of hydraulically driven tables, the force of the cutting action opposed the force of the table feed and so provided a means of stability. Up-cutting, however, requires very sturdy fixturing and good clamping since the cutter has a tendency to lift the workpiece out of the fixture. Down-cutting or climb milling on a hydraulically driven table machine would result in the cutter snatching the workpiece into the tool, creating instability, very poor finish, and poor tool life. The positive mechanical ball-screw drives on Computer numerical control (CNC) machines afford the use of either method with little to no compromise. When up-cutting, the work hardened layer from the previous cut scrapes across the entire rake face of the tool, whereas in down-cutting the individual tooth depth of cut may be calculated such that the tooth bites into the workpiece with sufficient individual tooth depth of cut to reach behind the work hardened layer and so machine within the softer, less abrasive material, resulting in increased tool life. Because of this, it is typical to find climb milling on more modern, positive drive machine tools.

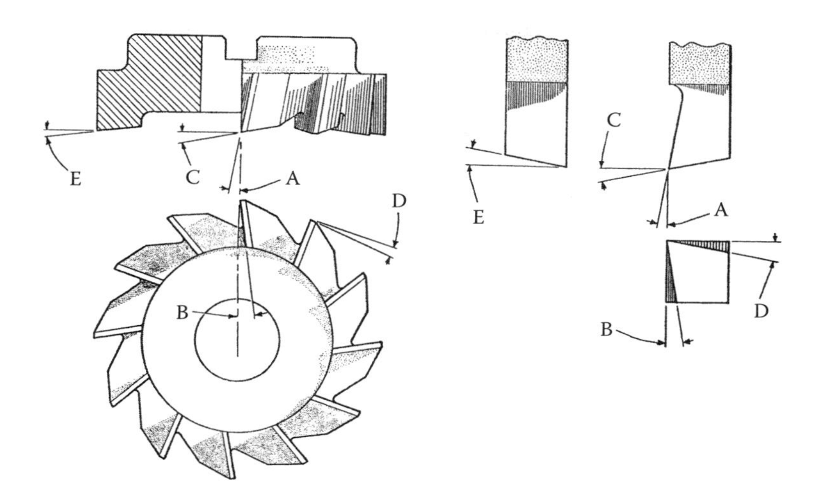

FIGURE 3.11 The tool geometry for both a milling cutter and a single point turning tool.

Feature	Milling Cutter	Turning Tool
A	Helix angle	Top rake angle
B	Radial angle	Side clearance angle
C	Relief angle	Front clearance angle
D	OD clearance angle	Side rake angle
E	Dish angle	Relief or trail angle

E. Broaching

The broaching process is analogous to milling but taking the cutter and rolling it out flat. Each tooth of the broach, as seen in Figure 3.13, takes the same tooth depth of cut but the broach tool is designed to have a set "rise" per tooth. The tool is designed such that it removes all the necessary material in one pass. If the rise per tooth is 0.05 mm (0.002 in.) and the total depth of cut is 5 mm (0.200 in.), then there will be 100 teeth in the broach. Allowing for a gullet for chip clearance, chip breaking, and the cutting tooth geometry, there may be a distance of 10 mm (0.400 in.) between each tooth. The broach would therefore be 1 m (39.4 in.) long.

Broaching can be carried out on external features such as cutting spur gears, as well as internal features such as ring wrenches. Broaching is a slow cutting speed process compared with drilling, turning, and milling, and it demands a more lubricious fluid, generally a straight oil, than one that is more of a water-based coolant. There are applications, however, particularly in internal broaching,

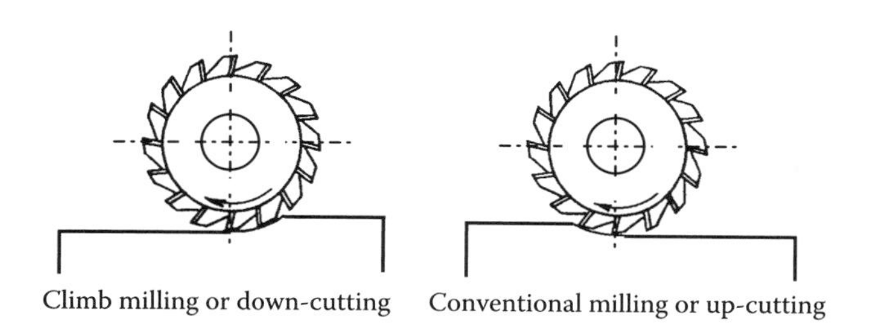

FIGURE 3.12 Climb milling and conventional or up-milling.

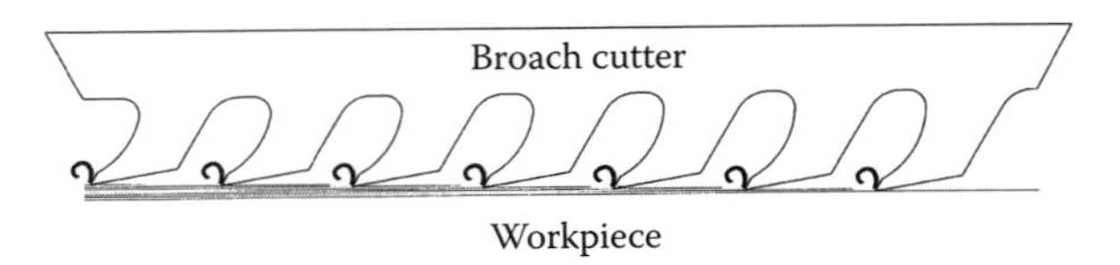

FIGURE 3.13 The basic operation of a broach tool.

where the frictional heat generates sufficient thermal distortion in the part to jam the broach cutter in the part and cause tool breakage. In that case, a water-based semisynthetic or emulsion would be preferable over the straight oil.

II. GENERATION OF CHIPS AND HEAT

From this somewhat simplistic overview of the machining processes and the influence of the cutting fluid, we may now delve deeper into the understanding of chip formation and the impact of the fluid properties. Real world metal cutting operations are made in three dimensions, described by the term oblique cutting, where the cutting edge is inclined at an angle to the direction of cut, thus facilitating a sideways curl to the chip. That third dimension complicates the theoretical modeling process, so a simplification was made to make the analysis only two dimensional, described by the term orthogonal machining where the cutting edge is at right angles to the direction of cut. The tool cutting geometry looks like that shown earlier in Figure 3.3. It can be seen that the rake angle determines the angle of the shear plane, and therefore the chip thickness. Due to the mode of shearing that takes place, the chip thickness is larger than the tool depth of cut. The shear angle directly affects the cutting forces, friction, and power. The combination of tool geometry, cutting speed, and cutting fluid affect the "partition of energy." The partition of energy is the distribution of the cutting energy, which will determine the surface integrity of the workpiece surface. The heat generated along the shear plane, as well as the frictional heat from any flank wear, will be split or partitioned into fractions that go off into the chip, into the tool, into the fluid, and into the workpiece surface. In general, it may be said that approximately 75% of the heat generated in the machining process comes from metal deformation, whereas the remaining 25% comes from friction. Though somewhat broad brush of all of that heat generated, approximately 80% goes off into the chip, 10% into the tool, and 10% into the fluid. Hence, the proper tool geometry has a critical impact on productivity, as well as the surface integrity of the process.

III. CONTRASTING PROCESSES

Cutting tools may be made from carbon steels, high-speed steels, cBN and diamond, cermets and ceramics. The proper choice of tool and geometry is paramount to the overall success of the machining process; however, each process has its own nuances for which special consideration needs to be made, not just in the tooling, but also the fluid and how it complements the overall performance.

The casual observance of chips being formed may not give a full appreciation for how different each of the chip making processes are:

Turning: The chip is generally continuous and the chip thickness constant. The cutting speed is medium to fast.

Milling: The cut is intermittent and the chip thickness is variable, either increasing from zero or decreasing from a maximum, depending on whether or not the tool is climb- or up-cutting (see Figure 3.12). The cutting speed is medium to fast.

Drilling: The chip is generally continuous, but the cutting speed varies from zero at the drill point to a maximum at the drill periphery.
Broaching: The chip is generally continuous and the chip thickness per tooth is constant. The cutting speed is slow.
Tapping: The chip is generally continuous. The cutting speed is slow to medium.

From the processes listed above, it is only turning where chip making is continuous, and the chip thickness and cutting speed are constant. This makes turning an ideal process to be used for academic research. Indeed, turning has been used extensively in research, especially due to the ease with which fast cutting speeds may be achieved. These studies do not, however, represent the likes of milling or drilling, so much care needs to be taken in extrapolating results from cutting tests made with one process and then relating them to another.

Generally, all the slow processes (broaching, tapping, and reaming) will require more lubricity in the cutting fluid, whereas all the fast processes (turning and milling) will require more cooling. Drilling is a special case due to the dramatic change in cutting speed across the rake face from the periphery to the drill tip.

With the advent of mini-computers, Klamecki,[11] Rowe,[12] and Childs[13] completed finite element analyses of chip formation to the point where, today, software is available to visualize and describe the machining process both mechanically and thermodynamically[14] to surprisingly accurate levels of predictability.

With both academic and empirical foundations firmly set, the refinement of machining technology continues and evolves, based on the solid rocks of the fundamental knowledge. Once understanding has been reached, however, one has to be careful not to view the rocks as icebergs that get in the way of radical advances. There is a certain comfort in knowing what we think we know, and there is no better illustration than in the ultra-conservative machine tool industry. Nevertheless, machine tools have evolved over time into stiffer, more mechanically and thermally stable structures. With better controls, using CNC, and faster spindle speeds with magnetically and hydrostatically levitated bearings and faster table speeds with linear motors, the way is being forged towards even higher productivity and more consistent part quality. Cutting tools are lasting longer — which brings us to the point of this chapter: the role of the cutting fluid in machining.

IV. ROLE OF THE CUTTING FLUID

It is going to be important to have our fluid nomenclature understood. "Coolant" is the slang term used to describe cutting fluid. Indeed, the cutting fluid does much more than just cool. In today's world of Internet search engines and keyword searches, "coolant" will generally bring about references to refrigerants. Though the name "metalworking fluids" has been a popular name, these fluids may also be used for machining plastics, ceramics, cermets, composite fiber reinforced materials, and glass — none of which is a metal. The broader term to use is "cutting and grinding fluids." In fact, Silliman edited a good reference book entitled *Cutting and Grinding Fluids — Selection and Application*.[15]

It appears that the key role of the cutting fluid in machining is to cool and then to lubricate, but it also serves many other very important functions. We will deal with those first, and then come back to the effects of cooling and lubricating. There are eight specific areas to mention.

A. Transportation of the Chips

A major role of the cutting fluid is that it transports the chips away from the cutting zone, at the same time cooling the chips and keeping dust and small particulates in the liquid rather than in the air. Hot chips are not the best things to have collecting around the machine base, the cutter or the part, so the fluid not only cools those chips but also washes them away from the machine tool into a filtering system for separation from the fluid. After processing, the separated and dried chips may even be recycled. Were it not for the dampening down effect of the liquid cutting fluid, small particulates and dust would otherwise be blown into the air, creating an unpleasant and dirty environment, as well as a respiratory health hazard.

B. To Arrest Rewelding

The cutting fluid helps to prevent rewelding. This is the reaction of material, at high temperature, to stick back onto itself at the tool edges and surfaces, as seen in the built-up-edge that occurs and is more pronounced in slower speed machining. It is also prevalent in terms of wheel loading when grinding soft materials. The chemistry of dissimilar materials works here. Copper compounds, in particular, may be added to a fluid when machining ferrous materials to prevent the rewelding. Work by Frost at the University of Bristol in England[16] showed how a copper de-loading pad prevented the build up of steel on a grinding wheel and reduced the propensity of the system, in cylindrical grinding, to go into regenerative chatter. To realize the full effect of some of these compounds and additives, the bulk fluid may have to be carefully monitored as they are often temperature and/or pH dependent.

C. Corrosion Protection

The cutting fluid should offer a level of corrosion protection to the machined workpiece. The "just machined" nascent metal surface is chemically active and will readily oxidize or react with the surroundings. Whereas most fluids will provide some corrosion protection, others may do just the opposite and cause some staining or discoloration due to their high surface active properties. The cutting fluid should not only protect the workpiece but also the machine tool, fixtures, and tooling. When designing fixtures and tool packages, it is important to appreciate the effects of galvanic corrosion that will occur between dissimilar materials. Some cutting fluids may have a strong surface activity, penetrating the surface of the material being machined. If the fluid can penetrate the interstitial fissures in the surface of the metal, then it can also penetrate quite readily between the surfaces of the fixture and the machine table or base. Capillary action will allow the cutting fluid to find its way between surfaces and set up galvanic cells. Such small spaces between surfaces may squeeze out any emulsified oil creating a system ripe for corrosion. Monitoring the electrical conductivity of the fluid and minimizing the mineral contaminants by using pure water will always help. Should bacteria begin to run rampant, they will consume the oil and change the corrosion protection properties of a fluid into that of a corrosive, as their waste products are acidic.

D. Power Reduction

Most cutting fluids reduce friction, and in so doing reduce the power required to machine a given material. Not only is this energy saving, but also if less power is consumed then less heat is generated. It will generally follow that if less heat is generated, the tools will last longer and the surface integrity of the workpiece will be protected. Overall, the system will tend to be more stable. The closer the system can be kept to ambient temperature, the more thermally stable the process, which impacts the integrity of the workpiece — both metallurgically and dimensionally. Thus, refrigeration of the cutting fluid may play a beneficial role in certain cases.

E. Extend Tool Life and Increase Productivity

The cutting fluid should be designed to first and foremost assist in the machining operation, maximizing stock removal rate and maximizing tool life. Surface active fluids with enhanced wetting agent chemistry will penetrate the surface of the workpiece and chemically react with the surface and subsurface to reduce shear stress. In so doing, it will reduce power consumption and reduce heat generation. This not only allows faster cutting rates, but also increases tool life for a given cutting speed.

F. Create a Certain Type of Chip

According to Ernst,[8] there are three types of chips: discontinuous or segmented chips, continuous chips with a built-up-edge, and continuous chips without a built-up-edge. Depending on the material, chips may be long and stringy, tightly curled, or virtually dust-like particles. All chips are characteristically smooth on the side that passes over the rake face. On their "inner" side however, they are quite rough as a result of the shearing action that takes place in the shear zone. The surface exhibits a serrated, saw-tooth appearance that is typical of a sheared material. Long stringy chips are generally produced when machining medium to somewhat softer materials. Tool geometry in combination with the fluid chemistry can produce either the continuous (not preferred) or the discontinuous chip (preferred). Chip breakers may be fitted behind the cutting point to curl the chip as it leaves the shear zone and break it into more manageable pieces. Hard, brittle materials, due to their lack of ductility, tend to produce dust like chips. The fluid application method will also affect the chip formation. High-pressure systems tend to act as a liquid chip breaker. The high pressure blast, onto the back of the chip, not only cools the chip along with the tool face, but also causes the chip to break into smaller particles. High pressure, through-the-tool applications, for drilling in particular, show a twofold benefit. Not only is the chip cooled and broken into small pieces, but it is quickly evacuated from the hole to minimize or even prevent scoring of the bore and reducing any additional frictional heat generated from chips compacting and clogging in the flutes of the tool. The evacuation of the chips in both drilling and tapping operations will also prevent frequent tool breakages by eliminating the possibility of tool jams.

G. Cooling

The cooling property of a fluid is one of the major contributions made to the machining operation. Though cooling of the tool and the chip at the tool/chip interface is key, the fluid also bathes the workpiece, the fixture, and the machine tool. The fluid, therefore, helps to establish thermal stability of the system and assists in better size control. Not only are there motors, pumps, and control cabinets emitting heat around a machine tool, but a person standing in front of a machine emits around 100 W of heat. A fixture or a part may be brought into the machine enclosure from another area, or even from outside where the temperature might be quite different from that inside the workshop and inside the machine enclosure. The fluid, particularly water-based fluids, have a tremendous capacity for heat and help to quickly bring the components and the machine tool to one temperature. Refrigeration of the cutting fluid may be particularly important if there are often large temperature gradients between the machine, the workpiece, and tooling. Refrigeration not only keeps the fluid cool and thermally stable, but also increases the longevity of the fluid by reducing the bacterial growth that typically occurs at warm or elevated temperatures.

It is important to keep the cutting tool cool in order to avoid it exceeding the temperature where softening will occur and so lead to rapid wear. Attempts have been made to cool the tool using cryogenics. Liquid nitrogen was used as a cutting fluid in a research application at Wright State University in 1995, particularly in the milling of titanium.[17] The application of a cryogenic fluid resulted in a substantial increase in tool life, but of course there are other concerns with this method, such as environmental and the overall economics of the process. The University of Nebraska-Lincoln reported the use of cryogenics as a heat sink, built into the tool holder of a turning tool. There was no release of liquid nitrogen to the atmosphere here, and again the tool life was extended substantially, this time by a factor of three to five times.[18]

In an attempt to move away from the use of liquid cutting fluids, air has been used through-the-tool at low volume and high pressure, as well as high volume and low pressure, neither of which made any improvement in part quality nor tool life. Air at high volume and high pressure did show signs of improvement; however, the compressor necessary to deliver the air through the tool had to be almost the size of an average machining center. Add in the cost of the maintenance of the compressor along with the noise pollution, and use of compressed air only is generally not a viable option.

H. Lubrication

Cooling and lubrication are the two main reasons to use cutting fluids. There is always heated (no pun intended) discussion among machinists as to which is the most important — the water vs. oil argument. Water is the best coolant with a heat capacity far in excess, and a latent heat of evaporation an order of magnitude greater than a typical straight oil. No matter how much heat is generated, the water will take it away from the cutting zone. Oil, on the other hand, lubricates where water does not. Lubrication reduces friction so that heat is not generated, and therefore does not need to be taken away.

There really is no argument, since the properties of both oil and water are important! For some processes, as we shall see, the emphasis may shift more to cooling than lubrication and vice versa.

Lubrication will reduce the generation of frictional heat, but there are other areas of heat generation that need cooling, such as the shearing action of chip formation. This is more pronounced in low speed machining operations. Low speed machining operations such as broaching and tapping tend to generate higher localized cutting forces and will form built-up-edges on the tool. The build-up of material is due to the softening of the workpiece material at the tool point. When localized pressure is so high that material becomes welded to the tool, it gradually builds up to a limit where the build-up becomes so large, that the force of the chip moving across the rake face breaks it away, sometimes taking some of the tool material with it. The build-up is cyclical, increasing and decreasing, but not necessarily at regular intervals. This is evidenced by occasional sharp "picks" of material on the back of the otherwise smooth chip. The built-up-edge will effectively change the tool geometry and so generate more frictional heat, longer and stringier chips, with heavier crater wear. Low speed machining operations benefit more from lubricity than cooling.

As the cutting speed increases and the built-up-edge effect decreases, there is less localized cutting force but more heat is generated in the shear zone. Also, as the cutting speed increases, the need for cooling takes over from the need for lubricity. It is generally true, however, that for better surface finishes an oil is preferred over water. The oil allows a certain amount of "smearing" of the material due to a larger area of contact between the tool and the workpiece surface, whereas the water will cause galling of the surface and result in a matt finish. It is perhaps better to think of the cooling vs. lubrication (water vs. oil) as complementary and with a different emphasis for different processes. At lower speeds (broaching, planning) adhesion wear is prevalent and at higher speeds (turning and milling) it is abrasion wear that is predominant. Softer materials benefit more from lubrication, too. With the advent of synthetic oils and compounds, it may not always be an oil that is used for lubrication. Synthetic chemical lubricants may be used in a full synthetic fluid, which may have a surprising effect on lowering the coefficient of friction for certain materials at certain machining rates. Some of the chemistry is temperature dependent, and so at low speeds the coefficient of friction may be quite high. Yet at higher speeds when perhaps an active sulfur compound is activated, the coefficient of friction then decreases. It can be seen that the effect of sulfur and chlorine on the coefficient of friction is temperature dependent, hence the use of sulfo-chlorinated fluids as "universal" chemical lubricants (see Figure 3.14). Chlorine is generally of little help in grinding where the interfacial temperature is high. Sulfur has better high temperature tolerance.

The cutting fluid also performs a secondary and very useful function of helping to lubricate the machine tool by keeping sliding surfaces both clean and oiled. Way covers benefit from

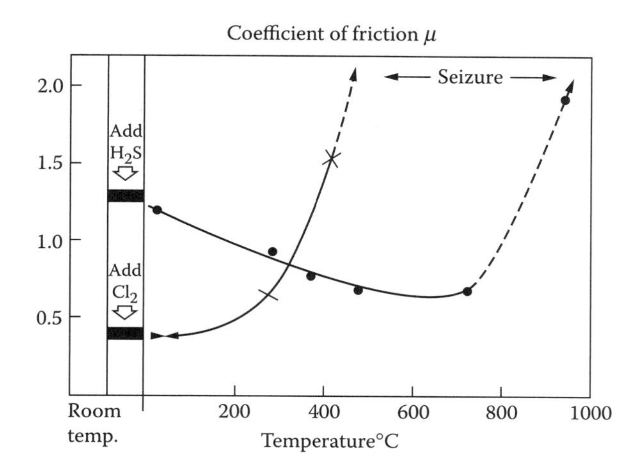

FIGURE 3.14 The impact of sulphur and chlorine compounds on the coefficient of friction for interfacial surface temperatures typical for machining and grinding.

a lubricating fluid, which prevents them from binding up. However, the cutting fluid needs to be chosen carefully to ensure that it is compatible with the machine lubricants, lube oils, and seals.

V. FLUID APPLICATION

There is a great deal of chemistry involved in producing a cutting fluid, not only for its beneficial effects in the metal cutting process, but also in maintaining the longevity of the fluid, oil droplet size, pH balance, bacterial resistance, corrosion control, wetting ability, and foam control. Such premium fluids offer assistance in achieving high productivity with economical advantage, however they can be very expensive. One of the great fallacies, when dealing with metalworking fluids, is to look at the cost of the fluid per liter as an indication of its cost effectiveness. It is critical to look at the overall benefits over time. Clyde Sluhan, the founder of Master Chemical Corporation and a pioneer in the industry, often said during his seminars, "If a cutting fluid costs you $25/gallon and it saves you money, you're a fool not to use it!" So imagine that a very expensive fluid has been purchased based on its proven performance in instrumented tests. It is critical that the fluid is applied properly in order to take advantage of the chemistry. Too often, cutting fluids are misapplied and their effectiveness and potential to enhance productivity is lost. Tool life suffers, surface finish is poor, and the parts are expensive to make. Once that occurs, a cost-cutting measure is usually instigated in an attempt to reduce the overall cost of producing the part. The first thing to go is the expensive cutting fluid, and now the potential for productivity gain has been lost forever. If only the fluid application method had been optimized. For some reason, cutting fluids are thought of as secondary and somewhat insignificant — the cheapest fluid will generally do. If that is the case, then why not just turn it off? The proper fluid application is essential to reap the full benefit.

It was a group of Russian investigators, the more prominent of which was Rebinder,[19] who discovered that chemicals are adsorbed into the surface of a metal, reducing its effective hardness and reducing shear stress. Fluid properties, such as surface tension, were important in order to penetrate the fissures of the machined surface (see Figure 3.15). Rebinder saw that boundary lubricants like oleic acid could penetrate the surface interstitial cracks and fissures to a certain depth and so weaken the strength of the material, an effect termed "the Rebinder effect." Later studies have suggested that the mechanism is a chemical reaction that takes place on the surface of the metal. Hard particle compounds made from the metal constituents along with sulfur, chlorine, and phosphorous, form on the surface and subsurface of the workpiece. These hard particles interfere

FIGURE 3.15 The rough surface of a specimen of 4140 steel after being reamed dry.

with the bulk metal matrix and provide a weaker path for shear to take place, reducing the effective shear stress. A loose, but perhaps illustrative analogy might be drawn from the comparison of digging in a pure clay soil vs. digging in a soil composed of clay mixed with small rocks and pebbles. It is far more strenuous digging in the pure clay than the soil with the interspersed rocks that seems to move with comparative ease. This theory was confirmed using carbon tetrachloride with ferrous materials, forming ferric chloride in the grain boundary and so increasing plasticity by facilitating slip. It is, therefore, important to apply the fluid properly on a macro-level so as to ensure that the chemicals make their way to the surfaces and subsurfaces where they are most effective on a micro-level.

Indeed, the proper application of a cutting fluid is critical. Looking at the two examples in Figure 3.16, it is easy to see how one may be fooled as to benefits of cutting wet vs. dry. Cutting dry, the swarf and chip will move across the rake face of the tool and so take the point of maximum heat a way back from the tool tip. The tool will get hot, but there is a larger bulk of the tool in which to dissipate the heat. Applying a fluid improperly to the top of the chip only will quench the "inner" surface of the chip and curl it much tighter, bringing the point of maximum heat much closer to the

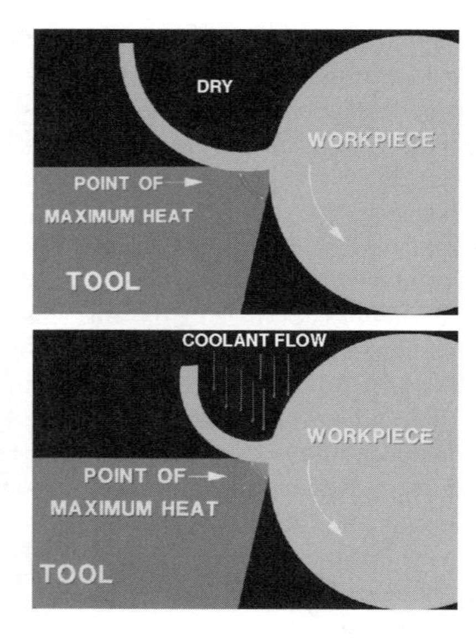

FIGURE 3.16 These two cartoons show how the point of maximum heat, for a dry application, moves closer to the point of the tool when a coolant is applied to the top of the chip only.

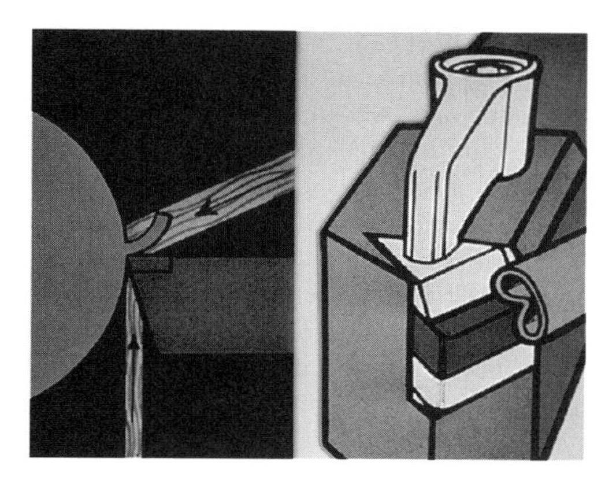

FIGURE 3.17 The drawing shows fluid being applied to the top and bottom of the chip, as well as being forced up into the cutting zone from beneath to form a film of fluid between the surfaces. This is for a lubricious fluid, whereas the nozzle arrangement on the right would apply fluid along the tool edges when a light-duty cooling fluid is being used.

point of cut, where there is less material to conduct away the heat; the tool life decreases as compared to dry cutting, despite the presence of a fluid. This gives the impression that cutting dry is better and yields longer tool life. If the fluid were applied properly to the under side of the chip and between the chip and flank face, the benefits of the fluid chemistry would be forthcoming.

The cutting fluid performs many functions, as has been described. However, unless the fluid is applied properly and carefully, the benefits of using a cutting fluid will be compromised (see Figure 3.17). If the fluid is a lubricating type (that means that the cutting fluid was chosen for lubrication over cooling), it must be directed into the cutting zone in an attempt to form a film between the tool, chip, and workpiece. If the fluid is merely flooded over the workpiece and machine bed, it will most definitely offer corrosion protection and lubrication to the machine, but it will not necessarily improve the cutting process. It is important, therefore, to direct the fluid into the nip between the rake face and the underside of the chip, at the point of cut. When a fluid is chosen for its cooling properties, it is more important to direct the fluid toward the edges of the tool. Forcing the fluid along the tool edges may be achieved using high-pressure fluid application. Not only will there be a chip breaking action, but the cutting fluid will provide longer tool life by maintaining tool hardness.

As cutting speeds increase and the use of carbide and ceramic cutting tools become more prevalent, it is critical that the flow of fluid be maintained so that the risk of thermal shock, due to intermittent fluid application, is eliminated. Care should be taken to maintain a constant flow of fluid in order to maintain a constant tool temperature. However, the tool and process itself may cause intermittent flow of the fluid, as in the milling process. Also, look to the part fixture with respect to clamps that may obstruct the flow of the fluid; and particularly when turning on a lathe, ensure that the chuck jaws do not block the fluid nozzles and interrupt the flow of fluid to the tool point.

One of the most difficult fluid applications is that for drilling. The tool is encapsulated by the part, which prevents the cutting fluid from getting to the cutting point. If through-the-tool fluid application is not available, then a steady stream of fluid should be directed at and down the drill. The drilling operation should be set up so that the drill pecks through the part rather than making one continuous pass. In this way, when the drill is withdrawn from the hole, the chip is broken, and fluid will typically enter and fill the hole thereby maintaining some fluid presence at the point of cut. The fluid should also be applied with sufficient force to blow away any random chips that might otherwise block the area and become trapped between the drill and the edge of the bore. Similarly, during a tapping operation, because of the large area of contact and the delicate nature of the tap, a highly lubricating type of fluid should be used. It is important to ensure the compatibility of any extraneous tapping oils, compounds, waxes, or any other concentrate used to supplement the tapping process, with the bulk fluid being used in the machine sump. Cross-contamination of chemicals may result in a decrease in the life of the cutting fluid, and may cause serious corrosion problems — even dermatitis and skin irritation for the operator.

Now, let us contrast turning and milling. During the turning process, the tool is generally in continuous contact with the workpiece and sees a relatively constant energy from friction and shear; whereas during the milling process, the cutting point is cutting intermittently, the chip thickness is ever changing and suffers from both mechanical and thermal cycle fatigue as an inherent condition of the process. That brings us to minimum quantity lubrication (MQL). It has been written and reported that dry machining is the wave of the future. There are obviously pluses and minuses to this argument, however, one area where MQL is often an acceptable method is milling. MQL has been shown to improve tool life significantly in milling operations. MQL is the technique of applying a very small volume of liquid into a fast moving air stream, generally the shop compressed air supply. The fluid atomizes to some extent and is transported in a very light "breath" into the cutting zone.

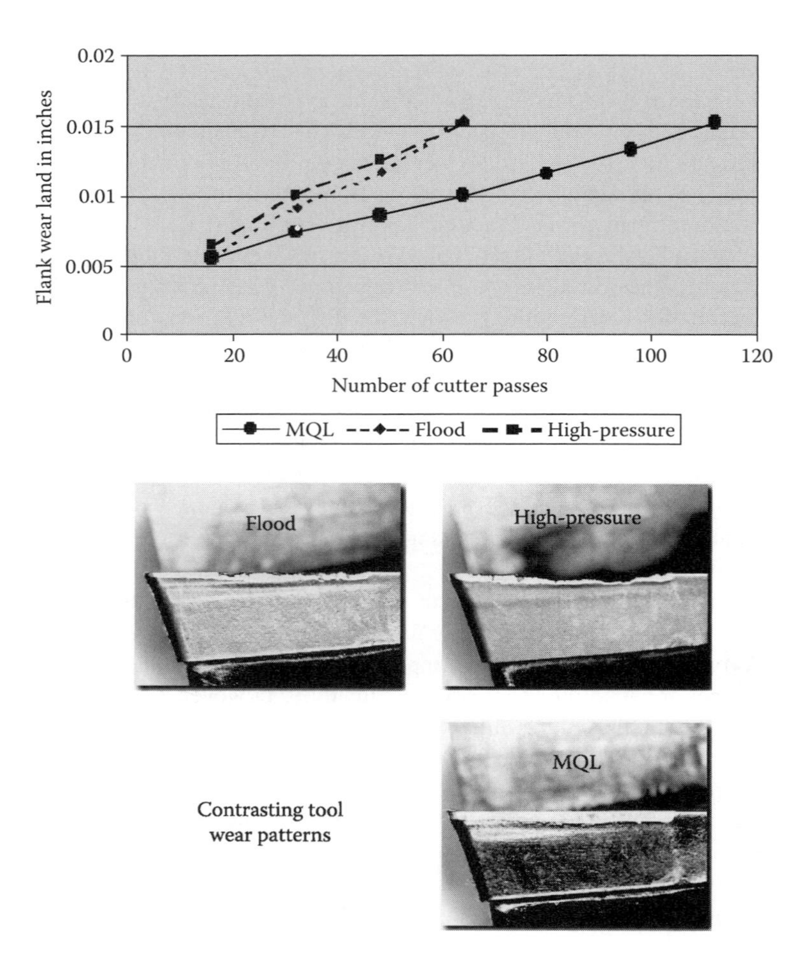

FIGURE 3.18 The graph shows how MQL leads both flood and high-pressure fluid application 2:1 in tool life all the way to 0.015 in. wear land. The photographs show evidence of the chipping that occurs with the flood and high-pressure application, whereas the flank wear land for MQL exhibits purely abrasion wear.

There is no visible fog or mist. The amount of fluid to air flow is very small, so contamination of anything is minimal, though over time a sticky residue can build up on part fixtures and guarding. The effect on tool life, however, is dramatic. It is believed that because of the intermittent cutting operation, for flood and high-pressure applications, the milling cutter is severely quenched every time it leaves the surface of the material. There is a large temperature gradient between the conditions of the cutting point in the cut, under the heat and pressure of the cutting action, and the noncutting action, when the cutter exits the cut. The cutter experiences cyclically, high temperatures and high loading for a very short time and then, under no load, a severe quench by the flood of cool fluid for a significantly longer period. During the MQL application method, the cutter still sees the cyclical force and temperature changes, however those changes are not nearly as severe. It has been demonstrated in many laboratories, as well as industrial settings, that MQL will reliably produce longer tool life than with either flood or high-pressure applications.

The graphs and pictures in Figure 3.18 show that the flood and high-pressure through-the-tool systems, though quite different with respect to the effectiveness of the fluid application, followed much the same path of flank wear. The MQL application, on the other hand, under all the same conditions of insert type, geometry, feeds and speeds, material and fluid, showed significantly better tool life. The wear land on the flanks of the tool inserts show evidence of chipping for flood and high-pressure, whereas for MQL, the land shows only abrasion wear. It should be noted that the surface temperature of the workpiece will be significantly higher when employing MQL over flood or high-pressure applications. It is therefore important to consider the dimensional precision and surface integrity, which results from each of those processes.

VI. WET VS. DRY DEBATE

The three major ways to apply cutting fluid, therefore, are flood, high-pressure, and MQL. MQL is not the same as machining dry. Machining dry is without any fluid at all, other than the atmospheric air surrounding the cutting zone. There is a push, by environmentalists in particular, to remove the cutting fluid altogether, viewing it as an unnecessary nuisance, not just because of the expense and maintenance of a fluid system, but also the ecological aspect of spent fluid disposal and operator acceptability. The pros and cons of essentially wet or dry will help make the decision as to which avenue to pursue.

The advantages of eliminating or dramatically reducing the use of cutting fluids are as follows: first, no fluid tanks to take up valuable floor space. An MQL tank holds a liter or less of fluid vs. hundreds or tens of thousands of liters for flood application systems. There are no large conventional coolant pumps to run and maintain, no fluid to have to filter, along with the maintenance of a filter media and filtration system, and no liquid contamination for hazardous waste removal. The chips will be drier and of higher financial value, since they may be recycled directly and more efficiently with less of a negative impact on the environment.

The disadvantages are: low tool life for the existing tools; however with multi coated or ceramic tools, although more expensive to purchase initially, the tool life might be equaled or even exceeded. There will be no way to evacuate the chips from the work zone unless a high-powered air vacuum is employed around the cutter. The vacuum system needs to be carefully designed to be sure that it does not become clogged with the chips, swarf, and debris. There will be very hot chips in the machine enclosure and around structural members of the machine tool that could give rise to thermal distortion, unless the machine is designed specifically for the evacuation of chips, directing them away from those critical areas using special guarding to funnel and move the chips out and away from the machining area. Dry machining of certain metals generates fumes. These fumes are similar to those found in welding booths. There is a fine submicron particulate and smoke along with a strong metallic odor that occurs when dry machining. The air in the machine enclosure will therefore need to be filtered or exhausted to the outside. There is no corrosion protection of the freshly machined surface, unless a system is employed to spray the surfaces directly after machining. There is no lubricity of the cutting process to reduce the frictional energy, heat and power, and no synergistic lubrication of the machine tool parts under the normal flow of a cutting fluid. The fine particles of dust generated in dry machining tend to infiltrate areas like in-process gauging equipment and the machine ways, no matter how well they are sealed or covered. There are many electrical motors with magnets as part of a machine tool that attracts the fine dust, particularly when machining ferrous materials. Over time, the infiltration of this electrically conductive dust will cause system failures. The last aspect of near-dry or MQL machining is that not all processes are performed best near-dry; some are better performed wet — drilling and tapping for instance. This means that machine tools will have to be either dry or wet or somehow a hybrid.

There is enough of an environmental consciousness within industry today that this question begs an answer. Perhaps a price will have to be paid for environmental protection to the sacrifice of productivity. This may be acceptable to some, however, in the global picture it must be put into perspective with other nations and their industries where environmental rules and regulations are not as strict as those in the U.S. Some of the new high-speed machining centers are designed specifically for MQL and compressed air/fluid application. In this case, there will be no flood or high-pressure through-the-tool alternative.

There are occasions when materials supply their own lubricant. Free-machining steels at one time contained lead, and machined far easier due to the lubricating quality of the lead content. However, due to environmental rulings, lead has been substituted with calcium, which produces a similar effect. There are also low shear stress aluminum and magnesium alloys that machine quite well dry.

Another prime example of a material with its own built-in lubricant, and somewhat of a special case, is cast iron. Chip formation in cast iron does not adhere to the principles of shear stress diagrams. The chips from cast iron are torn out of the surface rather than sheared. Relatively low power microscopy may be used to see on an etched surface of cast iron that, while the pearlitic structural constituents are stressed by the cutting tool and show the onset of plastic deformation, the imbedded graphite flakes are squeezed out as a soft mass and thereby act as a lubricant. The tearing action synonymous with machining cast iron is far removed from the shearing action and chip formation typical of a mild steel. The "tear chips" formed when machining cast iron are literally torn out of the surface and give rise to the matt surface finish typical of a machined cast iron surface. Though cast iron is easily machined dry, the presence of a fluid generally improves surface finish, dampens down the dust, and washes away the swarf.

VII. SURFACE INTEGRITY AND FINISH

Surface integrity is a term coined from the Metcut studies of the 1960s where, under a Department of Defense contract, they were able to produce the *Machinability Data Handbook*.[20] Surface integrity relates to features of the surface such as cracks, micro-cracks, grain growth, precipitation of carbides in grain boundaries for superalloys, chipping in ceramics, subsurface damage due to Hertzian stresses, and work hardening and residual stress. All these features affect the functionality of the part aside from the topographical surface finish characteristics and each may be affected by the cutting fluid and/or its application.

Surface finish problems generally arise with respect to the finish being too rough. Poor finish will result from chip welding, tool seizure, and the built-up-edge phenomenon, but it is important to separate the problems associated with the tooling from those associated with the cutting fluid. Generally, if the surface finish is adequate, when the tool is new and sharp, but deteriorates as the tool wears, then approaching the situation from the cutting fluid viewpoint could yield a solution. However, if the workpiece finish is unsatisfactory at the outset, with sharp tools, then both the tool and the fluid need to be reviewed. An increase in chemical activity of the fluid may help to prevent the built-up-edge. An increase in viscosity of a cutting oil may help to better hold the fluid in the cutting zone. Should excess heat be evidenced by highly discolored chips, as well as a hot workpiece, then some lubricity may need to be sacrificed for more cooling, in order to cool the tool and keep it sharper for a longer period, and thus prevent excessive rubbing of the wear land. However, there are many combinations of events that can produce a rough finish, including machine chatter, tool geometry, machine way errors, tool type, fluid type, and fluid filtration. Each case needs to be evaluated on the merits of the entire system. A mere change in fluid concentration or a change in fluid concentrate will rarely cure a surface finish problem that suddenly arises.

Stop-action studies have been carried out to measure the microhardness of the machined surface and the formed chip. Such a study requires a special tool holder that, during the machining operation, can be very quickly withdrawn from the cut, leaving the deformed chip intact and attached to the surface for metallurgical inspection. Such a tool holder will contain an explosive device that ejects the cutting tool backwards and out of the cutting zone, faster than the forward cutting speed. Studies have shown that the hardness of the chip is significantly harder than that of the bulk material, due to work hardening. Indeed, the chip is often as much as three times harder than the bulk material, as illustrated in Figure 3.19.

In addition, there is a thin layer on the machined surface, perhaps 0.025 to 0.050 mm thick, which is termed the "work hardened layer." This layer may be as much as twice the hardness of the bulk material. Interestingly, if the same microhardness measurements are taken around the neck of a tensile test specimen made of the same material, there will be a very close resemblance to the hardness measurements taken around the stop-action tool point. It is good to realize that there is a surface layer on the workpiece that has effectively been loaded to failure. Thus, certain machining

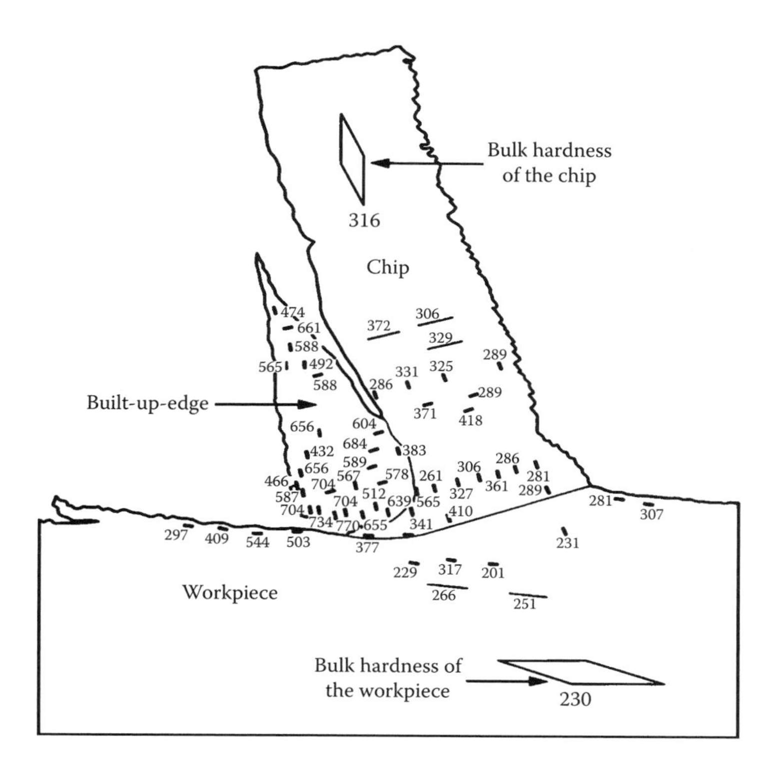

FIGURE 3.19 The Knoop microhardness numbers are shown around the cutting action where the bulk hardness of the material is 230 and the chip is 316. Hardnesses as high as 770 can be seen in the built-up-edge and 565 in the chip.

operations thought to be "gentle," may not be that gentle after all! The surface integrity is not representative of the overall properties of the bulk material.

Particularly with critical components, a post machining process may be required to create compressive stresses or remove micro-cracks, creating a more structurally sound workpiece. Such processes might be glass or shot peening, grit blasting, abrasive media tumbling, electro-chemical etching, or polishing, all of which will condition or remove the disturbed and work hardened layer.

VIII. MODES OF TOOL WEAR

Fluid filtration can have a major impact on the surface finish of the machined part and must be vigorously monitored. The swarf and chips contained in the fluid, as it performs its job of flushing and cleaning the work area and cutting zone, are very hard particles (as described earlier) that can spoil the surface finish and cause excessive tool wear leading to eventual tool failure. A dirty fluid contaminated by the hard particles of metal dust and swarf will act almost like an abrasive slurry in increasing tool wear. So, it is important to understand the wear mechanism of cutting tools, be able to identify the mode of wear, and look to the best solution when a failure occurs (see Figure 3.20). There are five main modes of tool wear and usually one is dominant.

A. Abrasion Wear

Abrasion wear is the wear that takes place, mainly on the flank face, due to the rubbing and abrading of the tool as it moves across the machined surface, abraded by hard particles in the matrix of the material (see Figure 3.21). These particles could be sand from castings, carbides, precipitates in the grain boundaries, or the chemical constituents formed as described by the Rebinder effect. Generally, it is abrasion wear that is the major mode of wear in cutting tools. It is observed as a flat that forms on the flank face of the tool. As the flank wear land grows the frictional heat increases, the tool softens and the surface finish deteriorates. Most precision cutting tools are considered worn out when the wear land reaches 0.250 mm. Some rough machining operations may take the land to

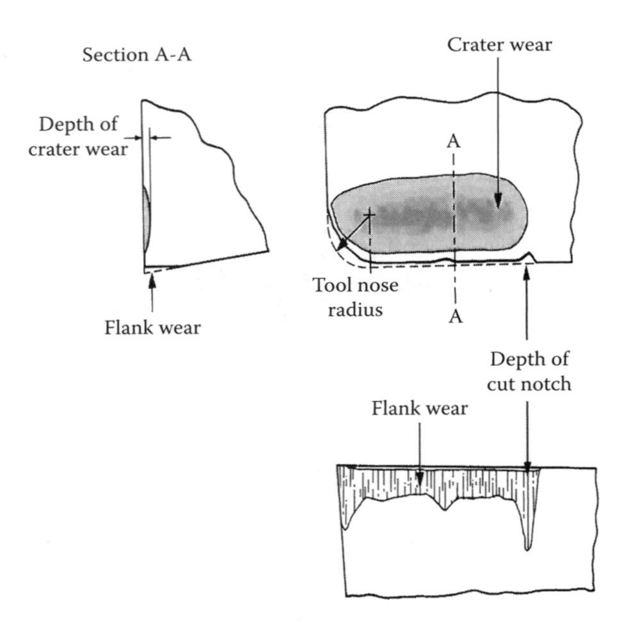

FIGURE 3.20 The nomenclature for the wear patterns on a tool.

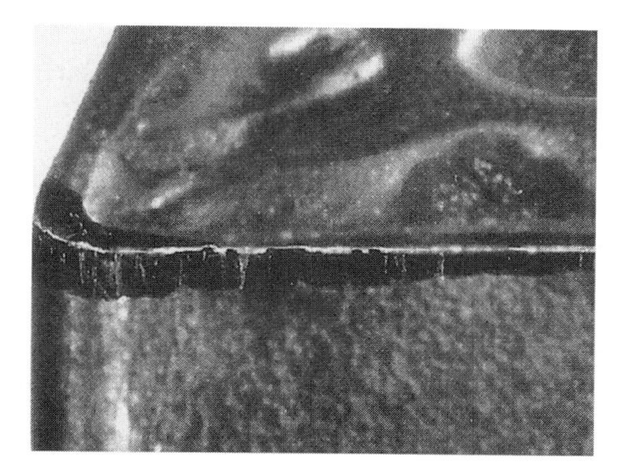

FIGURE 3.21 Abrasion wear on the flank face of the tool.

as much as 0.400 mm, but by that time there is a great deal of heat being generated, the surface finish is poor and chatter and vibration may even occur.

B. Adhesion Wear

Adhesion wear is wear that occurs from the generation of frictional heat, which softens the tool and creates a high temperature at the tool tip. The chip is hot and soft at the tool point and the conditions are ripe for welding. A small amount of material welds to the tool tip and material may build up on that, hence the name "built-up edge." That build-up on the tool tip will change the chip geometry, making the chip become longer and stringy. At some point the build-up will be such that it breaks away and moves off with the chip. When that breakage occurs, not only is the workpiece material removed, but it also takes a part of the tool with it, forming a crater on the rake face of the tool (see Figure 3.22), and changing the tool geometry. The build-up and breaking away of the built-up-edge occurs sporadically. The crater wear does not progress linearly. The combination of the built-up-edge and the crater wear are such that adhesion wear is not predictably linear, making tool life predictions very difficult in situations where this occurs.

C. Chemical Wear

Chemical wear will take place due to reactions between the tool material and chemicals in the cutting fluid or with the workpiece being machined. Some materials, such as magnesium, titanium,

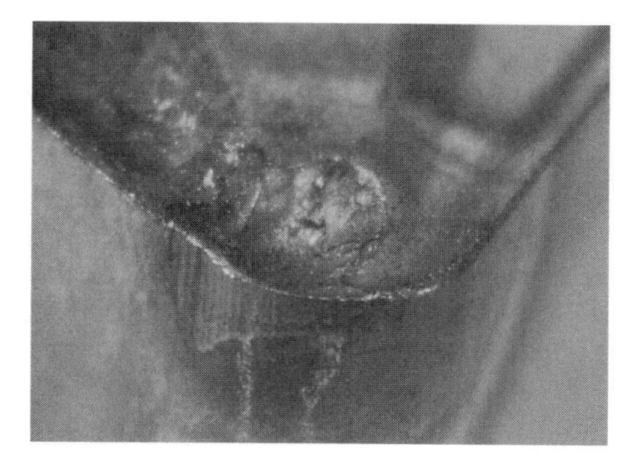

FIGURE 3.22 Crater wear on the rake face of the tool.

and aluminum, are more chemically reactive than say a nickel based superalloy or even a carbon steel. So, although this is generally a relatively insignificant wear mechanism, it can become significant depending on the nature of the material being machined.

D. Fatigue Wear

There are two types of fatigue wear: mechanical and thermal. Mechanical fatigue is analogous to bending a paper clip back and forth until it eventually breaks. Machine tools are neither infinitely rigid, nor are they completely vibrationally stable. The cutting action the tool sees is a mechanically regular oscillating force, which, like bending the paper clip, will eventually set up chatter and cause failure. Once the tool begins to vibrate, the tool depth of cut changes and that can lead to regenerative chatter if the frequency of vibration coincides with the frequency of the tool depth of cut changes; over time, the amplitude of vibration can increase until there is a catastrophic tool failure.

Intermittent cutting is inherent in the milling process. As previously mentioned, the improper application of the fluid across chuck jaws in turning will thermally cycle the tool as it moves from hot to cold temperatures. Thermal, as well as mechanical cycling, will cause tool failures evidenced by chipping rather than wearing of the cutting edge as seen in Figure 3.23.

E. Notch Wear

It was described earlier that the hardness of the chip and a thin layer of the machined surface were significantly harder than the bulk material. It may be visualized that in turning, the tool will have its tip in the bulk of the material; but at the distance equaling the depth of cut, the tool will be cutting through some significantly harder material (the work hardened layer) causing a notch to appear on the flank face, called the depth of cut notch (see Figure 3.24). Depending on the shape and geometry of the tool, the notch wear can be highly influential on tool life or be completely insignificant compared with other modes of wear.

Knowing that there is a work-hardened layer on the surface of a machined material that can cause a deep notch in the side of a turning tool, it is worth considering the effect on a milling cutter. If the cutter is used in the up-cutting mode, the tool has to penetrate the work hardened layer, and the edge of the layer then scrapes across the entire flank face of the tool. If the cutter is used in the climb milling mode and the tooth depth of cut is set to be below the work hardened layer, then the tool works within the softer material and never sees the work hardened layer, minimizing any crater wear.

The cutting fluid is present to extend tool life by minimizing the wear described above. The fluid is present for cooling and lubricating the tool also, but today the tool may also have some additional help by way of protection from just a few microns layer of a special coating.

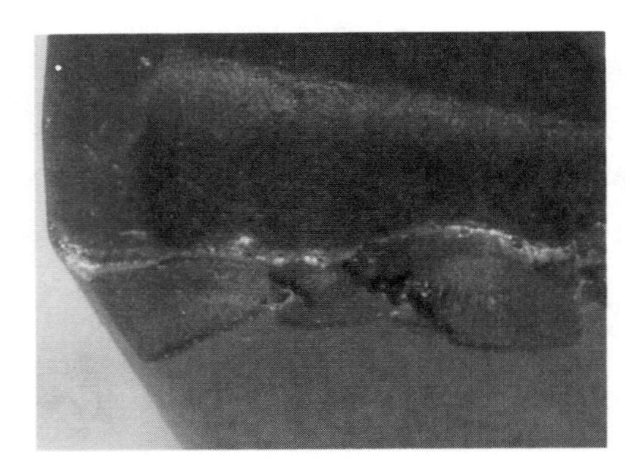

FIGURE 3.23 A badly chipped tool caused by fatigue from vibration and/or thermal cycling.

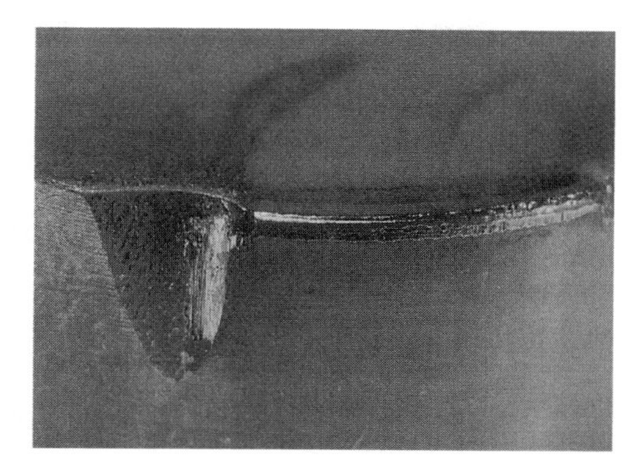

FIGURE 3.24 The depth of cut notch on tooling, next to a significantly narrower band of abrasion wear on the flank face.

IX. COATINGS

The first physical vapor deposited (PVD) coating to be put onto carbide tools was titanium nitride (TiN), but more recently developed PVD technologies include titanium carbonitride (TiCN) and titanium aluminum nitride (TiAlN), which offer higher hardness, increased toughness, and improved wear resistance. TiAlN tools in particular, through their higher chemical stability, offer increased resistance to chemical wear and thereby increased capability for higher speeds. These coatings were primarily developed for wear resistance, however other coatings offer other benefits. TiN and titanium carbide (TiC) are wear resistant coatings whereas TiAlN offers oxidation resistance, particularly important where cubic boron nitride (cBN) superabrasive tools are concerned. Aluminum oxide (Al_2O_3), a refractory coating, provides heat resistance, and molybdenum disulphide (MoS_2) acts as a hard lubricant.

Such is the effectiveness of just a few microns of a coating on tool life and performance, that the combination of the latest powder metal high speed steels with PVD coatings is proving to be more cost effective than carbide and ceramics in many applications.

X. HIGH-SPEED MACHINING

In the late 1920s to early 1930s, "large chip" process experimentation was being conducted by Dr. Saloman using higher than "normal" speeds. It seems odd, however, that abrasive/grinding engineers consider the "conventional" speed of a grinding wheel to be around 30 m/sec, whereas a "large chip" process such as milling, turning, and planning would see 30 m/sec as ultra-high-speed. Salomon showed that there comes a point where if the cutting speed is fast enough then the temperature of metal cutting decreases dramatically and the tool life becomes theoretically infinite. He was granted a German patent #523594 in 1931 for his work.[21]

High-speed machining needs to be defined. How fast is fast? Dr. Scott Smith, at the University of North Carolina (Charlotte) says that high-speed machining occurs when the tooth-pass frequency approaches a substantial fraction of the dominant natural frequency of the machine and tool system. This may sound complicated and has no mention of speed, however one of the critical components of high-speed machining is that the machining system is vibrationally stable, so rather than glibly state a spindle rev/min, the vibration frequency definition is best. The key is to match the chatter frequency with the tooth-pass frequency. For example if a machine system has a chatter frequency of 2200 Hz, then for a two tooth cutter the optimum spindle speed would be 2200 Hz $\times$ 60 sec/min/2 teeth = 66,000 rev/min. If that machine had a 45,000 rev/min spindle then significant vibration would occur, with accompanying poor tool life and poor surface finish. The next best speed would be an integer difference (66,000/2), namely 33,000 rev/min or (66,000/3) 22,000 rev/min. In order to enjoy the benefits of high-speed machining, methods are needed to be able to reach very fast speeds with high rates of acceleration. Hence, most high-speed machines have linear motors driving the ways.

High-speed machining is quite different from conventional speed machining. Conventional machining does not normally exceed 3 m/sec, with carbide cutters, and feed rates are typically up to 15 mm/sec. The conventional process, as it has been described, requires that strict attention be paid to the tooling and the type and application of the cutting fluid. In high-speed machining, the cutting speed ranges 5 to 15 m/sec with feed rates in the order of 30 to 40 mm/sec. When cutting fluid is used and applied as MQL or when compressed air or cryogenics are used, the feed rate may go as high as 425 mm/sec. High speed machining can also machine materials as hard as Rc72. Special tool holders are required for high-speed machining. The tools need to be held securely under the very high centripetal forces which would otherwise open a conventional collet or chuck. Where does it end? European programs, running at the University of West England and London South Bank University,[22] are centering on aerospace materials, both engine, nickel-based superalloys and airframe, titanium and aluminum alloys. The latest test rig for the RAMP program has been designed to demonstrate high-speed machining with spindle capabilities of 100 kW at 100,000 rev/min.[23]

XI. WHERE NEXT?

We have not been metalworking very long in the whole scheme of things; from the Ptolemaic period's first wood lathe it has been just over 2000 years. Since Taylor, however, during the post-industrial revolution era, a mere 100 years ago, we have witnessed an unprecedented acceleration in the technology. Taylor's passion and his struggle to find answers infected others with a fire and a quest for knowledge to increase productivity and enhance manufacturing engineering, yet with a social conscience.[4] It has brought us to this point in our understanding of machining metals. We still strive for productivity improvement, but now ultra-precision, rapid response (without sacrificing the economy of production), the environment, and ecological interactions have all entered the equation.

REFERENCES

1. Hodges, H., *Technology in the Ancient World*, Barnes and Noble, New York, p. 187, 1992.
2. Finnie, I., Review of the metal cutting theories of the past hundred years, *Mech. Eng.*, 78, 715–721, 1956.
3. Taylor, F. W., On the art of cutting metals, *Trans. ASME*, 28, 31–248, 1907.
4. Currie, R. M., *Work Study*, British Institute of Management, London, 1972.
5. Mallock, A., *Proc. R. Soc. London*, 33, 127–139, 1881.
6. Moon, F. C., *Franz Reuleaux: Contributions to 19th C. Kinematics and Theory of Machines*, Sibley School of Mechanical and Aerospace Engineering, Cornell, Ithaca, NY, *Trans. ASME J. Appl. Mech.* 2003.
7. Piispanen, V., Theory of chip formation, *Teknillinen Aikaausleni*, 27, 315–322, 1937.
8. Ernst, H. and Merchant, M. E., Chip formation, friction and high quality machined surfaces, *Trans. Am. Soc. Met.*, 29, 299–378, 1941.
9. Lee, E. H. and Shaffer, B. W., Theory of plasticity applied to problems of machining, *Trans. ASME J. Appl. Mech.*, 18, 405–413, 1951.
10. Oxley, P. L. B. and Hastings, W. F., Predicting the strain rate in the zone of intense shear in which the chip is formed in machining from the dynamic flow stress properties of the work material and the cutting conditions, *Proc. R. Soc. London*, A356, 395–410, 1977.
11. Klamecki, B. E., Incipient chip formation in metal cutting — A 3D finite element analysis, Ph.D. dissertation, University of Illinois at Urbana Champaign, 1973.

12. Rowe, G. W. et al. *Finite Element Plasticity and Metalforming Analysis*, Cambridge University Press, Cambridge, 1991.
13. Childs, T. H. C. and Dirikolu, M. H., Modeling requirements for computer simulation of metal machining, *Turkish J. Environ. Sci.*, 81–93, 2000.
14. Marusich, T. D. and Ortiz, M., Modeling and simulation of high-speed machining, *Int. J. Numer. Methods Eng.*, 38, 3675–3694, 1995, Third Wave Systems, Inc. AdvantEdge v3.6 Machining Simulation Software, Minneapolis, MN, 2001.
15. Silliman, J. D., *Cutting and Grinding Fluids: Selection and Application*, SME, Dearborn, MI, 1992.
16. Frost, M. F., A model of the loading process in grinding Ph.D. thesis, Department of Mechanical Engineering, University of Bristol, Bristol, UK, 1981.
17. Ding, Y. and Hong, S., A Study of the Cutting Temperature in Machining Process Cooled by Liquid Nitrogen. Technical Paper of NAMRC XXIII, May 1995.
18. Rajurkar, K. P. and Wang, Z. Y., Beyond cool, *Cutting Tool Eng.*, 48(6), 52–58, 1996.
19. Rebinder, P. A. and Likhtman, V. J., Effects of surface-acting media on strains and ruptures in solids, *Proceedings of the Second International Conference of Surface Activity*, Butterworths, London, pp. 563–580, 1947.
20. *Machining Data Handbook*, 3rd ed., TechSolve, Cincinnati, OH (http://www.techsolve.org/prodserv/Machining/ Mechining%20Data%20Handbook. htm).
21. King, R. I., *High Speed Machining Technology*, Chapman & Hall, New York, 1985.
22. Ezugwu, E. O., High speed machining of aero-engine alloys, *J. Brazil. Soc. Mech. Sci. Eng.*, 26(1), 1–11, 2004.
23. Jocelyn, A., *RAMP — Revolutionary Aerospace Machining Project*, University of West England, Bristol, UK, 2003.

4 Performance of Metalworking Fluids in a Grinding System

Cornelis A. Smits

CONTENTS

I. INTRODUCTION

Metalworking fluids have been applied for many reasons to enhance grinding as a metal removal operation for precision components. The importance of metalworking fluids and proper fluid application is proved every day in shops where precision components are produced.

Various publications on the performance of metalworking fluids, their application, and their secondary functions have increased the knowledge of fluid applications in the industry. The wide range of grinding operations and specific machine tools involved, as well as the environmental and health issues, creates a very complex technology with many constraints. Other considerations include the maintenance and ongoing contamination of the metalworking fluid by grinding chips, grinding wheel particles, machine tool and other oil leaks, make-up water minerals, elevated temperatures caused by grinding energy, and other possible pollutants which can cause a significant change in metalworking fluid behavior. Also, the carry-off of certain chemicals which stick to ground workpieces and metal fines will cause changes by depletion.

One can list the variables that are of importance under machine tool, grinding wheel, or metalworking fluid-related variables. The machine tool-related variables change over time due to wear of machine tool elements. However, this change is very slow and is measured in years instead of hours. The grinding wheel variables will be constant depending on duplication of manufacturing and the availability of facilities for maintaining a constant circumferential speed over the wear range of the grinding wheel. The metalworking fluid is the most dynamic element in the grinding system as it is continuously under change due to carry-off and the input of pollutants as mentioned earlier.

All of these conditions must be considered when selecting, applying, and maintaining metalworking fluids. In this context, the variables affecting performance will be discussed and the proper engineering criteria defined in order to choose the best selection of metalworking fluids and to correctly design its application.

II. METALWORKING FLUIDS AS AN ELEMENT OF THE TOTAL GRINDING SYSTEM

It is essential to know the grinding system that will be served by the metalworking fluid. This system includes the conditions of the workpieces before beginning the grinding operations. Figure 4.1 shows the most essential workpiece variables. The workpiece dimensions and static stiffness will define the grinding contact area with the grinding wheel and level of grinding force that can be applied. (Workpiece stiffness can be a significant constraint.) Also, the dimensions and shape of the part, combined with its thermal conductivity, will define the heat sensitivity of the workpiece — a constraint for maximum energy input.

Grinding is an energy inefficient method of shaping components and it is obvious that the thermal balance in the grinding zone is of significant importance. The thermal conductivity and diffusivity of the workpiece material has a significant impact on the grinding results. The flow of heat from the contact zone of workpiece and grinding wheel is very important in order to limit the thermal damage to the workpiece material. Application of the metalworking fluid can have a significant influence on this heat flow. The fluid speed leaving the nozzle, the flow rate, and the direction of the flow (nozzle position), are important factors. Other factors such as grinding wheel composition, wheel and workpiece speed, and the amount of energy input will influence the heat flow.

The metalworking fluid will carry away most (96%) of the input energy by its contact with the workpiece, chips, and grinding wheel. The energy input will end up in the metalworking fluid where it will be transferred to the surroundings by evaporation, convection, or in a forced manner, by a chiller. Evaporation is the dominating factor in obtaining a balance between energy input and output. The metalworking fluid will warm up until a temperature is reached that balances energy

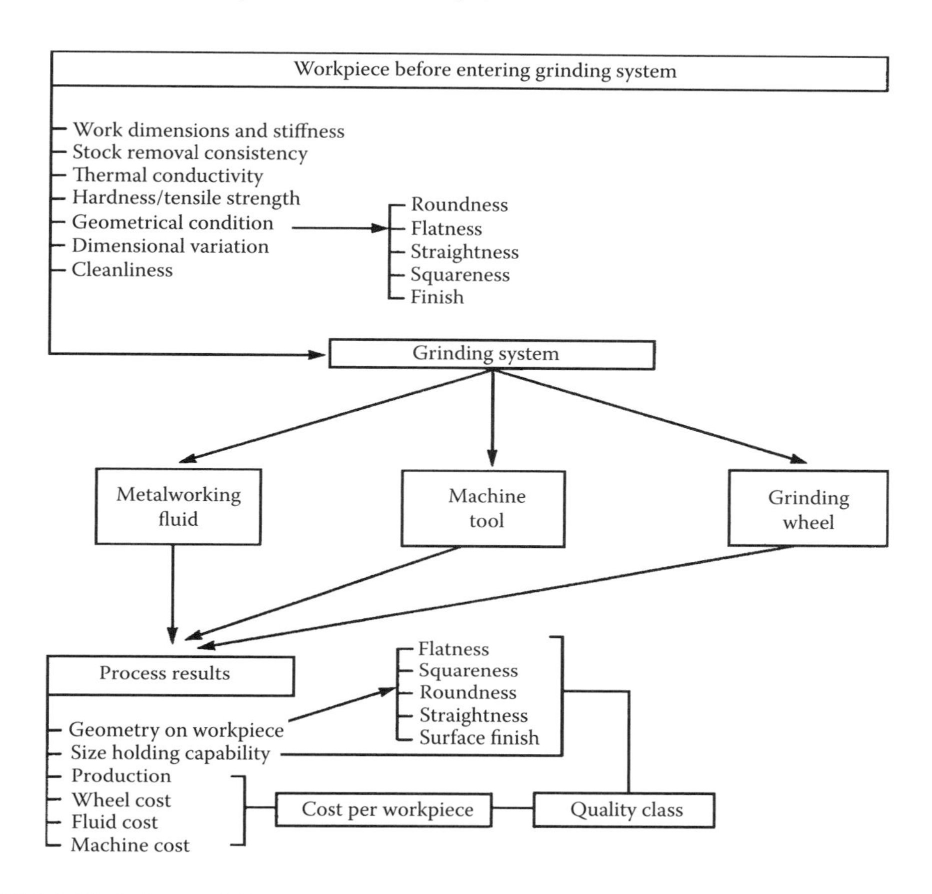

FIGURE 4.1 Workpiece variables.

input and output to the ambient through convection and evaporation. With a very high input of energy, a high make-up rate can be expected which causes a possible high rate of contamination if nontreated water is used.

Another workpiece material characteristic is its hardness (R_c). When hardness increases, higher forces are required to penetrate abrasive grains into the material surface. This will result in higher grinding forces and the normal force (F_n) will increase. Figure 4.2 shows that as you approach 50 to 60 R_c, it is very difficult for the grinding abrasive to penetrate into the surface of the workpiece. This results in very thin chips and considerable machine deflection, particularly for weak grinding systems such as internal grinding. Some of this behavior is due to the fact that the wheel itself flattens out in the contact areas, similar to the way a tire flattens against the road due to the weight of the car.

The normal force F_n is also dependent upon the lubricant capability of the metalworking fluid, the grinding wheel grade, the truing technology, and the aggressiveness of the cut (see Figure 4.3). The stock removal consistency, geometry, and dimensional variations are mechanically related properties that cannot be influenced by metalworking fluids.

Cleanliness of workpieces entering the system can affect fluid performance. Carry-over of oil and grease from other operations will be an influence on the grinding operation and present an ongoing contamination of the fluid.

Figure 4.4 shows the most important variables that affect manufacturing cost and quality class. The consistency of these listed variables is very important. Some will remain constant when implemented, others change over time. The metalworking fluid represents an essential element of these variables.

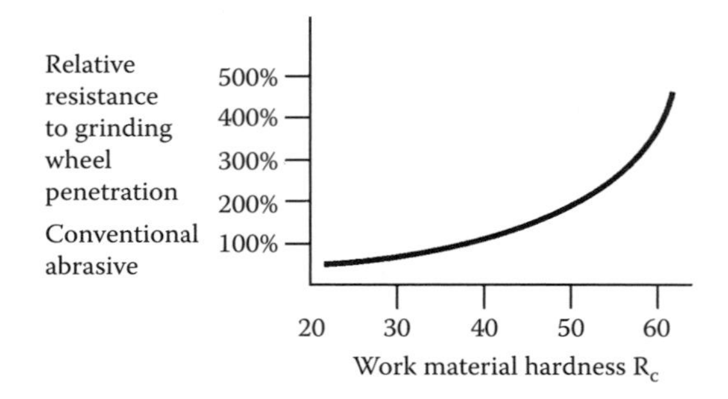

FIGURE 4.2 Resistance to grinding wheel penetration increases with increased work material hardness. As you approach 50 to 60 R_c, it is very difficult for the grinding abrasive to penetrate into the surface of the workpiece. This results in very thin chips and considerable machine deflection, particularly for weak grinding systems such as internal grinding. Some of this behavior is due to the fact that the wheel itself flattens out in the contact area — similar to the way a tire flattens against the road due to the weight of the car.

Grinding wheels depend on duplication capability when the optimum composition (grade) has been obtained. The truing tool and truing conditions, however, can change the cutting action of the grinding wheel which can lead to significant performance changes. Consistency in truing technology is essential.

III. PERFORMANCE OF METALWORKING FLUIDS

To express performance of a metalworking fluid, a wide range of criteria are applied. The criteria can be grouped as shown in Figure 4.5.

The environmental criteria will not be discussed in this section. Some attention will be given to the criteria for maintainability of the metalworking fluid, but only where an effect on the grinding operation is measurable.

The overall performance of a metalworking fluid is a result of the criteria listed. Most items listed under the environmental and maintainability categories are areas where attention has been focused and are well known. However, the reason for the application of a metalworking fluid is its ability to aid the grinding performance, which can be expressed by:

Productivity
Tool life
Energy consumption
Quality

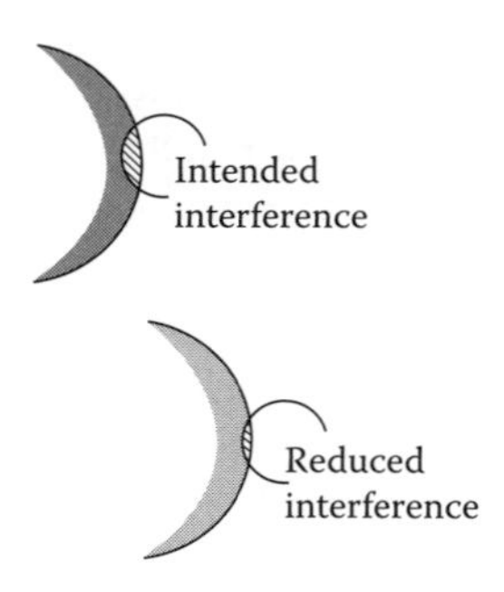

FIGURE 4.3 Factors affecting normal force.

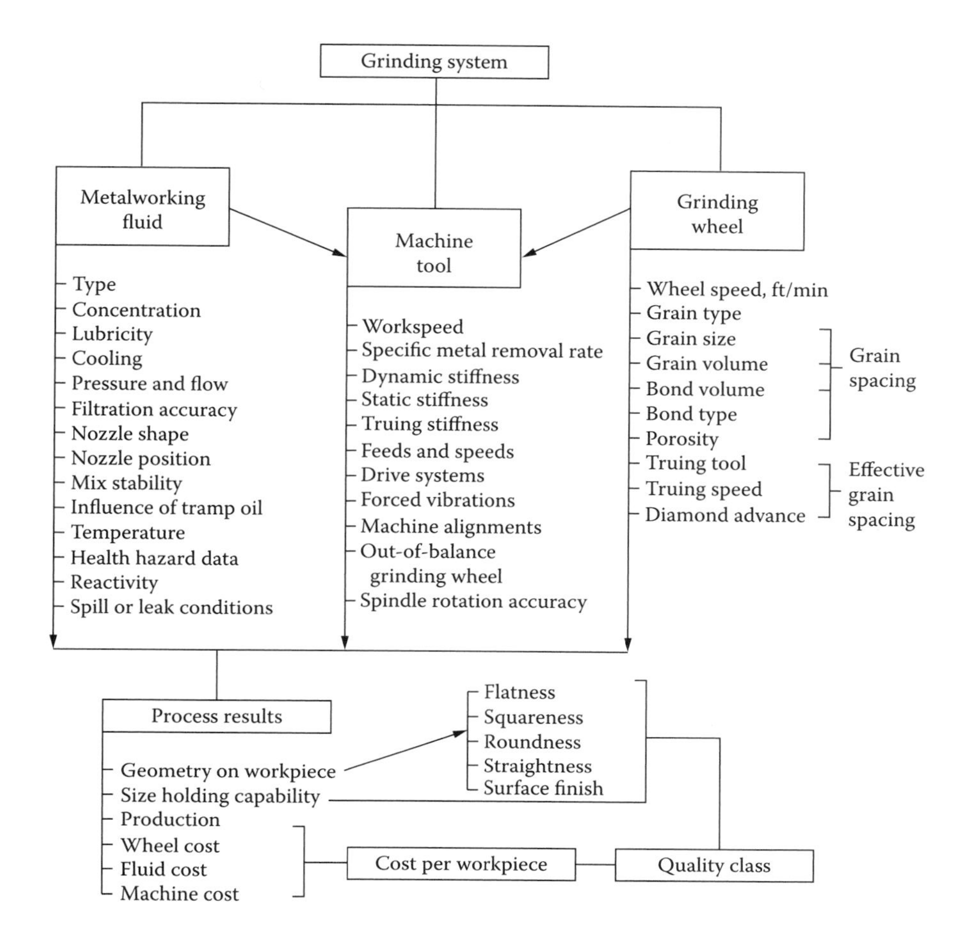

FIGURE 4.4 Factors affecting cost and quality of the finished part.

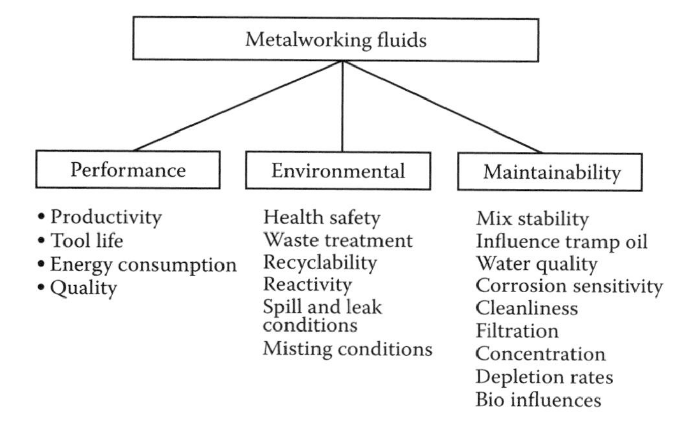

FIGURE 4.5 Metalworking fluid selection criteria.

These are the four significant process parameters that we have defined in cylindrical grinding operations.

Optimizing the operation must be the ultimate objective in any study of a metal cutting process. Optimization first requires a thorough understanding of the interrelationships of the significant variables; then these variables must be measured to find their quantitative influence on quality and cost.

A. Productivity

The productivity of a grinding process is expressed by the specific metal removal rate, Q' defined as the volume of metal removed per unit of time per unit of effective wheel width. Its units are in inches squared per minute (in.^{2}/min). This is the most important of all parameters since it determines production rate. Q' may be determined as follows.

For cylindrical grinding (internal or external):

$$Q' = \frac{\pi}{2}\frac{DSL}{WT} = \frac{\text{in.}^2}{\text{min}}$$

For surface grinding:

$$Q' = A \times V_w = \frac{\text{in.}^2}{\text{min}}$$

where:

D is the diameter or bore of workpiece (in.).
S is the stock removal on diameter (in.).
L is the work length (in.).
T is the grinding time (min).
W is the effective wheel width (in.).
A is the down feed increment (in.).
V_w is the table speed (in./min).

These equations represent the cubic inches of metal removed from a workpiece in 1 min by 1 in. of usable wheel width.

The formula for cylindrical grinding has been derived as follows.

The volume of metal removed per minute is:

$$V_m = \frac{\pi(D_1^2 - D_2^2)L}{4T}$$

where:

D_1 is the work diameter before grinding (in.).
D_2 is the work diameter after grinding (in.).
L is the length of workpiece (in.).
T is the grinding time (min).

Then:

$$Q' = \frac{V_m}{W} = \frac{\text{in.}^3/\text{min}}{\text{in.}}$$

$$Q' = \frac{\pi(D_1^2 - D_2^2)L}{4WT} = \frac{\text{in.}^3/\text{min}}{\text{in.}}$$

where W is the effective wheel width (in.).

Since $(D_1^2 - D_2^2)$ can be written $(D_1 + D_2)(D_1 - D_2)$ and substituted, then:

$$Q' = \frac{\pi(D_1 + D_2)(D_1 - D_2)L}{4WT}$$

Since $(D_1 + D_2)/2$ is the average diameter, or D; and $(D_1 - D_2)$ is the stock on the diameter, or S; then:

$$Q' = \frac{\pi DSL}{2WT} = \frac{\text{in.}^3/\text{min}}{\text{in.}}$$

This means that Q' equals the cubic inches of metal removed from a workpiece in 1 min by 1 in. of effective wheel width.

In plunge grinding, where the workpiece length L always equals the effective width W, the formula is:

$$Q' = \frac{\pi DS}{2T}$$

B. Tool Life

Tool life in grinding is expressed as a grinding ratio (G) defined as the volume of material removed with one volume unit of grinding wheel. This is an important parameter because wheel wear determines wheel costs and machine down-time in heavy metal removal operations, and it greatly affects quality when finish grinding. The expression for the G-ratio is:

$$G = \frac{DSLP}{2qdW} = \frac{\text{metal removal}}{\text{wheel wear}} = \frac{\text{in.}^3}{\text{in.}^3}$$

where:

q is the radial wheel wear (in.).
$d = (d_1 + d_2)/2$ is the average wheel diameter.
P is the number of workpieces ground.
L is the length of one part (for plunge infeed, L and W cancel).

What this expression really describes is the volume of metal removal for each volumetric unit of grinding wheel worn or dressed away.

C. Energy Consumption

The energy consumption in grinding is described by the specific energy (U) defined as the horsepower required to remove one unit volume of material per unit of time. The importance of specific energy lies primarily in its role as a process limitation or evaluation, since it expresses the energy required to perform the metal removal operation, and must be known and controlled in order to stay within the power available. It is also a very important parameter in predicting the behavior of nonrigid parts under grind. The expression for specific energy is:

$$U = \frac{N}{Q'W}$$

where:

N is horsepower (hp, kW, or Y).
Q' is the specific metal removal rate (in.^{2}/min).
W is the effective wheel width.

This represents the power required to remove one cubic inch of metal in 1 min.

D. Quality

The quality of a part is often measured as surface finish (f). This is one of the most important parameters in a finish grinding operation, and it is very closely related to specific metal removal rate and G-ratio. It can be described by $f = R_a$, where R_a is the average surface roughness value in use today throughout the world. The following definitions apply:

R_a, the arithmetic mean of departures of roughness profile from the mean line
R_q, the root mean square (RMS) parameter corresponding to R_a
R_z, the ISO ten-point numerical average height difference between the five highest peaks and the five lowest valleys within the measuring length

IV. Interrelationship of Grinding Parameters and the Influence of Metalworking Fluids

In 1907, F.W. Taylor reported that, in metal cutting, both tool wear and specific cutting force were exponentially related to cutting speed.[1] Approximately 50 years later, after a large mass of grinding data had been collected and recorded, it was discovered that the G-ratio and specific metal removal rate relationship for grinding also holds true in the Taylor equations for metal cutting processes. This application of the Taylor formulas to grinding has undoubtedly been the most important breakthrough in the quantitative understanding of the process. The formulas for G-ratio and specific energy as related to specific metal removal rate are shown below.

Productivity and tool life relationship:

$$Q'^{n_1} G = C_1$$

where:

G is the grinding ratio.
n_1 is the grinding ratio exponent.
C_1 is the grinding ratio at $Q' = 1$ (see Figure 4.6).

Productivity and energy consumption relationship:

$$Q'^{n_2} U = C_2$$

where:

U is the specific energy.
n_2 is the specific energy exponent.
C_2 is the specific energy at $Q' = 1$ (see Figure 4.7).

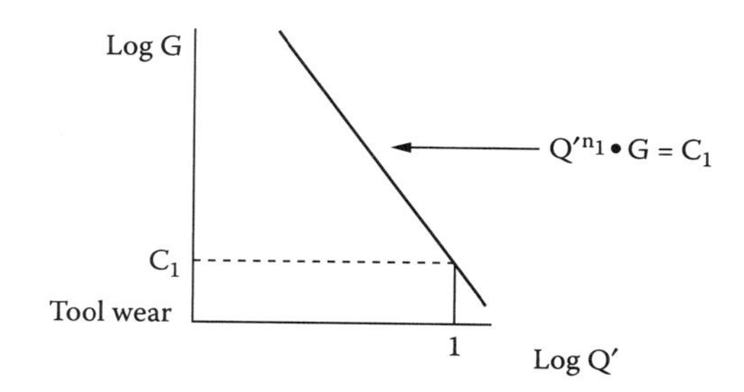

C_1 = G ratio at $Q' = 1$
C_1 = influenced by: wheel grading
Metalworking fluid
n_1 = influenced by: workmaterial
Metalworking fluid
Type of abrasive material (conventional)

FIGURE 4.6 Productivity vs. tool life.

An additional parameter that fits the exponential equations is surface finish (f).
Productivity and quality relationship:

$$Q'^{n_3} f = C_3$$

where n_3 = surface finish exponent and C_3 = surface finish at $Q' = 1$. This parameter is tied very closely to G and Q', and varies inversely with G and Q', as seen in Figure 4.8.

A. Grinding Performance Diagram

One of the most important qualities of the Taylor formulas as working instruments for optimizing the process is the simplicity and usefulness of their graphic representation. When

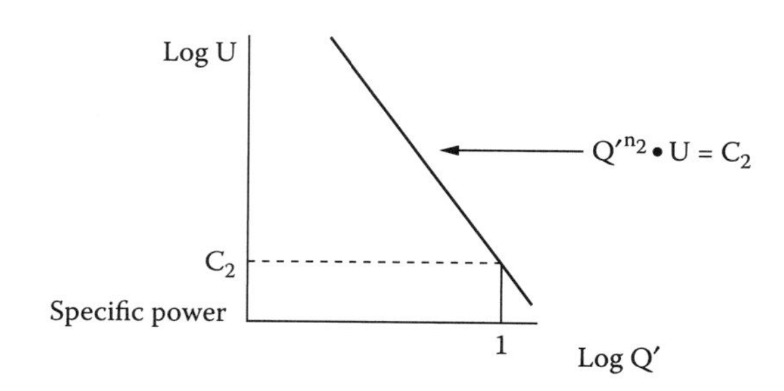

C_2 = U value at $Q' = 1$
C_2 = influenced by: wheel grading
Metalworking fluid
n_2 = influenced by: workmaterial
Metalworking fluid
Type of abrasive material (conventional)
grain size

FIGURE 4.7 Productivity vs. energy consumption.

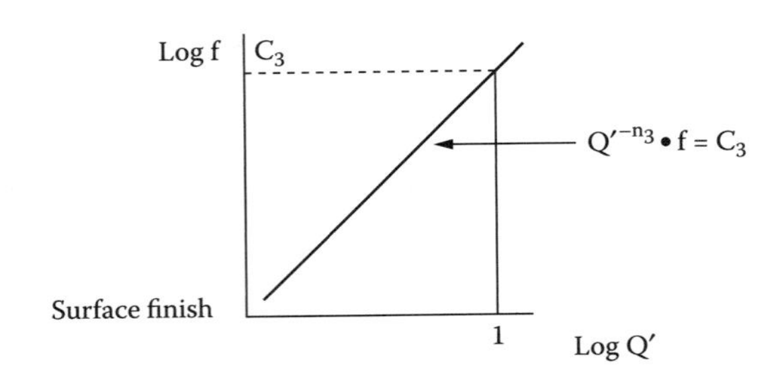

C_3 = Surface finish at Q' = 1
C_3 = influenced by: grain size
Metalworking fluid
Workmaterial hardness

n_3 = influenced by: hardness-structure wheel
workmaterial
Metalworking fluid

FIGURE 4.8 Productivity vs. quality.

a certain material is ground with a certain wheel at several different specific metal removal rates, and the resulting *G*-ratio and specific power data points are recorded on log-graph paper, the data lines drawn through the points are straight. The slope of the lines then represents the exponents (see Figure 4.6, Figure 4.7, and Figure 4.8).

To optimize the process, we attempt to find the wheel grading and cutting fluid combination that gives us the highest C_1 value (or *G*-ratio at $Q' = 1$) and the lowest possible exponent, n_1. This would mean the least possible influence of increased metal removal rate on the *G*-ratio. At the same time, we would like to have the lowest possible C_2 value (specific power at $Q' = 1$) and the highest possible exponent — or steepest slope of *U* — so that the increased metal removal rate would require the least possible increase in horsepower.

The *f* or finish line is always sloped in the opposite direction to both *G* and *U*, because finish values become larger as Q' is increased.

The development of grinding wheels and metalworking fluids in recent years has shown considerable influence on the values for C_1 and n_1 as well as on C_2, n_2, and C_3, n_3. CBN grinding technology has shown very flat *G*-ratio lines and also very flat specific power lines. The values for C_1 are 100 to 1000 times higher while the values for C_2 are 50 to 75% of the values for conventional abrasives.

B. The Influence of Grinding Wheels on Grinding Performance

The influence of a grinding wheel on both wheel life and horsepower consumption has been measured in many applications and research activities. Changing grain type, grain size, grain volume, bond type, and hardness of a grinding wheel will have an influence on the *G*-ratio line as well as the specific energy line and consequently on the surface finish line. The changes can be expressed in the values for the constants and the exponents.

The key is to select wheel gradings and generate a wheel manufacturing process that produces the following:

A high value of C_1 and a flat *G*-ratio line or a small value for n_1 which results in high productivity with controlled wheel wear. In general, a CBN grain will produce high values for C_1 and very low values for n_1.

A low value of C_2 and a value for n_2 of near 1. This means an increase in specific metal removal rate that does not result in an excessive increase of horsepower requirements and consequential grinding forces.
A low value for C_3 (good surface finish) and a low value for n_3. This means that an increase of the specific metal removal rate only has a small to moderate effect on surface finish.

Because of the many variables possible on grinding wheel compositions and manufacturing, an enormous databank is needed to predict or analyze the influence of the grinding wheel. Such a databank is essential, however, to generate a successful grinding system.

C. The Influence of a Metalworking Fluid on Grinding Performance

To measure the performance of a metalworking fluid, an empirical test is required. From the test data over a range of specific metal removal rates, the values for C_1 and the exponent n_1 as well as C_2n_2 and C_3n_3 can be defined. Figure 4.9 shows the G-ratio lines obtained with various cutting fluids x, y, and z.

To compare performance based on G-ratio at only one value of Q' is very misleading, as can be seen from Figure 4.9. Significant changes in value for C_1 and n_1 can be observed. Numerous times, performance has been measured on the basis of energy consumption. Figure 4.10 shows the relationship between U and Q' for the three fluids tested in Figure 4.9. Here again, significant changes can be seen in values for C_2 and n_2. For a true performance factor both tool life and energy consumption must be considered. The ideal performance would produce a very high G-ratio with a very low-energy requirement.

The performance factor (efficiency) can be expressed as:

$$E = \frac{G}{U} \qquad Q'^{n_1}G = C_1, \qquad Q'^{n_2}U = C_2$$

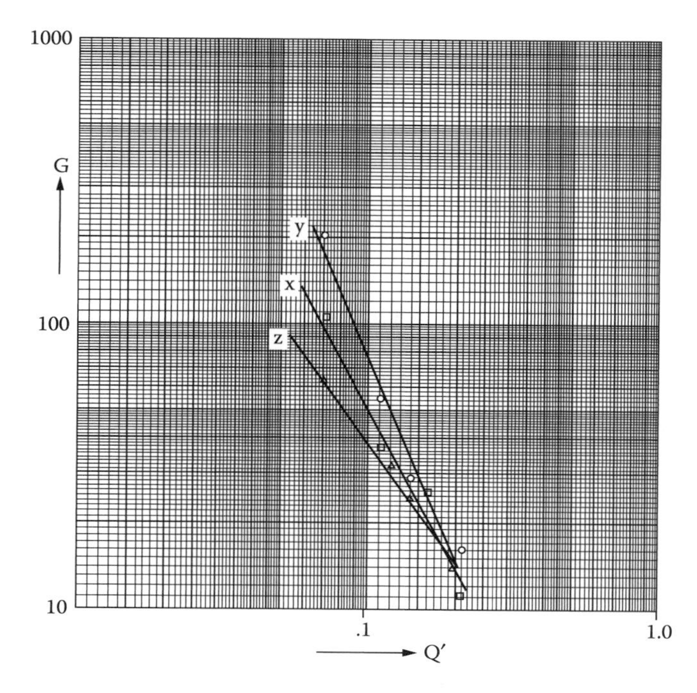

FIGURE 4.9 Wheel life with three grinding fluids at various specific metal removal rates.

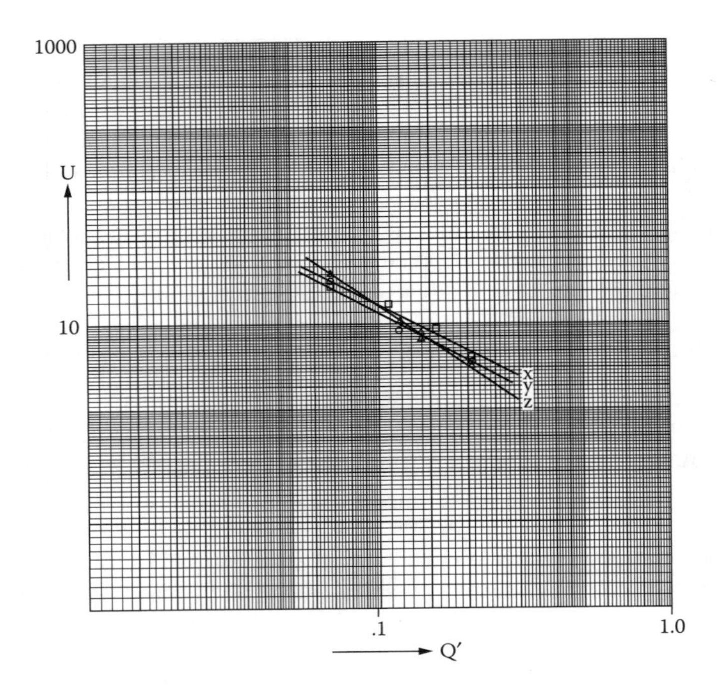

FIGURE 4.10 Energy consumption with three grinding fluids at various specific metal removal rates.

$$E = \frac{C_1}{C_2} Q'^{(n_2 - n_1)} \qquad \frac{C_1}{C_2} = C_E$$

$$E = \frac{C_E}{Q'^{(n_1 - n_2)}}$$

where E is the volume of material removed with one volume unit of grinding wheel per unit of specific energy.

The data from Figure 4.9 and Figure 4.10 are obtained in Figure 4.11 using the performance factor $E = G/U$. Now it can be clearly seen that product y outperforms products x and z up to a Q' value of 0.2 in.^{3}/min/in. Over a Q' value of 0.2, product z outperforms both y and x. For performance in grinding operations, product x can be considered as an underperforming product relative to product y, but relative to product z below $Q' = 0.12$, it outperformed z.

Using this method, fluid performance can be evaluated on the basis of two criteria:

1. The level of the E value
2. The rate of change of the E value as a function of Q'

The ideal fluid has a very high efficiency (E value) and increases with increasing specific metal removal rate (Q'). However, this is very unlikely. Usually, the E value decreases with an increase of the specific metal removal rate Q' (see Figure 4.12).

Under ideal conditions, the efficiency is sometimes nearly consistent and independent of Q'. This has been the case with a combination of superabrasives and straight oil applications. In these cases, n_2 is approximately equal to n_1 which means that the slopes of the G-ratio and U value were near, or equal to one ($n_2 = n_1 = 1$). Note the following statements:

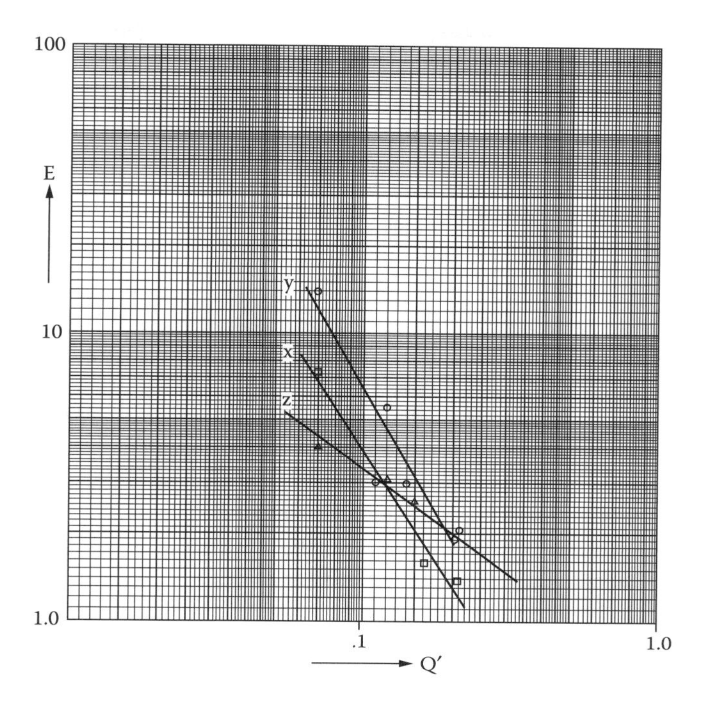

FIGURE 4.11 Grinding efficiency with three fluids at various specific metal removal rates.

$n_2 = n_1$: Obtained with superabrasives and special metalworking fluid or straight oil. On ball bearing steel material some synthetic abrasives are near this condition.

$n_2 < n_1$: Most applications are in this range of reduced efficiency at increasing specific metal removal rates.

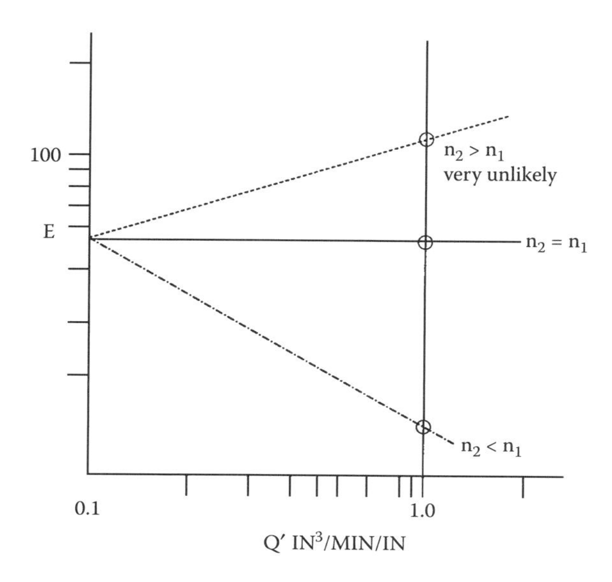

FIGURE 4.12 Grinding efficiency, three scenarios.

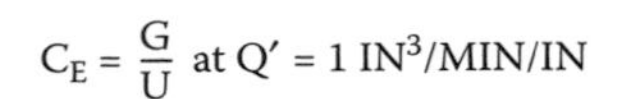

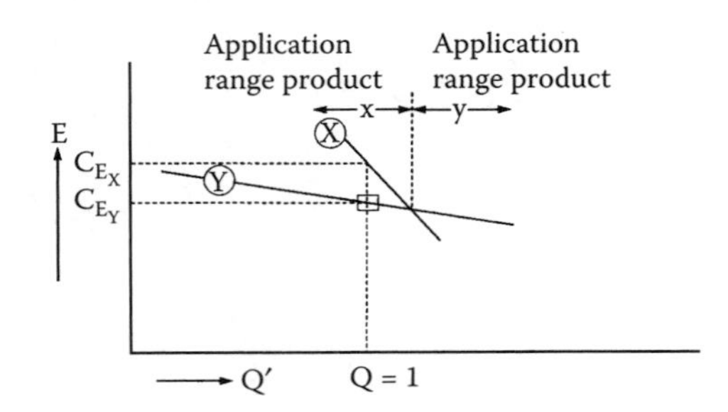

FIGURE 4.13 Selecting the proper fluid for the application (the higher the C_E value the better the overall performance).

See Figure 4.12 for a graphical representation.

To compare metalworking fluid performance, the E value at a specific metal removal rate $Q' = 1$ can be used. The most favorable application range can be selected as shown in Figure 4.13. Product x shows higher E values up to the intersection of the E value line for product y. As Q' values increase beyond the intersection point, product y has the best performance.

The primary function of the grinding performance factor E is in its expression of performance level as a function of the specific metal removal rate Q'. It can predict the best application range for metalworking fluids as far as grinding performance is concerned. Other factors related to the environment and to maintenance of the fluid must also be considered for the selection of fluids to be used in a specific metal removal rate range.

V. METALWORKING FLUID APPLICATION IN THE GRINDING ZONE

The application of metalworking fluids to the grinding zone has been discussed in many publications, but is still a "many times ignored" aspect. The application of the fluid to the grinding zone has to serve four basic functions: lubrication, cooling, chip removal, and workpiece surface protection. Lately, a study on flow mechanics resulted in a comprehensive flow model.[2]

Various other models for lubrication and flow have previously been presented. Hahn[3] discusses the application of lubrication between grain and work surface. A model developed by Powell[4] discussed the fluid flow through porous wheels. The local lubrication between grain and work surface, however, has not been modeled to our knowledge. All these studies point to the importance of flow rate, fluid speed entering the flow gap, fluid nozzle position, and grinding wheel contact with the workpiece. For our purposes we will focus on the practical area of fluid application.

Various fluid nozzle design concepts have been applied on grinding operations. The function of a coolant nozzle is to supply fluid to the cut zone and to clean the grinding wheel. There are several problems in accomplishing the task. One of the principal problems is the air barrier generated by a rotating porous grinding wheel. The grinding wheel acts like an impeller in a centrifugal pump. Air is drawn in from the sides and forced out around the circumference. As wheels get wider, there is less of a problem with air.

It is possible to classify the coolant nozzles into the following categories:

- Dribble
- Acceleration zone
  - Bourgoin Fluid Inducer
  - Wedge
  - Combined

Fire hose
Jet (medium pressure and super jet)
Wrap around

A. The Dribble Nozzle

The dribble nozzle is commonly used in most applications. About all that is accomplished is that some lubricant is deposited on the part's surface and some splash is sucked into the wheel. This does help in lubricating the cut zone. The major problem with this nozzle is that coolant cannot penetrate the air barriers and the fluid is not delivered efficiently into the cut zone. This can result in dry grinding despite the application of fluid.

B. Acceleration Zone Nozzle

The acceleration zone nozzle attempts to use the existing fluid systems to get the fluid up to the wheel speed, to cool the grinding wheel surface, and to wet the cutting grains. Several types exist.

1. Bourgoin Fluid Inducer

A nozzle (Figure 4.14) was developed by the French Research Center for Mechanical Industries (CETIM). Information on this device was first published in 1976, and assigned a U.S. patent in 1979.[5] The Nozzle requirements were:

1. 60 to 70° of wrap around the wheel were required to accelerate the fluid.
2. The gap between the nozzle had to be held at 1 mm (0.04 in.).
3. There had to be serrations inside the nozzle body to keep the liquid from separating tangentially.

The published results show evidence of improved wheel cleaning and reduced wheel wear. The device is no doubt much better than a dribble nozzle. The device is complex in that it must be adjusted for wheel curvature change as the wheel wears.

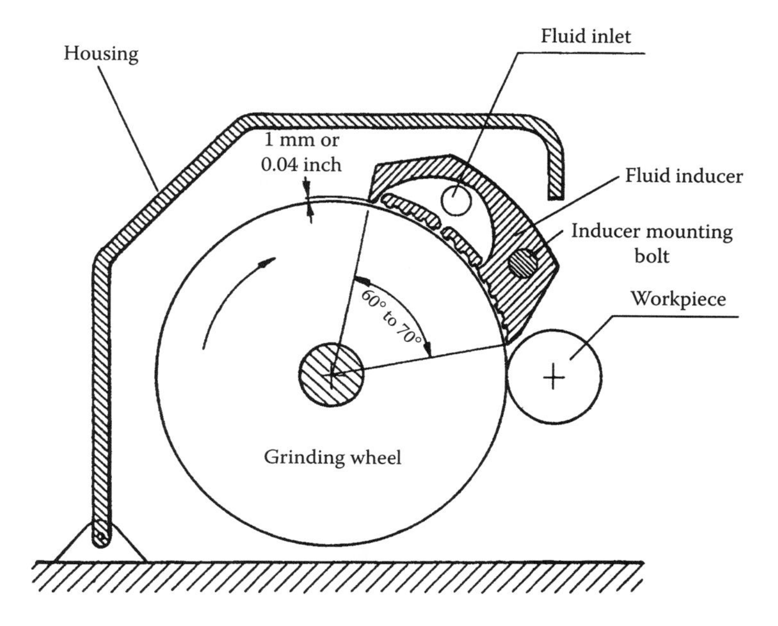

FIGURE 4.14 Bourgoin fluid inducer (from Ref. [5]).

The true value in wheel cleaning was probably the wetting of the cutting grits which led to the prevention of loading and smearing of metal on the wheel's surface.

2. Wedge Type

The wedge type nozzle is illustrated in Figure 4.15. The incompressible metalworking fluid is forced into a grinding wheel and consequently forced up to wheel speed. Complete development of this nozzle was never finished. The amount of wrap around the wheel no doubt should be the 60 to 70° worked out by the French. The nozzle wrap shown here was between 20 and 30°. The gap was controlled by maintaining a 30 to 60 psi nozzle pressure.

The sides of the wheel were covered with "horse blinder"-looking pieces. This forced the fluid alongside the wheel surface so it could be sucked into the wheel. This fluid would emerge out of the wheel's face, putting fluid in the cut zone.

This nozzle was very effective and data collected for ultrahigh speed (speeds above 16,000 sfpm) illustrated this fact.

3. Combined Inducer-Wedge

Probably the second most common nozzle to appear on the production floor is the combination nozzle. The nozzle is illustrated in Figure 4.16 and its configuration is somewhat empirical. This nozzle has a gap between the wheel and surface of approximately 0.04 in. for a certain number

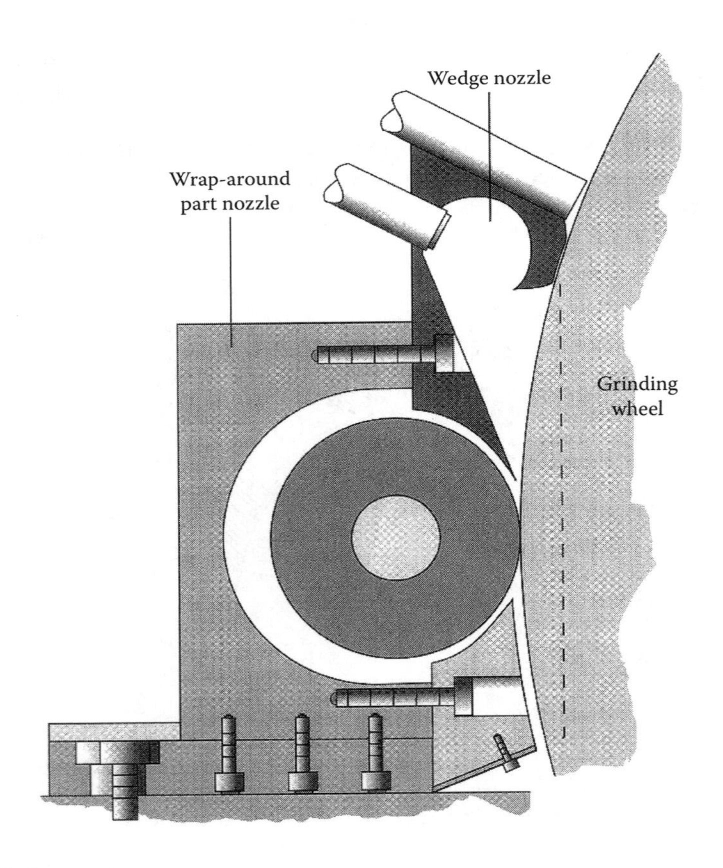

FIGURE 4.15 Wedge nozzle.

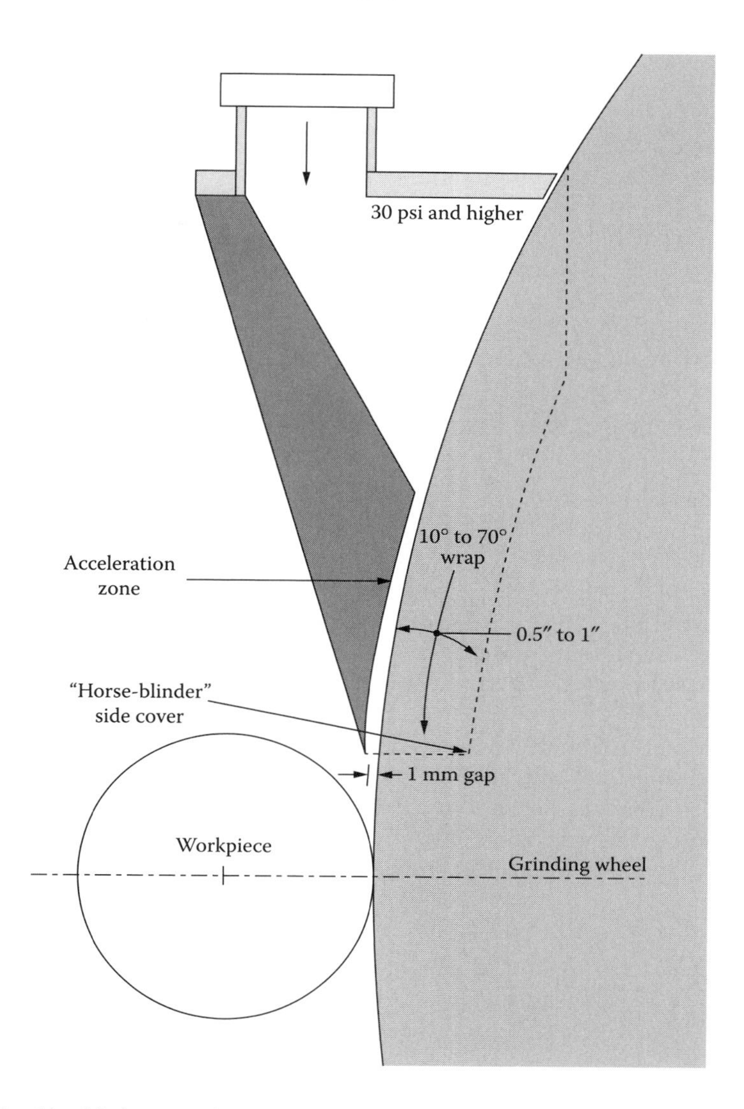

FIGURE 4.16 Combined inducer-wedge nozzle.

of degrees of wrap, usually a minimum of 10 to 20°. There is the problem with wheel curvature between new and worn wheels in maintaining the small gap to force the acceleration of the coolant. This nozzle will wet the cutting grains, remove heat, and lubricate the cutting zone. It is a fairly effective nozzle.

C. Fire Hose Nozzle

The idea behind this nozzle is to force the fluid to slightly above the grinding wheel speed, thus having the fluid present in the cut zone. One of the first applications of this nozzle was on the Micro Centric grinding machines for bearing grinding. It did not become very popular due to the splash and mist and it required an enclosed machine. A recent version of this nozzle is shown in Figure 4.17. Since liquid metalworking fluids are incompressible, the nozzle opening area can be calculated

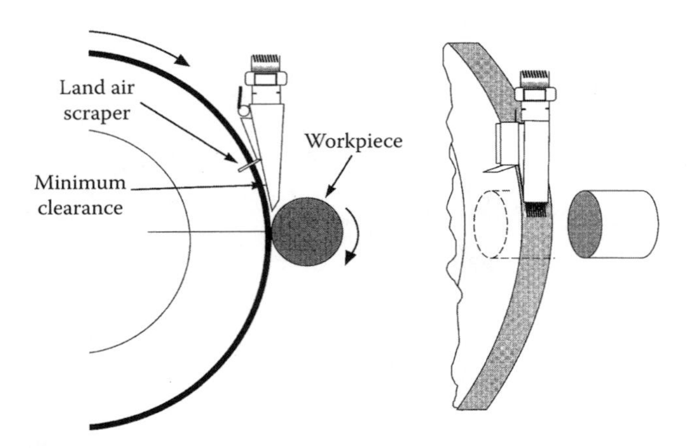

FIGURE 4.17 Fire hose nozzle.

by the formula $Q = AV$, where Q is the fluid flow, V is the desired speed of the fluid, and A is the area of the nozzle opening. This type of nozzle again must overcome the air pressure present on the wheel's surface. An air scraper can be used to achieve this. However, some wheels do not have an air barrier because the wheel has no porosity (such as very fine grit resin wheels and hot-pressed resin wheels).

The benefits of this nozzle are that it wets the grinding wheel and lubricates the cut zone. It does not remove heat as effectively, because the coolant is not forced around the wheel surface for as long as the acceleration zone nozzle, for example. The fire hose nozzle is an effective nozzle, however, and worthy of consideration for many applications.

D. Jet Nozzle

There are two jet nozzles applied in industry. The high-pressure version has been in use for approximately 40 years, and the medium-pressure was introduced in the early 1980s.

Sheffield introduced their super jet wheel cleaner in the early 1960s, in conjunction with abrasive machining and crush dressing. The system is illustrated in Figure 4.18. A number of these

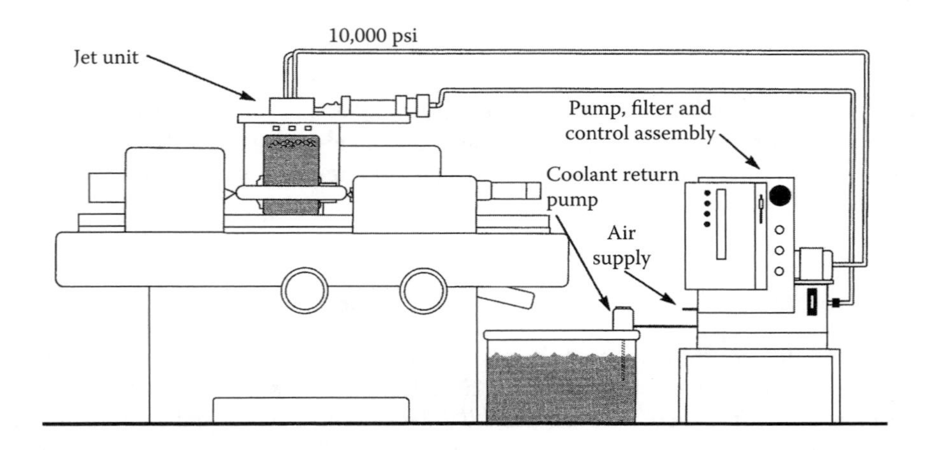

FIGURE 4.18 Jet wheel cleaning speeds grinding of high alloy materials.

units are in use today. High-pressure coolant (10,000 psi) is forced against the wheel at about an 1/8 in. (3mm) gap and moved back and forth across the wheel surface. The blast dislodges metal and swarf embedded in the wheel. Also, the fluid wets the grain surfaces. The jets are so powerful they easily overcome any air barrier.

One problem with the system is that the jet nozzles have very small openings and clog easily, but also wear fast, becoming ineffective. A good filter system is necessary, usually with a back flushing capability. The mist generated by this system usually requires an enclosure.

The system is effective in cleaning a wheel and wetting the grains on the wheel surface. The system does very little for cooling the wheel.

E. Wrap-Around Nozzle

A wrap-around nozzle is illustrated in Figure 4.15. In the case shown, the part is surrounded by metalworking fluid essentially submerged in the cutting zone. This both wets the part and extracts heat. It does little for wetting or cooling the grinding wheel. The scheme is of limited value, but is probably as good as the dribble nozzle systems discussed earlier.

A similar scheme could be used on the wheel side. This has not been attempted to the best of my knowledge. It may just act like a water brake and add heat to the wheel surface. The need is to extract heat from the wheel over a distance, especially for superabrasive grits.

F. Fluid Application for Superabrasive Grinding

A very interesting property of cubic boron nitride (CBN) is the thermal conductivity. A simple finite element model was presented by Glenn Johnson of General Electric[6] and is shown in Figure 4.19.

The grain is simulated by a cross and the metal surface by a square. The initial metal temperature was assigned the value believed to be the interface temperature when a chip is being formed. The temperature of the CBN grit was set at room temperature. Only conduction between the two elements was permitted. The contact time between the two was 80 μsec. (This simulated a grinding condition.) Note the large difference between CBN and aluminum oxide grain temperature in this short period of time.

This suggests several things:

1. Heat extraction into a superabrasive wheel is much more significant than with conventional abrasive.
2. Cooling a superbrasive grain as soon as it has completed its chip-making task is extremely important.
3. Thermal damage of a workpiece is less likely with superabrasives than with a conventional wheel.
4. Heat extraction from resin-bonded superabrasive is very critical for good *G*-ratios.

Schemes similar to that shown in Figure 4.20 have credibility.[7] Note the cooling high-pressure nozzle for the exiting wheel surface to drop the CBN grain temperature, the high-pressure jet nozzle to clean the film or load from the wheel, and the main supply for wetting the part and wheel before entering the cut zone.

VI. SELECTION OF FILTRATION SYSTEMS

Various parameters need to be defined for selecting the proper filtration system. The advantages and disadvantages of various systems will be discussed in Chapter 10. An important element that is

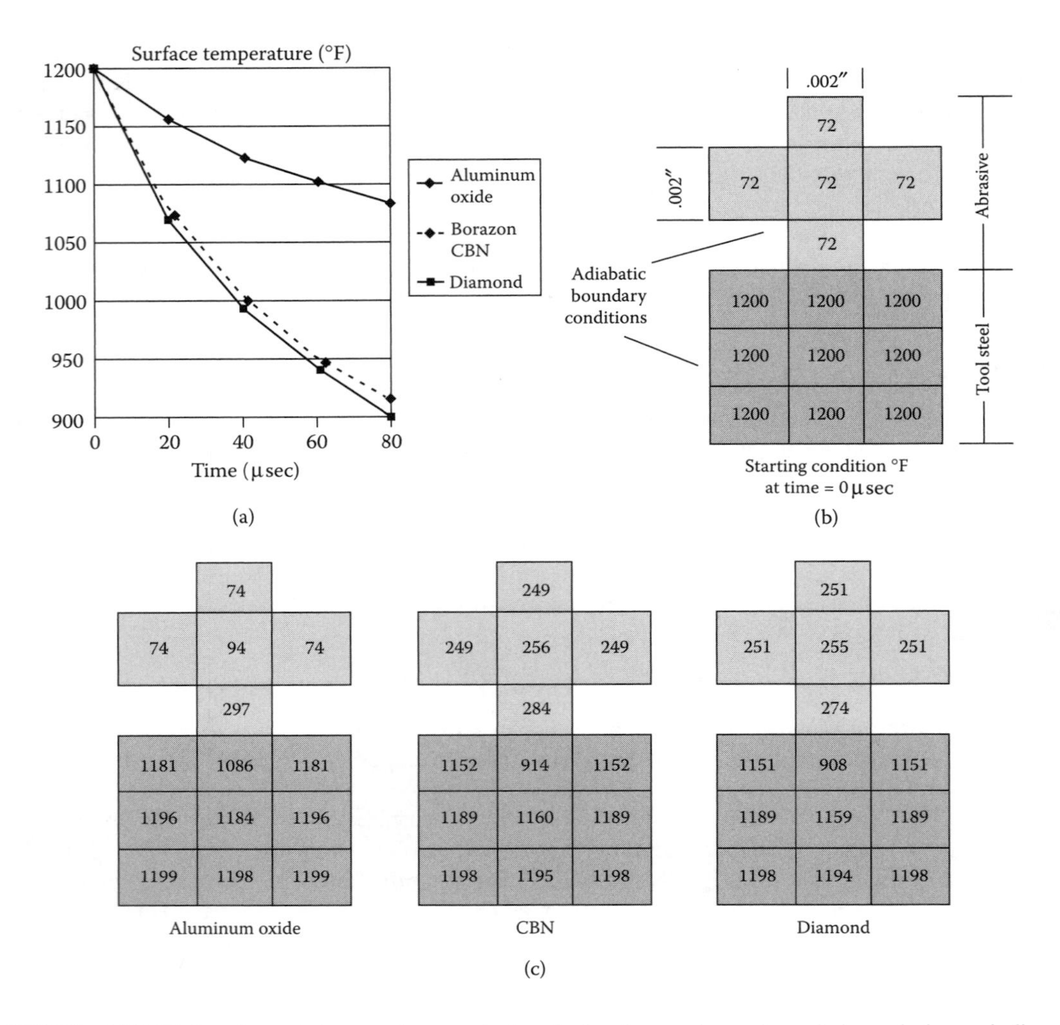

FIGURE 4.19 Finite element temperature analysis of abrasive grain and workpiece during grinding. (a) Thermal model using finite elemental analysis on effect abrasive thermal properties have on cooling the surface in the pass of one abrasive grain during the formation of a chip. (b) Finite elemental analysis model of heat transfer between workpiece and abrasive. (c) Comparison of temperature (°F) distribution and 80 μs of contact between workpiece and abrasive for three types of abrasive. *Source*: From Johnson, G.A., *Beneficial Compressive Residual Stress Resulting from CBN Grinding*, Society of Manufacturing Engineers, Dearborn, MI, MR86-625, 1986. With permission.

generally unrecognized will be reviewed in this chapter: the chip geometry and its effect on filtration systems.

The chips generated by the grinding process are defined by the following grinding parameters

Specific metal removal rate	Q'	in.^{3}/min/in.
Wheel speed	V_s	in./min
Work speed	V_w	in./min
Equivalent diameter	De	in.

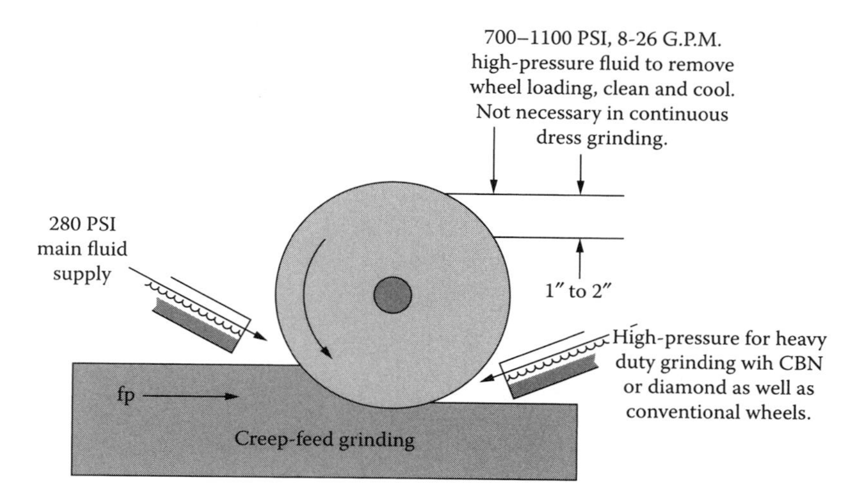

FIGURE 4.20 Use of high-pressure fluid supply in precision CBN grinding.

$$\text{De} = \frac{D_{\text{Workpiece}} \times D_{\text{Grinding Wheel}}}{D_{\text{Workpiece}} \pm D_{\text{Grinding Wheel}}}$$

Use + for OD grinding, use − for ID grinding, and for surface grinding use $\text{De} = D_{\text{Grinding Wheel}}$.

The dominating parameters for chip thickness (dt) and chip length (L) are Q' and De. Wheel speed is of importance for the chip thickness, but has no influence on chip length. Work speed affects both chip length and thickness. Figure 4.21 shows the influence of Q' and De on chip length (L) and thickness (dt). The relationship for the average undeformed chip thickness is as follows:

$$\text{d}t = \text{Cx} \sqrt[2+2a]{\frac{Q' V_{\text{w}}}{\text{De} V_{\text{s}}^2}} \qquad L = \sqrt{\frac{\text{De} Q'}{V_{\text{w}}}}$$

where:

Cx depends on wheel grade (grain volume) (in.).
a depends on grit size in the grinding wheel.
V_{w} is the work speed (in./min).
V_{s} is the wheel speed (in./min).

It is essential to understand that for most external grinding operations with work diameters below 2 in., the chips will be short and thick, while for larger ODs the chips become longer and thinner. For surface grinding operations, long wire-type chips are formed. This means that cake-type filters may not be applicable at low De and Q' values. Thus, for external grinding on finishing conditions ($Q' < 0.2$ and $\text{De} < 1.0$) cake-type filters will not function well. In this area, cartridge filtration or speed-related methods (cyclones, centrifugal) will be preferred.

Thus far, all discussions on chip form and size have concentrated on the average chip thickness. However, the chip size varies in thickness and length depending on the grinding wheel grain depth of penetration. At any time, in any grinding operation, there will be a variance in chip dimensions and very small chips will be generated.

Figure 4.22 shows a sampling of chips with various values of Q' for hardened, as well as soft, 1045 steel material. The value of De was 1.0 in. during the test. The dimension on the horizontal axis shows the chip size. The vertical axis shows the number of chips in the sample taken during

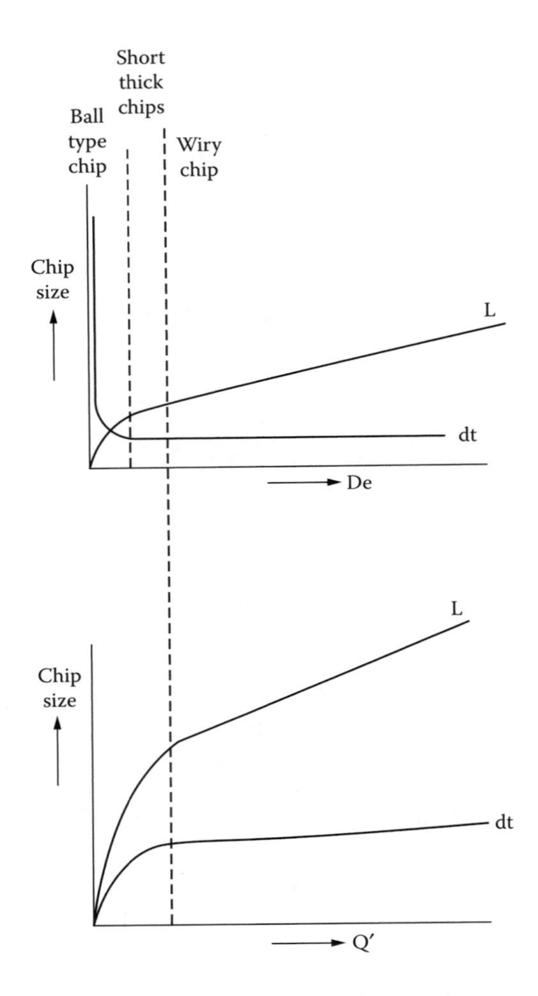

FIGURE 4.21 Effect of equivalent diameter and specific metal removal rate on chip size.

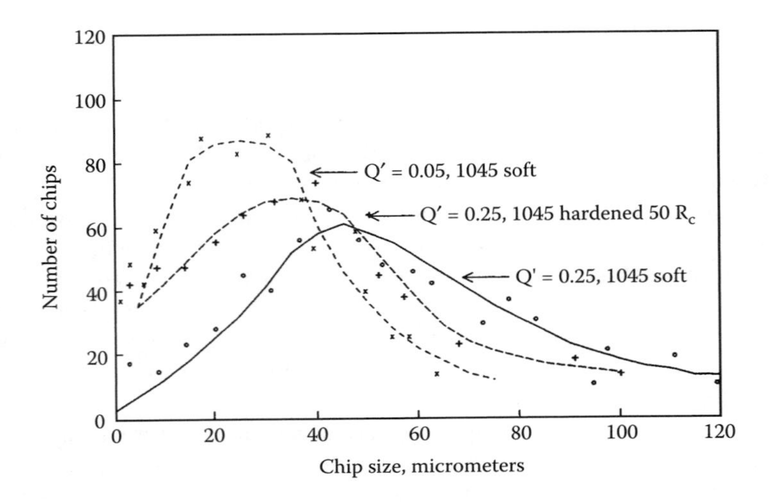

FIGURE 4.22 Chip size distribution for three grinding conditions.

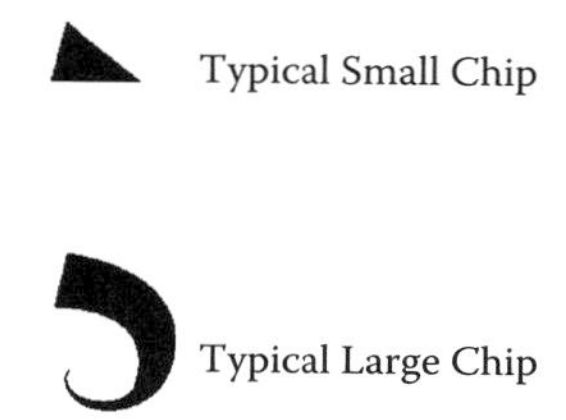

FIGURE 4.23 Typical chip shapes.

the grinding operation. The sample was collected directly after the grinding by means of a magnetic block, 1 in. below the grinding zone, attached to the workrest blade on a centerless grinder.

In Figure 4.22, one can observe that larger as well as smaller size chips are formed. It can be observed that when Q' reduces to typical finish grinding conditions, the content of small size chips increases and the average chip size changes modestly. It is interesting to see that the higher Q' values will produce, on average, larger chips, but will also produce a wider spread of chip size. Chips of 5 μm and smaller were still found and the volume was still significant.

We can conclude that, at the major grinding operations, smaller chips will be produced at a level of 5 μm or smaller. Filtration systems are difficult to design to filter these chips out of the fluid. Slowly, but surely, contamination will occur as a result of the build-up of small chips which remain in circulation. Filtration accuracy will dictate what size and what volume of chips will recirculate.

All the chips generally had a form varying from ball shape up to a wire type, with a 16:1 length-to-thickness ratio for the larger chips. However, these chips were mainly in the form of a curl, see Figure 4.23. With larger values for De it is expected that the typical large chip shape will dominate, while for small De values (De $\leq$ 1.0) the small chip form will be dominant.

For selection of a proper filtration system it is essential to have an understanding of the chip form and size range, as well as volume of grinding chips, grinding wheel wear particles, and other foreign materials which will pollute the metalworking fluids. Other important factors need to be considered, however, and this will be covered by other authors. The purpose of this contribution was to focus on chip geometry as a function of grinding conditions.

VII. KEEPING THE METALWORKING FLUID COOL

At high horsepower cuts, the temperature of the metalworking fluid will rise to a level that is no longer acceptable for work diameter tolerances, burning, etc. The main problem is how to dissipate the heat developed by the grinding process. Several methods can be used to realize this dissipation of heat. The methods used are:

1. Evaporation and convection
2. Cooling by air condensers
3. Cooling by forced evaporation
4. Cooling by refrigerating systems

A. Evaporation and Convection

If no forced cooling is available, the input energy must be transferred to the surroundings by evaporation and convection. Evaporation is usually dominant due to generally poor conditions for convection. Convection becomes significant only when a larger difference between fluid and ambient temperatures exists. In general, temperatures of metalworking fluids are 5 to 30°F

(3 to 17°C) over ambient and the convection surface is a limited area. In this range, 80% of the input energy will be transferred by evaporation when ventilation of the ambient air exists. In a totally enclosed environment, the evaporation is low due to saturation of the air with water, and convection will take over. In such cases, very high metalworking fluid temperatures can be expected.

Most of the evaporation, however, will take place when the fluid is applied in the cutting zone. For an estimate of the volume of water evaporated to balance the input energy, the following formula can be used:

$$V_e = \frac{Q}{2300} = \text{pounds of water per hour}$$

where Q is the energy input (BTU/h).

The warm-up curve of a fluid system depends on the volume of fluid in the system, convection, ambient temperature, and the humidity of the surrounding air. Only an empirical temperature-time plot can provide the information needed.

A warm-up curve is shown conceptually in Figure 4.24, while in Figure 4.25 an empirically measured warm-up curve is shown. In this example, the peak fluid temperature was up to 40.7°C (105°F) while the ambient temperature was 21.6°C (70°F). In this test the peak temperature was never reached due to termination of the energy input after 5 h. The evaporation will be maximum if this peak temperature has been reached.

B. Cooling by Air Condensers

Air coolers are all based on heat convection by blowing air through a condenser. For efficient convection a good temperature differential between ambient and fluid is needed. With a 15°F (8°C) temperature difference between fluid and ambient, a cooling surface of nearly 2000 yd^2 (1670 m^2) is needed. This method is basically not applied because of its inefficiency at the required small difference between fluid and ambient temperatures.

C. Cooling by Forced Evaporation

Another method is the evaporation system of a cooling tower. The dimensions are much smaller than for air coolers, but other disadvantages are introduced with this method. The cooling capacity

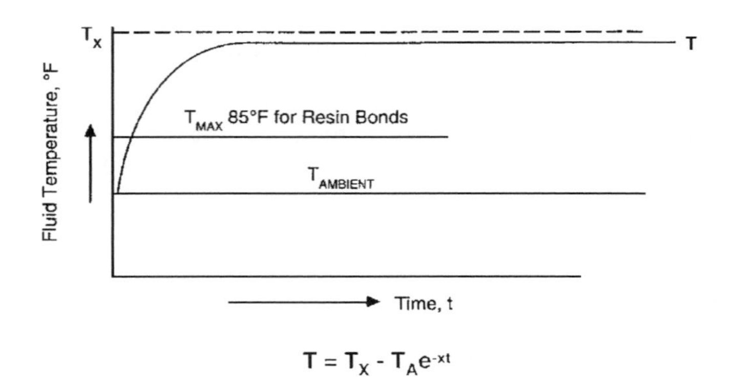

FIGURE 4.24 Generalized grinding fluid warm-up curve over time. Increased coolant temperature affects dimensional accuracy of the workpiece, performance (apparent hardness) of the wheel, and the capacity to remove heat from the cutting zone. The steady-state temperature (T_X) of the fluid depends on energy input to the system. Convection and evaporation to ambient, as well as intentional chilling of the fluid supply, can reduce fluid temperature. T_X is generally 15 to 20°F over ambient. Approximately 80% of the grinding power contributes to a rise in temperature.

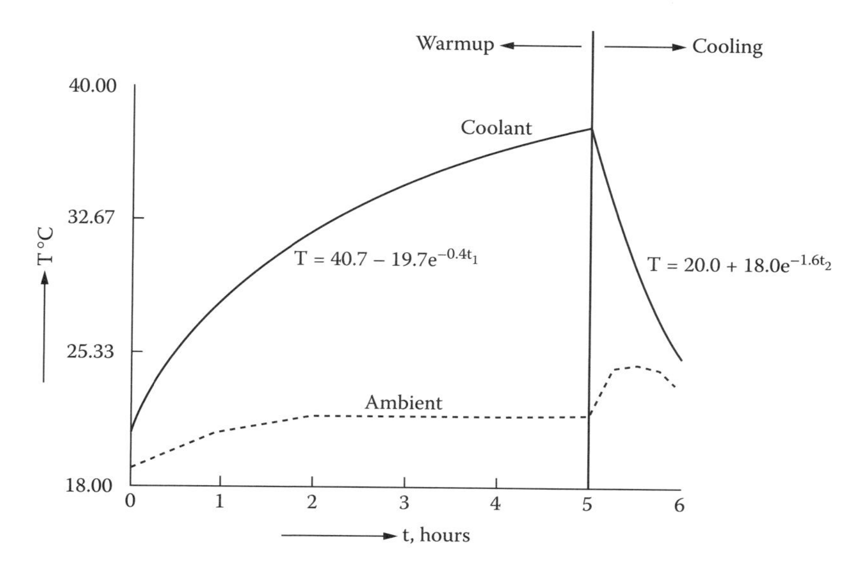

FIGURE 4.25 Actual grinding fluid warm-up curve.

of a cooling tower and the outlet coolant temperature depend on the ambient temperature and the humidity. The humidity under the conditions that the cooling tower is applied can be close to 90% if the cooling tower is installed near the machine in a badly ventilated area. One of the most important cost factors is the loss of concentrate together with the spray mist. Therefore, applying a direct circulation of a metalworking fluid through the cooling tower is not recommended. The best application of a cooling tower is to install the tower outside so that the relative humidity will always be the lowest possible and to use a heat exchanger combined with the cooling tower.

The required capacity of the cooling tower can be calculated with the following formula:

$$Q = \mu N \times 3406 = \mathrm{BTU/h}$$

where N is the energy input, kW h, and μ is the power efficiency.

μ depends on many variables but is generally between 0.8 and 0.90 depending on machine type and power. The loss counts for bearings and drive system consumed for the grinding operation.

If a filtering system is used, the energy input of the pumps must also be considered. This means that the total capacity should be:

$$Q = 3406[\mu N + (1 - f)Nf] = \mathrm{BTU/h}$$

where f is the energy efficiency of the filtering system and Nf is the power input by the filtering system.

For a cyclone filter, $f = 0.25 - 0.4$ (depending on cyclone type, piping, pressure, and differential over the cyclones). For nonspeed filtration systems (cake-type filters), $f = 0.65 - 0.7$.

If the cooling tower has this capacity, the coolant temperature then depends on humidity and ambient temperatures. For resinoid-bonded wheels it is essential to hold the coolant supply temperature below 28°C (81°F). See Section E.

D. Cooling by Refrigeration and Heat Exchangers

For calculating the required capacity, the same formula can be used as for the cooling tower, thus:

$$Q = 3406[\mu N + (1 - f)Nf] = \mathrm{BTU/h}$$

A refrigerating system is not as sensitive to ambient temperature, but the ambient temperature cannot be ignored. This means the cooling capacity must be available at the temperature that occurs in summer. In some cases, the energy from the grinding system can be utilized for purposes such as heating water or cleaner fluids or other heating purposes. In cases of larger systems with very high-energy input from grinding and machining operations, the savings could be considerable. For energy-saving purposes this concept should be seriously studied. Only by refrigerating systems can the energy be saved for other purposes.

E. Effect of Fluid Temperature on the Grinding Parameters with Resinoid-Bonded Wheels

A lot of work has been carried out on laboratory scale to determine the influence of the fluid and fluid temperature on the static hardness of the resinoid-bonded wheels. The actual effect on grinding performance was unknown. One of the applications where resinoid-bonded wheels are used is bar grinding. With the available equipment, data have been produced under production conditions. For these tests 50CrV4 steel has been used. At $Q' = 0.56$ in.^{3}/min/in. the temperature of the coolant has been varied from 20 to 48.5°C (68 to 118°F). The grinding ratio (G) and specific energy (U) have been measured, as well as the surface finish.

For each test, the wheel was under the specific temperature condition with grinding for 1 h. Then, after this conditioning time, a G-ratio measurement was carried out over 10 min grinding time and repeated three times. This measuring method was completed for all tests except for the first test at 20°C, due to problems with the available cooling capacity. The data are shown in Figure 4.26. After the test from 20 to 48.5°C fluid temperature, the wheel was brought back down to 20°C, and the grinding parameters were again measured. This test was then repeated again after a wear of 0.2 in. on the radius from the wheel. The data shows that the depth of influence is more than 0.2 in. into the resinoid wheel. (See data points marked *.)

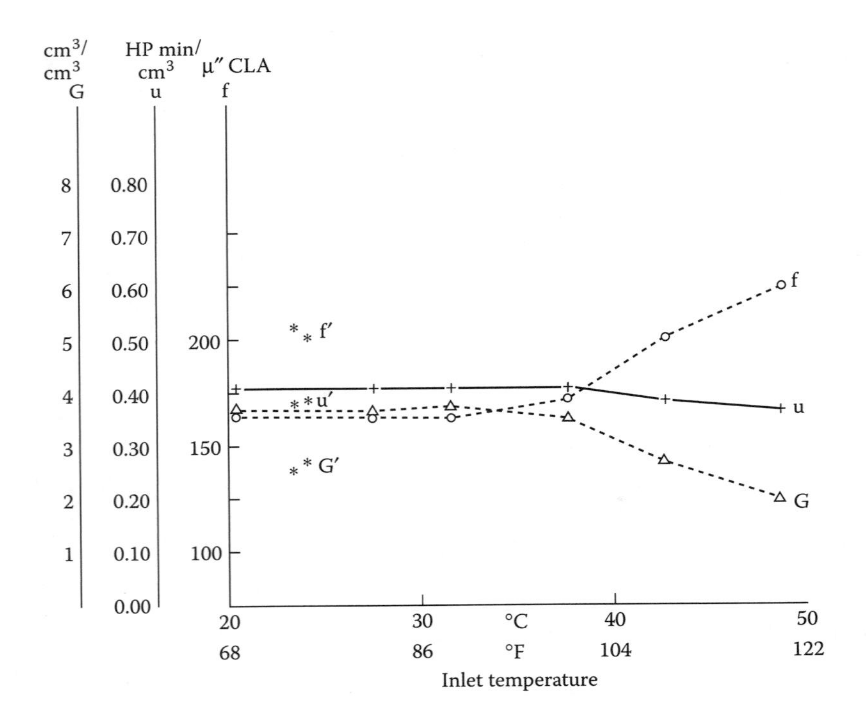

FIGURE 4.26 Effect of fluid temperature on the grinding parameters with resinoid-bonded wheels.

Figure 4.26 shows the grinding ratio (*G*), specific energy (*U*), and surface finish (*f*), as a function of the fluid inlet temperature. Up to 32°C (90°F), no effect could be determined, but over 32°C fluid temperature, the *G*-ratio decreased rapidly with the increase of fluid inlet temperature. The surface finish grew rapidly worse above a 32°C fluid temperature. The close relationship between *G*-ratio, surface finish *f*, and the specific power *U* is quite clear.

All tests were carried out with a 21SC46Q9B2 (46 grit) grinding wheel. The effect of grain size on the temperature sensitivity is unknown but laboratory research showed coarser wheels to be less sensitive to temperature than finer wheels. Therefore, it might be considered that the critical temperature for coarser resinoid-bonded wheels will be higher. This means that a coolant temperature of 32°C (90°F) might be the acceptable maximum for 36 to 60 grit size wheels.

The last test data showed that the depth of penetration was more than 0.2 in. into the resinoid bonded wheel. (The data are shown with *G*, *U*, and *f* indexes.) This means that when a resin-bonded grinding wheel is under a temperature condition of over 32°C, the wheel hardness will change. Under more favorable conditions — temperatures less than 32°C — the previous high-temperature conditions will determine the results (at least over 0.2 in. of depth into the wheel).

REFERENCES

1. Taylor, F. W., *On the Art of Cutting Metals*, Society of Mechanical Engineers, New York, 1907.
2. Schumack, M. R., Chung, J., Schultz, W. W., and Kannatey-Asibu, E., Analysis of fluid flow under a grinding wheel, *J. Eng. Ind.*, 113, 190–197, 1991.
3. Hahn, R. S., Some observations on wear and lubrication of grinding wheels, In *Friction and Lubrication in Metal Processing*, Ling, F. F., Whitely, R. L., Ku, P. M., and Peterson, M., Eds., ASME, New York, 1966.
4. Powell, J. W., The application of grinding fluid in creep feed grinding, Ph.D. thesis, University of Bristol, 1979.
5. Bourgoin, B., Centre Technique des Industries Mecaniques (France), Sprinkling device for grinding wheels, U.S. Patent, 4,176,500, 1979.
6. Johnson, G. A., *Beneficial Compressive Residual Stress Resulting from CBN Grinding*, MR86-625, Society of Manufacturing Engineers, Dearborn, MI, 1986.
7. Satow, Y., *Use of High Pressure Coolant Supply in Precision CBN Grinding*, MR86-643, Society of Manufacturing Engineers, Dearborn, MI, 1986.

5 Metalforming Applications

Kevin H. Tucker

CONTENTS

I. INTRODUCTION

The working of metal into useful objects has been suggested as the oldest "technological" occupation known to mankind; metal has been formed into usable utensils, weapons, and tools for over 7000 years.[1] It would be hard to imagine the social metamorphosis that would have occurred had it not been for the increases in metalforming technologies throughout history. It is equally difficult to imagine how metalforming would have advanced without the aid of lubricants. Only recently has there been any historic record of lubricant use in metalworking.

In *Tribology in Metalworking*, John Schey describes the historical evolution of metalforming.[1] Most historians believe that forging was the first metalforming operation. Malleable metals, such as "native gold, silver, and copper were hammered into thin sheets, and then shaped into jewelry and household utensils as early as 5000 B.C.," probably without lubrication. The making of a coin by driving metal into a die with several punch strokes was recorded in the seventh century B.C., again with no reported lubricant addition, other than possibly from "greasy fingers." However, materials used as lubricants were in evidence as early as the fifth century B.C. when Herodotus wrote about the extraction of light oil from petroleum. The manufacture of soap was recorded in 600 A.D., so a variety of lubricants would have been available throughout the history of metalworking. An eighteenth-century practice of forming rifle parts with a lubricant mixture of sawdust and oil was still used as recently as 60 years ago.

Metalforming lubricants are as varied as the many operations in which they are used. Animal fats were possibly the first lubricants used in primitive operations, as they were readily available from the rendering of animal carcasses. Stampers quickly learned that a coating of lard or tallow allowed the metal to be formed more easily, with more deformation, and with less tool wear.[1]

Other metalforming processes, such as wire drawing and rolling, have been dated to the Dark Ages where lubricants such as lard oils or beeswax were used with success. One final historical note: The first sketch of a rolling mill is credited to Leonardo da Vinci![1]

II. FACTORS AFFECTING FLUID REQUIREMENTS

A. TYPES OF METALFORMING FLUIDS

1. Oils

Petroleum mineral oil is probably the most widely used lubricant for metalforming. In light duty stamping, blanking, and coining operations, mineral oil provides the necessary separation of tool from metal. Many drawing operations can be performed with no lubricant application, relying solely on the ductility of the metal and part geometry. However, the application of mineral oil often enhances the speed at which parts can be produced. In order to optimize part-tool separation, the viscosity of the mineral oil can be varied to match draw severity.

Much work has been completed regarding the structure of mineral oil and its effect on performance. John Schey documented some of these effects.[1] Paraffinic oils are better suited for some operations due to their relatively high-viscosity index values. Paraffinic oils are less likely to be oxidized than naphthenic oils. Antioxidants, used to prevent oxidation, are more effective in paraffinic oils than naphthenic oils. Naphthenic oils though, are more easily emulsified and have more affinity for metal surfaces. However, most mineral oils contain some distribution of both straight and branched-chain hydrocarbons.

Synthetic oils are also of significance. Polybutene has been used as a rolling oil with good results for some time. Vegetable oils are also finding favor in some applications, although chain length affects viscosity, melting points, and suitability for specific operations. Cost is much higher than with mineral oils, although this may vary with market conditions.

While viscosity is often the determining factor in application and performance, additives designed for specific purposes are often included in straight oil formulations. Fats for rolling operations, extreme-pressure (EP) additives, and even solids such as molybdenum disulfide and talc, are often used as lubricants.

2. Soluble Oils

Fluids based on mineral or synthetic oils that contain emulsifiers to allow for the dilution of the product into water are called soluble oils. They are sometimes sold in their diluted form and referred to as preformed emulsions. These products are generally mixed at dilutions of 10 to 50% in water. They may be formulated with fats or fatty acids for light duty applications, or may contain EP additives for severe forming. Seldom are solid lubricants used because they are difficult to suspend in an emulsion.

3. Semisynthetics

Fluids containing a lesser amount of mineral oil, usually under 30% of the total concentrate volume, are called semisynthetics or chemical emulsions. Compared to the volume of oil used in a soluble oil fluid, the mineral oil content in a semisynthetic is much lower. The mineral oil may have been replaced by hydrocarbons such as glycols and esters used as oil substitutes, by emulsifiers such as soaps and amides, and even by water. Generally, semisynthetics will mix into water more easily than soluble oils, but other than subtle differences in mix appearance and use-dilution ratios, they are similar to soluble oils.

4. Synthetics

Synthetic fluids are generally of two types. One is the "solution" group in which water serves as the carrier. In this class, also referred to as water-based synthetics, all additives are either water soluble or are reacted with some other component to be water soluble. Generally these fluids will range in mix clarity from clear to hazy. A second type is a synthetic fluid that contains no mineral oil, but uses a hydrocarbon, such as a polybutene or glycol, as an "oily" replacement for mineral oil. The mixes of these fluids will generally range from cloudy to milky.

B. Lubricant Additives Used in Metalforming Fluids

With the exception of solid lubricants or "pigment" materials, the additives used in metalforming fluids are very similar to those used in the metalworking industry as a whole.

Fats derived from animals and vegetables are very effective boundary lubricants. Tallow, lard, and wool grease are extremely good boundary lubricants. Oil from the sperm whale, which had many uses, was such a good lubricant that the species became endangered. Its use in lubricants today is prohibited. Vegetable oils such as coconut, rapeseed, and tall oils are good sources of boundary lubrication. Fatty acids, esters, waxes, and alcohols derived from natural materials are also frequently used lubricants.

Soaps formed from a reaction between a metal hydroxide and fatty material are excellent lubricants. For instance, in the working of aluminum, aluminum soaps are often formed as reaction by-products of the operation. These aluminum soaps will lower the coefficient of friction. However, they are not water soluble. While excellent lubricants, aluminum soaps are often difficult to remove from the tooling and can actually cause increased wear of the punch and die.

Without a doubt, chlorine, phosphorus, and sulfur have been the most frequently used extreme-pressure additives. Of these, chlorine is most commonly used because it addresses the widest variety of operations. It is often used in conjunction with the other two, such as in sulfochlorinated additives.

Solid lubricants are suspended into pastes or thick emulsions. Graphite, mica, talc, glass, and molybdenum disulfide are just a few examples. This type of additive is generally reserved for the most severe of drawing operations.

C. Physical Properties

One of the biggest physical differences between metalforming fluids and other metalworking fluids is the physical appearance of the fluids. A typical machining or grinding fluid will typically be diluted in water at a dilution up to 10%. A stamping fluid will rarely be diluted at less than 10%. A machining and grinding fluid will have a viscosity near that of water, since water is its principal component. A stamping fluid often resembles paint in its ability to "coat" metal with a viscous covering.

The physical appearance of stamping and drawing fluids varies with the type of lubricant used. Straight oil stamping fluids range from very low viscosity, as in vanishing oils, to very high-viscosity "honey oils," so named because of its similar viscosity to honey. Pastes, which may be thick oil-in-water emulsions, or water-in-oil invert emulsions, are often used. These are generally used for very severe drawing operations in which the lubricant is painted onto the surface to ensure maximum carry-through of fluid through the die. Suspensions generally look like milky emulsions, although they may take on the appearance of the finely dispersed pigment material used as the lubricant, as in the case of graphite. Stamping fluids, especially those used for general-purpose stampings, may also be thin, milky emulsions. Finally, new water-based stamping fluids have the viscosity of water.

D. Lubricant Functions

1. Separation

Unlike metal removal fluids that have the primary responsibility of cooling, the primary function of a metalforming fluid is to separate the part from the tooling, preventing metal-to-metal contact. To do this, the fluid must provide a barrier, either physical or chemical, to prevent the contact of punch or die to metal. The main objective for providing this separation is the protection of the tooling. The cost of the tooling in most forming operations is quite significant when compared to all other components in the process. A lubricant that improves tool life can pay for itself through reduced tooling changes.

2. Lubrication

Along with part–die separation, a metalforming fluid must provide lubrication sufficient to make the part. Good lubrication can be defined as the reduction of friction at the part–die interface.

Lowering friction results in lower energy required to make a part. Further, lower friction results in less drag on the metal, which yields a more uniform flow of metal through the forming dies. Lower friction also benefits the operation by reducing the amount of wear, which in turn reduces the amount of metal fines and debris. With fewer metal fines in the punch zone, metal pickup or deposition on the punch is less likely. Although not as important as in cutting and grinding fluids, a lubricant may also provide cooling to the part and the tooling. In some metalforming applications, however, heat is necessary.

3. Corrosion Control

Another function of lubricants is to prevent corrosion. While a grinding or machining fluid may be expected to control corrosion for one or two days, stamping fluids are often required to provide corrosion protection for several months while parts are held in storage. Specialized testing equipment such as acid-atmosphere chambers, fog chambers, or condensing humidity chambers, are often used to predict fluid capabilities under a variety of storage conditions. A corrosion test involving a hydrochloric acid atmosphere is used in the automotive industry. Salt spray, ultraviolet light, condensing humidity, and elevated temperature chambers are all used to determine corrosion control of metalforming fluids.

Besides controlling corrosion in the finished part, a good fluid must also be safe for equipment. The fluid must prevent attack of metal ways, slides, and guide posts. Of even more importance, the fluid must be compatible with tooling material. Carbide tooling made with nickel or cobalt binders is often used for high-speed production punches and dies. The fluid must be formulated so that these binders are not leached from the metal matrix, which would result in punch or die degradation.

4. Cleanliness

After the part is formed, a fluid must be easily removed. An objective of a clean fluid should be the prevention of buildup of metal debris or fines in the forming die. Likewise, there should be adequate cleanliness to prevent residue deposition on the punch. The fluid should also be compatible with the selected cleaning equipment, whether it be vapor degreasing or alkaline wash. Along with part cleanliness, the fluid must also contribute to a clean working environment. Floors, walls, presses, and even operator's clothing need to be free of fluid residues.

5. Other Requirements

There are other aspects to a metalforming fluid that determine its acceptability for a specific operation. Operator acceptance factors such as ease of mixing, application method, and operator health and safety issues are of utmost importance. Press operators using reasonable safety practices must be able to work with the fluids, without the risk of dermatitis, respiratory distress, or other health problems. Fluid appearance, product odor, and rancidity control if the fluid is recirculated are aesthetic factors that need to be considered.

E. Lubrication Requirements

1. Boundary Lubrication

Boundary lubrication is defined as "a condition of lubrication in which the friction between two surfaces in relative motion is determined by the properties of the surfaces, and by the properties of the lubricant other than viscosity."[2] Boundary lubricants can be defined as thin organic films that are physically adsorbed on the metal surface. There are several theories on how boundary lubrication occurs. One explanation is that the polar molecules of the boundary lubricant are attracted to the metal surface. Under practical conditions, layers of the lubricant are formed.

As sliding friction occurs, some molecules are removed, but others take their place. This boundary film prevents metal-to-metal contact. If the film is broken, and metal-to-metal contact is made, part failure can occur. In practice, the majority of lubrication occurs as boundary lubrication.[2]

2. Extreme-Pressure Lubrication

Another type of boundary lubrication is classified as extreme-pressure (EP) lubrication. EP lubricants are those lubricants that will react under increased temperatures and undergo a chemical reaction with the metal surface. In metalforming lubricants, chlorine, sulfur, and phosphorus have been the traditional EP lubricants. It was often believed that these compounds simply reacted with metals, to form chlorides, sulfides, and phosphides. However, it has also been shown that reaction species other than simple metallic salts are formed between the metal surface and the chemical lubricant. EP lubricants also work synergistically with other lubricant regimes. They appear to adsorb onto the metal surface as do organic-film, boundary lubricants.[3]

Regardless of the activation mechanism, EP lubricants perform by chemically reacting with the metal substrate. Lubrication occurs through the sloughing off of the chemically reacted film through contact. Metal-to-metal contact is prevented by the chemical film. As removal occurs, regeneration of the film must occur through the availability of more lubricant, or failure occurs.

Even among tribologists there is some disagreement as to the activation temperature of extreme-pressure additives. Some lubrication engineers attribute the efficacy of EP additives to the melting points of the iron salts of chlorine, phosphorus, and sulfur. If this is the case, chlorine would be the earliest to be activated, followed by phosphorus and sulfur.[4] Other guidelines for the necessary reaction of EP additives have been established based on the observed process temperatures. Following this practical approach, phosphorus is effective in operations where temperatures do not exceed 400°F (205°C). For operations generating temperatures above 400°F (205°C), chlorine is selected. Chlorine will maintain effectiveness up to 1100°F (700°C). Sulfur is used in severe operations where temperatures exceed 1100°F (700°C), and will be effective up to 1800°F (960°C). This compares to fatty acid soaps which lose most of their effectiveness at 210°F (100°C).[5,6]

3. Hydrodynamic Lubrication

Hydrodynamic lubrication can be described as a "system of lubrication in which the shape and relative motion of the sliding surfaces cause the formation of a fluid film having sufficient pressure to separate the surfaces."[7] The viscosity of the lubricant system plays a significant role in determining the capability of the hydrodynamic lubricant film. In fact, some refer to this lubrication regime as "pressure-viscosity" lubrication.[8]

4. Solid-Film Lubrication

Solid-film lubrication involves the use of solid lubricants for part–die separation. Ideally, the solids serve as miniature ball bearings on which the sliding surfaces ride during part formation. This prevents metal-to-metal contact. However, the solids will sometimes attach onto the surface of the tooling or the part. Removal of solid films from finished parts can be a major problem. Examples of this type of lubricant are mica, talc, graphite, and polytetrafluoroethylene (PTFE).

F. Methods of Application

Many fluid application techniques are available. Selection of a method should be based on optimized lubricant delivery for performance, and on fluid conservation. Over-application of a fluid will not increase lubrication performance, but may make the workplace so messy that operators will dislike the fluid.

1. Drip Applicators

The application of drawing fluids by drip applicators is as simplistic as the name implies. In many shops, this method may involve a coffee can with a nail-hole in the bottom, which allows fluid to drip on a blank prior to the metalforming operation. This technique is one of the hardest to control for accurate delivery. It can also be very sloppy, but has the advantage of being one of the cheapest methods.

2. Roll Coaters

An application using roll coaters has two rollers, one above the other, in which the spacing and pressure between the two rolls are controlled by air pressure. The rolls are partially positioned in a fluid bath. As the metal source is fed through the rollers, the fluid is "rolled" onto the metal prior to the stamping operation. Fluid application can be controlled to milligrams/surface area of metal.

3. Electrodeposition

This fluid application procedure is becoming more prominent. Once reserved for the rolling mills, small units are being tested on the shop floors of metalforming plants. An electrical charge is passed through the fluid, with the opposite charge applied to the metal. As opposite charges attract, the fluid is deposited onto the metal surface. Advantages are the accurate application of a small amount of fluid per surface area, little or no waste, and high-speed application. The major disadvantage is that the high cost associated with an electrodeposition unit puts it out of the reach of the average shop.

4. Airless Spray

Equipment that sprays measured amounts of fluid to precise locations in the drawing press without mixing with air is known as an airless spray unit. Tubing is placed into fluid reservoirs which feed the fluid through a pumping system onto the selected delivery site. Advantages include little or no waste, fluid savings, and a cleaner work environment. Disadvantages include limitations on the viscosity of the lubricant that is applied.

5. Mops and Sponges

Unfortunately, many small stamping shops still use mops and sponges to apply drawing fluids. While these are obviously low-cost application devices, expenses related to this poor fluid delivery system generally overshadow any possible benefits. Disadvantages include poor control of the amount and location of fluid delivery, excessive fluid waste, and sloppy work environment.

G. Removal Methods

In almost all cases, the metalforming fluid will eventually need to be removed from the formed part. In some cases, it is the ease with which this occurs that actually determines which fluid is selected for the operation. Often, problems with plating, painting, or electrocathodic deposition of primers (E-coat) can be traced to poor cleaning or contaminated cleaner reservoirs. There are several basic types of removal methods.

Straight oil-type products have historically been removed through vapor degreasers. A vapor degreasing process uses vapors of a solvent such as boiling 1,1,1-trichloroethane. The vapors solubilize and remove the oily residue on the part. The presence of water in the residue will decrease the efficiency of a vapor-degreasing solvent. Today, the use of vapor degreasers is being drastically reduced due to environmental and operator safety concerns. The future availability of these cleaning systems is doubtful.

As an alternative to solvent degreasing, alkaline cleaners are being used. These cleaners can be applied through a dip bath, impingement spray, or a tumbling parts washer, among others. One advantage is that, unlike vapor degreasers, alkaline cleaners are also effective for water-diluted lubricants, even those that contain oil. Alkaline cleaners are often warmed to speed up the processes that cause residue removal.

Finally, many companies install multistage washers designed to clean off stamping fluid residues, acid-etch the surface, and react with the metal part to form a conversion coating (often a phosphate) for subsequent plating or painting operations. Once installed, these washers are very economical to run and are controlled through specific ion monitoring devices to control washer-chemical concentrations.

H. Pre-Metalforming Operations

In today's stamping plants, there are a variety of experimental processes being evaluated that may be the norm for stamping plants of tomorrow. One of the most interesting is the trend towards "prelubes." This involves the application at the rolling mill of a thin film of lubricant, prior to rolling of the coil. This lubricant may either remain a liquid on the coil, or through either evaporation or chemical reaction convert to what is known as a dry-film lubricant. When this coil arrives at the stamping plant, no further application of lubricant occurs. The operation is completed using only the thin film of mill-applied lubricant. The use of pre-lubricated metal is in its infancy, and has had many growing pains. Since no application of fluid in the press occurs, buildup of metal debris and other residue material is difficult to wash off.

Similar applications are being tried with experimental paints. The paints are applied at an off-site facility prior to forming. When stamped, the paint serves not only as the lubricant, but the part is already prepared for shipping with no cleaning costs. This operation is already in practice for very small parts.

Many times, metal coils arrive with residues of the rolling oil. Any subsequent application of lubricant must be compatible with these mill-applied oils.

In some very rare applications, a metal will be "pretreated" with a conversion material, such as phosphoric acid on cold-rolled steel. The acid will react with the steel to form a phosphated dry-film lubricant. This film serves as an extreme-pressure lubricant and will perform very well on deep-drawing operations.

I. Post-Metalforming Operations

Once the part has been formed, other processes may follow. Cleaning to remove lubricant and mill oil residuals must occur before any finishing operations such as painting or plating. For automobile body parts, E-coating of primer paints is completed before final paint coats are applied. Craters that are visible in final paint coats are attributed to poor surface quality of the metal substrate prior to E-coating. The part must also be free of any contaminants that may affect glue adhesion, welding, and application of modern sound deadeners in automotive applications.

In the electrical industry, post-forming operations include the application of resinous insulating materials. If a compatibility problem exists, cracking or removal of the insulation may occur. Most magnet wire is coated with a protective varnish. The drawing lubricant must be compatible with the varnishes to prevent wire surface quality defects.

The can-drawing industry has one of the most involved postforming processes in metalforming. Following the making of a can, the can body will be cleaned, dried, decorated, sprayed with internal and external coatings, and baked. Next, the neck of the can, and a flange for the lid attachment, will be formed. After all of these procedures, the lubricant applied early in the original can-forming process may contribute indirectly to mechanical problems that occur in the final necking or decorating operations.

The making of a part is only the beginning of the metalforming operation. From the original mill-supplied metal to the final manufacture of a formed part, stamping fluids must not only provide lubrication, but ensure final part quality and acceptance for use.

III. METALFORMING PROCESSES

A. Blanking Operations

One of the most common metalforming operations is the blanking process, as shown in Figure 5.1. Nearly all metalforming operations are preceded by a blanking stroke. In this operation, the metal blank is cut from a sheet or strip into a shape desired to affect the final part. The remaining scrap is in the form of a "skeleton" with a hole where the blank had been. The blank can be round as in the formation of metal cans, square as in the making of medicine cabinets, or patterned if needed for an intricate shape or design.

Rough blanking uses a die set with a cut edge. The metal is clamped tightly to a specified hold-down force. The punch serves as a moving knife edge that cuts its way through the metal much like a cookie-cutter through dough. The stationary die also serves as a cut edge to complete the cutting operation. In rough blanking the metal is only cut from 35 to 70% through the depth of the metal. Because of the metal flow, the remaining metal edge is torn away. As can be seen in Figure 5.2, the top edge of this blanked part is much smoother than the lower edge, which has been broken off.

In fine blanking, the process is very similar to rough blanking. However, fine blanking provides a "finished" quality appearance and size control. As such, the metal is forced 100% through the die cut edges, resulting in a smooth, even appearance over the entire surface edge (see Figure 5.3).

There are many types of blanking classifications. Cutoff is a class of blanking that describes the use of a punch and die to make a straight or angular cut in the metal. This operation is basically a shearing operation completed in a forming press. Parting (Figure 5.4) is the opposite of blanking in that the scrap is "parted" from the original metal strip, leaving the desired finished piece.[9] Punching (or piercing) is the making of a hole in metal, leaving a round slug of metal as scrap. Notching is similar to punching, only a slit is formed rather than a hole. Shaving uses a cut edge to remove rough edges for precision finishes. Trimming is also a type of blanking. Trimming removes excess metal from a formed part giving the final desired dimension. The *Tool and Manufacturing Engineers Handbook* lists over ten different blanking operations, and over 30 uniquely different metalforming operations.[9]

B. Drawing

The drawing of metal can be simply described as a punch forcing a flat metal blank from a blankholder, into and through a forming die, reshaping the flat piece into a three-dimensional shape such as a cup or box (see Figure 5.5). There are degrees of severity of drawing operations, and

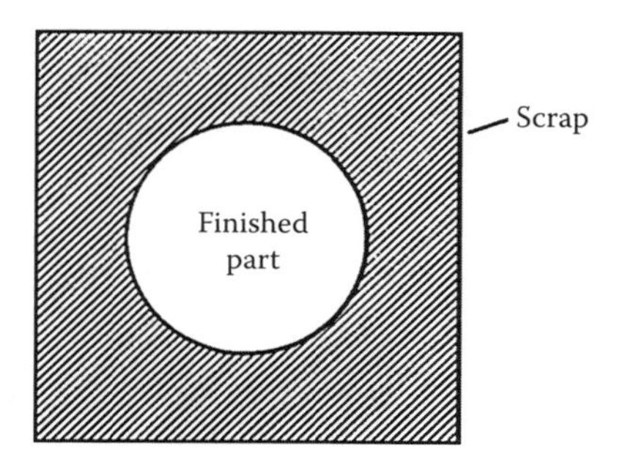

FIGURE 5.1 Blanking.

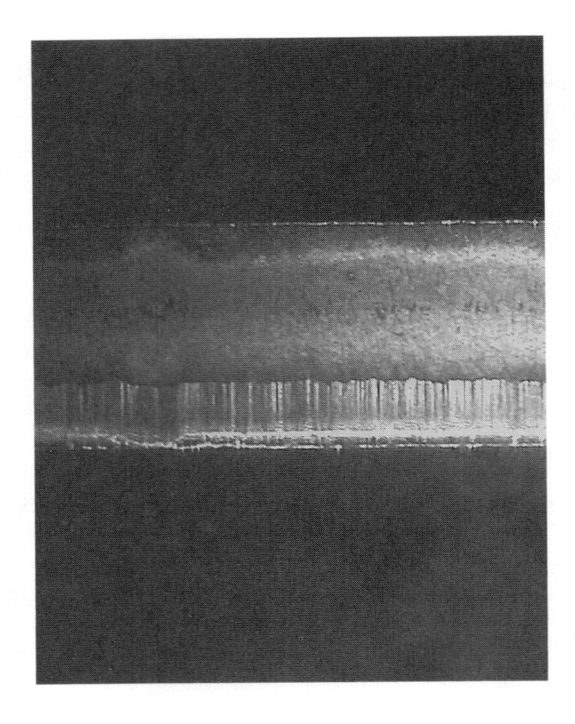

FIGURE 5.2 Photograph showing edge of a rough blanked part.

FIGURE 5.3 Photograph showing edge of a fine blanked part.

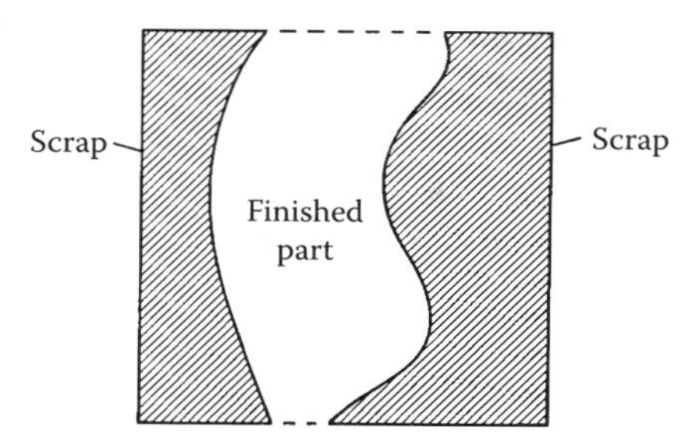

FIGURE 5.4 Parting.

numerous definitions used to define such. A practical and easy way to define the severity of a drawing operation is to look at the draw ratio. A draw ratio is determined by the ratio of the depth of draw to the original blank diameter[10]:

$$\text{draw ratio (DR)} = \frac{\text{depth of draw}}{\text{blank diameter}}$$

Based on this relationship, draw severity is commonly classified as:

Classification	Draw Ratio
Shallow draw	<1.5
Moderate draw	1.5–2.0
Deep drawing	>2.0

In low and moderately severe operations, there is little or no wall thinning. In deep drawing, wall thinning may occur to a greater degree than in shallow drawing. While draw ratios can be used as an indication of difficulty, the severity of a drawing operation is affected not only by the draw depth but also by the actual amount of wall thinning that occurs within the sidewall.

While the severity of the drawing process can be expressed as a draw ratio, the "drawability" of the metal used can also be expressed as the ratio of the blank diameter (D) to the punch diameter (d).[10] This relationship is referred to as the limited draw ratio. The metalforming operation is controlled, to a large extent, on the limits of the metal to be drawn to a desired depth.[11]

$$\text{Limited draw ratio (LDR)} = \frac{\text{blank diameter } (D)}{\text{punch diameter } (d)}$$

For practical use, the LDR is also a measure of the maximum reduction possible in a drawing operation. The LDR is sometimes expressed as the percentage reduction from the blank diameter to

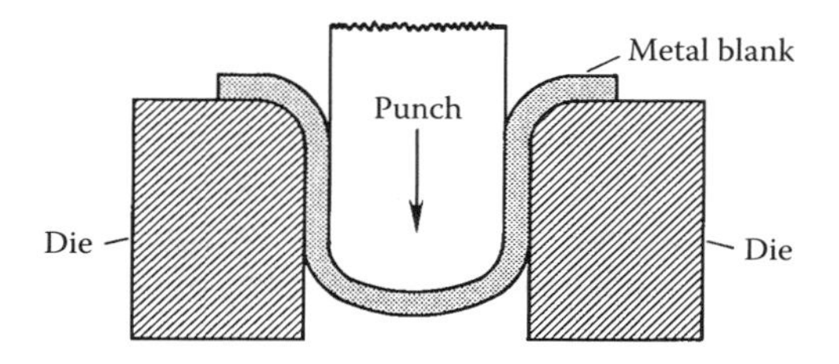

FIGURE 5.5 Deep drawing.

the cup diameter as $100(1 - d/D)$.[12] The variety of commercial qualities of metal must be considered when a drawing operation is designed. For instance, cold-rolled steel is available in several grades including draw quality aluminum killed (DQAK), which is designed for improved drawability.

As a rule of thumb, press operators and die engineers use a generic definition for deep drawing that is based on a visual analysis of the part geometry. If the "cup" depth is equal to or greater than half the "cup" width, then the operation is considered to be a deep draw. This is perhaps the definition for deep drawing most often used on the manufacturing floor.

Success of a drawing operation is also affected by the ratio of metal thickness to the blank diameter ratio. This ratio can be expressed as t/D, where t is the thickness and D is the blank diameter to which the shell is being reduced.[13]

This ratio is used to approximate maximum allowable reductions for draw and redraw depths. Because the thickness remains constant, and the blank diameter decreases, each successive draw reduction is decreased accordingly. Obviously, other factors including corner radii and metal quality need to be considered.

Blankholder pressure, also known as hold-down force, holds the metal in place under the proper pressure to prevent deformities. With too little hold-down force, the metal may twist in the die, causing wrinkles. Too much hold-down pressure will prevent smooth, even metal flow, causing tear-offs or part rupture.

C. Drawing and Ironing

1. Background

The use of canned foods dates back to Napoleonic days when a French candy maker, Nicolas Appert, conducted food preservation experiments by canning soups and vegetables in champagne bottles.[14] By the early 1800s, the French navy carried bottled vegetables on their ships with reportedly good results. Soon after, a patent for canning in "bottles or other vessels of glass, pottery, tin, or other metals or fit materials" was granted to an Englishman, Peter Durant. In 1811, Durant sold the patent, and the new owners soon were making the first "tin cans" at the phenomenal rate of ten cans per man per day! With the tin mines in England, tin-plated steel was soon being produced in commercial quantities. In spite of the high cost of the canned food, not to mention that the can needed a hammer and chisel to open, canning became a booming industry.[14]

In the late 1800s, Luigi Stampacchia received a patent for a process using a double-action press to produce a "double-drawn" can, and the first "ironed" can was produced.[14] A U.S. patent was granted in 1904 to James Rigby for producing cans with the wall thickness reduced by burnishing.[14] Even though other developments took place, and many more patents were issued for the making of cans, it was not until the 1960s that the first commercial beverage cans were produced by the drawing and ironing process first developed in the 1800s. The introduction of the "easy open" end in the early 1960s led to wide consumer acceptance, and now close to 100 billion cans are produced throughout the world each year.[15] In fact, each person in the U.S. accounts for 547 beverage cans used each year![16]

Since 1959, when Coors Brewery made the first two-piece aluminum can,[17] nearly all beverage cans, as well as many similar types of food containers, are manufactured through the operation known as drawing and ironing, sometimes referred to as the D&I process. As described earlier, drawing is the reshaping of a flat metal piece into a cup. Ironing involves a thinning of the metal side-walls, as shown in Figure 5.6. Beverage cans used for soft drinks and beer are called "two-piece" cans because the complete container actually consists of two separate portions. One portion of the two-piece can is the lid, which is produced using a stamping operation. The major section of the two-piece can is the body. The can body is manufactured using the D&I process and will be the focus of this section. The can body and the lid are made independently of each other, and often

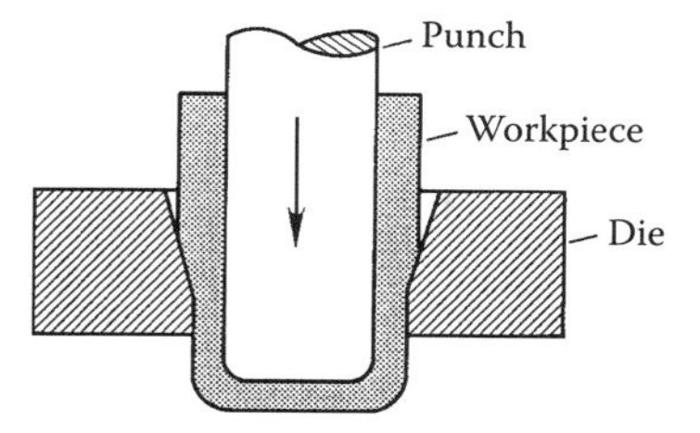

FIGURE 5.6 Ironing.

at completely different locations. The can bodies and lids are then shipped to a bottling plant, which adds the beverage and seals the can.

Steel has the largest marketshare of cans produced. By the mid-1970s, steel cans had dominated the beverage can marketplace. Presently however, aluminum cans have an almost exclusive market share for what is referred to as the "beer and beverage" container. In fact, carbonated beverage cans now represent the largest single use of aluminum in the world.[18] This preference for aluminum beverage cans has been supported not only by years of qualitative taste and flavor success, but also from an ecological standpoint due to the recyclability of aluminum. Today, with increased consumer awareness of environmental issues, and with improved collection of scrap aluminum, over 50% of newly manufactured aluminum cans are from recycled aluminum.[19]

In recent years, the use of recycled aluminum has kept the cost of manufactured aluminum cans much lower than the increase observed in the cost of aluminum as a raw material. In fact, during the decade from 1980 to 1990, manufacturing costs for an aluminum can unofficially decreased by approximately 10% in spite of increases in raw material costs and inflation!

Steel producers have devoted extensive research and development efforts in making their coil stock more attractive to can manufacturers as an alternative raw material to aluminum. Because of large subsidies by steel makers, it is actually cheaper to produce a steel can than to make an aluminum beverage can. However, the recycling efforts have generally not been as successful with steel cans as they have with aluminum. In spite of extensive advertising efforts promoting steel as being "recyclable," the low returns per can paid to collectors have so far failed to interest consumers in recycling cans. Recycled steel cans amount to much less than the recycling rate for aluminum cans. In addition, long-term storage of beverages in steel cans may impart what is described as an "iron" taste to products. Technology improvements in interior can coatings have helped to alleviate this concern, however.

For food containers such as those used for fruits and vegetables, steel is the preferred metal. Aluminum cans, however, are making significant inroads into this marketplace and have recently been used for soups, vegetables, fruits, and even wines.

2. Definition

The two-piece can is made in a two-stage process. Whether making an aluminum can or one from tin-plated steel, the process is very similar. The first step is the cupping operation. During the cupping operation, a blank is produced from the coil stock and, in the same punch stroke, is formed into a shallow cup. The cupping operation can be compared to the drawing operations described earlier.

The formed cup is then conveyed to a horizontal metalforming press called a "bodymaker," or "wall-ironing machine." What makes the D&I operation unique is that within the bodymaker, two metalforming procedures are carried out using one ram stroke of the punch. This is accomplished through the use of a series of forming dies. The first stage of making a can body is the punch forcing the cup through a redraw die. In this die, the cylindrical cup is merely reshaped or redrawn into a longer cup having a smaller diameter.

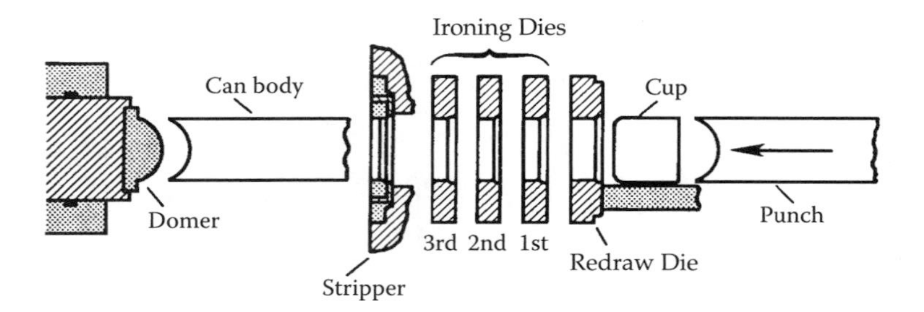

FIGURE 5.7 Bodymaker tooling.

As the punch is extended, the cup is forced through a series of ironing dies, called a tool pack (see Figure 5.7). The tool pack usually consists of three ironing dies, and is used because no single die can make the necessary forming reductions. In the ironing dies, the can sidewall is reduced to the desired wall thickness. The total reduction in thickness going from thickwall to thinwall may be in excess of 70%. As the punch forces the can through the first ironing die, the inside diameter of the can is now the exact diameter of the outside of the punch. The cup is now lodged tightly onto the punch. As the cup goes through the successive ironing dies, the outside can diameter is reduced, and the sidewall thickness is reduced. Controlling the amount of reduction, as well as the volume of metal moving through the tool pack, determines the shape of the can, and to a large extent, the length of the can. Because of the tooling setup in this ironing process, the bottom and top walls of the can are considered as thickwall, while the middle of the can is thinwall.

The nose of the punch is hollow. As the punch continues through the tool pack, it strikes a positive-stop at the end of the ram stroke. This stop is a piece of tooling called the "domer" which is shaped to coincide with the hollow nose. The punch nose forces the can into the domer tooling, shaping the bottom of the can.

As the punch begins to retract from the newly formed can, the can is removed from the punch by segmented pieces of spring loaded tooling known as stripper fingers. The fingers open to allow the can to exit the tool pack, but then close around the outer punch diameter which allows it to be withdrawn from the can as the ram is returned.

Subsequent operations include trimming the can to a specified length, washing to remove drawing lubricants, decorating the exterior of the can with the desired brand label, and applying an interior coating to prevent contact of the food or beverage product with the metal substrate.

3. Lubrication Requirements

Lubricant for making aluminum can bodies is applied by flood application through a series of lubricant rings placed between each ironing die in the tool pack. The D&I lubricant is also directed at the cups entering the redraw die. Once the punch has entered the tool pack, only the outside of the can, in contact with the ironing dies, is affected by the bodymaking coolant. The interior of the can, in contact with the punch, must rely solely on the residual cupping lubricant along with any bodymaking coolant retained prior to entering the redraw housing.

An interesting variation in lubrication application allows for a pattern such as crosshatching or longitudinal grooves to be machined onto the punch surface. The lubricant is more easily carried into the work zone. This practice is rarely used with oil-containing or emulsifying products, but is frequently recommended for water-based synthetics.

The ironing process has been compared to similar processes such as wire drawing, extrusion, and tube drawing that involves heavy plastic deformation. S. Rajagopal, from the IIT Research Institute, claims "the reduction in thickness and the resulting generation of new surfaces made wall ironing more severe from the tribological standpoint than deep drawing wherein the surface area remains nearly constant."[20] He described four basic modes of lubrication encountered in the D&I operations:

1. Thick film, in which the surface of the sheet metal is separated from the tooling surface by a film of lubricant which is thick in comparison with the peaks and valleys of the two surfaces
2. Thin film, in which a continuous film of lubricant is interposed between the sheet metal and tooling surfaces, but whose thickness is similar in magnitude to the peaks and valleys of the surfaces
3. Boundary, in which the lubricant film continuity is disrupted by contact between the roughness peaks of the sheet metal and the tooling, with the valleys thereby serving as the lubricant carriers
4. Mixed, which is intermediate between the thin-film and boundary regimes, and in which the contact load is carried partly by asperity contact (roughness peaks) and partly by a thin film of lubricant

Greater demands are placed on bodymaking coolants than perhaps any other metalworking fluid. Not only must the coolant provide sufficient lubricity for what has already been shown to be one of the most severe forming operations in commercial application today, but secondary fluid requirements may actually be the determining criteria on which the fluid selection is made.

These secondary requirements include adequate detergency to remove dirt, metal fines, and other debris from the punch, dies, and other tool pack surfaces. There must be sufficient cleanliness to carry dirt in the fluid to the filtering system for removal, but not so much so as to prevent lubricant "plate-out" on the punch and can surfaces. Too much detergency will cause removal of residual cupping fluids, resulting in poor interior can quality.

A fluid must also protect the metal surfaces it contacts. For instance, in aluminum can drawing, the coolant must not stain aluminum, should protect slide ways and bodymakers from corrosion, and must not attack cobalt or nickel binders used in carbide tooling.

A D&I lubricant may be required to meet specific chemical guidelines established by the end-user. For instance, cans intended for direct contact with foods must be formulated with raw materials that meet the Food and Drug Administration (FDA) criteria for this application. Several breweries require coolants to pass stringent taste and flavor tests before being approved for use in making cans that will ultimately contain their beer.

Bodymaking coolants must be compatible with a variety of cupping fluids, way lubricants, greases, cleaners, and water conditions. They must often handle copious amounts of gear oils or way lubes that leak into the system. While these tramp oils are very effective at helping to remove metallic soaps formed in the tool packs, exorbitant levels may create problems with microbial control, waste treatment, or excessive filter media usage. Any evidence of incompatibility will show up as lower can quality, or decreased production.

Good coolants must also be compatible with some of the antimicrobial additives used to control odors and reduce demands on the filter system. Finally, bodymaking coolants are sometimes even called on to ease waste treatment requirements.

There are as many types of coolants for can drawing as there are varieties of metalworking fluids. Soluble oil emulsions used in the early days of can drawing have given way to synthetic emulsions, which used esters or synthetic base-stocks such as polyalkyleneglycols (PAGs) or polyalphaolefins (PAOs) as replacements for the petroleum mineral oils. The synthetics quickly gained acceptance during the oil shortages of the 1970s. As expected, the synthetics are more expensive, but can save money in the long term through better lubricity, improved cleanliness, and less fluid carry-off per can. Improved work environments are another benefit of the synthetics, as mist levels are lower than for soluble oils.

Water-based, "solution synthetics" have gradually gained acceptance as lubricant technology for can drawing continues to evolve.[21] These fluids have clear to hazy mixes, which contrast with the milky appearances of their soluble oil and "synthetic emulsion" predecessors. They are often formulated to reject tramp oils, reduce friction, and lower the need for antimicrobial additives. While cleanliness is very good, there is a fine balance between too much and too little detergency, and overall performance seems to be more difficult to control. In general, greater attention must be paid to selection of filter media, concentration control, tooling surface finish, and tramp or leak oil removal methods. A small amount of tramp oil, in the 1 to 3% range, seems to improve overall performance dramatically with water-based, solution synthetics. This level appears ideal for optimum fines removal, which dramatically reduces friction between the metal contact surfaces within the tool pack. With reduced friction, tool life is improved.

Lubricity requirements for tin plate are rather unique. Whereas coolants for tin plate must meet the same process requirements as those for aluminum, tin serves as a solid lubricant in addition to the lubricating fluid. Rajagopal showed that under deep-drawing conditions, tin performed as a boundary lubricant, serving as a solid film to inhibit contact between the steel substrate and the punch or dies. This reduces the chance for galling.

Since tin is relatively expensive, any reduction of tin coating can result in considerable savings. Mishra and Rajagopal showed that the load required to draw a cup decreased with increasing tin-coating thickness. In addition, increasing the viscosity of lubricants had the same effect on deep drawing as increasing the tin-coating thickness. They concluded that any attempt to decrease tin-coating thickness must be accompanied by an increase in the viscosity of the liquid lubricant. They determined that total elimination of tin plate could only be accomplished by using lubricants that rely on thick-film regime lubrication.

Byers and Kelly reported success with a synthetic lubricant at not only improving can plant efficiency rates, but in lowering tin weights as well.[22] In their study, the fluid viscosity was relatively low. However, the synthetic fluid did contain lubricants that had very high viscosities and perhaps relied on a thick-film regime. One can-making plant documented in their study was able to reduce tin coatings by 60%, which resulted in a saving of over a million dollars per year in tin-plated steel costs alone.

Figure 5.7 shows the tooling progression from initial cup stage to finished can during the drawing and ironing process.

D. Draw–Redraw

Redrawing can describe several operations. The process of lengthening an already drawn part, without sidewall reduction (as occurs in an ironing operation) is known as redrawing. Redrawing can also refer to inside-out reforming in which a cup is placed in the press so that the punch moves through the outside bottom of the cup, through the length of the cup, and out the top, which literally turns the cup inside out.

Redrawing a part to obtain the proper part height or diameter can occur any number of times. Many times, successive stages in transfer presses are simply to redraw the part to a selected depth. However, each draw that occurs limits the successive draws to a percentage of the first draw reduction. The percentage reduction that can be obtained for any successive draw can be calculated by the equation[23]:

$$R = \frac{100(D - d)}{d}$$

where R is the percent reduction, D is the initial diameter being reduced, and d is the final diameter to which the blank or shell is reduced.

The equation is for a round cup. For square or rectangular parts, the successive draw reductions are controlled by the depth of draw desired, and radii on the punch and die rings at the corners.

Since metal flow at the corners is multidirectional for a rectangular part, metal-thickening occurs, which is a major limiting factor in how much reduction can occur in one die.[23]

The draw–redraw process has become a very popular method for making containers for food items. A large majority of pet food containers found on grocery shelves are made by the draw–redraw process. Sardine cans, as well as stackable fruit and vegetable cans, are also being made. Due to the noticeable absence of ironing in the process, these cans are generally not extended to the draw depths possible in the drawn and ironed cans.

The lubricant of choice for draw–redraw cans is generally a paraffinic wax. Often, these are food-grade lubricants with FDA approval. This enables the food product to be packed with few after-draw handling processes.

E. Wire Drawing

Donald Sayenga of the Cardon Management Group presented a technical report for the Wire Association International, Inc., at Interwire 91.[24] The report was an excellent summary of the history of wire. Mr. Sayenga traced the roots of the word wire to the Latin verb, *uiere*, meaning "to plait." He also attributes to the 1771 *Encyclopedia Britannica* the following description: "Wire, a piece of metal drawn through the hole of an iron into a thread of a fineness answerable to the hole it passed through." Two hundred years later, one would be hard-pressed for a better definition. More importantly, however, the report showed not only wire's impact on history (where would the phone company be without wire?), but somewhat romanticized the men and women responsible for the development of wire, used as a source of inspiration in the making of products such as jewelry, tools and musical instruments, chains, fencing, cables, and even suspension bridges.

The drawing of wire is considered to be one of the most difficult metalforming operations.[25] Wire is made by pulling metal bar stock through a series of reduction dies until the correct shape and size are reached (see Figure 5.8). Most drawn wire is cylindrical, but it can also be drawn to flat or rectangular shapes. Wire can be made from steel, aluminum, copper, or other ductile metals. In addition, copper-clad or tin-coated alloys are used to provide solid-film lubrication. Finished wire can be used in a variety of applications from magnetic wire and electrical applications to high-strength reinforcing cables. In fact, the earliest application of drawn wire recorded is in making chain link armor plate in 44 A.D.[26] It is believed that this wire was drawn through dies using old-fashioned muscle power. The first lubricant used in wire drawing was developed by Johan Gerdes, who used an accidental combination of urine and urea to draw steel wire.[27] As the story goes, Gerdes, somewhat frustrated, threw several steel rods out a window "where men came to cast their waters."[27] After "reacting" with the urine, the steel rods were easily drawn.

Since copper wire has the largest volume of use, we will look specifically at copper wire drawing. The rolling mill is where the wire begins. Molten metal alloy is poured into molds and after cooling is cold-rolled into slabs. These slabs are then shaped into continuous thick rods and gathered into large coils ready for the rod-breakdown process.

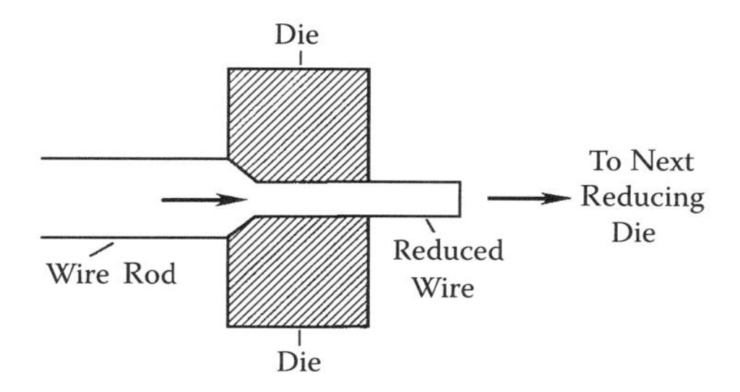

FIGURE 5.8 Wire drawing.

Wire drawing can be described in three general classes: rod breakdown, intermediate, and fine wire. Rod breakdown is the first wire-drawing process. The rod diameter is reduced through a series of forming dies. A take-up reel pulls the wire through the dies, with the wire tension controlled by wraps around a wheel-like device called a capstan. During the plastic deformation process, the volume of wire remains constant. As the wire diameter is reduced, the speed of the wire must be increased through each subsequent die to account for the amount of wire produced. As an example, several miles of extra-fine gauge wire can be drawn from a 6 ft section of 100 mm rod. Jan Kajuch of Case Western Reserve University summarized this process.[28] The relationship of the initial volume of wire to the final length of wire after drawing can be shown by some basic principles:

$$V_1 = V_2$$

$$A_1 L_1 = A_2 L_2$$

$$\frac{\pi (D_1)^2}{4} L_1 = \frac{\pi (D_2)^2}{4} L_2$$

$$\frac{(D_1)^2}{(D_2)^2} = \frac{L_2}{L_1} \text{ or } \frac{D_1}{D_2} = \sqrt{\frac{L_2}{L_1}}$$

where $\pi = 3.14159$, D is the wire diameter (in. or mm), L is the wire length (in. or mm), A is area, and V is volume.

This relationship can then be used to determine the basic die parameters for elongation percentage as follows[28]:

$$\text{percent elongation} = \frac{L_2 - L_1}{L_1} 100 = \left(\frac{L_2}{L_1} - 1 \right) 100$$

$$\text{percent elongation} = \left[\frac{(D_1)^2}{(D_2)^2} - 1 \right] 100$$

Now that we can calculate the percent elongation of the wire, we can determine the amount of reduction that occurs in the dies. We can see from the relationship of elongation to area that[28]:

$$E = \frac{100 A_r}{100 - A_r}$$

where E is die elongation (%) and A_r is reduction of area (%).

The reduction of area (A_r) can be expressed as:

$$A_r = \frac{100 E}{100 + E}$$

and reduction of wire diameter (D_r) in the die is determined by[28]:

$$D_r = 100 \left(1 - \frac{A_r}{E} \right)$$

The preceding equations are elementary for the engineer responsible for making sure the dies are set up properly to draw wire. For the lubrication engineer who must determine if the lubricant is functioning properly, the dies must be set up so as not to exceed the maximum or minimum value for wire elongation, while the machine setup must account for the increased wire speed through each subsequent die. This is all based upon the understanding that the wire volume remains constant.

1. Lubricants

Fortunately for everyone, considerable advances in lubricant technology have been made since Gerdes used urine. As recently as 25 years ago, wire-drawing lubricants were the original combinations of solid lubricants such as metallic stearates, lime (CaO), and calcium hydroxide ($Ca(OH)_2$). These dry powders were applied by drawing the wire through a box containing the lubricant. This box was referred to as a "ripper box." An advantage of the stearate powder was that the lubricant stayed on the wire through the die.

The lubricants used today are complex blends of esters, soaps, and extreme-pressure lubricants. In general terms, the need for lubrication is highest in the rod breakdown operation where the greatest reductions occur. As the wire becomes finer, the need for lubrication decreases, but the requirement for detergency increases. A wide variety of fluid types are used in wire-drawing operations, but are typically oil or polyglycol-based lubricants diluted in water at 10% or higher concentrations in the rod mill. In fine wire drawing, a typical dilution may be as low as 1% of a high-detergency fluid.

2. Metal

The quality of the rod supplied obviously affects the wire-drawing process. Care should be taken by the plant metallurgist to ensure that the metal is the correct quality and alloy. Surface defects caused by extraneous alloying agents or other foreign particles may alter the drawability of the wire. Pockets of air in the continuous cast meal rod may result in wire breaks. Surface contaminants such as rolling oil or dirt are also of concern.

3. Dies

The advent of man-made diamond dies revolutionized the dies used in wire drawing. These artificial diamond dies, known in the industry as *compax* dies, hold their tolerance much better than previous natural diamond. The compax dies are much harder, and last many times longer than natural dies, which reduces downtime. Provided the setup for the reduction profile has been correctly made, there are seldom problems with manmade diamond dies.

4. Other Factors

The equipment used in the wire industry ranges from "high-tech" multihead wire machines able to complete much of the rod breakdown and subsequent wire gauges, to small, antiquated but efficient machines that have drawn millions of miles of wire. Regardless of age or ability, the equipment is always well cared for so as to maintain the extremely tight tolerances permitted in wire applications. Wire-drawing rolls and capstans should be periodically checked to make sure that the proper ratios and roll grooves are used. At the same time, a visual inspection for cracks or chipped capstans can be completed. Lubricating nozzles should be placed for optimum application. The take-up system should be inspected for balance and proper tension.

Operators are generally well educated through on-the-job training. It is imperative that an operator knows how to "string" a machine quickly and safely. A wire company invests a lot of money to ensure that operators understand their role in the wire operation.

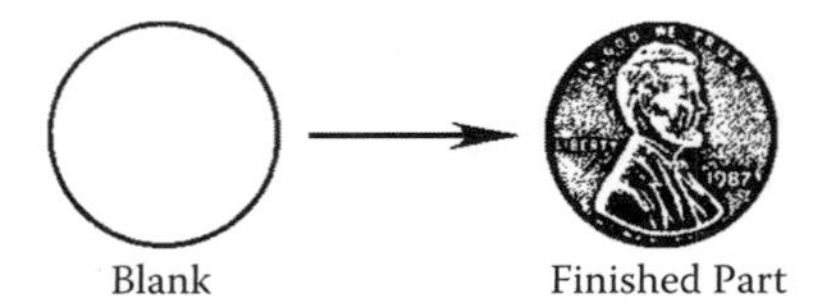

FIGURE 5.9 Coining.

F. Other Metalforming Processes

1. Coining

The world's monetary system would be drastically different without the ability to coin metal. In this operation, metal is thinned or thickened to achieve the desired design.[29] For a coining operation, there is a set of dies, which may or may not be dissimilar, in which the reverse pattern of the final design has been engraved. During the stroke, the metal is forced into the dies, and the defined pattern is imprinted into the metal, as in Figure 5.9. Very beautiful and intricate patterns can be created within the dies. This operation is different from embossing, in that there is no change in thickness during embossing.

2. Punching, Piercing, and Notching

The previously mentioned operations are all very similar both in function and in process. When holes are needed, the hole is "punched" out using a punch and die (see Figure 5.10). Piercing is often used interchangeably with punching, but is really distinguishable in that piercing produces holes by a "tearing" action as opposed to punching's "cutting" action.[29] Notching (shown in Figure 5.11)

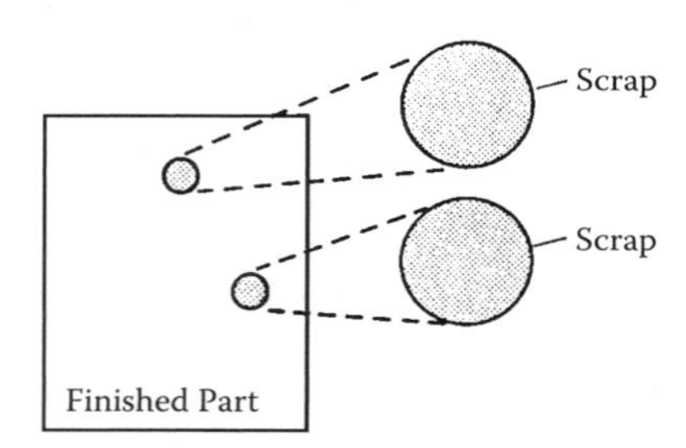

FIGURE 5.10 Punching.

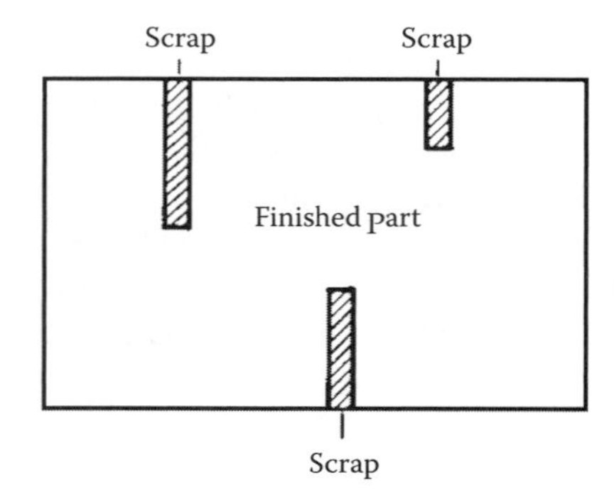

FIGURE 5.11 Notching.

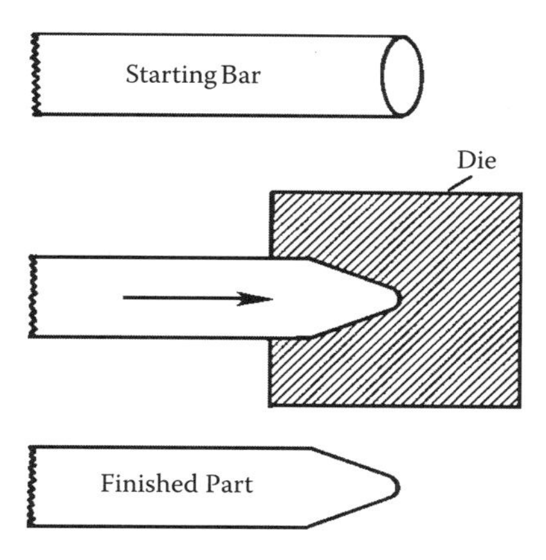

FIGURE 5.12 Swaging.

is actually very similar to punching, but the final shape is a slice in the edge of a finished part. Because part flatness is often important, notching is often carried out in a progressive die, with multiple strokes.[29] From a lubrication standpoint, most punching, piercing, and notching operations in low-gauge metals are done dry or with a light viscosity mineral oil. In thicker-gauge metals, or in more brittle metals, formulated lubricants may be used to increase tool life.

3. Swaging

Swaging is defined as "a metalforming process in which a rapid series of impact blows is delivered radially to either solid or tubular work. This causes a reduction in cross-sectional area, or a change in geometrical shape."[29] Perhaps an easier definition compares swaging to the squeezing of tube stock through a die in order to affect its appearance. Cone shapes and tapers on the ends of tooling or tubing are made through swaging, as shown in Figure 5.12. Jet spray devices and welding points are good examples of parts produced through swaging.

4. Hydroforming

A metalforming process that is becoming more popular is hydroforming. Originally, hydroforming was a way to make a small number of deep drawn parts relatively cheaply. Hydroforming is the forming of hollow parts through the application of internal hydraulic pressure to a tubular blank. This results in the plastic deformation of the blank, forcing it to take the shape of the die cavity. In hydroforming, fluid pressure replaces the punch in conventional deep drawing.

Although there are many different part configurations possible, perhaps the best analogy is to compare hydroforming to making a jelly donut. In hydroforming, the "donut" (or hollow blank) would be placed in a die cavity and sealed. When the jelly (or liquid) is injected under intense pressure, both internal and external to the blank, the donut expands and assumes the shape of the die cavity. Depending upon the shape of the die cavity, you can make a short, fat donut; or you can make a long, thin donut; or you can make any combination of jelly donut configuration that is desired.

This method of forming is being used in making a variety of parts in different industries, but particularly in applications where structural components such as rectangular tubes are being formed. In automotive applications, some companies are finding that another type of hydroforming has a niche in body panels. One advantage of hydroforming over conventional deep drawing is that it is very versatile, especially in making relatively intricate part geometries and irregular shaped parts.

Lastly, hydroforming is suggested to have benefits over conventional forming presses such as:

- Parts can be made in fewer steps or operations
- Part finish is often superior to drawn parts
- Material is more evenly distributed (reduced thinning)

Disadvantages are usually limited to the speed at which a great number of parts can be made, and the initial costs of establishing a hydroforming operation when compared to the amount of capital already tied up in conventional forming presses.

5. "Near-Net Shape" Forming

Much discussion recently has centered on how to reduce the number of metalworking operations in the manufacture of a completed part, while at the same time improving surface finish and appearance. One approach to this dilemma is something referred to as "near net shape" forming. During World War II, scientists observed that piercing armor plates at high velocities produced extremely clean holes with little distortion to the surrounding material. This process became known as "adiabatic softening phenomenon."[30] Over time, this observation led to the development of the near net shape forming process.

In its simplest version, near net shape forming can be defined as what happens if you hit metal so hard and so fast that, through the resulting increase in temperature and pressure, the metal instantaneously exceeds its melting point at the localized point of contact. Theoretically, the adiabatic phenomenon occurs at a small, localized area at the contact point for only an instant. Under this condition, metal is more easily formed. Since it is more easily formed, net shape forming allows parts to be made without the need for subsequent deburring, grinding, and other "finishing" processes.

This process is most commonly used as an alternative to blanking. Blanks created in this manner are more uniform, and demonstrate less strain hardening and deformation. It is actually a "cutting" operation. The near net shape operation often combines the initial act of "cutting" with a power stroke to complete the forming process so that the softened metal is actually pushed into the forming die. This creates "near-net" shapes in the cold state.[31]

This process is not widely used at this time, but the advantages of near net shape forming when compared to conventional forming appear to be:

- Less blank variation
- More consistent part geometry
- Better finish
- Fewer post-forming finish operations

Disadvantages of the process include high initial costs and the applications are limited, as most formed parts do not require the precision afforded by near net shape forming.

IV. CONCLUSION

This chapter has reviewed some of the more common metalforming applications. It would be impossible in this short summary to address all of the operations, so focus was given to a few of the more common techniques. There are excellent reference books available that give deeper insight into very specific technical aspects of lesser known metalforming procedures.

Included in this discussion have been many examples of the history and tradition of metal-forming. These have been presented primarily because of the desire of the author not only to chronicle the significance of the event, but to show that there is more to lubrication topics than merely "making parts."

Also included were practical guidelines for various metalforming operations. These sensible descriptions were not aimed at the engineer responsible for the highly detailed, mechanical setups necessary to form metal into parts. Rather, they were intended as pragmatic descriptions to help better understand metalforming processes in general, and, in turn, obtain a better appreciation for all of the benefits gained in our daily lives from someone shaping metal into formed parts. From opening car doors to opening beverage containers, we see the intrinsic value of lubrication knowledge and practice. Perhaps this chapter will lead to an even greater awareness of the necessity to continue to make advancements in the science, and in the art, of metalforming lubrication.

REFERENCES

1. Schey, J. A., *Tribology in Metalworking*, Vol. 134, American Society for Metals, Metals Park, OH, pp. 1–3, 1983.
2. O'Connor, J. J., Ed., *Standard Handbook of Lubrication Engineering*, McGraw-Hill, New York, p. 2-1, 1968.
3. Lyman, T., Ed., *Metals Handbook: Forming*, 8th ed., Vol. 4, American Society for Metals, Metals Park, OH, p. 23, 1979.
4. Weast, R. C., Ed., *Handbook of Physical Chemistry*, 50th ed., The Chemical Rubber Company, Cleveland, OH, B118–B119, 1969.
5. Hixson, D. R., Pressworking lubricants, *Manuf. Eng.*, 82(2), 56, 1979.
6. Lyman, T., Ed., *Metals Handbook: Forming*, 8th ed., Vol. 4, American Society for Metals, Metals Park, OH, p. 24, 1979.
7. O'Connor, J. J., Ed., *Standard Handbook of Lubrication Engineering*, McGraw-Hill, New York, p. 3-1, 1968.
8. Lyman, T., Ed., *Metals Handbook: Forming*, 8th ed., Vol. 4, American Society for Metals, Metals Park, OH, p. 25, 1979.
9. Wick, C., Benedict, J., and Veilleux, R., Eds., *Tool and Manufacturing Engineers Handbook: Forming*, 4th ed., Vol. 2, Society of Manufacturing Engineers, Dearborn, MI, 4-1–4-9, 1984.
10. Lyman, T., Ed., *Metals Handbook: Forming*, 8th ed., Vol. 4, American Society for Metals, Metal Park, OH, p. 163, 1979.
11. Wick, C., Benedict, J., and Veilleux, R., Eds., *Tool and Manufacturing Engineers Handbook: Forming*, 4th ed., Vol. 2, Society of Manufacturing Engineers, Dearborn, MI, p. 4-41, 1984.
12. Lyman, T., Ed., *Metals Handbook: Forming*, 8th ed., Vol. 4, American Society for Metals, Metal Park, OH, p. 163, 1979.
13. Wick, C., Benedict, J., Veilleux, R., Eds., *Tool and Manufacturing Engineers Handbook: Forming*, 4th ed., Vol. 2, Society of Manufacturing Engineers, Dearborn, MI, p. 4-34, 1984.
14. Langewis, C., *Two-Piece Can Manufacturing: Blanking and Cup Drawing*, Presented at the Society of Manufacturing Engineers (SME) Conference, Clearwater Beach, FL, pp. 1–12, 1981.
15. News, *CanMaker*, 4(8), 1991.
16. News, *CanMaker*, 4(3), 1991.
17. Conny, B. M., Extracts from Coors: A catalyst for change, *CanMaker*, 4, 33–38, 1991.
18. Church, F., Productivity gains head off beverage can shortage, *Mod. Met.*, 41(9), 68–76, 1985.
19. Golding, P., Aluminum drinks can recycling in Europe: Five successful years of consumer recycling, *CanMaker*, 4, 42–45, 1991.
20. Rajagopal, S., *A Critical Review of Lubrication in Deep Drawing and Wall Ironing, from a Conference on Metal Working Lubrication*, American Society of Mechanical Engineers, New York, pp. 135–144, 1980.
21. Tucker, K., *A Solution Synthetic for 2-Piece Cans*, STLE Annual Meeting, Non-Ferrous Session, Atlanta, GA, 1989.
22. Byers, J., A fluid and tooling system for the production of high-quality two-piece cans, *Lubr. Eng.*, 42(8), 491–496, 1986.

23. Wick, C., Benedict, J., and Veilleux, R., Eds., *Tool and Manufacturing Engineers Handbook: Forming*, 4th ed., Vol. 2, Society of Manufacturing Engineers, Dearborn, MI, p. 4-34, 1984.
24. Sayenga, D., *Wonderful World of Wire*, Presented at "Interwire-91", Wire Association International, Inc., Atlanta, GA (Cardon Management Group, USA), 1991.
25. Geiger, G.H., Copper drawing agents — some new ideas, *Wire J. Int.*, 23, 60, 1990.
26. Gielisse, P., *A Seminar on Copper Wire Drawing*, Cincinnati Milacron, June, 1983.
27. Schey, J. A., *Tribology in Metalworking*, American Society for Metals Park, OH, p. 3, 1983.
28. Kajuch, J., *Basic Concepts of Wire Drawing Process*, *Cincinnati Milacron*, March, 1992.
29. Wick, C., Benedict, J., and Veilleux, R., Eds., *Tool and Manufacturing Engineers Handbook: Forming*, 4th ed., Vol. 2, Society of Manufacturing Engineers, Dearborn, MI, pp. 4-4–4-5, see also pp. 4-8, 14-1, 1984.
30. Lennart Lindell, Using adiabatic process technology to cut precision blanks, *The Fabricator*, pp. 1998, August.
31. Lincoln Brunner, Adiabatic forming: The answer to competing with china?, *Mod. Met.*, http://www.Imcpress.com/ModernMetals_files/Modern%20Metal%20Articles%202004.htm

6 The Chemistry of Metalworking Fluids

Jean C. Childers

CONTENTS

I. INTRODUCTION: FLUID TYPES

Throughout the twentieth century, metalworking chemistry evolved from simple oils to sophisticated water-based technology. The evolution of these products is shown in Figure 6.1. Between 1910 and 1920, soluble oils were initially developed to improve the cooling properties and fire resistance of straight oils. By emulsifying the oil into water, smoke and fire were greatly reduced in factories, thus improving working conditions. With the presence of water in the fluid, tool life was extended by reducing wear since the fluid kept the tools cool. However, water-diluted fluids caused rust on the workpiece, thereby creating the need for rust inhibition.

Synthetic fluids were first marketed in the 1950s because of better cooling and rust protection compared with soluble oils in grinding operations. In the early 1970s, oil shortages encouraged compounders of cutting fluids to formulate synthetic oil-free products that could replace oil-based fluids in all metalworking operations. Synthetic fluids offer benefits over soluble oil technology. These benefits include better cooling and longer tank life due to good hard-water stability and resistance to microbiological degradation. However, soluble oils, while indeed more susceptible to bacteria growth, provide better lubricity and easier waste treatability than synthetic fluids. These trade-offs encouraged the developed of semisynthetic fluids. This class of water-based fluids contains some oil and oil-based additives emulsified into water to form a tight microemulsion system. These semisynthetic fluids are an attempt to reap the benefits of oil-soluble technology while retaining the good microbial control and long tank life of synthetic fluids.

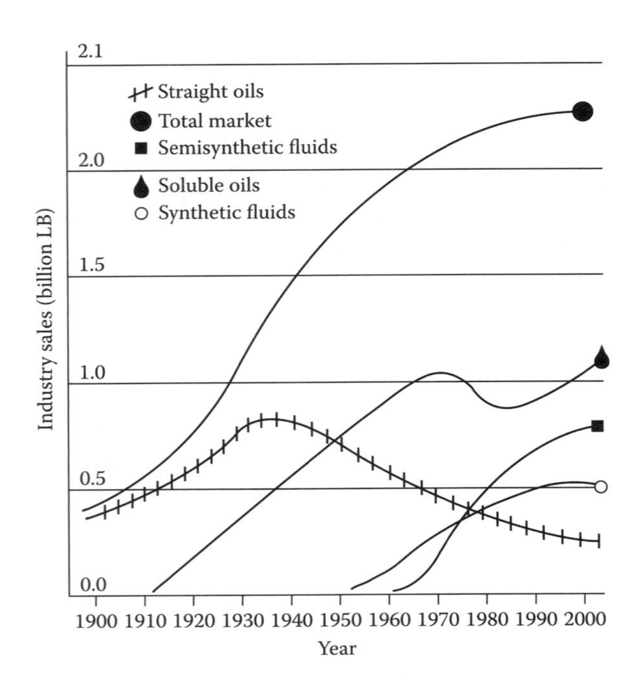

FIGURE 6.1 Evolutionary product life cycle.

In the 1980s, synthetic and semisynthetic fluids were growing in a mature market, displacing high oil content technology. Then, in 1985, changes were made to three key components common to many metalworking fluids throughout the industry. Sodium nitrite disappeared as a rust inhibitor due to concerns over nitrosamine formation, certain chlorinated paraffins were eliminated, and oil companies changed their refining processes to produce either severely hydrotreated or severely solvent refined oils having better toxicological profiles.[1]

In the early 1990s, oil prices dropped, placing oil technology at the forefront in pricing. With increasing waste treatment costs, easier to waste-treat soluble oils gained market share over synthetics. Additionally, hazard regulations on ethanolamines commonly used in synthetic fluids for corrosion control further encouraged the use of soluble oils. Therefore, mature straight oil and soluble oil technology held its 65% market share.

Oil prices rose sharply again in 2004 and 2005, but so did the cost of other chemicals used in all types of metalworking fluids. It remains to be seen whether this will affect preferences for fluid types.

The chemistry of metalworking fluids is as diverse as a library of cookbooks. Each formulating chemist will develop his own fluid formula to meet the performance criteria of the metalworking operation. However, similar to lasagna, each "recipe" will have common ingredients or raw materials (i.e., noodles, cheese, meat, sauce, spices). That is why fluids are sometimes called black box chemical blends. No user is fully aware of the exact composition of the fluid used, but the user knows whether it meets certain performance criteria (lasagna that tastes good). There are many additive blends that will function as metalworking fluids and there is no assurance of the perfect fluid for an operation. Misapplication of that perfect fluid could render it unacceptable.

This review of the chemistry of metalworking fluids will identify the building blocks of metalworking fluids, the reasons for utilizing them, and the key parameters for additive selection.

II. FUNCTIONS OF FLUIDS

A metalworking fluid's principal functions are to aid the cutting, grinding, or forming of metal and to provide good finish and workpiece quality while extending the life of the machine tools. The fluids cool and lubricate the metal-tool interface while flushing the fines or chips of metal away from the work-piece. The fluid should also provide adequate temporary indoor rust protection for the workpiece prior to further processing or assembly. Water-based fluids should resist the growth of microorganisms and the development of objectionable odors.

III. ADDITIVE TYPES

The chemical additives used to formulate metalworking fluids serve various functions. These include emulsification, corrosion inhibition, lubrication, microbial control, pH buffering, coupling, defoaming, dispersing, and wetting.

Most of the additives used are organic chemicals that are anionic or nonionic in charge. Most are liquids, used for ease of blending by the compounder. Some of the basic chemical types utilized are fatty acids, fatty alkanolamides, esters, sulfonates, soaps, ethoxylated surfactants, chlorinated paraffins, sulfurized fats and oils, glycol esters, ethanolamines, polyalkylene glycols, sulfated oils, fatty oils, and various biocide/fungicide chemical entities. Many of these chemicals are also used in common household and personal care products found in our own homes.

The functional additives used in metalworking fluids each contribute to the total composition. The effect of the addition of an additive is tested by the chemist to ensure that optimal properties of a fluid are maintained. In general, a fluid should be stable, low foaming, and waste treatable. Many of the properties of additives are mutually exclusive. Typically, if a fluid has excellent biological and hard-water stability, it may be difficult to waste treat. If it provides excellent lubricity, it may be difficult to clean. The following reviews the typical properties of additives and the significance to the formulator and user.

A. Stability

The fluid concentrate must be stable without clouding or separating for a minimum of six months storage. The fluid may be tested in cold and hot atmospheres to assess the effect of shipment or storage in winter and summer climates. Some chemists check for gelling, freezing, or "skinning" of the fluid, which may signify handling problems.

B. Oxidative Stability

Some consider the oxidative stability of additives important. Aerating and heating the coolant can accelerate any destructive oxidation of the chemical additive.

C. Emulsion Stability

In soluble oils, emulsion stability is the most critical property. The emulsifier system must be balanced, based upon its alkalinity, acidity, and hydrophylic–lipophylic balance (HLB) to ensure a stable emulsion with no cream or oil separation on the surface of the fluid. It is useful to understand the HLB system for efficient selection of emulsifiers. The HLB system was developed in 1951 by William C. Griffin of the Atlas Powder Company, and may be used to match the HLB of the emulsifier blend to the HLB requirement of the oil to be emulsified.[2] A lower number means the emulsifier is more lipophylic (oil soluble), while a higher number means it is more hydrophilic (water soluble).

Petroleum oil typically has an HLB requirement in the range of nine. To form a water-in-oil emulsion (called an invert emulsion where oil is the continuous phase) will require an emulsifier blend with an HLB of four to six, while an oil-in-water emulsion (where water is the continuous phase) will require emulsifiers having an HLB in the 8 to 12 range. The bottles shown in Figure 6.2 demonstrate how emulsion stability is affected by changes in the emulsifier HLB.

It is possible to purchase emulsifiers with known HLB values, but the HLB can be estimated by stirring some into water and observing the mixture.[2]

FIGURE 6.2 Sample bottles containing oil-in-water emulsions made with emulsifiers having different HLB values: HLB = 11 on the left, 12 in the center, and 13 at the right of the picture. (Photo courtesy of R. Bingeman, Uniqema.)

HLB = 1 to 4:	Emulsifier does not disperse in water.
HLB = 3 to 6:	Emulsifier has very poor stability in water.
HLB = 6 to 8:	Emulsifier forms a milky mix that separates with time.
HLB = 8 to 10:	Emulsifier forms a stable milky mix.
HLB = 10 to 13:	Forms a translucent mixture.
HLB > 13:	Forms a clear solution in water.

Emulsifiers used in metalworking fluid formulations include anionic compounds, such as sulfonates, fatty acid soaps (fatty acids neutralized by amines or caustic), and derivatives of polyisobutylene succinic anhydride (PIBSA). Nonionic emulsifiers are also used, including alkylphenol ethoxylates, alcohol ethoxylates, fatty-alkanolamides (products of a condensation reaction between a fatty acid and an alkanolamine), and esters of polyethylene glycol (PEG esters).

Natural sodium petroleum sulfonate was one of the major emulsifiers used in metalworking emulsions until the autumn of 2003 when Shell closed its plant in Martinez, California.[3] This closure caused major shortages, and formulators switched to synthetic sulfonates, based upon either straight-chain or branched-chain alkylbenzenes (alkylates). Derivatives of PIBSA have also been proposed as suitable alternatives.[4]

D. Hard-Water Stability

All fluid types are tested for hard-water stability because of the progressive increase in hard-water salts in the used fluid. As the fluid evaporates, only deionized water is removed, leaving behind water salts, such as calcium and magnesium. Carry out of the fluid on the parts also depletes the fluid volume. As more water and fluid concentrate are added, more salts accumulate in the tank. Calcium and magnesium cations build up in the fluid. Therefore, in soluble oils, the sodium sulfonate emulsifier is changed to calcium sulfonate, an additive that is not an emulsifier. The destabilization of the emulsion causes oil separation and loss of fluid concentration. In synthetic fluids, hard-water stability problems are visible as soap scum formation on the surface of the fluid and machines. Typically, anionic additives may have hard-water stability problems, whereas nonionic-type additives are stable to hard-water salts. Chelating agents, such as EDTA, also may be used to tie up calcium and magnesium water hardness ions, making them unavailable to react with anionic emulsifiers.

E. Mixability of Fluid Concentrate

The ease of dilution of the fluid concentrate is important from a practical perspective. The oil must "bloom" into the water without gelling to ensure fast and complete mixing. Often fluid concentrate is not premixed and is added at a point in the tank where there may be little agitation. Without good mixability, the fluid concentrate could sink, thereby not contributing to increasing the fluid concentration as intended. High soap components and high concentrate viscosity can cause mixability problems.

F. Foam

Owing to constant agitation, spraying, and recirculation of metalworking fluids, foam can easily form in the tank. Besides being a nuisance, foam interferes with the lubricity and cooling functions of the fluid. Air does not lubricate, so air entrained in the fluid renders a fluid ineffective. Foam also interferes with the worker's view of the workpiece, affecting machining accuracy and measurements. Many emulsifiers and lubricity additives may serve their function very well in a stagnant system but may be only marginally useful if they foam excessively. Antifoam agents may be incorporated to reduce foaming.

G. Residue/Cleanability

The fluid should not leave a sticky or hard-to-clean residue on the parts or equipment. Some boron-based corrosion inhibitors can leave a sticky residue, as can soap-based products in hard water. Chlorinated paraffins and pigmented lubricant additives can be difficult to remove in cleaning operations.

H. Corrosion Inhibition

Fluids are tested for their corrosion-inhibition properties. Since water is the diluent for the majority of fluids, corrosion inhibition is critical. Some additives are film forming (amine carboxylate), some are more like vapor phase inhibitors (monoethanolamine borates), while others actually form a matrix with the metal surface to provide protection (azoles). Consider both ferrous (iron) and nonferrous (aluminum, copper) metal alloys.

I. Lubricity

Additives tested for lubricity can be combined to obtain various types of lubricating properties, depending on the fluid requirements.

Boundary lubricants, such as lard oil, overbased sulfonates, esters, soaps, and sulfated oils provide a boundary between the workpiece and tool. This slipperiness is ideal for all systems, especially when machining aluminum. Soft metals need boundary lubricant to allow metal removal with good tolerance by inhibiting the tool from welding onto the aluminum workpiece.

Extreme-pressure additives, such as sulfur, chlorine, and phosphorus actually form metal complexes with the metal surface at elevated temperatures. Chlorinated additives are the most effective with typically 40 to 70% chlorine in the additive compared with sulfurized additives with 10 to 15% sulfur, or phosphate esters with 5 to 15% phosphorus. Each has its problems. Chlorinated additives, in general, are under scrutiny owing to concerns about health hazards. Sulfurized materials can stain metals and can quickly cause rancidity. Phosphate esters, the least effective of the three as a lubricant, can cause fungus and mold growth because phosphorus is a good nutrient.

Hydrodynamic lubricity additives provide a variation on boundary lubricity through high viscosity in the fluid. Typically, this term is used when describing straight oils with added viscosity improver, although some synthetic fluids are formulated with high-viscosity polymer additives that give the fluid a slippery, thick appearance. These elastic polymers may drop in viscosity under shear and heat, but if they are true rheological additives they will regain their viscosity when cooled.

J. Microbial Control

Just as water in a swimming pool will require chemical additions to control microbial growth, water-based metalworking fluids require one or more EPA-approved microbial control agents (microbicides) to help keep bacteria and mold counts as low as possible. Formaldehyde condensate biocides are commonly used to control bacteria, while sodium pyridinethione is often used for controlling mold. Orthophenylphenol or para-chloro-meta-cresol may be used when the presence of phenolic materials does not cause waste disposal issues. So-called bioresistant or biostable fluids are sometimes formulated without registered microbicides through the careful selection of ingredients that have low biodegradability. This subject will be addressed in greater detail in Chapter 9.

K. Chemical Structures

The chemical nature of most metalworking fluid additives is organic. Figure 6.3 to Figure 6.7 show the chemical structures of some of these additives. Emulsifiers, corrosion inhibitors, and lubricant

Mineral oils

Paraffinic 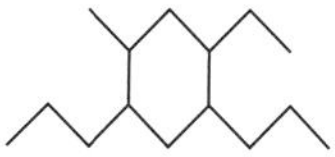 Naphthenic

Animal & Vegetable oils

$$\begin{array}{l} H_2C-O-R_1 \\ \quad | \\ HC-O-R_2 \\ \quad | \\ H_2C-O-R_3 \end{array}$$

where at least one of R_1, R_2, or R_3 is a fatty acid group of carbon chain length 12 to 22 and the remainder (if any) are hydrogen

Oleic acid

$CH_3(CH_2)_7CH=CH(CH_2)_7COOH$

FIGURE 6.3 Oils and fats.

additives are all of importance in formulating metalworking fluids, as described in Section IV to Section VIII.

IV. STRAIGHT OILS

A straight oil is a petroleum or vegetable oil that is used without dilution with water. It can be alone or oil compounded with various polar and/or chemically active additives. Light solvents, neutral oils, and heavy bright and refined stocks are among the petroleum oils used.

Petroleum sulfonate

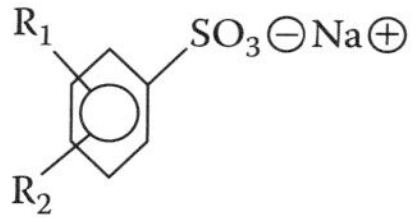

R_1, R_2, $SO_3^{\ominus}Na^{\oplus}$

where $R_1 = C_{15}$ to C_{30} alkyl chain and $R_2 = C_{15}$ to C_{30} alkyl chain or H molecular weight from 380 to 540

Fatty acid soap

O, $O^{\ominus}\oplus$

where chain length is $C_{12} - C_{22}$ and $\oplus$ is ethanolamine or alkali

Super (1:1) alkanolamide

$R-C(=O)-N(CH_2CH_2OH)_2$

Kritchevsky (2:1) alkanolamide

$R-C(=O)-N(CH_2CH_2OH)_2$ + 25% free diethanolamine

Glycerol monooleate

$C_{17}H_{33}COOCH_2CH(OH)CH_2OH$

Sorbitan monooleate

$C_{17}H_{33}COOCH_2CH(OH)-CH-CH(OH)-CH(OH)-CH_2$ (ring closed through O)

Fatty acid ethoxylate

$R-C(=O)-O[CH_2-CH_2-O]_nH$

Fatty alcohol ethoxylate

$R-CH_2-O[CH_2-CH_2-O]_nH$

Alkyl phenol ethoxylate

$C_nH_{2n+1}-C_6H_4-O-[CH_2-CH_2-O]_xH$

FIGURE 6.4 Emulsifiers.

Tolyltriazole salt

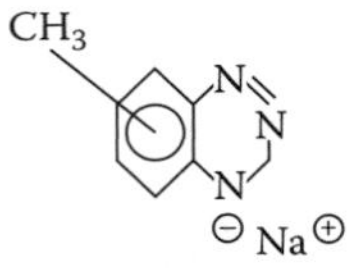

Calcium sulfonate

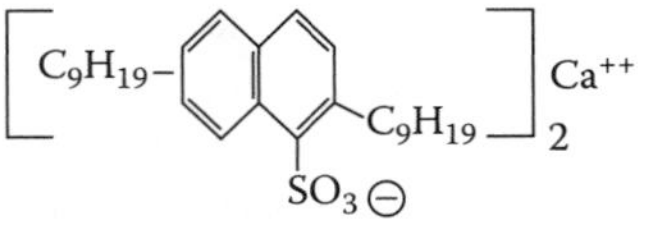

Triethanolamine

$(HOCH_2CH_2)_3N$

Monoethanolamine borate

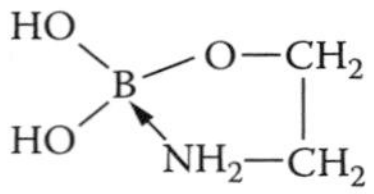

Amine dicarboxylate

where acid chain length is C_9-C_{12}
and ⊕is $NH_2(CH_2CH_2OH)_2$

FIGURE 6.5 Corrosion inhibitors.

Paraffinic oils offer better oxidative stability and less smoke during cutting than naphthenic oils. However, most compounded oils contain naphthenic oils because the lubricant additives are more soluble and compatible in naphthenic oils.[5]

For environmentally favorable requirements, vegetable oils are the oils of choice. Although considerably more expensive than petroleum oils, they are easily biodegraded for disposal. It follows then that they are more prone to biological deterioration than petroleum oils. Nondrying oils, such as rapeseed, castor, and coconut oils are best. Rapeseed oil, being the lowest in saturated fatty composition, is the best in lubricity because of its long C_{22} carbon fatty chains. It burns clean and is smoke free, which is a great advantage over petroleum oils that are frequent fire and smoke hazards.

Straight oils provide hydrodynamic lubrication. When compounded with lubricant additives, they are useful for severe cutting operations, for machining difficult metals, and for ensuring optimal grinding wheel life.

A. Compounded Oils, Mineral Oils, and Polar Additives

One of the basic compounded straight oils is a naphthenic oil with 10 to 40% boundary lubricants added. These may include animal oils, such as lard oil or tallow, or vegetable oils, such as palm oil, rapeseed oil, or coconut oil.[6] Oil-soluble esters of these oils are beneficial because they reduce the inherent biodegradation of fatty oils. Examples are methyl lardate and pentaerythritol esters. *Blown oils*, oxygen-polymerized vegetable and animal oils, increase the affinity of the additive for the metal surface, thereby providing added slip between the tool and workpiece.

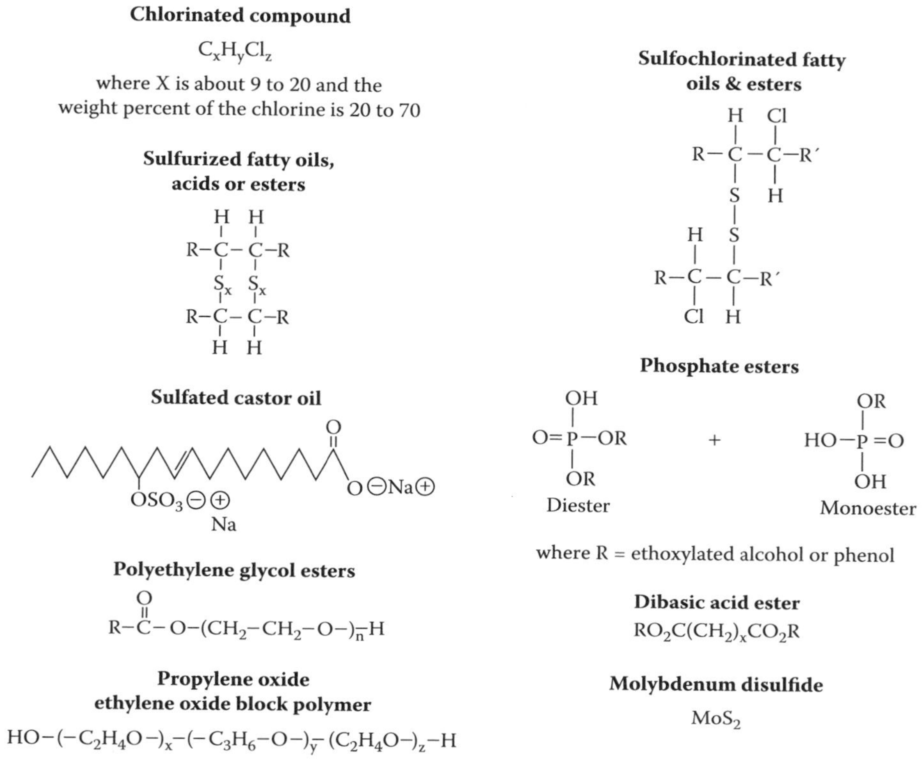

Chlorinated compound

$C_xH_yCl_z$

where X is about 9 to 20 and the weight percent of the chlorine is 20 to 70

Sulfurized fatty oils, acids or esters

$$\begin{array}{c} H \quad H \\ R-C-C-R \\ S_x \quad S_x \\ R-C-C-R \\ H \quad H \end{array}$$

Sulfated castor oil

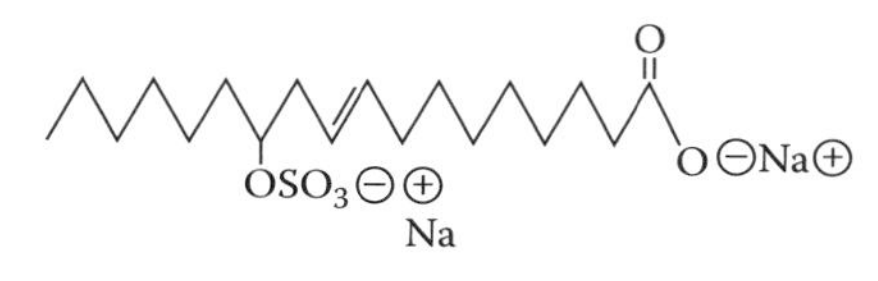

Polyethylene glycol esters

$$R-\overset{O}{\overset{\|}{C}}-O-(CH_2-CH_2-O-)_n-H$$

Propylene oxide ethylene oxide block polymer

$$HO-(-C_2H_4O-)_x-(-C_3H_6-O-)_y-(C_2H_4O-)_z-H$$

Sulfochlorinated fatty oils & esters

$$\begin{array}{c} H \quad Cl \\ R-C-C-R' \\ S \quad H \\ H \quad S \\ R-C-C-R' \\ Cl \quad H \end{array}$$

Phosphate esters

$$O=P(OH)(OR)-OR \quad + \quad HO-P(OR)(OH)=O$$

Diester + Monoester

where R = ethoxylated alcohol or phenol

Dibasic acid ester

$RO_2C(CH_2)_xCO_2R$

Molybdenum disulfide

MoS_2

FIGURE 6.6 Lubricants.

These polar additives increase the wetting ability and penetrating properties of the oil and provide a slippery boundary lubricant film. The keys to choosing these additives are oxidation resistance, oil solubility, and gumming properties. Petroleum oil fortified with these polar lubricant additives are used in machining nonferrous metals where staining by other additive systems are problematic. The polar additives provide a rust-inhibiting barrier film from the atmosphere thereby providing excellent indoor rust protection. These fortified oils are primarily used for light-duty cutting operations.

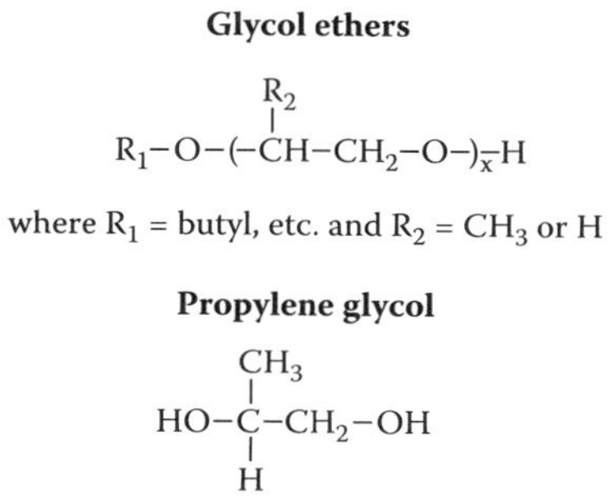

Glycol ethers

$$R_1-O-(-\overset{R_2}{\overset{|}{C}}H-CH_2-O-)_x-H$$

where R_1 = butyl, etc. and $R_2 = CH_3$ or H

Propylene glycol

$$HO-\underset{H}{\overset{CH_3}{C}}-CH_2-OH$$

FIGURE 6.7 Couplers.

B. Chemically Active Lubricant Additives

For more difficult machining operations, extreme-pressure additives, such as sulfurized, chlorinated, or phosphated additives are added to the mineral oil. These additives are surface reactive and form metallic reaction product films on the tool surface, thereby acting much like a solid lubricant at the metal-tool interface.

These additives are used alone or in combination with one another and paired with polar additives to give a lubricating oil with a wide range of effectiveness at various temperatures and pressures. An oil that contains lard oil chlorinated paraffins, and sulfurized lard oil can bridge the lubrication needs as follows: At low temperatures and pressures, the lard oil provides good boundary lubrication until temperatures reach 570 to 750°F. The chlorinated paraffin then takes over, forming an iron chloride film. Then, as temperatures climb to approximately 1300°F, the sulfurized fat takes over, forming a metallic sulfide lubricant film.[7]

The chlorinated additives could be chlorinated waxes, paraffin, olefin, or esters. Chlorinated additives are nonstaining but they can be corrosive, since small levels of hydrogen chloride can be released. Therefore, inhibitors such as epoxidized vegetable oils are often used to inhibit corrosion on the workpiece.

The sulfurized additives are either active or inactive. A sulfurized mineral oil is an active additive in that there is free unbound sulfur that easily reacts as the EP lubricant. However, this free sulfur can stain yellow nonferrous metals. Sulfurized fats such as lard oil have a stronger chemical bond with the sulfur and may be less likely to stain metals. Typically, a straight oil which contains sulfurized oils is dark in color and has a pungent odor. There are, however, other sulfurized additives, such as trinonylpolysulfide (TNPS), that is light yellow in color and ideal for water- and amine-free metalworking fluids. The simplest sulfurized mineral oil formulation would contain approximately 1% sulfur, but a fluid for difficult tapping or threading operations would contain approximately 5% sulfur.

There are sulfochlorinated additives where both sulfur and chlorine are reacted onto one molecule. These are good for machining low carbon steel and nickel-chrome alloys.

Phosphate esters provide both boundary lubricity from the ester component and phosphorus extreme-pressure lubricity at low temperatures. The effects are less dramatic than with sulfur and chlorine. Phosphate esters must be oil soluble and can be used "as is" in their free acid form or can be neutralized with an alkaline material. Neutralized phosphate esters are nonstaining and non-corrosive, and can provide rust protection properties to the oil blend.

Solid lubricants are used to a limited extent in nonrecirculating systems. Molybdenum disulfide (MoS_2) and graphite are dispersed or suspended into the oil. These additives form metallic sulfide films and flat lubricant structures that provide excellent lubricity for very difficult machining operations.

C. Straight Oil Formulations

Oil Formulation	% By Weight
Naphthenic 100 s mineral oil	90
Lard oil	2
Chlorinated paraffin	6
Sulfurized lard oil	2
	100

Straight oils are used in difficult machining and forming operations. They are ideal in recirculating systems with a lot of downtime and where rancidity of the water dilutable fluid is a problem. Straight oils are very stable to degradation, provide good rust protection, and with regular removal of metal chips, are the most trouble-free metalworking fluids from a service aspect. Their limitations are higher cost, smoke and fire hazards, operator health problems, and limited tool life through inadequate cooling.

In drawing and forming operations, oils of high viscosity are valuable. The thicker, more viscous oils provide a tougher hydrodynamic lubricant barrier film. In chip removal operations, however, high-viscosity oils will not clear the chips very well and will act as an insulator, thereby further reducing the cooling properties on the tooling. The viscosity of a finished cutting oil should be low enough to clear the chips and not insulate the heat from the operation, but high enough to control oil misting, a common health concern associated with the use of straight metalworking oils.

V. SOLUBLE OILS

With the changeover to carbide tooling and increased machine speeds, water-diluted metalworking fluids were developed. Soluble oils or emulsifiable oils are the largest type of fluid used in metalworking. The product concentrate, an oil fortified with emulsifiers and specialty additives, is diluted at the user's site with water to form oil-in-water emulsions. Here the oil is dispersed as little droplets in a continuous phase of water (see Figure 6.8).

Dilutions for general machining and grinding are 1 to 20% in water, with 5% being the most common dilution level. Drawing compounds are diluted with less water — typically 20 to 50%. At rich 50% dilutions, an invert emulsion is often purposely formed with the oil as the continuous phase. This thickened lubricant has superb lubricating properties and clinging potential on the metal to avoid run-off prior to the draw.

A. OIL

The major component of soluble oils is either a naphthenic or paraffinic oil with viscosities of 100 SUS (Saybolt universal seconds) at 100°F, sometimes termed a 100/100 oil. Higher-viscosity oils can be used but with greater difficulty in emulsification, although with possibly better lubricity. Naphthenic oils have been predominantly used because of their historically lower cost and ease of emulsification. Today, naphthenic oils are hydrotreated or solvent-refined to remove potential carcinogens known as polynuclear aromatics. However, fewer refineries are producing naphthenic oils. A soluble oil concentrate will contain up to 85% oil and little or no water.

Vegetable-based oils may also be used to prepare a water-dilutable for metalworking, although the choice of emulsifiers may need to be different. Drawbacks to the use of vegetable oils in such

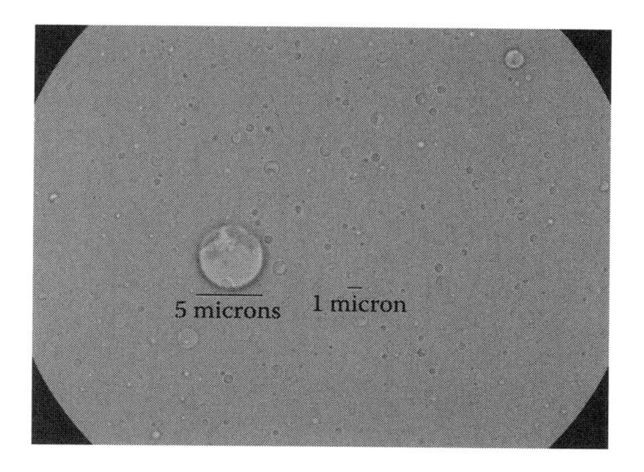

FIGURE 6.8 Microscopic view of a freshly prepared soluble oil mix. Notice that the individual oil droplets suspended in water vary in size. (Photo courtesy of Milacron, Inc.)

applications are increased cost, tendency to undergo oxidation and hydrolysis reactions, and microbial growth issues due to the fact that these oils are more biodegradable. Biodegradability is good for waste treatment, but may be a problem if long sump life is required.[8]

B. Emulsifiers

The next major class of additives in a soluble oil is the emulsifiers. These chemicals suspend oil droplets in the water to make a milky to translucent solution in water. The size of the emulsion particle determines the appearance. Normal milky emulsions have particle sizes approximately 0.002 to 0.00008 in. in diameter (2.0 to 50 μm), whereas micro-like emulsions with a pearlescent look have emulsion particle sizes of approximately 0.000004 to 0.00008 in. (0.1 to 2.0 μm)[9]. Some compounders relate the effectiveness of the two types of emulsions to comparing basketballs with small ball bearings. One can visualize more ball bearings entering a tight metal-tooling interface for lubrication than basketballs. Others claim that biostability can be enhanced with a microemulsion. Advantages of a standard milky emulsion are large oil droplet size for forming operations, ease of waste treatability, and lower foam than with microemulsions.

The predominant emulsifier is sodium sulfonate, which is used with fatty acid soaps, esters, and coupling agents to provide a white emulsion with no oil or cream separating out after mixing with water. Nonionic emulsifiers, such as alcohol or nonylphenol ethoxylates, PEG esters, and alkanolamides are also used when hard-water stability or microemulsion systems are desired. Many basic soluble oils are complete with this combination of oil and emulsifier system.

C. Value Additives

Many specialty compounders include other additives to add further value to the product. Since the fluid will be diluted with water, the possibility of rust formation is introduced. Normal rust control is usually satisfactory, but this depends on the emulsifier. Some added rust inhibitors include calcium sulfonate, alkanolamides, and blown or oxidized waxes. To impart biostability along with rust inhibition, boron containing water-soluble inhibitors may be coupled into the formulation.

The pH of the diluted fluid should be 8.8 to 9.2 to ensure rust protection and rancidity control. This pH should be buffered so the pH is maintained upon recirculation of the fluid. This is more attainable with amines as alkaline sources rather than caustic soda or potash.

To control rancidity of the fluid from bacteria growth further, biocides are often added to the oil. Further tankside additions will be necessary to prolong bacteria control.

The lubricity of a soluble oil comes from the oil emulsion. Since the viscosity of water-dilutable fluids is almost equal to that of water, the film strength or hydrodynamic lubrication potential is negated compared with straight oils. Lubricant additives are commonly added for medium- to heavy-duty operations. Boundary lubricants such as lard oil, esters, amides, soaps, and rapeseed oil are used just as they were in straight oils. Likewise, chlorinated, sulfurized, and phosphorus-based extreme-pressure additives, discussed previously, are popular value lubricant additives.

Defoamers are sometimes added if the product foams excessively due to the emulsifier system's properties. Both silicone and nonsilicone defoamers are used, silicone being the most effective at low doses. However, many plants forbid the use of silicone where plating, painting, and finishing surfaces will be affected because of "fish eyes" forming in the painted surface.

The advantages of soluble oils over straight oils include lower cost, since they are diluted with water, heat reduction, and the ability to run at higher machining speeds. Soluble oils are also cleaner, cooler, and more beneficial to workers' health because oil mists are no longer inhaled.

The advantages of straight oils over soluble oils include no rancidity, good wettability of the metal surface, good rust protection, and no destabilization problems from emulsions oiling out due to hard-water buildup and bacterial attack.

The following is a typical formulation showing the proportions of the additives in a soluble oil product:

Function	Component	% By Weight
Oil	100/100 naphthenic hydrotreated oil	68
Emulsifier	Sulfonate emulsifier base	17
EP lubricant	Chlorinated olefin	5
Boundary lubricant	Synthetic ester	5
Rust inhibitor	Alkanolamide	3
Biocide	Phenol-type	2
		100

VI. SEMISYNTHETIC FLUIDS

A. Oil/Water Base

Semisynthetic fluids are similar to soluble oils in that they are emulsions, and similar to synthetic fluids in that they are water-based fluids. The product concentrate usually appears to be a clear solution of additives. However, there is usually 5 to 30% mineral oil emulsified into the water to form a microemulsion. The emulsion particle size is 0.000004 to 0.0000004 in. (0.1 to 0.01 μm) in diameter.[9] This is small enough to transmit almost all incidental light.

B. Emulsifiers

The emulsifiers used to achieve this microemulsion will disperse oil into water to form a clear concentrate. Most of these are the same types used for soluble oils, although a higher emulsifier-to-oil ratio is necessary. Alkanolamides are the most commonly used emulsifiers, along with sulfonate, soap, esters, and/or ethoxylated compounds as coemulsifiers. A good, waste-treatable fluid would contain an amide and sulfonate base, or soap package. A hard-water stable product would use a nonionic-type emulsifier, along with the amide.

C. Value Additives

Couplers such as fatty acids and glycol ethers may be required to regulate the clarity and viscosity of the fluid. Both oil- and water-soluble rust inhibitors are used, keeping in mind that oil-soluble additives must also be emulsified. Alkanolamines such as triethanolamine are added to help buffer the pH to a good alkaline level for rust protection.

Lubricant additives can also be either oil or water soluble. Boundary lubricants and extreme-pressure additives based on sulfur, chlorine, or phosphorus can fortify a semisynthetic fluid for more difficult machining operations. Water-soluble chlorinated fatty acid soaps or esters are an example of this type of additive that need not be emulsified into the microemulsion.

It is useful to understand that some chlorinated lubricants formulated into metalworking fluids must be reported under SARA 313 in the U.S., while others having different carbon chain length or chlorine content do not need to be reported.

Commonly used rust inhibitors are amine-carboxylates or amine-borates. Amines in general are critical for good rust control.

Many compounders also add a biocide/fungicide package to protect the product from microbial growth. As an excess of emulsifiers is required (typically, two parts emulsifiers to one part oil), a defoamer may be necessary. However, selection of defoamers for semisynthetics can be difficult because if the defoamer can be coupled or emulsified into the microemulsion, it will no longer defoam the fluid. If it separates in the drum of product concentrate, it is effective only if totally removed with product concentrate.

Owing to the abundance of emulsifiers, the semisynthetic fluids will also emulsify tramp oil. To some users this is a plus, because they have no means of tramp oil removal and their system stays cleaner with this fluid. After time, the once translucent fluid will appear milky, much like a soluble oil. Many feel they are creating an *in situ* soluble oil. Others believe this acceptance of foreign oil deteriorates the quality of the fluid. Should a formulator want to make a semisynthetic that rejects tramp oil, the formulator might carefully emulsify the oil with alkanolamide. The alkanolamide must be a 2:1 amide with no fatty acid present in order to neutralize the excess mole of amine, which forms a soap.

D. Semisynthetic Formulation

Function	Component	% By Weight
Emulsifier	Sulfonate base	5
Emulsifier	Alkanolamide	15
Oil	100/100 naphthenic oil	15
Corrosion inhibitor	Amine borate	6
Coupler	Butyl carbitol	1.5
Biocide/fungicide	Triazine/pyridinethione	2
Diluent	Water	55.5
		100

The oil and chemical additives must be mixed together first, then the water should be slowly added to obtain a clear microemulsion. The product should be quality controlled before adding the water. All adjustments should be made at this point to ensure a stable and clear product. Instability will result in a separated product that cannot be reconstituted without removal of the water.

Many users like the "semi" nature of these fluids because of the advantages of both soluble oils and synthetics without many of their individual disadvantages. The advantages of semisynthetic fluids are rapid heat dissipation, cleanliness of the system, resistance to rancidity, and bioresistance. The bioresistance is due to the small emulsion particle size and small amount of oil in the fluid for anaerobic bacteria to feed on. Rust protection and lubricity are better than in a synthetic fluid because the oil and oil-soluble additives provide a barrier film that protects from corrosion and adds lubricity. The disadvantage is foam in grinding operations, acceptance of tramp oil, and less lubricity than soluble oils.

VII. SYNTHETIC FLUIDS

Synthetic metalworking fluids are water-based products containing no mineral oil. The particle size of synthetic fluid is typically 0.000000125 in. (0.003 μm) in diameter.[9]

A. Water Base

The water in the products provides excellent cooling properties, but no lubricity. Water also causes corrosion on metal surfaces. Synthetic fluids are formulated with multiple rust inhibitors and lubricant additives in heavier duty products to reproduce the machinability properties of oil-based products.

B. Corrosion Inhibitors

Synthetic fluids usually contain an ethanolamine for general corrosion inhibition and pH buffering capability. Synthetic corrosion inhibitors are amine borates, commonly termed borate esters, and amine carboxylate derivatives. These low-foaming additives are replacements for amine plus nitrite combinations, which were discontinued from use due to potential carcinogenicity from nitrosamine formation. Nonferrous inhibitors include benzotriazole, tolyltriazole, and mercaptobenzothiazole. An amine-free inorganic inhibitor is sodium molybdate. Basic amine-fatty acid soaps and alkanolamide also provide excellent rust protection for synthetic systems, and are good lubricants.

C. Lubricant and Other Value Additives

Other synthetic lubricants include polyalkylene glycols and esters, both of which are low-foaming lubricants with good hard-water stability. Owing to their nonionic water solubility, however, they are difficult to waste treat, which results in high chemical oxygen demand (COD). Boundary and extreme-pressure lubricants used in synthetics must be water soluble. Boundary lubricants include soaps, amides, esters, glycols, and sulfated vegetable oils. Chlorinated and sulfurized fatty acid soaps and esters and neutralized phosphate esters provide extreme-pressure lubricity.

A fungicide is added to protect the synthetic fluid from yeast, fungus, and molds that are prevalent in these fluids. Bacteria are nearly nonexistent due to the high pH and oil-free nature of the synthetic system. Defoamers, wetting agents, and dyes are auxiliary additives found in many synthetic fluids. The wetting agents, or surfactants, reduce the surface tension of the fluid thereby promoting good coverage of the metals for lubrication.

D. Synthetic Formulation

Function	Component	% By Weight
Diluent	Water	70
Rust inhibitor	Amine carboxylate	10
pH buffer and inhibitor	Triethanolamine	5
EP lubricant	Phosphate ester	4
Boundary lubricant	PEG ester	5
Boundary lubricant	Sulfated castor oil	4
Fungicide	Pyridinethione	2
		100

Much new product development is centered around synthetic products in order to produce additive systems that provide optimal lubricity and rust protection in an easily disposed fluid. One such concept is the marriage of semisynthetic technology with synthetic chemistry. By using multiple emulsifiers to couple synthetic water-insoluble lubricants into water, a waste-treatable system is created with petroleum oil absent from the formula.

Synthetic fluids have found widespread use in multiple machining, grinding, and forming operations. They are the products of choice where clean fluids with long tank life and modest lubrication are needed.

VIII. BARRIER FILM LUBRICANTS

A. Drawing, Stamping, and Forming Compounds

In stamping, drawing, cold forming, and extrusions, barrier film-type lubricants are used as the metalworking compound.

The emulsion products used in cutting operations are often formulated differently for drawing operations. The emulsifiers used have a lower HLB value (are more oil soluble), enabling them to emulsify high levels of lubricity additives like chlorinated paraffin. In addition, a thickened emulsion can be formed with amides and esters to give the fluid a higher viscosity enabling it to cling to the metal part during the drawing operation. Blown vegetable oils and lard oils are often used as boundary lubricants because these high-viscosity oils chemically adhere to the metal surface, providing optimal boundary lubrication. Methyl lardate is added to ensure total coverage of the metal prior to the draw. Biocides are not typically used in once-through stamping and drawing applications because bacterial colonies do not grow out of control since the fluid is not recirculated.

Honey oils are used in very difficult high-stress draws of heavy gauge metals. These are essentially chlorinated paraffin with surfactant added in order to aid in the subsequent cleaning of parts.

Vanishing oils are an evaporative-type lubricant used to stamp or draw where parts will not be washed. These are typically mineral spirits with a flash point of approximately 140°F with lubricant additives including lard oil, methyl lardate, chlorinated paraffin, or chlorinated solvents. After the draw the mineral spirits evaporate leaving a dry invisible residue.

These vanishing oils have a very high volatile organic compounds (VOC) value and are coming under scrutiny due to environmental concerns over clean air. Unfortunately, it is not possible to reduce the VOC content of these oils without leaving more residue on the part.

Before chlorinated paraffins became widely used, pigmented pastes were popular drawing lubricants. They are still used in difficult operations or where the use of chlorinated lubricants is not preferred. These are calcium carbonate/fatty acid/oil-based pastes. They may also contain mica or graphite for added lubricity. They are difficult to clean and may contain a surfactant to aid in its removal.[10]

B. Wire Draw Lubricants

Solid calcium stearate and other metal stearate soaps are used in wire drawing and cold heading or forming operations. Hydrated lime is mixed with tallow, hydrogenated tallow, fatty acids, or stearic acid to form flake soaps. Borax, elemental sulfur, MoS_2, and talc are added to supplement the lubricity properties.

Dispersions of MoS_2 and graphite in mineral oil are used in cold- and warm-forming operations. After zinc phosphating a metal part, sodium stearate is applied, thereby forming a zinc stearate film on the blanks. MoS_2 will then adhere to the stearate film providing an excellent solid-film lubricant up to 750°F.

Typical Wire Draw Lubricant Formulation	% By Weight
Aluminum stearate	10
Calcium stearate	20
Hydrated lime	66
MoS_2	4
	100

C. Prelubes

Prelubes are rust preventatives applied to coil steel that also contain a drawing lubricant package so parts can be formed without cleaning and applying the drawing compound. Polymers and dry-film lubricant packages are used without any extreme-pressure additives, which would be activated under the extreme weight of a coil of steel to cause staining.

With thickened emulsions, solid lubricant dispersions, and pastes, product stability and dispersion properties are important, as are ease of cleaning and high levels of lubricity.

IX. WASTE MINIMIZATION[11]

The waste disposal of metalworking fluids is an issue affecting the choice of metalworking fluid additives. There are three criteria that can be used in assessing the waste minimization parameters. They are waste treatability, hard-water stability, and biostability.

The rising cost of waste disposal and environmental concerns drive the need for waste-treatable additives. Additives that are stable to bacteriological degradation and hard-water salts will promote the long tank life of a fluid, thereby requiring less frequent disposal.

A. Waste Treatability

In general, anionic additives — those with a negative charge — are the easiest to waste treat because acidification or reaction with cationic coagulants makes removal chemically possible. Nonionic additives — additives with no charge — are difficult to treat because chemical treatment methods are ineffective.

The relative water solubility of the additive also affects its relative waste treatability. The more oil soluble an additive, the more likely it will be removed from the waste stream. For example, a soluble oil that contains oil, an emulsifier base, and a chlorinated paraffin will be easy to treat as long as the emulsifier is anionic. The oil and chlorinated paraffin, having no water solubility, will be removed with the partly water-soluble emulsifier. This phenomenon explains why soluble oils are easier to treat than semisynthetic fluids, which are easier to treat than synthetic fluids (Table 6.1).

B. Hard-Water Stability

Many additives will react with the calcium and magnesium salts in the water used to dilute a fluid. These calcium complexes are not usually soluble in water, so they separate from the fluid, thus destabilizing and reducing the effectiveness of the fluid. It can be seen in soluble oils as an oiling out or creaming of the emulsion. It shows up as scum or froth in synthetic fluids. By formulating

TABLE 6.1
Waste Treatability of Additives

	Easy	Moderate	Difficult
Emulsifiers	Sulfonates, soaps, sorbitan esters, glyceryl monooleate, alkanolamides, octylphenolethoxylate (HLB 10.4)	Sulfonate base	Nonylphenolethoxylate (HLB 13.4)
Corrosion inhibitors	Calcium sulfonates, amine borates	Triethanolamine	Amine dicarboxylate
Lubricants	Amphoteric	Sulfated oils, phosphate esters	Polyalkylene glycols, PEG 600 esters, block polymers, imidazolines

TABLE 6.2
Hard-Water Stability

	Clear	Stable	Precipitate or Scum
Emulsifiers	Nonylphenolethoxylate (HLB 13.4)	Octylphenolethoxylate (HLB 10.4), alkanolamides	Sulfonates, soaps, sulfonate base
Corrosion inhibitors	Amine dicarboxylate, amines, amine borates		Soaps
Lubricants	Block polymers, polyalkylene glycols, phosphate esters, PEG 600 esters	Sulfated castor oils, amphoteric salt	

metalworking fluids with additives that are not destabilized by these salts, tank life can be extended, thereby lessening the frequency of fluid disposal. The additives that are easiest to waste treat are usually the most sensitive to hard-water salts (Table 6.2).

C. Biostability

The third criterion for determining which additives contribute to waste minimization is the bio-resistance of the additives. This is the ability of an additive to slow the growth of micro-organisms in the fluid. The additive, essentially, does not act as a food source for bacteria or mold, or it may interfere with other food sources.

A study was undertaken to evaluate the biostability of key water-soluble metalworking fluid additives. The test used recirculating aquariums of each additive that were periodically inoculated with bacteria, yeast, and molds from typical fluids. Microbial growth was monitored to determine which additives were biosupportive, biostable (neither supportive nor resistant), or bioresistant (see Table 6.3).

Bioresistant chemical additives are those that contain boron, are cyclic or saturated, and are branched chained fatty acids or amine-based compounds. These include amine borates, rosin fatty acids, ethoxylated phenols, neodecanoic acid, and monoethanolamine.

Biosupportive or biodegradable chemical additives are typically fatty acids, natural fats and oils, anionics, straight-chained additives, or phosphorus-containing additives. These include soaps, amine carboxylates, sulfonate bases, lard oil, and phosphate esters.

TABLE 6.3
Biostability of Metalworking Additives

	Bioresistant	Biostable	Biosupportive
Emulsifiers		2:1 Tall oil amide, natural sodium sulfonate, nonylphenolethoxylate (HLB 13.4)	Alkali fatty acid soap, octylphenolethoxylate (HLB 10.4), 2:1 fatty amide, sulfonate base
Corrosion inhibitors	Amino methyl propanol, amine borate	Triethanolamine	Amine dicarboxylate
Lubricants		600 PEG ester, polyalkylene glycol, block polymers	Sulfated castor oil, amphoteric

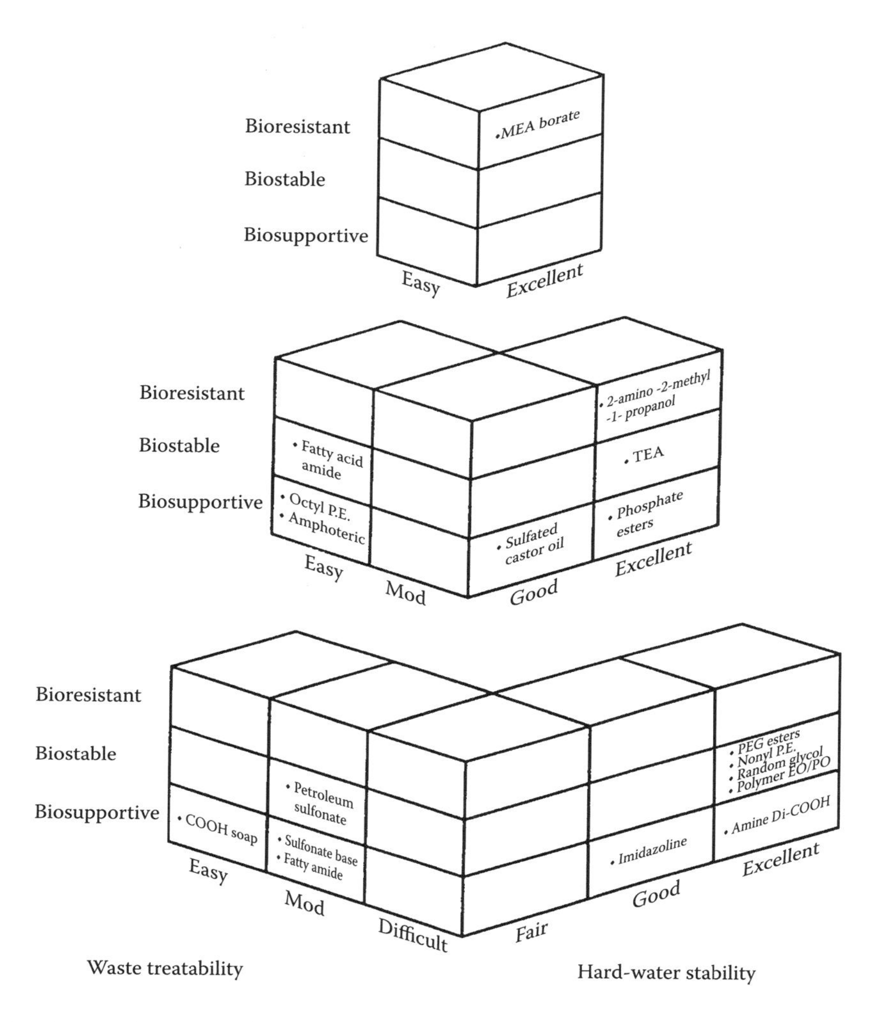

FIGURE 6.9 Waste minimization in three dimensions.

Bioresistant additives are difficult to waste treat, and conversely, additives that are biosupportive or biodegradable are relatively easy to waste treat. These mutually exclusive parameters make it difficult to have the best of both worlds. By combining the waste treatability, hard-water stability, and bioresistance properties of metalworking fluid additives, a matrix is formed (Figure 6.9) that directs a formulating chemist to the best choices for a system.

For overall waste minimization, the following semisynthetic bioresistant fluid formulation guide applies:

Corrosion inhibitors	Amino methyl propanol, monoethanolamine borate ester
Coemulsifiers	2:1 DEA rosin fatty acid amide, sodium sulfonate
Coupler	Branched diacid
Oil	Napthenic oil
Microbiological aids	Biocide/fungicide
Diluent	Water

X. CONCLUSION

The needs of the consumer, e.g., lubricity, tank life, or water disposability, are paramount in fluids development. For this reason, there are many variations of fluid types within any metalworking fluid compounder's product line. Custom formulations are the nature of the metalworking fluids industry.

Regulatory reporting requirements have opened the doors to metalworking fluid formulations. Once proprietary blends are now identified on safety data sheets and drum labels. New instrumental methods of chemical analysis have unveiled what was once closely held, confidential technology. This has placed even more emphasis on the right choice of fluid for an application.

Having developed some formulations designed for a specific task, the next chapter describes laboratory test methods for evaluating the performance and acceptability of the fluid.

REFERENCES

1. Anon, Metalworking fluid health and safety — History, chronology and future, *Compoundings*, 101, 11, 1996.
2. Bingeman, R., *Using the surfactant HLB system to save time while optimizing emulsion performance, presented at STLE Annual Meeting, Las Vegas, May 17*, 2005.
3. Lege, C. S., Meeting the need for alternatives to natural sulfonates, *Compoundings*, 54(1), 13, 2004.
4. Tocci, L., Sulfonate rivals proliferate, *Lubes'N'Greases*, 9(6), 14, 2003.
5. Decraen, L., Who needs naphthenic base oils?, *Lubes'N'Greases*, 11(2), 24, 2005.
6. Foltz, G., *Definitions of metalworking fluids*, *Waste Minimization and Wastewater Treatment of Metalworking Fluids*, Independent Lubricant Manufacturers Association, Alexandria, pp. 2–4, 1990.
7. Drozda, T. J., *Cutting fluids and industrial lubricants*, Society of Manufacturing Engineers, Dearborn, MI, pp. 5–7, 1988.
8. Woods, S., Going green, *Cutting Tool Eng.*, 57, 2, 2005.
9. Silliman, J. D., *Cutting and Grinding Fluids — Selection and Application*, 2nd ed., Society of Manufacturing Engineers, Dearborn, MI, pp. 35–47, 1992.
10. Olds, W. J., *Lubricants, Cutting Fluids and Coolants*, Cahners, Boston, MA, 1993, pp. 100–104.
11. Childers, J. C., Huang, S. J., and Romba, M., Metalworking additives for waste minimization, *Lubr. Eng.*, 46(6), 349–358, 1990.

7 Laboratory Evaluation of Metalworking Fluids

Jerry P. Byers

CONTENTS

I. INTRODUCTION

According to a survey conducted in 1989 by *American Machinist* magazine, there were 1,870,753 metal-cutting machines and 456,028 metal-forming machines in the United States.[1] (Unfortunately, more recent information is not available.) Some of these machines will use no metalworking fluid, some will use straight oil, but the majority will use water-based metalworking fluids. Although the metalworking fluids significantly affect both the part quality and the productivity of the plant, they account for less than 1% of the manufacturing cost of the end product.

Manufacturers will spend hundreds of thousands of dollars on the machines, tens of thousands of dollars on the skilled operator, hundreds or thousands of dollars on the cutting tool or grinding wheel, and only pennies per mix gallon on the metalworking fluid. Yet, if the fluid is not correctly matched to the operation, the result will be scrapped or poor quality parts and the entire investment will have been wasted. It is precisely because the user receives value over and above the cost of the metalworking fluid that many, elaborate laboratory procedures have been developed to aid in the selection of the right fluid for the right application.

A laboratory test method must be meaningful. It must simulate the most important conditions of the metalworking operation. The results must be measurable and must be compared against a standard or a reference fluid. Variables that may affect the test results must be controlled. Unrealistic conditions contrived to accelerate the test may lead to false conclusions and should, therefore, be avoided.

A given fluid parameter can be measured using several valid methods. Selection will depend on which method best simulates the conditions to which the fluid will be exposed. This chapter provides an overview of the methods available and the meaning of the results. Complete, step-by-step procedures for many of these tests are provided in the references.

II. CHEMICAL AND PHYSICAL PROPERTIES OF THE NEAT FLUID

Tests that are generally conducted on the neat, undiluted fluid as sold will be considered in this section. The first two properties, specific gravity and viscosity, give the most fundamental information about any lubricating liquid.

A. Specific Gravity

The specific gravity of a material is the mass of a given volume divided by the mass of an equal volume of some reference material, usually water, at a standard temperature. Specific gravity is also called relative density. Density, a closely related value, is the mass of a material divided by its volume. Since the density of water is very close to 1.0 g/ml at normal temperatures, the specific gravity of a fluid is nearly identical to its density. These related properties can easily be determined with an electronic, digital read-out density meter. An equally valid approach is to read the level at which a calibrated, glass hydrometer floats in a cylinder filled with the liquid at a specified temperature (see ASTM method D1298[2]).

B. Viscosity

The single most important property of a lubricant, designed to be used neat, is viscosity. (This is of much less importance if the lubricant is to be diluted, for example, to 5% in water prior to use.) The viscosity of a fluid is its internal resistance to flow.

Kinematic viscosity is determined by using one of several types of glass viscometer tubes. The time, in seconds, is measured for a fixed volume of fluid to flow, under gravity, through a calibrated capillary tube. This procedure is described in ASTM methods D445 and D446.[2] During this flow down the capillary, the fluid is under a head pressure that is proportional to its density. Thus, kinematic viscosity is a function of both internal friction and density. The units of viscosity are often expressed as centistokes (cSt), defined as millimeters squared per second. An alternative unit in common use is Saybolt universal seconds (SUS). ASTM method D2161[2] provides equations and conversion tables for converting from cSt to SUS.

A Zahn cup is a device sometimes used to determine the kinematic viscosity of opaque fluids that would tend to coat a glass viscometer tube, making the end points difficult to detect. The Zahn cup is a small metal cup having a rounded bottom with a hole in the center. The cup is filled with fluid and the time required for the fluid to flow out of the cup in an unbroken stream is measured. The result is expressed in Zahn seconds. Thick stamping and drawing fluid mixes are often evaluated using this technique. ASTM method D4212 describes the procedure.

The viscosity of a liquid varies with temperature, increasing as temperature decreases. Kinematic viscosities for a liquid at any two temperatures can be used to predict the viscosity at another temperature. This can be done graphically according to ASTM method D341.

The relationship between temperature and viscosity can be expressed by a viscosity index (VI). Using ASTM method D2270, the VI can be calculated from kinematic viscosities at 40°C and 100°C. A high VI indicates a low rate of change in viscosity with temperature, whereas a low VI indicates a high rate of change.

Dynamic viscosity is a function of the internal friction of a fluid and is not related to density. It is reported either as centipoise (cP) or as Pascal seconds (Pa sec).

$$1 \text{ cP} = 1 \text{ mPa sec}$$

One method of determining dynamic viscosity is ASTM method D2983, using a Brookfield viscometer. This instrument measures the resistance of a fluid to the rotation of various shaped spindles at various rotation speeds. The viscosity of many lubricants will vary with the speed of rotation or the shear rate. True "Newtonian" fluids, however, have a constant viscosity regardless of shear rate.

Kinematic viscosities in centistokes may be converted to dynamic viscosities in centipoise or mPa sec by multiplying by the density (g/cm^3) of the fluid, where both are determined at the same temperature.

C. Flash and Fire Points

The flammability of an oil is extremely important when considering employee safety, manufacturing plant insurability, and transportation requirements. Several different methods exist, which will allow the comparison of products, under constant conditions, for their tendency to ignite.

An open cup method, such as ASTM method D92 (Cleveland open cup) or D1310 (Tag open cup) can be used to determine both a flash point and a fire point. An open cup of oil is slowly heated at a controlled rate, while a small flame is passed over the cup at prescribed intervals. The flash point is the temperature at which a brief ignition of the vapors is first detected. The fire point is at some slightly higher temperature at which a sustained flame burns for at least 5 sec.

Closed cup methods such as ASTM D56 (Tag closed cup tester) and D93 (Pensky-Martens closed cup tester) are run in a similar manner with the cup being opened periodically for introduction of the ignition source. The Pensky-Martens tester incorporates stirring of the sample. Only a flash point is determined with closed cup methods.

D. Neutralization Number

The acid or alkali (base) content of a lubricant can be determined by a simple titration procedure. Acids and bases may be present in the lubricant as supplied, or may develop during use through degradation of product components. Depending upon solubility characteristics, the sample will be diluted in either an organic solvent mixture or an aqueous solution. A colored indicator is then added in order to detect the neutralization end point. A simple acid such as hydrochloric (HCl) or a base such as potassium hydroxide (KOH) is slowly added until the indicator changes color. In the case of an acid number, the results are expressed as the number of milligrams of KOH required to neutralize a gram of sample. In the case of a base number, the titration is done with an acid but the results are expressed as if the base contained in the sample were KOH (again, milligrams of KOH per gram of sample).

E. Lubricant Content

It may be necessary to determine the content of materials that enhance the lubricity of oil, such as fats, chlorine, phosphorous, and sulfur. As a measure of the fat content, ASTM method D94 is used to determine a saponification number. In this procedure, any fat present is converted to a soap by heating the sample with a known amount of alkali (KOH). The excess alkali that is not consumed in the conversion process is then measured by titration with an acid. The saponification number is expressed as the number of milligrams of KOH consumed by one gram of the sample.

Chlorine, sulfur, and phosphorous compounds are extreme pressure lubricants. These can be measured by many wet chemical methods or by instrumental techniques, such as x-ray fluorescence spectroscopy.

III. STABILITY DETERMINATIONS

Oil in water emulsions account for a majority of all cutting and grinding fluids on the market, as well as a significant amount of the metal-forming fluids. These products are either emulsions as sold in the case of semisynthetics, or they become emulsions prior to use in the case of soluble oils. A dispersion of oil droplets in water is accomplished through the use of surfactants and emulsifiers that rely upon electrostatic or steric repulsive barriers in order to maintain stability. Some synthetic products, containing no oil, are actually microfine emulsions of sparingly soluble synthetic organic surfactants and lubricants. The long-term storage stability of these dispersions is critical, and must be carefully evaluated. Consideration must be given to the product as sold, as well as to the stability of the end-use dilution.

A. Neat Product Stability

Perhaps the best way to determine whether a product will be stable for 1 year is to set a sample on a shelf and watch it for 1 year. Since few formulators can afford that luxury, some means of accelerating the aging process needs to be devised. Heating a sample of the product is a common technique:

- Heating will accelerate most chemical reactions. Of particular concern is the potential hydrolysis of certain emulsifiers under aqueous, alkaline conditions. If this is likely to happen over time, it will happen much faster at elevated temperatures.
- Heating lowers viscosity, increasing the possibility that emulsion droplets will collide and coalesce during Brownian motion.
- Heating to a reasonable temperature will determine whether the "cloud point" of various surfactants use in the formulation is likely to be exceeded during typical storage and transportation conditions.

The temperature selected should not be unreasonably high. Martin Rieger states

> If massive separation of (an emulsion) occurs quickly at temperatures below about 45–50°C (113–122°F), the emulsion is clearly unstable. It should be reformulated. ... Similar breakdown at 75–85°C (167–185°F) is probably irrelevant.[3]

Dr. T.J. Lin determines emulsion stability by observing the degree of creaming or phase separation after storing samples at room temperature and at 45°C (113°F).[4] ASTM Method D3707 recommends that a 100-ml sample of the emulsion be placed at 85°C (185°F) for 48 to 96 h, but this high a temperature is probably too severe. Some laboratories report that products which are unstable at or above 72°C (160°F) will still have excellent long-term storage stability, but stability at 55°C (130°F) for 3 to 5 days is absolutely essential.

Cold-temperature stability should also be considered. Exposure to cold temperatures during winter shipment and storage is unavoidable. Refrigerator temperatures of 5°C (40°F) are not unreasonable. Stability under freeze-thaw conditions is beneficial, but few emulsions will withstand such treatment. ASTM method D3209 specifies three 16-h exposures to 20°F (− 7°C) temperatures, which is reasonable. ASTM method D3709 is more severe with nine freeze-thaw cycles over a 2-week period between 0°F (− 18°C) and room temperature.

Antifoaming agents can also affect neat product stability. Antifoams and defoamers function because of their sparing solubility — if they are too soluble, they do not defoam! If a product is formulated with an antifoaming agent, that will be the first material to separate out. Such separation does not necessarily mean that the emulsion itself is unstable, and may have very little effect upon product performance. Antifoaming agents need to be carefully selected to give the most stable product possible.

B. Dilution Stability

If the metalworking fluid is designed to be further diluted with water prior to use, then this mixture needs to be evaluated for stability. Dilution stability will depend upon both the quality of the metalworking fluid concentrate and the quality of the water used for dilution. Levels of dissolved calcium and magnesium salts are referred to as "hardness," usually expressed as ppm of calcium carbonate ($CaCO_3$). Total dissolved solids, including sodium chloride and sodium sulfate, will also have an effect. In addition to the initial water quality, consideration must be given to the unavoidable buildup of salts as the fluid is used and water evaporates. It is always best, therefore, to test a product in several different waters, which may be synthetically prepared in the laboratory.

Typically, a product will be expected to perform under a variety of water conditions, from soft (75 ppm $CaCO_3$ or less) to very hard waters (400 to 600 ppm $CaCO_3$). Water hardness may also be expressed in other units, as shown below:

1 grain hardness = 17.1 ppm $CaCO_3$
1 Clark degree hardness = 14.3 ppm $CaCO_3$
1 German degree hardness = 17.9 ppm $CaCO_3$

While water hardness (the calcium and magnesium ion content) is typically thought to be the component that deactivates anionic emulsifiers, rendering them insoluble in water and destabilizing the emulsion, there are other ions found in water (chlorides, sulfate, nitrate, etc.) that can affect emulsion stability and need to be considered. Measuring the electrical conductivity of the fluid is one technique for quickly monitoring the overall salt or electrolyte content of water-based fluids, and may be used to predict emulsion stability (Figure 7.1).

CNOMO[5] method 655202 describes a procedure for determining the ease with which a metalworking fluid concentrate can be dispersed in water, as well as determining the stability of that dilution. Using a 100-ml graduated cylinder, 5 ml of concentrate is added to 95 ml of water (200 ppm $CaCO_3$). The cylinder is stoppered, inverted 180°, and then returned to the upright position. The number of inversions is counted until the concentrated material completely disperses. Five inversions or less is considered very good, while 40 or more is bad. The cylinder is then allowed to stand for 24 h, and the amount of floating oil or cream layer is measured.

DIN[6] method 51367 is used to determine the percent emulsion stability by measuring the relative change in the oil content of the lower portion of a container of emulsion before and after a 24-h static stand. A liter of mix is prepared in a separatory funnel using 20° German hardness (GH) water (358 ppm $CaCO_3$). A special, narrow-necked bottle is used to perform an oil break by acidifying a sample of this freshly prepared mix. The remaining mixture is allowed to stand undisturbed for 24 h. The bottom 100 ml is then drained off and a second oil break is conducted. Percent emulsion stability is defined below.

$$\% \text{ Stability} = \frac{\text{24 hour oil break results}}{\text{Initial oil break results}} \times 100$$

ASTM method D1479 details a similar procedure, but calculates percent oil depletion.

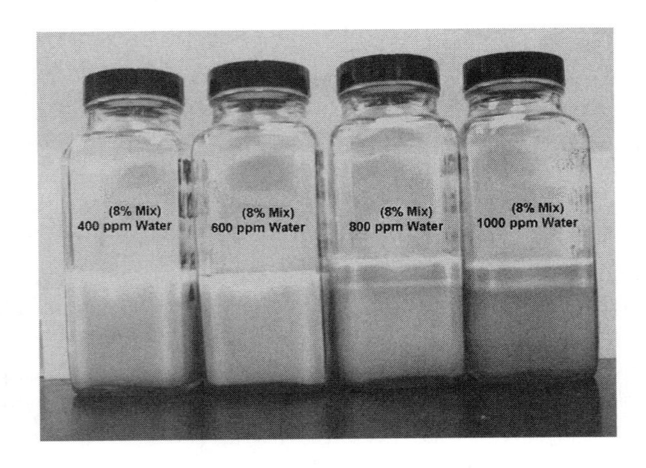

FIGURE 7.1 Effect of water hardness on emulsion stability.

Thus far, only 24-h dilution stability has been considered. D. Smith and J. Lieser have described a much longer stability test using a 5-gal aquarium equipped with an over-the-side filter and circulating pump.[7] This test is usually conducted for at least 30 days.

At the other extreme from a 30-day test is a 10-min emulsion stability test described by J. Deluhery and N. Rajagopalan, which involves turbidity measurements using a spectrophotometer with mixes containing various amounts of calcium chloride over a range of wavelengths.[8]

Another means of quantifying emulsion stability is to measure the oil droplet size. This can be done by taking measurements from a photomicrograph,[4] or using various instrumental methods. The Coulter Counter® measures electrical conductivity changes as oil droplets, suspended in a salt solution, passed through a small hole. Laser light scattering is another common technique. Particle size determinations are a useful measure of emulsion stability if the assumption is made that smaller oil droplets result in more stable emulsions. Dr. T.J. Lin questions that assumption, however.[4]

Zeta potential may also be used to help understand and control emulsion stability. An electrokinetic potential exists between the oil droplet and the surrounding liquid in which it is suspended, and is in the millivolt (mV) range. When an electrically charged field is applied, a charged particle will be attracted to one of the poles. The zeta potential can be measured by monitoring the movement of a particle through a microscope as it migrates in the voltage field. Ren Xu reports that "stabilization occurs when the zeta potential is at least +30 mV."[9]

IV. FOAM TESTS

Foaming in a metalworking fluid can lead to higher operating costs due to fluid loss, shorten the life of pumps due to cavitation, and reduce both cooling and lubrication at the chip–tool interface. The metalworking fluid formulator, however, is generally forced to use the same surfactants used by the household products industry, which equates foaming with cleaning action. It is important, therefore, to evaluate the foam control of the metalworking fluid being considered.

A. Factors to Consider

A number of factors influence the amount of foam generated in a cutting or grinding operation. These include:

- Quality or hardness of the water
- Fluid composition as sold
- Build up or depletion of fluid components with age
- Type and speed of metalworking operation
- Filtration system design
- Fluid return trench design
- Fluid pressures and flow rate
- Fluid temperature
- Contaminants such as leak oils, floor cleaners, etc.

With so many factors to be considered, it is obvious that no single foam test will predict the performance of every fluid in every application. Choose a foam test that seems to give the best correlation with past experience, and use the same water for the laboratory test that will be used in the manufacturing operation. With all foam tests, it is best not to run the test on freshly prepared dilutions. Some amount of aging is necessary for the mixture to equilibrate and for reaction of anionic surfactants with water hardness to take place. Aging for at least 1 h is recommended, although W. Niezabitowski and E. Nachtman have stated that "the true foaming characteristics of fluids become apparent after 1 week of standing."[10] Twenty-four hours of aging is more convenient and is probably sufficient in most cases. Gentle agitation using either a shaker or aeration will accelerate the aging process.

B. Bottle Test

Perhaps the simplest and most commonly conducted test procedure is a bottle test similar to ASTM method D3601. The bottle should be no more than half full of fluid, and shaking should be at some specified, reproducible rate. The initial foam height is noted immediately after shaking stops, and the time is recorded for the foam to collapse to some predetermined level (see Figure 7.2). The initial foam height and either the collapse time or the residual foam height after a specified waiting period should be used to compare foaming tendencies of various fluids. Figure 7.2 shows the results of a bottle foam test using a metalworking fluid in both soft and hard water (higher hardness equals lower foam).

C. Blender Test

The blender test is a very severe, although not unrealistic method, simulating the agitation a fluid will receive as it is whirled around by a grinding wheel, cutting tool, or pump impeller. ASTM method D3519 details the procedure. Two hundred milliliters of aged metalworking fluid dilution is placed in the jar of a kitchen blender. The mix is agitated at approximately 8000 rpm for 30 sec, and the foam height is measured immediately after the blender is switched off. The time is recorded for the foam to collapse to 10 mm in height. If more than 10 mm of foam remains after 5 min, the residual foam height is then recorded.

D. Aeration Test

ASTM method D892 describes a test method in which air is blown into the fluid to generate foam. The apparatus consists of a 1000-ml graduated cylinder and an air diffuser stone. A fluid volume of roughly 200 ml is aerated at 94 ml of air per minute for 5 min. The foam volume is measured immediately after discontinuing aeration and 10 min later. This procedure, designed for lubricating oils, has little relevance to metalworking applications.

E. Circulation Test

The CNOMO test D655212 describes a fluid circulation test using a centrifugal pump and a water-jacketed 2000-ml graduated cylinder with an outlet on the side, near the bottom (see Figure 7.3).

FIGURE 7.2 Bottle foam test — effect of water hardness.

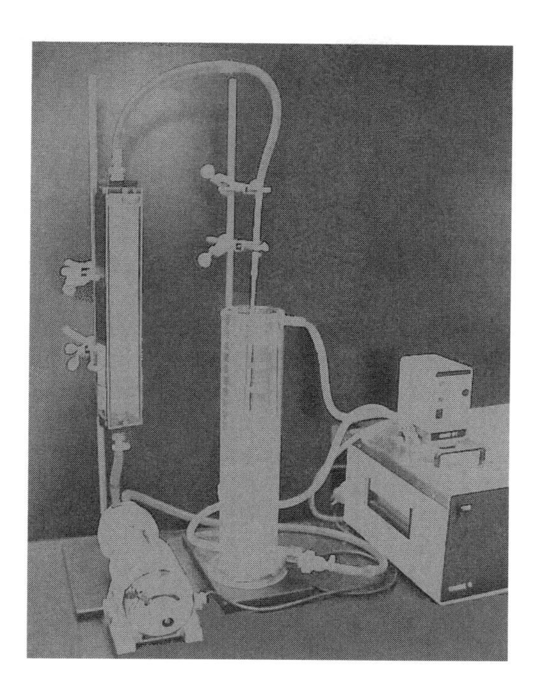

FIGURE 7.3 Pump foam test.

A fluid mix prepared in 200 ppm hardness water is added to the cylinder to the 1000-ml level. It is then pumped from the bottom of the cylinder at a rate of 250 l per hour and cascaded back upon itself from a height of 390 mm above the 1000-ml mark. The test is run for a maximum of 5 h or until the foam level reaches the 2000-ml mark. The volume of the foam above the 1000-ml mark is recorded immediately after the pump is stopped and 15 min later. This test simulates fluid flow in a machine sump or central system, but is much more severe due to the extremely high turnover rate. Observation of the sides of the graduated cylinder above the fluid level may also be useful as an indication of the cleanliness of the product.

F. Cascade Test

ASTM method D1173 is sometimes referred to as the Ross-Miles foam test, and is widely recognized in the soap and detergent industry. Although the procedure specifies 120°F (49°C) for the fluid temperature, lower temperatures could be used for metalworking fluids. A volume of 200 ml of fluid is allowed to drain at a controlled rate from a glass pipette over a distance of 90 cm into a receiving cylinder containing 50 ml of the same fluid. When all the fluid has drained out of the pipette, the foam height is measured initially, and then again 5 min later.

G. Air Entrainment and Misting

Air forced into metalworking fluids can be held at the fluid surface as bubbles of foam, or it can create two other phenomena: air entrainment or misting. Air that is held in suspension by the fluid and is slow to rise to the surface is called air entrainment. These extremely small, suspended air bubbles can cause a clear mix to become hazy or clouded, and can reduce the machine operator's view of the part being machined or ground. All metalworking fluids entrain some air, but it is more noticeable in very clear synthetics. Products relying on organic corrosion inhibitors tend to entrain more air than those relying on inorganic inhibitors.

Air that is quickly rejected by the metalworking fluid can be propelled above the surface of the fluid causing misting, effervescence, or the "cola effect." This phenomenon is only encountered with very low foaming synthetics and can be sufficient to cause a fog-like cloud to develop near the floor around return trenches or above the central coolant system.

The formation of particulates in the air is unavoidable during metalworking. Metal dust will be generated if metals are machined dry. Application of fluids during machining will reduce the amount of metal dust particles, but the fluids themselves become aerosolized. This is unavoidable, but it is significant that the application of fluids during metalworking can often result in lower levels of particulate in the air than when cutting metal dry![11] The misting properties of fluids may be studied using various techniques, and it is important to realize that the technique chosen can affect the mist size and the chemical composition of the mist.[12] One such study showed that the misting characteristics of fluids vary by fluid type, and that the presence of extraneous oil leakage from machine components will drastically increase the mist levels with fluids of all types.[13] Other studies have shown that the addition of polymers can be effectively used to reduce misting.[14,15]

V. LUBRICITY

There is a variety of tests for evaluating the lubrication properties of metalworking fluids. Each has its own inherent advantages and limitations. Lubricity tests can be broadly divided into three groups. One group is based upon simple rubbing or rolling action. Another group is based upon metal removal or chip-making processes. The final group incorporates forming or drawing of a metal sheet. Owing to the complexity of field conditions, no single test machine can simulate the lubrication requirements for all in-plant metalworking operations. That is why it is so difficult, or even impossible, to correlate bench test data with actual performance. Therefore, several different lubricity tests should be used to evaluate metalworking fluids. A broad overview of some of these methods is provided below.

A. Rubbing Surfaces

Bench tests that evaluate lubricity in rubbing processes are perhaps the most widely used, and yet of least value with respect to metal cutting and grinding. Evaluation of rubbing action may, however, be of importance in cutting and grinding applications where the workpiece or tool rubs against a support. Examples are blade wear in centerless grinders and tool guides in deep hole drilling or reaming. Rubbing tests are of greater value for stamping and drawing applications.

1. Pin and V-Block Test

The pin and V-block test is perhaps the most widely recognized of the rubbing tests. Two steel jaws having a V-shaped notch in them apply pressure to a rotating steel pin immersed in fluid (see Figure 7.4).

Two different tests can be run with this machine. ASTM method D3233 covers a technique of increasing pressure on the jaws until failure, in order to measure the load carrying properties of the fluid. ASTM method D2670 measures the antiwear properties of a fluid as a ratchet mechanism advances in order to maintain a constant load on the pin. The number of teeth advanced by the ratchet during the prescribed testing period is reported as the measure of pin wear.

Table 7.1 provides data on five metalworking fluids developed using these two ASTM methods. The fluids are arranged with the high oil products at the top of the table, synthetics and water at the bottom. Note that a very light-duty, clear synthetic gave the lowest number of teeth wear and was comparable to the heavy-duty soluble oil on failure load. This result is due to the incorporation of a

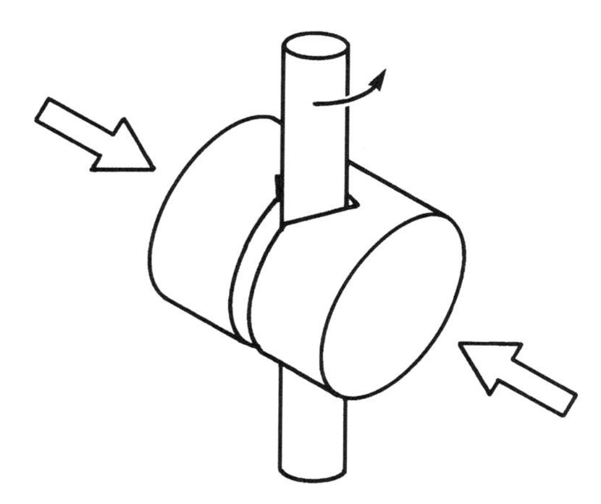

FIGURE 7.4 Pin and V-block lubricity test.

small amount of antiwear additive, which allows the light-duty product to pass the rubbing test, but will in no way assure heavy-duty cutting or grinding performance. Note, also, that in the case of the two soluble oils, chlorine, and sulfur additives improved the failure load, but did not eliminate the wear.

2. Four-Ball Test

The four-ball tester uses three steel balls held stationary in a cup-shaped cradle while a fourth ball rotates against the others under an applied load (see Figure 7.5). Using this basic concept, two different types of tests may be run. One test measures the size of the point contact wear scars on the three stationary balls after a specified time under a constant speed of rotation and load (ASTM D4172). This test is used to determine the relative wear preventive properties of various fluids. The second test measures extreme pressure capability by using a constant speed of rotation with increasing loads until welding occurs (ASTM D2783). D. Kirkpatrick has used both techniques to compare synthetic, semisynthetic, and soluble oil metalworking fluids.[16]

TABLE 7.1
Pin and V-block Results

Product Type	Dilution (%)	Failure Load		Teeth Wear
		(lb)	(N)	
Heavy duty soluble oil with chlorine and sulfur	5	4500 +	20,025 + [a]	5
Soluble oil	5	2100	9345	28
Moderate-duty semisynthetic	3	4500 +	20,025 + [a]	12
Heavy-duty synthetic	5	4400	19,580	100
Light-duty synthetic	5	4500 +	20,025 + [a]	0[a]
Water	100	300	1335	Failure

[a] Indicates the best values, best lubricity.

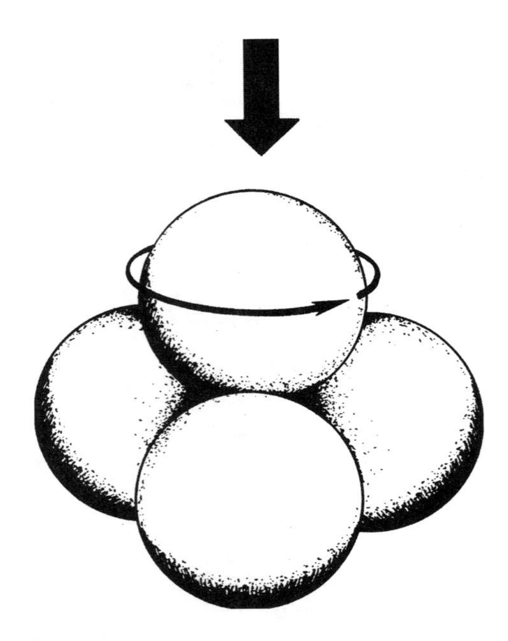

FIGURE 7.5 Four-ball lubricity test.

3. Block on Ring

A metal block under an applied load against a rotating steel ring has been used by R. Kelly and J. Byers to compare can drawing fluids[17] and by A. Molmans and M. Compton[18] to compare cutting and grinding fluids[18] (see Figure 7.6). Several measurements can be made from this test:

a. Frictional force
b. Wear scar measurements on the block
c. Weight loss measurements on the block
d. Failure load at which the lubricant film ruptures

ASTM methods D2714 and D2782 cover these procedures.

The Reichert test is similar, using a cylindrical steel roller pressed against the rotating steel ring. The lower third of the ring is bathed in lubricant residing in a cup-shaped reservoir. As the ring

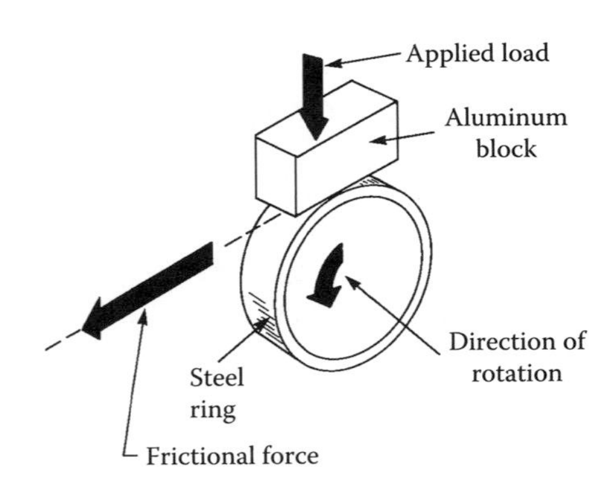

FIGURE 7.6 Block-on-ring lubricity test.

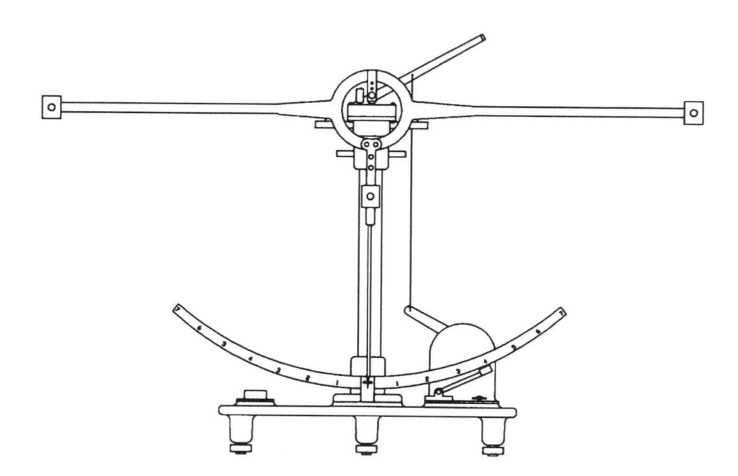

FIGURE 7.7 Friction pendulum. (*Source*: From Roehl, E. L., Sakkers, P. J. D., and Brand, H. M., *Cosmet. Toiletries*, 105, 79, 1990. With permission.)

rotates, it produces an elliptical wear mark on the roller. The size of the worn area is related to the load-carrying capacity of the lubricant.

4. Soda-Pendulum

The friction pendulum or Soda-pendulum can be used to measure the coefficient of friction over a wide range of temperature[19] (see Figure 7.7). The pendulum spindle is supported by four balls in a cup containing the test fluid. If no friction was present, the pendulum arm would swing constantly from side to side with no change in the width of swing. Friction, however, makes each swing shorter than the previous one. The coefficient of friction can be calculated from the amplitude of any two subsequent swings. Roehl et al. have used this method to compare the lubricity of materials such as isostearic acid and isopropyl myristate.[20] As the graphs in Figure 7.8 show, the isostearic acid is the better of the two lubricants.

B. Chip Generating Tests

The tests described in this section employ machines, which actually remove metal and generate nascent metal surfaces, that can interact with the lubricants. Some degree of rubbing action is also involved.

1. Lathe Tests

Dr. Charles Yang has described a lathe test using a single point, V-shaped tool that simulates chip crowding conditions found in heavy-duty machining operations. He has shown that the vertical cutting force provides a reliable method for predicting tool wear, which can be difficult to measure accurately. Using this lathe method, Dr. Yang demonstrated that the presence of 125 ppm calcium water hardness significantly reduced the cutting forces, indicating improved lubrication with a metalworking fluid mix, compared with the same fluid diluted with deionized water. Thus, water quality can have a significant effect on the lubricating properties of metalworking fluids. Low- to medium-water hardness can improve lubricity, but high-water hardness almost always leads to a loss of performance.[21]

Dr. L. DeChiffre has also developed a lathe test, in which he measures frictional force, tool wear, and chip–tool contact length.[22]

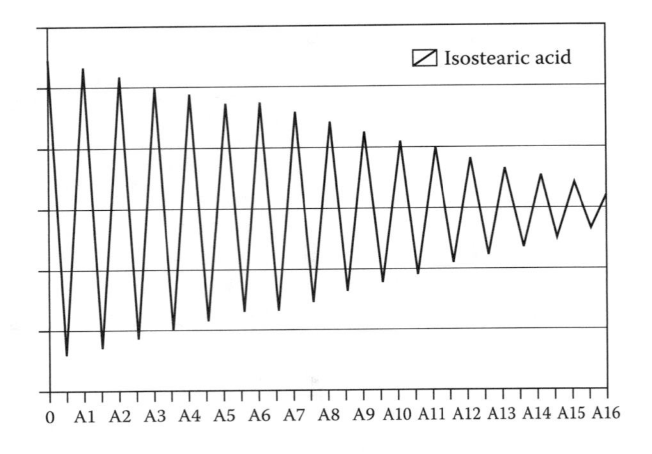

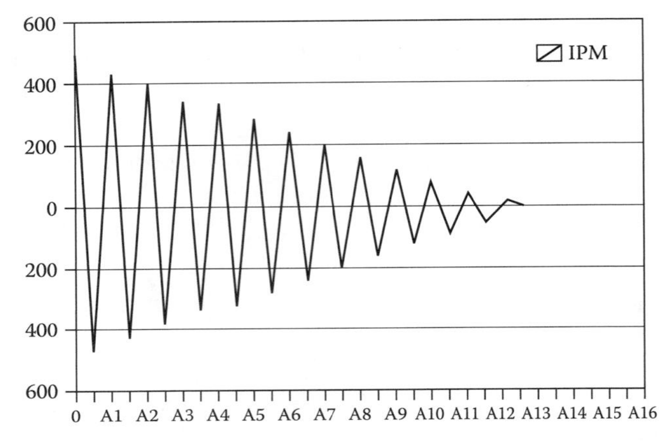

FIGURE 7.8 Damping of friction pendulum. (*Source*: From Roehl, E. L., Sakkers, P. J. D., and Brand, H. M., *Cosmet. Toiletries*, 105, 79, 1990. With permission.)

Using Dr. Yang's method and SAE 1026 steel, cutting force values were determined for the same five metalworking fluids shown in Table 7.1. Table 7.2 shows that water performed poorly on the lathe test, followed by the light-duty synthetic. A simple soluble oil and a moderate-duty,

TABLE 7.2
Lathe Test Results

Product Type	Dilution (%)	Cutting Forces	
		(lb)	(N)
Heavy-duty soluble oil with chlorine and sulfur	5	438	1948
Soluble oil	5	464	2065
Moderate-duty semisynthetic	3	460	2046
Heavy-duty synthetic	5	400[a]	1779[a]
Light-duty synthetic	5	480	2135
Water	100	530	2357

[a] Indicates the best values, best lubricity.

low-oil content semisynthetic gave almost identical results. These results show that oil alone does not provide the lubricity. A heavy-duty soluble oil with chlorinated and sulfurized additives performed better than the simple soluble oil. Finally, note that a heavy-duty clear synthetic gave the best results (lowest forces).

2. Grinding Tests

The grinding process also makes chips, but at temperatures and speeds that may be much higher than for a machining operation. A grinding wheel can be considered as a cluster of randomly oriented, negative rake cutting tools,[23] which are chemically very different from the tools used in machining. It is, therefore, important to evaluate metalworking fluids for their ability to reduce grinding wheel wear or increase metal removal rates.

A simple, horizontal spindle surface grinder can be used to evaluate the grinding ratio or G-ratio.[18,24] The G-ratio is obtained by dividing the volume of metal removed by the volume of wheel lost due to wear. High G-ratios indicate low wheel wear and good grinding performance. Surface finish and power consumption may also be measured.

Table 7.3 shows data from a moderate-duty surface grinding test on SAE 8617 steel using a vitrified bond, aluminum oxide wheel with the same five fluids from Table 7.1 and Table 7.2. Note that the heavy-duty soluble oil provided the best G-ratio, surpassing both the heavy-duty synthetic, which performed well on the lathe, and the light-duty synthetic that was best on the pin and V-block test. Each condition has a different set of fluid requirements for optimum performance.

Many other types of grinding operations may also be used for metalworking fluid evaluations. Ref. [25] describes a centerless grinding test on 52,100 steel, while Ref. [26] describes testing done with a cylindrical or center-type grinder and 52,100 steel.

3. Drilling Test

Several investigators have used drilling tests to evaluate metalworking fluids. Dr. Herman Leep compared drilling, turning, and milling test methods, and found that testing with high-speed steel drills was "the best method for discriminating between different cutting fluids."[27] The number of holes drilled, surface roughness, tool wear, torque, and cutting forces have all been used as discriminators by various investigators. W.R. Russell notes that

> there are definite performance variables that exist between manufacturing lots (of twist drills), as well as variables that exist in tool performance between tools of the same lot.[28]

His article gives several recommended metallurgical and mechanical considerations in the selection of drills for evaluating coolants.

TABLE 7.3
Surface Grinding Results

Product Type	Dilution (%)	G-Ratio
Heavy-duty soluble oil with chlorine and sulfur	5	8.0[a]
Soluble oil	5	5.0
Moderate-duty semisynthetic	3	4.0
Heavy-duty synthetic	5	5.7
Light-duty synthetic	5	2.9
Water	100	2.1

[a] Indicates the best values, best lubricity.

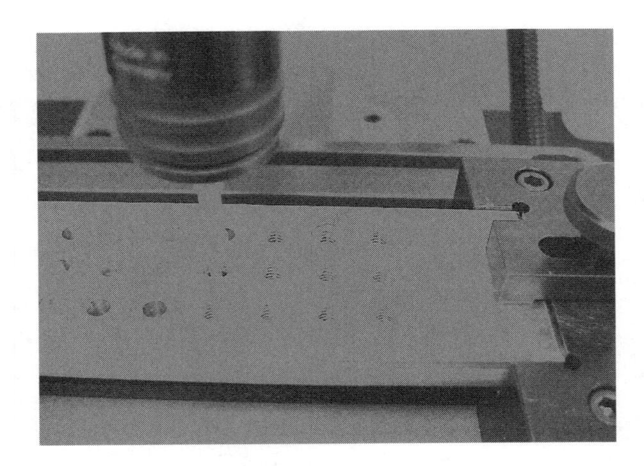

FIGURE 7.9 Tapping torque lubricity test using a predrilled aluminum test bar.

4. Tapping Torque Test

Much has been written in recent years about the tapping torque tester.[29–32] The interest in this test is due to the fact that it is perhaps the only bench-scale metal cutting test available. Torque values are measured as a tap cuts threads into a predrilled hole in a metal specimen, which can be made of various metals (see Figure 7.9). The average torque value of five runs is then calculated. Test results may be expressed either as a simple torque force value or as a percent efficiency, the ratio of the average torque value of a reference fluid to that of the test fluid. The same tap is used on both the reference fluid and the test fluid. L. DeChiffre states that an evaluation of surface finish is also necessary.[33]

Table 7.4 lists tapping torque efficiency values for four of the metalworking fluids used in previous comparisons. Two different cutting speeds were used with 1215 steel. At 400 rpm the data shows very little correlation with in-plant experience or with lathe test results. Note that the heavy-duty soluble oil and heavy-duty synthetic looked worse than the moderate-duty semisynthetic. At 1200 rpm, the light-duty synthetic, moderate-duty semisynthetic, and the heavy-duty soluble oil behave more or less as expected; but the heavy-duty synthetic was a complete failure. This may indicate that the lack of rubbing lubricity seen with this product on the pin and V-block test is an important factor in the tapping test. These data underscore the need for careful selection of the test conditions in order to generate reliable conclusions.

TABLE 7.4
Tapping Torque Results Using 1215 Steel

Product Type	Dilution (%)	Percent Efficiency	
		400 rpm	1200 rpm
Reference fluid (94% naphthenic oil + 6% lard oil)	100	100	100
Heavy-duty soluble oil with chlorine and sulfur	5	90.6	101.5[a]
Moderate-duty semisynthetic	5	103.2[a]	94.6
Heavy-duty synthetic	5	100.1	Failure
Light-duty synthetic	5	92.3	91.6

[a] Indicates the best values, best lubricity.

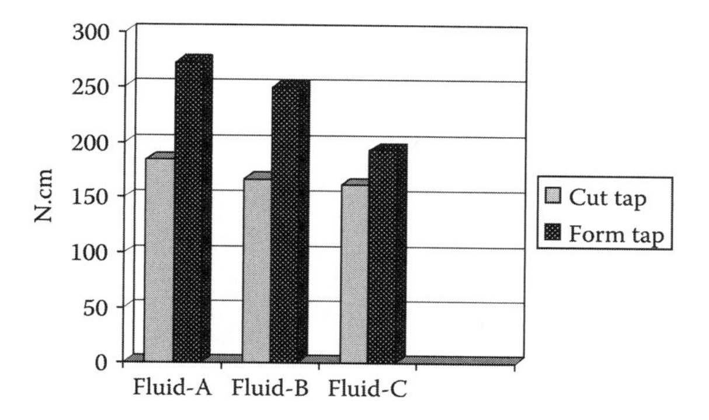

FIGURE 7.10 Bar chart of tapping torque results for 6061 aluminum.

Many factors besides the fluid composition can affect the tapping torque test results, including the quality of the tap, whether it is a "cut" or "form" tap, the exact size of the predrilled and reamed hole relative to the tap size (sometimes expressed as thread percentage), the metal alloy, and hardness of the metal (which can vary across the metal specimen). Figure 7.10 demonstrates the difference in data generated with cut taps vs. form taps in 6061 aluminum for three different fluids. Note that form taps require higher forces than cut taps, and show greater differentiation between the three fluids than do cut taps. Cut taps make threads by cutting into the wall of the hole and removing chips of metal during the process. Form tapping, unlike cut tapping, does not (or should not) generate any chips. Form taps push the metal and force it to flow into the required shape.

5. Tech Solve Machinability Guidelines

The U.S. EPA awarded Tech Solve in Cincinnati a 3-year grant to develop the *Pollution Prevention Guide to Using Metal Removal Fluid in Machining Operations*, which may be found at the organization's web site, *www.techsolve.org*. Tech Solve assembled a 60-member industrial council called the International Working Industry Group (IWIG) to accomplish the task. This group decided that it was necessary to develop some test methods for evaluating metal removal performance. Since no single machine test would adequately predict fluid performance, the group agreed upon four different metal cutting tests, each examining a different aspect of metal removal. Some of the critical parameters for each test are listed below.

Drilling — an operation utilizing a tool with two cutting edges, where the cutting speed varies along the edges and the chips must move up the flute:

- Half-inch diameter, oxide-coated high speed steel (HSS), 135° split point drill bit.
- One-inch hole depth.
- AISI/SAE 4340 steel (32–34 HRC).
- 420 rpm (55 SFPM), 0.007 ipr feed rate.
- Thrust force, torque and wear are measured.
- End point is 0.010-in. uniform drill wear.

End-Milling — a condition with interrupted cuts:

- One-inch diameter end mill cutter body with grade SM-30 uncoated carbide inserts.
- 400 SFPM speed, 0.005-in. feed per tooth, 0.5-in. axial depth of cut, 0.06 radial depth of cut.
- Climb milling.
- AISI/SAE 4140 steel (24–26 HRC).

- Cutting forces and tool flank wear are measured.
- End point is 0.010-in. uniform flank wear.

Turning (Plunging) — single point tool, plunge cut:

- Uncoated carbide inserts, grade K313.
- AISI/SAE 4340 steel (24–26 HRC) bar with dimensions 6-ft long and 1-in. diameter.
- 150 SFPM (574 rpm).
- Plunge width of 0.1 in.
- Plunge rate of 0.001 in./revolution.
- Test length 620 cycles (plunges).
- Cutting force and tool flank wear are measured.

Surface Grinding — multiple cutting points, high speed:

- 32A60-IVBE wheel, 12 in. × 1 in.
- 6000 SFPM
- AISI/SAE 4140 steel (32 and 56 HRC hardness)
- Measure cutting force, wheel wear, and calculate G-ratio

C. Metal Deformation Tests

There seems to be general agreement that no single bench test will provide all the information needed to evaluate a metal-forming lubricant. C. Wall,[34] K. Dohda, and N. Kawai,[35] and ASTM standard practice D4173 have all used at least four bench tests to study the various aspects of the metal forming process. Figure 7.11 illustrates six laboratory test methods commonly used.

Figure 7.11(a) is the flat bottom cup or deep draw test. In this procedure a lubricated metal disk or blank is forced through a circular die by a blunt-nosed punch, forming a cylindrical cup. The maximum drawing force during the test can be used as a measure of lubricity. Another measure is

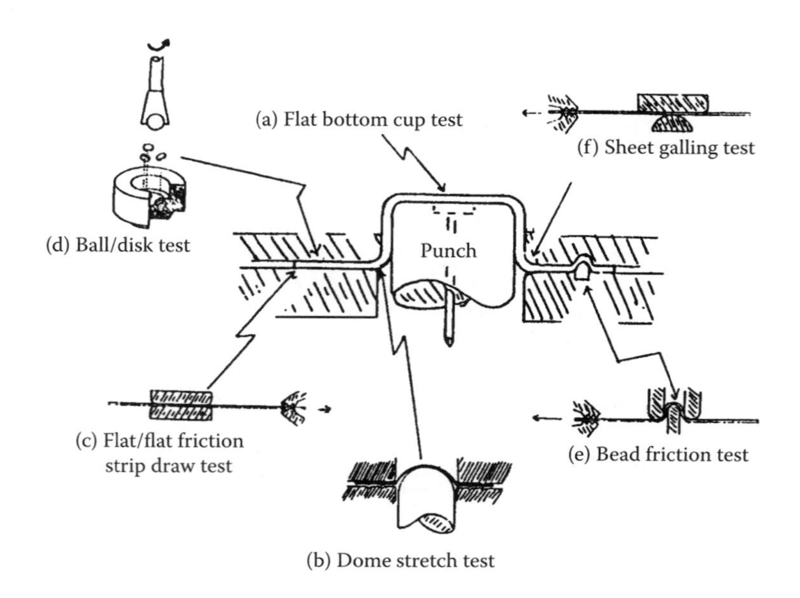

FIGURE 7.11 The metal-forming process separated into six areas of interest. (*Source:* From Wall, C., *Lubr. Eng.*, 40, 139, 1984. With permission.)

the limiting draw ratio or LDR, defined as the maximum successful blank diameter divided by the diameter of the punch.[36] The cup test combines all aspects of metal forming, including frictional forces and metal deformation forces.

Figure 7.11(b) illustrates the dome stretch test. A lubricated metal sheet is stretched over a domed punch with sufficient clamping force to prevent complete cup formation. The maximum drawing force and dome height are measures of lubricity. This test examines stretch forming and the metallurgical aspects of the process.

Figure 7.11(c) is the strip draw test, which uses flat dies and a metal strip to evaluate lubricants under conditions of pure sliding friction. The pulling force is measured at increasing clamping forces. The coefficient of friction is calculated by dividing the average steady state pulling force by twice the clamping force.[36]

The ball on disk wear test is a modification of the four-ball test described earlier. Figure 7.11(d) shows that the three stationary balls have been replaced with a cup holding three disks made of the metal to be evaluated. The cup also contains 5 ml of lubricant. The modified four-ball test can be used for evaluation of drawing lubricants, as well as aqueous rolling fluids.[37]

Draw beads are commonly used to control metal flow during stamping, particularly in the automotive industry. They aid in preventing wrinkling and maintaining wall uniformity. The draw bead simulator shown in Figure 7.11(e) evaluates lubricants by pulling a lubricated metal strip through a series of draw beads and grooves (hills and valleys) so that the metal experiences a number of bending and unbending operations. The pulling force is plotted vs. the length of travel. All strips from the same lot of metal are tested under the same clamping force.[36] A reference oil is used for a comparison standard.

Figure 7.11(f) shows a sheet galling test developed by Bernick et al.[38] It is used to evaluate the ability of a lubricant to prevent scuffing and improve die life. The test consists of a flat bottom die and a round top die with a radius of 1 in. A normal load is applied hydraulically. By plotting the pulling pressure against time, the static frictional pressure or peak pressure (P_s) and the dynamic frictional pressure (P_d) can be measured. The ratio P_s to P_d can be used to evaluate the ability of a lubricant to prevent galling. A slightly different test for galling is the compression–twist friction test described by Dohda and Kawai.[35]

Each of the basic tests described in this section addresses a different aspect of the total metal-forming process. Although the flat bottom cup draw test is, perhaps, the best simulation of a production stamping and drawing operation, no single test can be relied upon as the perfect predictor. It is necessary to select two or three tests that give reproducible results and include the most critical facets of the operation being considered. Only tests using sheet metal stock should be considered as realistic.[36]

D. Electrochemical Methods

Metalworking fluids function as lubricants by depositing a thin layer of molecules on metal surfaces that tend to prevent welding of the chip, tool, and workpiece. If the rate or degree of molecular adsorption can be determined, then the effectiveness of a fluid as a lubricant can be predicted. Naerheim and Kendig have used electrochemical impedence measurements as a means of quantifying this chemical adsorption and have shown a relationship between such measurements and metal cutting forces for three cutting fluids.[39] They anticipate that great time and cost savings could be realized from the use of electrochemical techniques instead of machinability testing.

VI. OIL REJECTION

Leak oil is an unavoidable contaminant to metalworking fluids and may build to significant levels. The actual amount of oil present may never be known if a refractometer or total oil determination is used as the only measure of metalworking fluid concentration. With these methods, all oil present is assumed to have been contributed by the fluid. In some plants, the leak or tramp oil level may actually exceed the amount of metalworking fluid concentrate present!

Tramp oil will affect such performance properties as chip settling, foam, misting characteristics, microbial control, wetting action, cleanliness, lubricity, ability to cool, residue character, and skin irritation. Low levels of tramp oil can actually improve some aspects of fluid performance, but high levels will almost always damage performance.

Metalworking fluids may be formulated to either emulsify leak oil or reject it. If a coolant sump does not have some means of removing oily contaminants, it is best that the fluid is capable of emulsifying it. Complete oil rejection is probably an unreasonable expectation for products formulated with high oil contents and, hence, a high level of emulsifiers. It is important to note that many hydraulic oils have antiwear agents and detergents, which may cause the oil to be somewhat self-emulsifying. Lubricating oils may also contain additives designed to help the oil reject water. Such additives can get into the metalworking fluid and cause emulsion instability.

Emulsion products may be formulated to reject oil by using the minimum amount of emulsifiers necessary to hold the product oil in suspension. Thus, there is no excess emulsifier present to pull leak oil into the emulsion. However, this also means the product may not be robust enough to withstand loss of critical emulsifiers to water hardness or extraction into oil floats or onto metal fines, resulting in a split mix.

CNOMO test method 655203 offers a procedure for estimating a fluid's tendency to emulsify oil. Ninety milliliters of metalworking fluid dilution plus 10 ml of oil are stirred at 10,000 rpm for 15 sec. The mixture is transferred to a graduated cylinder and allowed to stand for 24 h. The volume of floating, unemulsified oil is then read from the markings on the cylinder.

VII. CONCENTRATION CHECKS

Concentration control is extremely important. Every metalworking fluid is designed to perform relatively trouble free within a specific range of dilutions with water. Too weak a mixture can lead to one set of problems (rust, microbial growth, mix instability, lack of cleanliness), while too strong a mixture can lead to another set (foam, skin irritation, high cost, heavy residues). Metalworking fluids are mixtures of ingredients, each performing a definite function. It is wise to check the level of several of these components during use in the plant. Several methods of checking these concentrations are discussed below.

A. Refractometer

Perhaps the most widely recognized concentration test method is the hand-held refractometer (see Figure 7.12). A drop of fluid is placed on one side of a glass prism and exposed to a light source.

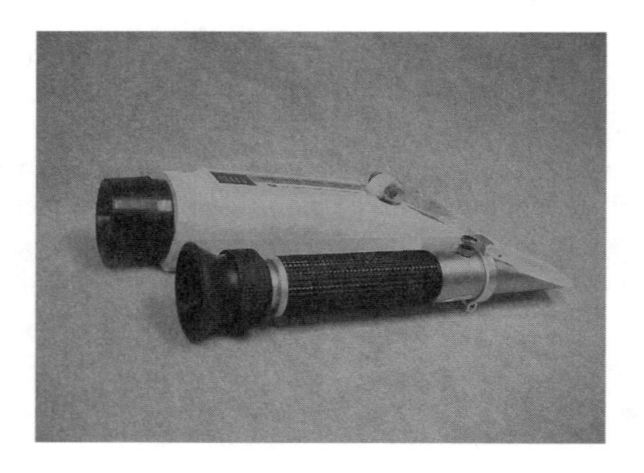

FIGURE 7.12 Two models of handheld refractometers.

By looking through the eyepiece, one can observe a band of light falling across a number scale. The position of the band of light is determined by the amount of material dissolved or dispersed in the water. The higher the concentration, the higher the band of light will appear on the scale. The value from the refractometer scale is converted to a product concentration value by using a previously prepared graph relating metalworking fluid concentration to refractometer reading. This is a very rapid method, but it has several shortcomings:

1. The refractometer cannot distinguish metalworking fluid components from contaminants. It measures anything that gets into the fluid as if it were the product of interest.
2. As the system ages and contaminants increase, the band of light becomes blurred, and its exact position difficult to determine.

Refractometers are very reliable for freshly prepared mixtures, but the accuracy decreases as the fluid is used. Used mixes will tend to give refractometer readings stronger than the actual value. In short, the refractometer is not very accurate, but it is better than no concentration checks at all.

B. Oil Content

If a soluble oil or a semisynthetic product is being used, then oil content is sometimes used as a means of concentration control. A volume of 10 ml of emulsion is added to a Babcock ice cream test bottle graduated on the neck to 20%. It is then filled with 30% sulfuric acid and centrifuged for 10 min at 1000 rpm. The volume of floating oil is determined from the graduations. This value is compared to a previously prepared chart or graph showing metalworking fluid dilution vs. oil content in order to determine the concentration. The one shortcoming of this method is that leak oil from hydraulics, spindles, and machine ways is indistinguishable from the oil in the metalworking fluid. This causes used mixes to give falsely strong concentration values. Some users try to resolve this concern by allowing the mix to stand quietly or even centrifuge the mix prior to conducting the acid split. This may provide more accurate results, but is not totally effective at eliminating the interference.

C. pH Measurements

Water-based fluids have a chemical property known as pH. Pure water has a pH value of 7.0. Water containing an acid will have a lower pH, while water containing a base has a higher pH. The pH of a solution may be determined by using pH paper and observing a color change, or with an electronic pH meter. Most metalworking fluid mixes are basic or alkaline and have pH values between 8.0 and 9.5. The pH is a very useful number, but it cannot be used as a measure of concentration for metalworking fluids since the pH can remain relatively constant despite wide variations in the product concentration. Generally, aeration of a metalworking fluid during use will cause the pH to drop slightly from its initial value.

D. Alkalinity

The concentration of basic or alkaline components is determined by a free alkalinity titration.[7] A 10-ml sample of metalworking fluid is placed in a beaker or flask with a few drops of methyl orange indicator. Dilute (0.1N) hydrochloric acid is slowly added with stirring until a color change signals the end point. The volume of acid added is related to the concentration of the metalworking fluid. The alkalinity will generally increase as the fluid is used due to accumulation of carbonates from makeup water and the buildup of alkaline materials from the coolant, which are somewhat resistant to depletion. Alkalinity and pH are related, but one cannot be used to determine the other. The concentration of weakly basic components in the metalworking fluid can vary greatly without affecting the pH, but will be readily detected by alkalinity measurements.

E. Emulsifier Content

Many emulsifiers, lubricants, and corrosion inhibitors used in the formulation of metalworking fluids carry a slight negative electrical charge. They are known as anionic surfactants. Examples of these are fatty acids, such as oleic acid and sodium petroleum sulfonates. Anionic surfactants tend to be depleted over time due to scum formation with water hardness, adsorption onto metal surfaces, and extraction into tramp oil layers. It is, therefore, of critical importance to monitor their presence in the fluid. One method is to determine the amount of cationic or positively charged surfactant required to neutralize all of the negative charges present.

Several such analytical procedures have been described in the literature.[40] Essentially, these procedures call for an exact volume of metalworking fluid containing anionic surfactants to be placed in a clear glass bottle. A colored indicator (bromophenol blue, methylene blue, bromocresol green, etc.) is added to the bottle along with a water insoluble solvent (carbon tetrachloride or chloroform). A dilute solution of a quaternary ammonium chloride salt (the cationic) is then added stepwise. The bottle is capped and shaken between each addition, and then allowed to stand until the solvent layer separates cleanly from the water layer. After all of the anionic charges have been neutralized and a slight excess of cationic has been added, a color change is observed in the solvent layer and the titration is stopped. The volume of cationic surfactant solution required to produce this color change is proportional to the amount of anionic surfactant in the sample of metalworking fluid.

These cationic/anionic titration methods are extremely accurate and do not require sophisticated or expensive laboratory equipment. The spent chlorinated solvents, however, are considered hazardous wastes and should be distilled for reuse. Instrumental methods of quantifying the anionic surfactant level include high-pressure liquid chromatography (HPLC), gas chromatography (GC), and ion-specific electrodes.

F. Boron Content

Boron compounds are used in some products for corrosion inhibition, and are easily measured using an instrument called an atomic absorption spectrophotometer (or AA). Borates are very water soluble and are not readily depleted. In fact, Dr. Giles Becket reports that in one fluid system the boron level rose to twice the level of other components in a 10-week period of time.[40] It cannot, therefore, be used as the primary concentration method for controlling a metalworking fluid system.

G. Microbicide Level

Microbicides may be present in the metalworking fluid as formulated or may be added tankside. Since most microbicides tend to be depleted in the process of killing bacteria or mold organisms, it is important to monitor their levels or their effectiveness. Most suppliers of these materials can recommend an analytical method for their product. Gregory Russ mentions procedures for triazine bactericides and for phenolic fungicides.[41] Dr. E.C. Hill has developed a dipstick method that measures the capacity for a fluid to control bacterial growth rather than measuring the level of specific compounds.[42] A pad carrying spores of a Gram-positive bacteria, dried nutrients, and a growth indicator is mounted on a plastic strip. This is dipped into the sample of metalworking fluid and then incubated at 37°C overnight. If sufficient microbicide is present, there will be no color change on the pad. If the microbial control is weak, the pad will turn a pink or red color indicating bacterial growth.

H. "What Is the Concentration?"

Clearly, there is no one answer to the question: "What is the concentration of my coolant?" A metalworking fluid may contain between 10 and 20 different ingredients, each selected to perform

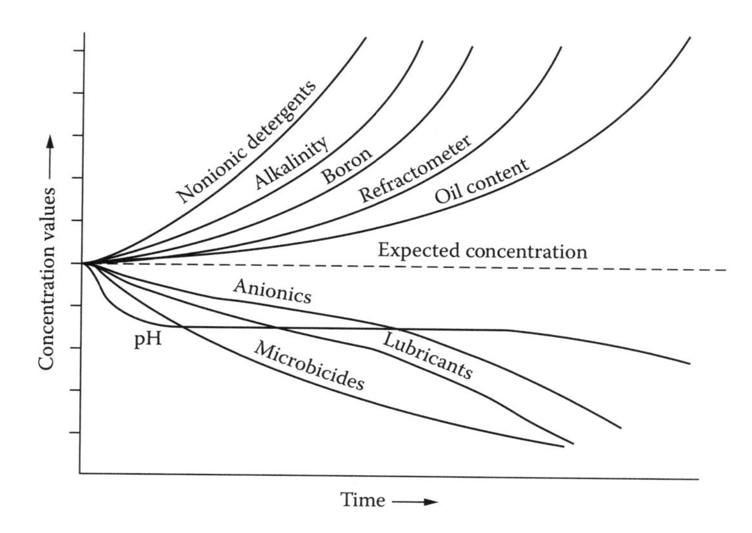

FIGURE 7.13 Metalworking fluid composition changes over time.

a certain function. Some components will increase in concentration over time (alkaline materials, borates, oil, and nonionic detergents) while others will be depleted (anionic surfactants, extreme pressure lubricants, and microbicides). Figure 7.13 attempts to depict this reality.

The only circumstance in which we can confidently speak of a single coolant concentration is when a dilution has been prepared in deionized, sterile water and stored in a capped bottle. All is not hopeless, however! An effective technique for monitoring a metalworking fluid is to select its most important functions and check the level of the components responsible for these functions. Frequently, the critical parameters will be alkalinity (responsible for rust control) and anionic surfactants (responsible for emulsion integrity, corrosion control, and lubricity). The useful or effective coolant concentration will be somewhere between these two values.

VIII. STAMPING AND DRAWING FLUID EVALUATIONS

Fluids used in automotive stamping and drawing processes must undergo several unique tests that require a great deal of specialized equipment. Four of the most common tests will be described here, although the exact procedures will vary from one account to another.

A. Phosphate Compatibility

Automotive body parts are given a zinc phosphate treatment before painting. The fluid used to stamp body parts must not interfere with these processes. To check for compatibility, cleaned metal test panels are coated with lubricant using a draw bar. The panels are aged for 1 week at 50°C and then sent through an eight-stage phosphating process. After drying, the panels are evaluated for uniformity of appearance and size of crystal formation. A small crystal structure is preferred. The phosphate coating weight, in grams per square meter, is also determined by x-ray fluorescence.

B. Electrocoat

The electrocoat or E-coat process uses phosphated test panels and a gallon of the selected E-coat paint. A cathodic charge is set on the panel, and an anodic charge is set on the paint container. A current of about one amp and 240 V is applied for 2 to 3 min in order to obtain a uniform coating thickness of about 1 mil (0.025 mm). The panel is rinsed and then cured for 20 min at 177°C. At this point, any foreign materials may flash off from the panel causing cratering of the paint. The panels are rated for size and quantity of craters.

Several variations of this test are used. A panel is first coated with lubricant using a draw bar and allowed to dry. The panel may then be placed in the E-coat bath, with stirring overnight, to determine if any fluid on the panel will be digested by the paint. On the following day, the panel is E-coated and evaluated as before. Alternatively, the E-coat bath may be contaminated with about 0.1% of lubricant and stirred before coating a panel. One final variation on the E-coat test is to prepare a sandwich of two panels with a few drops of fluid between. During the baking phase of the process, the lubricant may spatter out from the joint causing the coating to crater.

C. Adhesive Strength

Various adhesives and sealants are used in the manufacture of automobiles. It is important that any residual stamping lubricant does not interfere with these adhesives. One way to address this concern is to conduct an adhesive strength test. Clean, 1 in. × 4 in. coupons (2.5 cm × 10 cm) are coated with lubricant using a draw bar and allowed to dry. Two coupons are then glued together under carefully defined conditions of area and thickness of adhesive. The couple is cured for 20 min at 177°C and allowed to cool. The assembly is then placed in a tensile strength apparatus and pulled apart. Failure load and the amount of extension prior to failure are recorded.

D. Cleanability

Ease of removal from the stamped metal surface is a primary concern for stamping lubricants. Coated test panels are placed in a heated, agitated, alkaline cleaning bath for an appropriate length of time. The panels are then rinsed under a stream of water and observed for uniformity of the water film on the surface of the metal. A uniform film, called a break-free surface is the objective.

Other requirements for stamping and drawing fluids may include ability to weld lubricated steel panels, a staining test run at elevated temperatures (the bake/stain test), and various corrosion tests described in Chapter 8.

IX. MISCELLANEOUS

A complete discussion of every metalworking fluid bench test would be nearly endless. It is not the intent of this chapter to cover every one. Yet, there are a few more general topics that should be mentioned briefly here. Certain other properties are so important and complex that whole chapters have been devoted to them. Corrosion testing will be detailed in Chapter 8, while microbial resistance will be considered in Chapter 9.

A. Waste Treatment

Waste treatment is, usually, the first test required for fluid approval in a plant. Fluids that do not pass this test will not be used. Fluids can certainly be developed to be compatible with a particular plant's waste treatment process, but the term waste treatable means different things to different people. Some plants use an acid and alum process in which the pH is lowered with acid and alum is added. The pH is then brought back to neutral with caustic and the floc is separated. Other plants do not use acid, but rely instead upon polyelectrolytes to break the emulsion. A few plants use biological treatment systems with specially acclimated bacteria to digest the chemicals in the water. Some plants use ultrafiltration to separate organics from water, while others evaporate the water. It would be quite difficult to develop a metalworking fluid that could be treated successfully by all of these methods. It is, therefore, important to understand the end user's waste treatment process and the factors that make one product more easily treated than another. Chapters 6 and 13 are helpful in this regard.

B. Residue Characteristics

Fluids in use will splash and collect in areas where the water can evaporate, leaving behind product components and any contaminants. All metalworking fluids will leave some type of residue. Especially in hard water, the deposit can be sticky and will pick up metal fines if it cannot be redissolved by the fluid. Studying the kind of residue produced by the fluid in hard and soft water is desirable. This can be done by allowing some of the diluted fluid to evaporate in a petri dish or beaker. Testing to see how easily the residue will redissolve in the mix may be even more important than knowing how solid or sticky it becomes. An ideal residue would be one that is liquid, nonsticky, and quickly redissolves.

C. Chip Settling and Filtration

The ability of a metalworking fluid to settle metal particles or chips is an important property, necessary for good performance. The fluid carries the chips away from the work area to a sump or central system tank where they should then settle from the fluid and be removed. If a fluid suspends metal fines, chip recirculation may result, leading to scratched surface finishes. It is also possible for fines to settle too quickly, plating out on machines and filling up return trenches or overhead piping (see Figure 7.14). Chip settling characteristics may be studied by adding a weighed amount of metal fines to the bottle foam test described earlier in this chapter. After shaking, the time required for the bulk of the fines to separate from the fluid should be recorded. Whether the chips settle or rise to the top of the fluid should also be noted.

A filtration test can be conducted using vacuum filtration of a fluid/metal chip mixture through selected filter media. The time required for the fluid to pass through the media, and whether the flow is continuous or stops due to filter plugging, can be used to rate products. Plastic membrane filters should never be used in such tests, since these have no counterpart in industry.

D. Product Effect on Nonmetals

Rubber seals and plastic machine components, such as machine window panels, are frequently bathed in metalworking fluid. Major ingredients in the fluid, including the oil, water, and alkaline materials, are known to affect the integrity of these nonmetal components.[43] ASTM method D471 can be used to evaluate fluid compatibility with such materials. The elastomer is immersed in the

FIGURE 7.14 Section cut from overhead piping showing accumulated metal chips.

fluid for up to 30 days at room temperature or higher. At the end of that time, changes in specimen appearance, weight, volume, hardness, tensile strength, and elongation are recorded. With rigid plastics, effects may not be observed unless the material is stressed. While many water-based fluids will have similar effects, great differences have been noted in the resistance of various types of elastomers.[43,44]

Another nonmetal to be considered is the machine paint. Improperly prepared surfaces prior to painting and poor quality paint are the major causes of problems. The compatibility of the metalworking fluid and the machine paint may be checked using steel panels that have been properly prepared and painted. The panels may either be soaked in a dilution of the product being considered, or a few drops of the concentrated product may be placed on the painted surface. Some test procedures may require that the paint be scratched before the fluid is applied to observe the tendency of the fluid to get under paint edges and begin lifting it off the substrate. After several days of exposure, the surface is examined for signs of discoloration, softening, or bubbling of the paint.

E. Surface Tension

Surface tension is a measure of the inward pull of a liquid that tends to restrain the liquid from flowing or wetting a surface. It is related to such performance properties as cleaning action, lubrication, and foam. Two techniques are used most frequently for this determination:

- Du Nouy ring tensiometer, ASTM method D1331
- Dynamic or bubble tensiometer, ASTM method D3825

The du Nouy tensiometer is a torsion arm balance with a platinum wire ring in a horizontal position hanging from the end of the arm. The liquid to be tested is poured into a shallow cup and placed on an adjustable platform below the ring. The ring is submerged just below the surface of the liquid. By simultaneously adjusting the height of the platform and the torsion on the arm, a measurement is made of the force required to pull the ring away from the surface. Using this procedure, pure water has a surface tension of about 73 dyn/cm at 20°C. Addition of surface-active agents such as emulsifiers, soaps, and detergents will cause this value to decrease. The surface tension of a water-based metalworking fluid will depend upon the type and concentration of surface-active agents present.

Another method of determining surface tension is to measure the pressure required for bubble formation as a gas flows through a capillary tip immersed in a liquid. Such dynamic measurements can be important whenever the surface area of a liquid is changing rapidly, for example, when a metalworking fluid is pumped out of a relatively quiet reservoir and sprayed into the metal cutting zone. This technique is also unaffected by the presence of foam on the liquid surface.

X. CONCLUDING REMARKS

This chapter was intended to provide a broad overview of the many evaluation methods applied to metalworking fluids. In the interest of space and the reader's time, no attempt was made to give complete, step-by-step instructions. Instead, many references have been listed for those desiring further detail. None of these procedures should be considered to be the final word on testing methods. The reader should feel free to modify the procedures to meet his or her own needs. Performance in the manufacturing environment is, of course, the ultimate test of a metalworking fluid.

One final word about metalworking fluid evaluation. Never conduct a test simply because it is a published, standard method. If the procedure does not simulate the actual conditions of use, or if it was not designed for the type of product being considered (such as using a test designed for fuels on a water-based fluid), the data generated may be worse than having no data at all!

REFERENCES

1. 14th American machinist inventory of metalworking equipment, *Am. Machin.*, 133, 11, 91–110, 1989.
2. American Society for Testing and Materials (ASTM), 1916 Race Street, Philadelphia, PA 19103, USA.
3. Rieger, M., Stability testing of macroemulsions, *Cosmet. Toiletries*, 106(5), 59–69, 1991.
4. Lin, T. J., Adverse effects of excess surfactants upon emulsification, *Cosmet. Toiletries*, 106(5), 71–81, 1991.
5. Committee De Normalisation De La Machine Outiels (CNOMO), Service 0927 bat f24, 8–10 Avenue Emile Zola, 92109 Billancourt Cedex, France.
6. Deutsches Institut fur Normung (DIN), Burggrafenstrasse 4–10, D-1000 Berlin 30, Germany.
7. Smith, M. D. and Lieser, J. E., Laboratory evaluation and control of metalworking fluids, *Lubr. Eng.*, 29, 315–319, 1973.
8. Deluhery, J. and Rajagopalan, N., A turbidimetric method for the rapid evaluation of MWF emulsion stability, *Colloid Surf. A*, 256, 145, 2005.
9. Xu, R., How to disperse particulates, *R&D Mag.*, 45(7), 20, 2003.
10. Niezabitowski, W. and Nachtman, E., *Way and Gear Oil, Hydraulic Fluids and Greases as Contaminants in Water Base Metal Removal Fluids: Corrosion and Foam Effects*, *Strategies for Automation of Machining: Materials and Processes*, ASM International, New York, pp. 167–170, 1987.
11. Woskie, S. et al., Factors affecting worker exposures to metalworking fluids during automotive component manufacturing, *Appl. Occup. Environ. Hyg.*, 9(9), 612–621, 1994.
12. White, E. and Lucke, W., Effects of fluid composition on mist composition, *Appl. Occup. Environ. Hyg.*, 18, 838–841, 2003.
13. Turchin, H. and Byers, J., Effect of oil contamination on metalworking fluid mist, *Lubr. Eng.*, 56(7), 21–25, 2000.
14. Gulari, E., Manke, C., and Smolinski, J., Polymer Additives as Mist Suppressants in Metalworking Fluids: Laboratory and Plant Studies, *Symposium Proceedings of the American Automobile Manufacturers Association: The Industrial Metalworking Environment: Assessment and Control*, 294–300, 1995.
15. Kalhan, S. et al., *Polymer Additives as Mist Suppressants in Metalworking Fluids, Part 2A: Preliminary Laboratory and Plant Studies*, *Design and Manufacture for the Environment*, SAE International, pp. 47–51, 1998.
16. Kirkpatrick, D., *Trend to Synthetic Cutting Fluids*, *Conference on Lubrication, Friction and Wear in Engineering*, Institution of Engineers, Australia, 1980.
17. Kelly, R. and Byers, J., Synthetic fluids for high speed can drawing and ironing bodymakers, *Lubr. Eng.*, 40(1), 47–52, 1984.
18. Molmans, A., and Compton, M., Heavy duty synthetic metalworking fluids are a reality, *Synthetic Lubricants and Operational Fluids*, Fourth International Colloquium at Esslingen, Germany, pp. 401–405, 1984.
19. Thornhill, F., Other parameters and measurement advantages, In *Monitoring and Maintenance of Aqueous Metalworking Fluids*, Chater, K. W. A and Hill, E. C., Eds., Wiley, Chichester, 1984.
20. Roehl, E. L., Sakkers, P. J. D., and Brand, H. M., Isostearic acid and isostearic acid derivatives, *Cosmet. Toiletries*, 105(5), 79–87, 1990.
21. Yang, C., The effects of water hardness on the lubricity of a semi-synthetic cutting fluid, *Lubr. Eng.*, 35(3), 133–136, 1979.
22. DeChiffre, L., Laboratory Testing of Cutting Fluid Performance, *Lubrication in Metal Working*, Vol. 2, Third International Colloquim at Esslingen, Germany, pp. 74.1–74.5, 1982.
23. Springborn, R. K., *Cutting and Grinding Fluids: Selection and Application*, American Society of Tool and Manufacturing Engineers, pp. 10–11, 1967.
24. Mehta, A. K. et al., A test technique for the evaluation of grinding fluids, *Proceedings of the Institute of Mechanical Engineers Conference, Tribology, Friction, Lubrication and Wear*, 1, 517–522, 1987.
25. Yoon, S. and Krueger, M., A killer combination for ideal grinding conditions, *Am. Mach.*, 142(11), 96–102, 1998.

26. Krueger, M. et al., New technology in metalworking fluids and grinding wheels achieves 130-fold improvement in grinding performance, *Abrasives Magazine*, 8–15, October/November 2000.
27. Leep, H. R., Investigation of synthetic cutting fluids in drilling, turning and milling processes, *Lubr. Eng.*, 37(12), 715–721, 1981.
28. Russell, W. R., Cutting tools for cutting fluid evaluation, *Lubr. Eng.*, 30(5), 252–254, 1974.
29. Faville, W. and Voitik, R., The Falex tapping torque test machine, *Lubr. Eng.*, 34(4), 193–197, 1978.
30. Webb, T. and Holodnik, E., Statistical evaluation of the Falex tapping torque test, *Lubr. Eng.*, 36(9), 513–529, 1980.
31. Hernandez, P. and Shiraki, H., Comparison of aqueous extreme pressure cutting fluids on the no. 8 tap torque tester and other cutting methods, *Lubr. Eng.*, 43(6), 451–458, 1987.
32. Zimmerman, J. et al., Experimental and statistical design considerations for economical evaluation of metalworking fluids using the tapping torque test, *Lubr. Eng.*, 59(3), 17–24, 2003.
33. DeChiffre, L., Function of cutting fluids in machining, *Lubr. Eng.*, 44(6), 514–518, 1988.
34. Wall, C., The laboratory evaluation of sheet metal forming lubricants, *Lubr. Eng.*, 40(3), 139–147, 1984.
35. Dohda, K. and Kawai, N., Correlation among tribological indices for metal forming, *Lubr. Eng.*, 46(3), 727–734, 1990.
36. ASTM D4173, Standard practice for evaluating sheet metal forming lubricant.
37. Riddle, B. L., Kirk, T. E., and Kipp, E. M., Reactive additives improve aqueous aluminum foil rolling, *Lubr. Eng.*, 47(1), 41–45, 1991.
38. Bernick, L. D., Hilsen, R. R., and Wandrei, C. L., Development of a quantitative sheet galling test, *Wear*, 48, 323–346, 1978.
39. Naerheim, Y. and Kendig, M., Evaluation of cutting fluid effectiveness in machining using electrochemical techniques, *Wear*, 114, 51–57, 1987.
40. Becket, G. J. P., Knowing the true concentration is the key to longer cutting fluid life, *Lubrication in Metal Working*, Vol. 2, Third International Colloquium at Esslingen, Germany, pp. 105.1–105.12, 1982.
41. Russ, G. A., Coolant control of large central systems, *Lubr. Eng.*, 36(1), 21–24, 1980.
42. Hill, E. C., Biocide assays in metalworking fluids as an indication of spoilage potential, *Industrial Lubricants — Properties, Application, Disposal*, Sixth International Colloquium at Esslingen, Germany, Vol. 2, pp. 21.2-1 to 21.2-5, 1988.
43. Rolfert, E., The influence of metalworking fluids on common elastomers, *Lubr. Eng.*, 49(1), 49–52, 1993.
44. Moon, D. and Canter, N., The seal compatibility problem, *Manufacturing Engineering*, June 2001.

8 Corrosion: Causes and Cures

Giles J.P. Becket

CONTENTS

I. INTRODUCTION

Metalworking fluids can be divided into two basic types: water-free and water-mixed products. Those free from water are mineral oils that contain comparatively small amounts of oil-soluble chemicals to enhance the performance of the product. The second type are water-mixed fluids, which may either be solubilized or emulsified in water, the water being approximately 90 to 98% of the total material. Apart from being generally considered a safer material than mineral oil, water does a much more effective job of cooling the tool and the workpiece. Unfortunately, water has a great capacity to corrode the majority of metals. Moreover, it is not just a question of water causing corrosion, bacteria that can come to inhabit water-mixed fluids are capable of causing corrosion through a number of processes. Thus, it is extremely important to have a good understanding of the mechanisms that govern corrosion if it is to be avoided. This chapter also considers corrosion-testing techniques in order to evaluate cutting, grinding, stamping, and drawing fluids. Finally, suggestions are offered to help avoid or correct corrosion problems that can occur.

II. DEFINING CORROSION

Corrosion of a metal is the deterioration of the material because of a reaction with its environment. Looked at another way, corrosion occurs when a metal returns to one of its possible natural states, e.g., iron oxidizes back to iron ore (oxide) and copper can be corroded by sulfur-containing compounds and returned to its sulfide. Even aluminum corrodes to give a surface layer of oxide that is chemically similar to the bauxite from which it was originally won. Corrosion is a completely natural process, but one that does not suit our modern-day requirement of using metals for structural purposes. In fact, with the exception of only a few metals (notably silver and gold), metals never occur naturally. They are always in the form of compounds, because in this form they are chemically in a lower energy state, which is thermodynamically preferable. In short, if a metal atom can lose one or more electrons from its structure and then go on to combine with other (nonmetallic) elements (e.g., oxygen, sulfur, and chlorine), thereby losing some energy and reaching a more stable (lower energy) state, it will. We view this process as corrosion, which, while generally being a nuisance, is at least made use of in a battery when the freed electrons are channeled to some useful purpose. (A battery may be viewed as controlled corrosion in a container.) Corrosion is, above all, an electrochemical process, and anything that aids the flow of electrons invariably promotes corrosion. Thus, seawater, which conducts electricity well because of the dissolved minerals contained in it, is far more corrosive than pure water, which is a relatively poor electrical conductor.

Since corrosion is dependent on the metal(s) involved, the nature of the corroding environment, and physical forces such as temperature, pressure, and friction, it can vary from the mildest surface discoloration to total disintegration of the metal. It is therefore possible to categorize environmental attack not only by whether it is upon ferrous or nonferrous metal, but also by the severity of that attack.

A. Staining

Staining is defined as light corrosion resulting in discoloration or tarnish. This is distinct from more general corrosion in that it is only a surface effect and is unlikely to affect the structural strength of the metal. It is undesirable mainly because it degrades the metal's appearance or because it interferes with electrical contacts in switches and sockets. An interesting aspect of staining is that it does not need a wet environment to occur, which is a common requirement in other types of corrosion. Copper or silver, for example, will discolor even in a dry atmosphere of oxygen, sulfur, or halogen to give the resulting metal oxide, sulfide or halide. The layer formed acts as a solid electrolyte with nonhydrated ions migrating through the lattice. The staining, which is the result of the solid corrosion product, can build to form a coating (or scale) that is thick enough to crack under differential thermal stress, whereupon more intensive corrosion can occur in the fissures. Staining, nevertheless, is even more prevalent in a wet environment, for not only can the agents responsible for dry corrosion still degrade the metal, but many other corrosive process are brought into play.

B. Corrosion

Corrosion is environmental attack on a metallic surface causing changes in metallurgical properties. Whereas staining is a relatively thin layer over the metal surface, corrosion is usually considered to be a more extensive attack.

Aluminum or zinc (amphoteric metals) corrode to a white powdery material, whereas copper gives a typical green product. Low-alloy steels tend to show a brown granular oxide layer if the corrosion is brought about by the effects of water and oxygen. Although a corrosion layer does offer some protection against further attack, there is not a significant reduction in the rate of atmospheric corrosion until after about 15 months following the onset. By this time, however, the degree of corrosion, especially with low-alloy steels, can be substantial.

Bacterially induced corrosion (anaerobic), which is discussed later, is black and quite different from oxidative corrosion.

C. Rusting

Rusting is the corrosion of ferrous materials, a special case resulting from the importance of ferrous materials. Of all the metals used for structural purposes, iron and steel far outweigh all others. Unfortunately, these ferrous materials are particularly prone to corrosion, especially in moist air, which in a polluted environment is also frequently acidic. Iron ore, while being a reasonably concentrated source in nature, requires heating with about four times its own weight of coal or coke to be reduced to iron. The iron is then used in diverse locations until eventually much of it reverts (corrodes or rusts) back to an oxide state similar to that which was originally mined. Unfortunately, iron oxide is no longer localized, but well-distributed throughout the land. Furthermore, the acidic pollution created during the smelting of the original ore remains. Thus, rusting is more than just a concern for the loss of outward appearance or structural failure that inevitably results. Its control also serves to reduce atmospheric pollution and retain another diminishing resource.

III. MECHANISM OF CORROSION

A. Ferrous Metals

If we think of chemical reactions at all, it may be that we are used to considering them mainly as some chemical interchange occurring in bulk, for example, a fatty acid being treated with an alkali to form a soap (saponification), or iron ore (hematite) being reduced by heating it with carbon to produce pig iron. Incidentally, the term *reduction* was originally used since there is a distinct reduction in volume observed when going from the ore (oxide) to the metal. When the metal eventually oxidizes it will regain its former natural rusty bulk; iron in the metallic form is, quite literally, unnatural.

In these and all chemical reactions there is a movement of electrons between the reacting species. However, if the reactants are separated in some way, then the reaction can only proceed as long as there is a conducting pathway so the electrons (that form the "currency" of a chemical reaction) can move between them. This then is the situation with corrosion. Electrons and ions pass between a fixed metallic surface and the environment, or between two metallic surfaces through an environment. During this process, the metal is oxidized and some part of the environment undergoes chemical reduction while gaining some energy. The propensity with which any particular metal loses one or more electrons (oxidizes) determines how likely it is to corrode. A naturally occurring passivation layer can markedly retard this corrosion. (This will be discussed later.)

Before an electric current can flow, there has to be a potential difference between two points — more free electrons at one point than the other — and a conducting pathway. In the case of a single metallic surface, this potential difference arises from small changes in local environment. Such a situation occurs when a water drop comes to rest on a ferrous surface and leaves behind a ring of brown oxide, as illustrated in Figure 8.1. Although pure water is virtually noncorrosive (nonconductive), in practice, gases (O_2, CO_2) and ions (Cl^-, CO_3^{-2}) are likely to be present to increase conductivity and hence corrosivity.

The area in the center of the water droplet is lower in oxygen content than at the rim, thus an ionic differential exists — what is known as a concentration cell. The low oxygen area is termed *anodic* and it is here that the iron loses electrons which flow into the bulk of the metal.

$$Fe \rightarrow Fe^{+2} + 2\ \text{electrons}$$

Nearer the rim of the water droplet (an area termed *cathodic*) the electrons released from the above reaction combine with the water and then reduce some of the more plentiful oxygen atoms to hydroxyl ions (OH^-).

$$\frac{1}{2}O_2 + H_2O + 2\ \text{electrons} \rightarrow 2OH^-$$

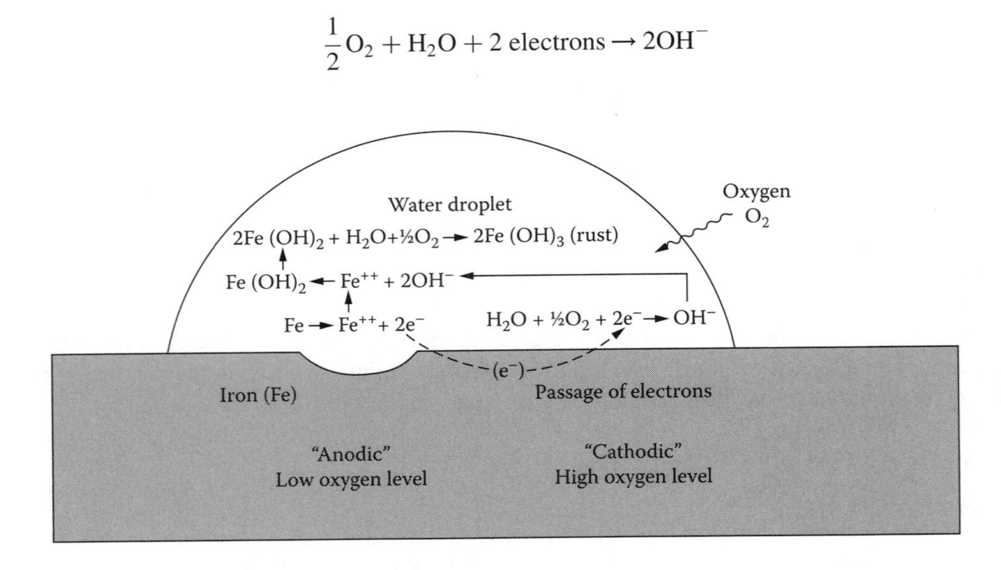

FIGURE 8.1 Rusting of an iron surface.

These hydroxyl ions readily combine with the ferrous ions (Fe^{+2}) produced in the anodic oxidation of the iron and iron hydroxide is produced.

$$Fe^{+2} + 2OH^- \rightarrow Fe(OH)_2$$

The iron hydroxide combines with more water and oxygen to form the somewhat more complex oxide we term *rust,* which is generally written as $Fe(OH)_3$ (ferric hydroxide), and can be seen as the brown granular material that rings the pit that had once been the original iron surface.

$$2Fe(OH)_2 + H_2O + \frac{1}{2}O_2 \rightarrow 2Fe(OH)_3(\text{rust})$$

B. Nonferrous Metals

Nonferrous corrosion is the reaction of any metal, other than cast iron or steel, with the environment. Only those metals that are frequently encountered in a metalworking operation will be covered.

1. Aluminum

If a piece of aluminum is cut, it forms an invisible oxide film across the freshly exposed surface almost instantly. This film is only about 0.005 μm thick, but will protect the bulk of the metal from further attack by the atmosphere. However, if the aluminum is in contact with a solution that is continuously able to dissolve away this protective layer, then clearly the attack proceeds and the metal is eventually lost. The advantage aluminum has of being corrosion resistant in air, is somewhat counterbalanced by its being attacked by both aqueous acids and alkalis. In normal use as a structural material aluminum is reasonably unlikely to encounter corrosive solutions. In a metalworking situation, however, the water-based coolant invariably has an alkaline pH, i.e., above pH 7.

The most common cause of aluminum staining with metalworking fluids is when the alkalinity (pH) of the mix is too high. High pH values will dissolve away aluminum's protective oxide layer and thus expose it to further corrosion. Very mild corrosion shows up as the alloy gaining a very pale yellow or golden appearance. This is not uncommon with synthetic fluids that are deemed suitable for use with aluminum alloys, but have been in the system for an extended time (probably more than a year or two), and by now the inhibitors are becoming depleted and the overall alkalinity of the mix has risen.

More aggressive corrosion is when the aluminum becomes stained a gray or even black color. This too is probably the result of high pH and alkalinity, just a more severe case of the golden stain mentioned above. This type of severe staining is seen, for example, when a synthetic fluid designed for ferrous metals, *but not nonferrous metals,* is filled into a machine working on aluminum parts. Such synthetics tend to have very high pH values (>9.5) since this retards corrosion on ferrous metals, but unfortunately promotes it on nonferrous metals. If, however, the aluminum is showing staining when a fluid designed for nonferrous metal is in use, then there may be several reasons. For example, additions of biocide or cleaning additives can raise the pH to a level when staining will occur. This is particularly likely with triazine-type biocides, which are not only alkaline, but often require the addition of caustic prior to their use.

Once a mix is causing aluminum to stain due to high alkalinity (pH), then it is not usually considered good practice to reduce it by adding acidic components. Never, for example, add phosphoric acid or any other type of mineral acid to a metalworking fluid in an effort to reduce the mix's alkalinity. This would almost certainly cause severe corrosion to the machine tool, and moreover, any phosphorus added to the mix can greatly promote microbial growth. In short, even a little phosphate introduced to a mix can result in a damaged machine and a stinking sump!

One method that can help with high pH-induced corrosion is to add a little tramp oil to the mix. This is not suitable for synthetics that are based on so-called cationic chemistry, because they reject tramp oil. However, a semisynthetic or soluble oil that is causing aluminum to corrode may be helped by adding a little (0.5 to 1%) tramp oil. This is certainly not an ideal method of preventing corrosion since adding tramp oil is deliberately contaminating the mix (not usually a good idea), but on occasions when drastic methods are sought, the tramp oil will tend to coat the parts and thus form a physical barrier between the mix and the aluminum. It is for this reason that soluble oils often provide good aluminum corrosion control. It is not so much the presence of inhibitors in the mix, but the fact that the emulsified product oil is able to plate out onto the aluminum and so protect it. (Of course, it plates out onto everything else too, including the machine and the work area, where its presence is not particularly beneficial.)

There are occasions when a white powdery coating appears on an aluminum surface. This is generally nothing to do with the metalworking fluid, but a result of plain water coming into contact with the metal. Zinc and magnesium present in the aluminum alloy are converted to the hydroxide form, which appears on the aluminum surface as this white powder. Generally, the powder is more prevalent on a rough-cast surface than one that has been machined. If this is seen (as it sometimes is, in say, the manufacture of aluminum automobile wheels), then suspect the cleaning process that usually follows the machining process. It is probably due to the final rinse water not being blown or dried off the casting, and it is this that causes the white staining.

2. Zinc

Zinc, like aluminum, is also attacked by both acids and alkalis, and such metals are termed *amphoteric*. Generally speaking, all that holds true for aluminum, holds true for zinc, but more so. Thus, conditions that can cause mild staining of aluminum frequently result in more severe staining of zinc. Low pH fluids are a good way to reduce staining (there are synthetics available now with pH values around 7.4), or use of a soluble oil that has a fairly coarse emulsion. The coarseness of the emulsion will ensure that the zinc component (for example, carburetor body) picks up a coating of oil that will protect from staining by the aqueous phase of the metalworking fluid.

3. Magnesium

Magnesium is even more prone to staining than zinc, simply because it is a highly reactive metal. However, because magnesium is so light, it is becoming more frequently used in automobiles and aircraft. Conventional wisdom has always been that magnesium should not be machined using water-based fluids. Straight oils should be used. This was not so much due to its propensity to staining, but because magnesium reacts with water (especially if it is hot or steam) causing the liberation of hydrogen gas. Hydrogen gas, in the presence of oxygen (air), makes an extremely explosive mixture and metal cutting can easily supply a spark as the source of ignition. There are water-based metalworking fluids available for machining magnesium, and generally they are slightly unstable oil emulsion products that plate out oil on to the magnesium surface, thus retarding any possible reaction with water. The oily layer also protects the metal from becoming stained during the machining.

4. Copper

Copper is only slowly corroded by acids and alkalis, though fatty acids present in many cutting fluids can react to form pale green soaps. However, fatty acids are unlikely to cause corrosion or discoloration of copper during the short time they are in contact while machining. Copper is discolored though, by the formation of copper sulfides due to a reaction between certain sulfur-containing compounds that can be present in cutting fluids. The problem is not insuperable however, since there are several extremely effective copper corrosion inhibitors that can be incorporated into a cutting fluid. These generally act by forming a molecular layer of an insoluble organic compound over the entire copper surface. Benzotriazole, and similar organic molecules, are widely used as copper corrosion inhibitors. Only small amounts are required to be effective, but they do deplete with time and may need to be replenished.

It is worth noting that even fairly low levels of dissolved copper (say 5 to 50 ppm) can accelerate the corrosion of aluminum being machined in the same fluid. Thus, what may first be taken as simple aluminum staining, due perhaps to elevated alkalinity, could be caused by copper ions in the mix. Adding a small amount of copper corrosion inhibitor (for example 1:10,000 of tolyl triazole), would "tie up" the copper ions and thus reduce or eliminate the aluminum staining. The source of the copper ions could well be from the aluminum alloy itself since many alloys (frequently those that begin with a 2, for example 2024 or 213) contain copper in their composition.

IV. TYPES OF CORROSION: BOTH FERROUS AND NONFERROUS

A. Uniform Attack

As discussed earlier, a drop of water on a metal surface results in areas that are either anodic (where electrons are given up) or cathodic (where electrons are used to reduce oxygen to hydroxyl ions, OH^-). If the water or other aqueous fluid is not just a drop, but a complete coating over the metal, then these anodic and cathodic areas are continually changing, resulting in a fairly uniform degree of corrosion. This may be considered to be the most common form of corrosion encountered in everyday life, the slow attrition of metallic (especially ferrous) materials. Since it is a long-term process, it is only of significance in a metalworking environment as far as the machine tool is concerned. The component is in contact with the metalworking fluid for far too short a period. Where it does impinge on the workpiece is when it is stored wet with coolant, and this is invariably a mistake since metalworking fluids (coolants) should never be thought of as long-term rust preventatives.

B. Effects of Electrolytes

Pure water ionizes only slightly to form hydrogen and hydroxyl ions (H^+, OH^-, although it is more correct to say that the H^+ goes on to recombine with a water molecule to form the hydroxonium H_3O^+). This degree of ionization is low, and as a result, pure water corrodes most metals at a very low rate. The addition of either an acid, an alkali, or a salt, greatly increases the ionic content of the water and promotes corrosion. The obvious example is seawater, which causes much faster corrosion than freshwater.

The electrolyte can play one of several roles. It can increase the electrical conductivity of the corroding fluid, thus bringing about faster dissolution of the metal. It may also react directly with the metal surface to form a soluble compound that is washed into the fluid. However, should the electrolyte react with the metal surface to form an insoluble film, which not only prevents dissolution of the metal but also further attack by the electrolyte itself, then we say the metal has become passivated and the rate of corrosion can be considerably decreased.

C. Differential Aeration

An example of differential aeration is the water drop on an iron plate, which we used earlier in describing the mechanism of corrosion. Here, areas low in oxygen are anodic to areas high in oxygen and it is in the anodic areas that the metal begins to corrode away. It is not just the simple example of the water drop where differential aeration occurs. Consider the slide way of a machine tool that had pools of oil lying over an area damp with water-based metalworking fluid. The area well under the oil film would be oxygen deficient compared with the more exposed aqueous material. As a result, it is quite possible for a differential oxygen cell to be set up and for there to be corrosion *under* the oil since that would be the anodic area.

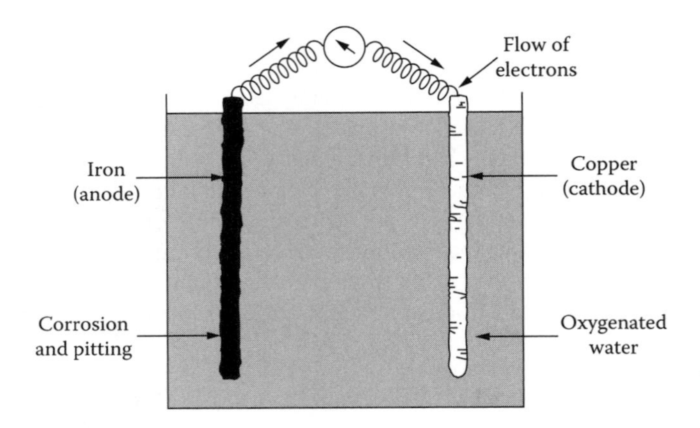

FIGURE 8.2 Bimetallic corrosion.

D. Bimetallic

If pieces of iron and copper are placed in water, it is possible to measure an electric current flowing (see Figure 8.2). In fact, a simple battery has been constructed, and in a short time it will be noticed that the iron begins to become discolored and soon pitting sets in. This is corrosion involving two metals joined together and wetted by one liquid — commonly known as bimetallic corrosion.

A typical example of this type of corrosion is where the steel rivet in a copper sheet is rapidly corroded. However, should the metals be reversed and the rivet is copper and the sheet steel, then, so long as a large section of the sheet is wetted, the corrosion occurring to the steel will be widely spread and thus hardly noticed. If it is only the area around the copper rivet that is wet, then naturally the steel in that localized part of the sheet will experience the full degree of the corrosive action and severe pitting (and a loose rivet) will result.

Since one metal will generally promote corrosion in another connected to it, it is important to know which couples are best avoided. In fact, metals can be listed to form a series, the electrochemical series. If two metals are brought in contact and wetted, then the metal that is *higher up the list* (more electropositive) will corrode in preference to the other. The metal is more electropositive, after all, since it is more inclined to lose electrons (which are negative). Referring back to the diagram of the water drop will make it clear that the pitting of the metal occurs where the metal loses its electrons.

Thus, in Table 8.1 when any two of the metals are in contact, and wetted, that which is above the other on the list will tend to corrode. Since pure metals are not generally used in practice, it makes more sense to include common alloys in the list as well.

E. Erosion Corrosion

Whereas erosion is just the mechanical wearing away of a surface by a fast-moving stream of fluid, erosion corrosion has the added dimension that the abraded material is not usually the metal but the protective film. Many metals and alloys develop a protective film, frequently an oxide layer, which when abraded away reforms from the parent metal below. Gradually therefore, the metal is lost if the film is continuously being worn away. Materials such as stainless steel, which depend heavily on a protective layer to maintain good corrosion resistance, are particularly vulnerable to this type of corrosion. The attack causes the formation of characteristic smooth grooves and holes in the surface of the metal. Laboratory corrosion testing of a metal under *static* conditions will not show this form of corrosion if the material is ultimately for use in an environment where, although it will be subjected to the same chemicals, it will be under dynamic conditions. Erosion corrosion is particularly important in pipe work and in pumping equipment.

TABLE 8.1
Electrochemical Series for Selected Metals

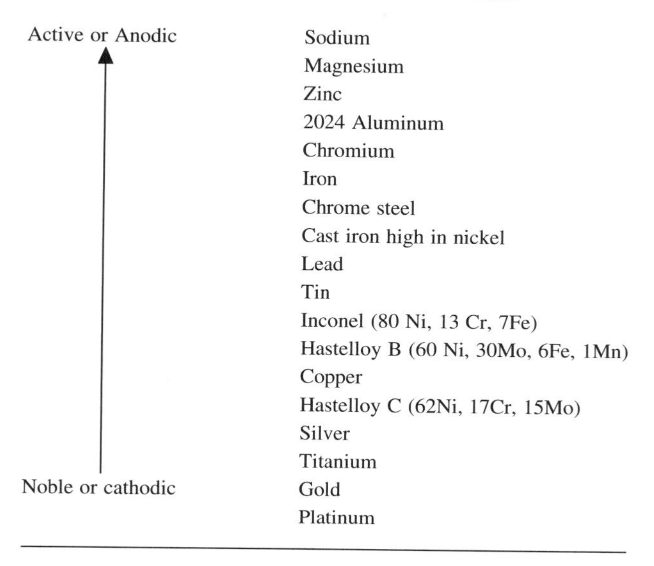

Active or Anodic ↑	Sodium
	Magnesium
	Zinc
	2024 Aluminum
	Chromium
	Iron
	Chrome steel
	Cast iron high in nickel
	Lead
	Tin
	Inconel (80 Ni, 13 Cr, 7Fe)
	Hastelloy B (60 Ni, 30Mo, 6Fe, 1Mn)
	Copper
	Hastelloy C (62Ni, 17Cr, 15Mo)
	Silver
	Titanium
Noble or cathodic	Gold
	Platinum

F. Pitting

Pitting, as the name suggests, is the localized attack of a metal surface that leads to small holes. The pit occurs when the corrosive site (the anodic area) remains small and in one spot. As discussed earlier, whenever there is a local difference on the metal surface, due to differences in relative oxygen concentrations, or indeed an impurity in the metal, or more or less any morphological difference, then a potential gradient will result between these two areas. This difference in electrical potential means that one area is anodic compared with the other (cathodic) area, thus a galvanic cell is set up and corrosion can begin. A typical example is when there is a slight scratch in the protective (oxide) film of a metal. The minute area of metal exposed becomes anodic to the (comparatively) enormous area of cathodic oxide surface, which is more noble, and as a result corrosion occurs in the scratch, forming a pit. It would appear that corrosion products resulting from the reaction prevent the metal in the scratch from reforming the protective film and thus stopping the corrosion. Chlorides (and the other halides) are particularly prone to causing pitting corrosion, especially in stainless steels.

G. Fretting Corrosion

Fretting corrosion is similar to erosion corrosion, except that in this case the mechanism by which the oxide or other protective layer is worn away is by direct contact of another solid material not the flow of a fluid. There has to be relative movement between the two surfaces, typically vibration, in order for fretting corrosion to occur.

H. Intergranular Corrosion

Although metals appear to the naked eye to be uniform, they consist, in fact, of grains of metal with boundaries between them. These are the intergranular boundaries. Should corrosion occur preferentially along the boundaries then the metal is weakened and can fail in service. Even though the boundaries are considered to be anodic (prone to corrode) compared with the grains proper, this fortunately does not result in significant corrosion. If it did, then certainly, metals could never have been used for structural purposes, and technology would have remained in the era of wooden wheels and ships! However, this insignificant corrosion between the grains can result in a real problem if the nature of the boundary is changed to make the interface considerably different from the parent grains. This can occur in austenitic steels (a solid solution of one or more elements in face-centered cubic iron) when they are heated to between 1000 and 1400°F. Under this condition the carbon in the steel tends to migrate to the boundaries, react preferentially with the chromium in the alloy, and precipitate out of solid solution. In effect then, the steel in the boundary area is considerably lower in chromium than it is in the interior of the grain. This is a situation where the boundary is much less noble (more anodic) than the grain and causes rapid corrosion. There are various ways to prevent this from occurring, for example, by ensuring that the carbon content of the steel is particularly low ($<0.02\%$) or by adding small amounts of exotic metals (columbium or tantalum) which react more strongly with the carbon than the chromium.

High temperatures present during welding can unwittingly cause this type of heat treatment and thus intergranular corrosion, which could lead one to believe that a weld had failed, whereas in fact this more involved process had occurred.

I. Stress Corrosion

There are two stages in stress corrosion: the initiation of a microfissure (crack), and then the propagation of the crack. The initiation can result from many causes, for example pitting corrosion or intergranular corrosion, whereupon a small, sharp, anodic area is surrounded by a large cathodic area. Under stressed conditions this could be enough to cause the crack to propagate since it has been shown that there is a point of maximum stress just ahead of the point of the crack into which the crack moves. Of course, this stresses area continues to move ahead with the crack following until either the metal parts, a soft area, or a hole is reached wherein the energy of crack propagation can be relieved. However, with stress corrosion there is good evidence that the propagation of the crack is accompanied by a corrosive action actually occurring within the crack.

Situations where stress corrosion is common are in parts that have been welded and not stress relieved, in heat exchangers where there can be a buildup of corrosive deposits, or in metals that show a strong chemical susceptibility to a particular material that was used when cutting the metal. Examples of the latter are brass cut with a coolant high in amines, or aluminum or titanium cut with a coolant high in chloride ions. Owing to the intermetallic nature of the corrosion, it is not usually possible to wash off the offending material with any degree of certainty.

One point worth noting is that sodium ions can cause stress corrosion cracking in nickel alloys. Thus, even if the metalworking fluid is free from chloride ions, but contains sodium ions from, say, a sodium sulfonate emulsifier package or the sodium salt of a copper corrosion inhibitor, nickel coupons undergoing stress corrosion tests may fail.

J. Bacterially Induced Corrosion

So far we have covered chemical corrosion, or more precisely, electrochemical corrosion, and although we have chosen to differentiate between, say, bimetallic and intergranular corrosion, when it comes down to it the process is essentially the same: an electric current flowing between two dissimilar regions eating away the surface that gives up its electrons most easily. Bacterial corrosion, however, is substantially different; to start with, it is biochemical in nature. A biochemical reaction, just as any other chemical reaction, depends on the movement of electrons, except in this case the initiating step is biological in nature.

In metalworking fluids mainly composed of water, there are two groups of bacteria that can flourish: first, those that require an oxygenated environment, termed *aerobic* bacteria (from the Greek *aeros* meaning air), and second, those that proliferate in the absence of oxygen, termed *anaerobic* bacteria.

1. Aerobic Corrosion

A water-mix cutting fluid provides a reasonably favorable environment for culturing bacteria. The warmth, water, dissolved or emulsified organic materials (oils, corrosion inhibitors — always a good source of nitrogen), and areas of high and low oxygenation, all encourage rapid bacterial (and mold) growth unless some material that retards microbial action is incorporated into the mix. Such antimicrobial agents are termed *biocides*.

Aerobic bacteria can influence corrosion in a number of ways, some indirectly, other directly. The most obvious way bacteria can affect the corrosion control of a metalworking fluid is simply to metabolize (destroy) the chemicals that were originally included in the mix to confer corrosion protection on machines and workpieces during usage. These anticorrosive agents are essential if rusting is to be prevented since the metalworking fluid may easily be 95% water. Moreover, anticorrosive agents are frequently rich in nitrogen, which is a prime source of energy for the majority of bacteria.

Whenever aerobic bacteria attack a water-based metalworking fluid, a set of clearly defined steps occur. Initially, it appears that the microbes break down the long chain molecules, which make up the lubricants, emulsifiers, and anticorrosive agents, into shorter sections. These shorter molecular sections are rapidly oxidized utilizing the oxygen dissolved in the mix. This utilization of the dissolved oxygen (DO) can easily be monitored using a DO meter. At room temperature, water (metalworking fluid), which is microbially uncontaminated, will have about 9 ppm of dissolved oxygen in it. (This is the oxygen that fish rely on to "breath.") However, once microbes begin to multiply and break down the mix, this DO level can quickly drop to 2 ppm or less. The majority of the oxygen has not been "used up" by the bacteria, but rather, has combined with the cutting fluid components that the bacteria are busy destroying. The oxygenated fragments of the metalworking fluid are, in fact, now short-chain acids, which have a sour odor and are volatile, thus making the fluid smell rancid. Moreover, being acidic, the pH of the mix begins to fall. Thus, by regularly monitoring the DO level of a mix, it is possible to prevent its bacterial destruction by noting when the DO begins to fall and adding a remedial amount of biocide before too much damage is done. In many ways this has the advantage over using dip-slides since a DO reading takes only 3 to 5 min, whereas the incubation time for a dip-slide is anywhere from 2 to 5 days, which can be far longer than the time needed for bacteria to cause rancidity in a metalworking fluid.

Adding biocide to a rancid fluid is nearly always too much too late, for now the damage has been done, and killing the bacteria will not restore the fluid components that have been destroyed. A program of continual monitoring and proper fluid maintenance are the keys to good fluid performance and corrosion control.

It is interesting to note that a jar of plain water inoculated with 100 million bacteria per milliliter (10^8 cfu/ml) has virtually no odor at all. However, add a little corrosion inhibitor or emulsifier and within 24 h the pH and the DO will have fallen and the contents of the jar will smell rancid. Bacteria have little smell; it is the damage they do to the product components that is responsible for the foul odors and subsequent corrosion.

2. Anaerobic Corrosion

The other type of bacteria which can cause corrosion was originally investigated early in the 20th century by two Dutchmen, Von Wolzogen Huhr and Van der Vlugt, while studying the corrosion of buried pipes in the polder regions of Holland. They noticed that black staining (iron sulfide) occurred not only as an adherent corrosion product on the pipes themselves, but also in the soil in the vicinity of the corroded pipes. From this they ultimately deduced the presence of sulfate-reducing (or anaerobic) bacteria.

Many microbes are able to reduce small amounts of sulfate for the synthesis of sulfur-containing substances, however, comparatively few are able to utilize sulfate reduction as their major energy-producing activity. By far the majority of living organisms derive most of their energy — to go on

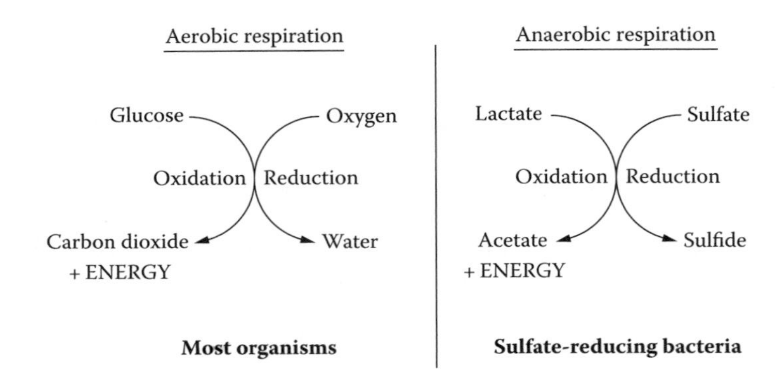

FIGURE 8.3 Bacterial respiration.

living — by oxidizing sugars, particularly glucose, and this can only be done if oxygen is taken in at the same time and reduced to water.

Not so with sulfate-reducing (anaerobic) bacteria. They do not use sugars, but a simpler type of chemical known as a lactate, which they oxidize to acetate, thus deriving the energy to live. However, instead of using oxygen (like most other organisms) to carry out their main energy-producing process, they reduce sulfates to sulfides (see Figure 8.3). Unfortunately for metalworking fluids, sulfides (especially iron sulfide) are black in color and smell sulfurous. Although there are comparatively few types of bacteria that reduce sulfate on a large scale, they are, unfortunately, very widely distributed, for example, in fine metallic swarf (especially cast iron) in a machine tool or system. With oxygen present there will be no problem from these sulfate-reducing bacteria — no discoloration or foul smells and no corrosion — but the bacteria will not die. Even a few parts per million of DO will prevent anaerobes from feeding and multiplying — which is about all bacteria can do with the energy they obtain — and thus they will not be noticed. However, other bacteria that require oxygen to live (aerobes) can rapidly deoxygenate static areas and inadvertently provide the necessary conditions needed by the anaerobes.

Where does the sulfate necessary for anaerobic life come from in a cutting fluid? First, from the emulsifiers. Many widely used emulsifying agents are based on sulfated and sulfonated long-chain molecules that are able to provide an excellent energy source. However, even without this source, many metals (and most mineral oils) have sufficient sulfur or sulfur-containing impurities to supply the bacteria.

Furthermore, water invariably contains a certain amount of calcium and magnesium sulfate dissolved out of the gypsum and other rocks that it percolates through before being held in a reservoir. The anaerobic bacteria, of course, are ubiquitous and are present in water, on rags, in dandruff. These bacteria wait in a dormant state for the oxygen level to decrease to allow them to flourish.

Thus, it appears that with or without oxygen, bacteria can cause corrosion. The problem is generally seen as rust if it is due to aerobic bacteria that have created conditions likely to cause corrosion, or as black staining if the culprit is the anaerobic type of bacteria which has actually taken some sulfur from the metal.

V. CORROSION PREVENTION METHODS

A. Inhibitors

1. Passivators

Look back to Table 8.1, listed in decreasing order of electropositivity — where the metals higher up are more likely to corrode than those below — it will be noted that chromium is, in fact, above iron.

Thus, it would be expected that when chromium is added to iron, it would render the resultant alloy *less* corrosion resistant; however, this is clearly not the case, since chrome steel is prized for its corrosion resistance. Indeed, the iron is now stainless steel. Plain and simple iron, or carbon steel, will react with damp air to form a surface oxide layer that is porous, thus more water and oxygen can penetrate through to the virgin metal underneath and continue the corrosive process. Chromium, on the other hand, also forms an oxide film, but it is not porous; it is impervious to further penetration and so the initial incredibly thin (transparent) oxide layer acts as a self-sealing barrier to continued oxidative corrosion. The alloy is unreactive to moist air; it has become passivated. Aluminum (even higher up the table) exhibits the same property of self-passivation, and even high silicon cast iron can form a film of protective silica (SiO_2). Sodium, however, does not form a passive layer. Indeed, a piece of sodium metal left in moist air is so reactive that it will probably burst into flames. Clearly, the nature of the film or corrosion product that forms on a metal is much more important than its relative position on the electrochemical series. Although the most likely film to form on a metal naturally is the oxide, the same principle of a thin protective film formed *from the substrate metal itself* applies to other passivating agents. For example, iron becomes passivated when immersed in solutions of nitrites (NO_2^-), chromates (CrO_4^{2-}), molybdates (MoO_4^{2-}), tungstates (WO_4^{2-}), or pertechnetates (TcO_4^-). In fact, iron can even be temporarily passivated by dipping into concentrated nitric acid solution. However, none of these methods are applicable to metalworking fluids (with the possible exception of molybdates) owing to toxicity considerations, even though nitrites were used extensively in the past. Thus, there are few, if any, options open to formulators of metalworking fluids if they are seeking to use oxidizing agents as a means of passivating a workpiece and thus preventing corrosion.

Sodium silicates have been used in metalworking fluids to reduce staining of aluminum alloys. Silicates have a very high pH (typically a 1% solution will be above pH 11), but nevertheless can be used as aluminum corrosion inhibitors. The disadvantage appears to be that a minimum level of silicate needs to be maintained (this level depends on the formulation of the mix as a whole), because if the amount falls below this threshold then the aluminum will stain far more than if the silicate had been completely absent. Moreover, many products (particularly semisynthetics) do not easily form stable concentrates if sodium silicate is included in the formulation.

2. Organic Film Formers

This method of corrosion control is much more akin to painting the metal surface. As shown in Figure 8.4, fatty acids and similarly configured molecules have a long water-repelling hydrocarbon "tail" and a "head" that has a strong affinity for the metal surface.

The long, thin molecules line up roughly parallel to each other and perpendicularly to the metal surface, forming a fatty monolayer that is essentially impervious to water and oxygen. This effect

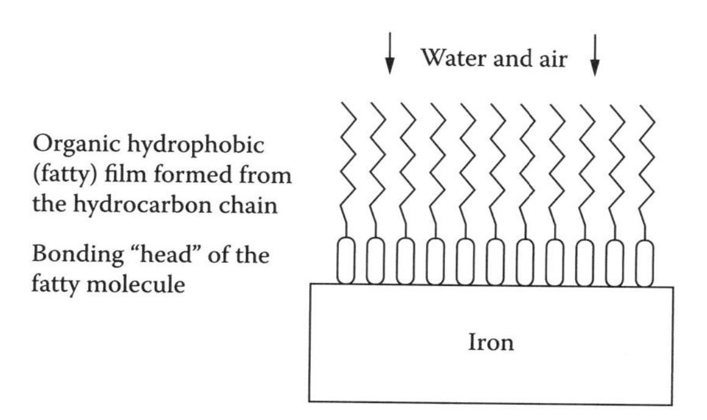

FIGURE 8.4 Organic corrosion inhibitor film formation.

can be easily demonstrated by noticing how a steel wool scouring pad remains rust free so long as there is still soap impregnated in it, soaps being simple salts of fatty acids. However, once the soap is depleted, the steel wool rapidly rusts away. Although these fatty molecules are excellent corrosion inhibitors for use in water-based metalworking fluids, they are not really tenuous enough to provide long-term corrosion protection during, say, storage. They were never designed with that in mind, and since it follows that any substance that can be deposited from an aqueous solution can almost certainly be washed away just as easily by water, long-term corrosion protection of machined parts usually requires an oil-based anticorrosive coating to be applied. However, the machine tool will be protected by the anticorrosion agents in a metalworking fluid, as long as it is kept at a sufficient concentration and in good condition.

B. Inhibitors in Microbially Induced Corrosion

If microbes are causing corrosion, then the remedy is to destroy the microbes. This can be done in several ways, depending on the severity of the problem.

If the metalworking fluid has had a long history of bacterial odors and the associated corrosion problems, then there is little that can be done except to throw away the fluid, clean the machine, and refill. To prevent the problem from reoccurring, several steps should be addressed, not least of which is ensuring that the fluid is maintained at the correct concentration. A mix that is too lean will soon fall prey to microbial contamination. Pollution of the mix with waste materials is another common cause of bacterial problems. If the rise in bacterial population is a new occurrence, then simply dosing the system with the correct biocide may be sufficient. However, a little detective work on why the problem occurred in the first place is important to prevent its reoccurrence.

With anaerobic contamination the problems of foul odors and black staining can frequently be relieved by oxygenating the mix, simply by keeping it *continuously* circulating, and checking to make sure there are no stagnant areas or heavy deposits of sludge in the base of the tank or system. These latter places are ideal breeding grounds for anaerobic bacteria, especially if the sludge is from cast iron or other low-grade ferrous material rich in sulfur. Removal of the sludge is essential. In short, more bacterial problems and the associated corrosion can be avoided if the mix concentration is held at the recommended level, the fluid is continuously circulated, stagnant areas are designed out of the system, and finally, the filtration and drag out equipment is working efficiently.

VI. CORROSION TESTING METHODS FOR METALWORKING FLUIDS

Before filling an expensive machine or central system with an unfamiliar metalworking fluid, it is advisable for engineers to ask for data from the supplier regarding the ability of the fluid to prevent corrosion as the final mix might consist of up to 98% water filled into what is essentially a cast iron structure. Many users do their own testing and may even have evolved their own test methods. However, unless care is taken, testing a water-based metalworking fluid to evaluate its anticorrosive properties is so fraught with problems that the customer may end up missing a good in-use product or selecting something that turns out to be unsuitable. Consider the following test results.

A semisynthetic metalworking fluid was tested to determine its break point using five different (but widely used) ferrous corrosion tests. The values shown below are corrosion break points for the fluid, or the minimum concentration in water that will prevent corrosion, according to that particular test. The test material in each case was cast iron.

Test method	**A**	**B**	**C**	**D**	**E**
Break point (%)	2.5	4.5	>5.0	1.25	5.0

Since the recommended use concentration of the fluid was 3%, and field trials showed that it did not cause corrosion at this level, it is probable that people using only test methods B, C, or E would reject the product as having poor corrosion control. Thus, slavish adherence to a corrosion test method will neither guarantee that the best fluid is finally selected nor that it will be used at the most economic dilution. However, it is equally obvious that no one can risk an unknown fluid in an expensive machine without at least some testing. It is as well therefore to have some knowledge of the various test methods so at least an informed interpretation of their results can be made. The three metal types that are usually involved in corrosion testing are:

Ferrous. Cast iron and steels
Aluminum alloys
Copper alloys

Testing against ferrous alloys is by far the most common.

A. Ferrous Metals

Ferrous corrosion tests generally originate from one of three sources:

Customer (end user)
Manufacturer (or supplier)
National or international testing organization (who tend to establish standardized tests)

Generally, customers' tests will be a variation on one or more of the standardized tests. Although metalworking fluid manufacturers will use these standard tests as well, they frequently have a number of self-devised methods to highlight factors they consider important, especially during the development of new products.

Any test method chosen *must* be cross-checked against a machine trial, preferably in a typical, yet noncritical, machine, using available water.

1. Chip Test

a. *Steel Chips on a Cast Iron Plate*

This method is the basis of the Herbert test (U.K.), The Institute of Petroleum IP 125 (U.K.), and the DIN 51360 part 1 (Germany).

The Herbert test was the forerunner, but the specifications are so imprecise that its value is somewhat questionable. Basically, with all these tests, a cast iron plate is cleaned and polished up with a fine abrasive paper. Four small piles of clean steel chips are positioned on the plate and are then wetted with the test mix. The four piles are treated with four different dilutions of the same mix, or conversely, with four different products. The plate and its chips are placed in a closed container for 24 h, after which the chips are removed and the degree of corrosion on the plate examined.

The DIN 51360 part 1 gives highly specific instructions, even to the point of specifying the level of salts in the water used to make the mixes. Therefore, it alone can be considered as an absolute test allowing comparison between different testers and localities.

b. *Cast Iron Chips on a Steel Plate*

This test method is analogous to that described above, but now the plate is steel and the chips are cast iron. The test procedure is essentially the same and is used as a standard in France (CNOMO) and is widely accepted in Italy.

c. *Cast Iron Chips on Filter Paper*

This test has gained wide acceptance since it can be quick (as little as 2 h), is reasonably reproducible, is simple (there is no metal plate to clean and polish), and the final paper (test result) can be fixed into a notebook for future reference. Typically, about 2 g of clean (dry cut) cast iron chips are spread onto a filter paper in a Petri dish. The diluted mix is pipetted onto the chips and the dish covered. After a set period of time the mix and chips are removed and the paper examined for staining; if there is any it is usually graded depending on its severity. Test methods that use this general technique are the IP 287 (U.K.), DIN 51360 part 2 (Germany), and ASTM D4627 (U.S.). A variation of this test is to soak the chips in the metalworking fluid first, and then drain the mix and pile the chips onto the paper. The damp chips then cause staining of the paper if the corrosion control of the fluid is insufficient.

2. Flat Surface Tests

a. *Open Cast Iron Cylinders*

Small cast iron cylinders, about 1 in. in diameter and 2 in. long, are ground flat and then lapped to a polished finish. These are stood in a 100% humidity cabinet and a film of the metalworking mix is pipetted onto the virgin surface of the metal. The cabinet is closed and usually left overnight. The next day the cast iron surface is examined for signs of corrosion. Generally, a series dilution of the produce will be used to determine the break point for that particular metalworking fluid.

b. *Stacked Steel Cylinders*

This test is analogous to that described above, except that steel cylinders are used and after the mix has been placed on the top surface of the steel, a second steel cylinder is mounted on top of the first. The majority of the mix is squeezed out, but a thin film remains between the disks. The 100% humidity cabinet is closed and left overnight. The next day the top cylinder is removed and the resulting corrosion of the steel (if any) examined and rated. Various finished steel parts, such as bearing races, may also be used.

3. Panel in Closed Cabinet

Although not normally applied to metalworking fluids used for cutting and grinding, there are special tests required for fluids used for stamping and drawing products. These tests take the form of dipping the component part in the tests fluid, allowing it to drain and then hanging or clamping the part in an environmentally controlled cabinet. The reason for these tests is that parts stamped or formed in some way from sheet metal are often placed in bins after being cut from the roll of metal and may wait a considerable time before being used. Thus, it is particularly important that the fluid used in the metal fabrication process leaves a coherent film of corrosion inhibitor over the metal surface. Typical parts would be automotive panels cut and formed from mild steel. Any subsequent corrosion to these panels would involve either scrapping them or expensive cleaning processes.

a. *Humidity Cabinet*

The simplest panel test involves a closed cabinet where the humidity is maintained at or near moisture saturation levels and at an elevated temperature. Often the heating circuits in the apparatus will be programmed to cycle through heating and cooling periods so that from time to time the moisture in the enclosed air has a chance to condense out onto the component. This falling below the dew point clearly mimics conditions that can occur with components stored in factory areas where the temperature varies throughout the day. Incidentally, the air in factories that use metalworking fluids in significant amounts is generally saturated with water vapor.

b. *Acid and Salt Atmospheres*

Since the damp air in factory environments is frequently contaminated by acid materials (exhaust from furnaces and heat treatment facilities), components can be subjected to dilute acid droplets condensing out on them. To ensure that the metalworking fluids used in stamping and drawing leave behind sufficient rust protection under these harsh conditions, humidity testing in an acid atmosphere is carried out. It is very similar to ordinary humidity testing, but with the added challenge of acid vapors in the chamber. Salts can also be introduced into the test atmosphere as a further variation of this test.

4. Stress Corrosion Tests

In-service failure of aerospace components generally has catastrophic consequences. Therefore, metalworking fluids used in these fields are checked to ensure that they will not induce any form of point corrosion that could lead to propagation, or initiation, of a crack through the component. A typical material that has been linked to metal cracking is chlorine, when present in the fluids used to work titanium. Since cracking can occur months or even years after exposure to the causative agent, more subtle testing is required than in simple ferrous corrosion testing.

Generally, the test for stress corrosion consists of bending a coupon, cut from the subject metal, into a U-form and holding it in a clamp. The metal, which is under considerable stress, is then soaked for a short time in the metalworking fluid under investigation. The metal, still held in the clamp, is then subjected to high temperatures in a special oven.

After removal from the oven, the component is etched and polished, and examined for metallurgical defects. Using this technique, it is possible to screen out those metalworking fluids that could cause stress cracking in particular metals.

B. Nonferrous Metals

Corrosion testing of nonferrous metals is usually a simple matter of taking a coupon of the metal in question, cleaning and polishing it, and then partially immersing it in the test fluid. After a time (typically 24 h) the component is removed and examined for discoloration, beneath, on, and above the fluid line.

Another test procedure, which is in essence similar, is to half fill a small clear glass bottle with nonferrous turnings, then measure in enough test fluid to half submerge the pile. The bottle is stoppered and left undisturbed. The turnings are checked, especially at the fluid–air interface, every day for corrosion or discoloration. Typically, such a test would be left to run for 5 to 7 days.

Even if no corrosion is observed, it is good practice to filter the test fluid and to measure the concentration of dissolved metallic ions in it. A metalworking fluid suitable for nonferrous metal should show little or no corrosion of the material and should not dissolve more than a few parts per million of it over a period of about a week at room temperature.

C. Multimetal Sandwich

Many of the components used in aircraft components are fabricated from several metals, and it is important to know whether during the machining of these multimetal components there could be corrosive interaction brought about by the metalworking fluid. A test for this relies on taking pieces of the various metals and clamping them together and bringing them in contact with the metalworking fluid. Some tests simply require that the pieces are clamped and then soaked (probably at an elevated temperature) in the fluid. Other tests are more severe and involve clamping the metals and then drilling and possibly reaming through them with the test fluid.

D. Galvanic Test Methods

Since corrosion is always accompanied by the passage of electrical current, considerable efforts have been made to try to utilize this phenomenon as a means of testing for corrosion. The method differs from those so far discussed because in conventional stain or rust tests the degree of corrosion observed on the test piece is taken as a direct measure of what is likely to occur in real usage. Thus, if a particular cutting fluid causes significant rusting of cast iron test pieces or even chips on a filter paper, then one is naturally hesitant about filling an expensive machine tool with the product, since the machine tool is made largely from just an alloy of iron. However, galvanic and similar electrochemical tests try to *predict* possible corrosion problems by making a voltaic cell out of two (usually dissimilar) pieces of metal and measuring the current that flows between them when immersed in the test fluid. Once again we come back to the idea of a corrosion cell being a battery, except here we are trying to relate the magnitude of the current generated to future possible corrosion problems. There are numerous variations on this theme, such as first joining the two coupons together and then soaking them in the fluid, before separating them, and then measuring currents produced by the now corroded components. However, no matter how much use one makes of these corrosion tests, they still have to be related to field results to be really useful.

E. Other Electrochemical Techniques

More involved electrochemical techniques than that discussed under galvanic test methods are potentiostatic polarization and AC impedance spectroscopy. (Figure 8.5 shows a potentiostatic polarization plot of a single fluid, and Figure 8.6 compares two different fluids using the same technique.)

Potentiostatic polarization is where a known DC potential is applied between a metal electrode and a standard (for example, a calomel) electrode. The electrolyte, for our purposes, would be the metalworking fluid under test. The magnitude of the current that flows is dependent on the applied potential and the corrosion-inhibiting characteristic of the metalworking fluid. By applying a potential that increases from typically −2 V (a high cathodic potential), through zero to an anodic potential of typically +2 V, and measuring the absolute current flow, it is possible to distinguish at least four distinct electrochemical regions. These are (going from cathodic to anodic potentials) reduction of water, reduction of oxygen, passive region, and oxidation of water. The passive region is the area that shows little or no change in current flow for an increase in applied potential.

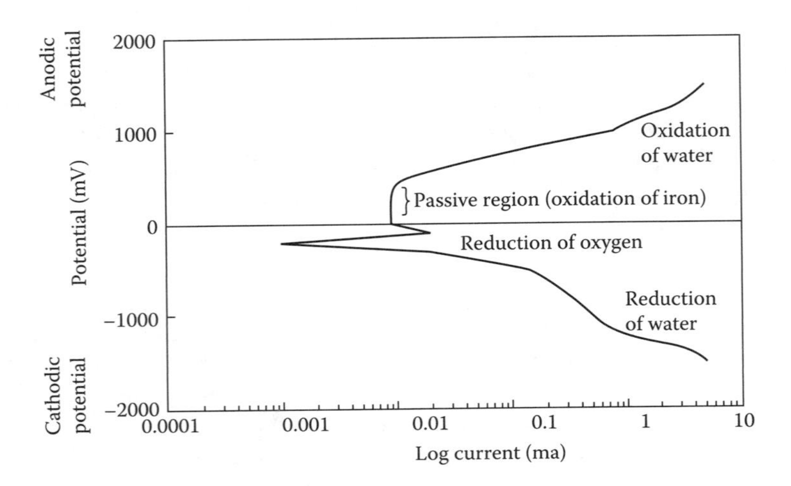

FIGURE 8.5 Typical potentiostatic polarization plot.

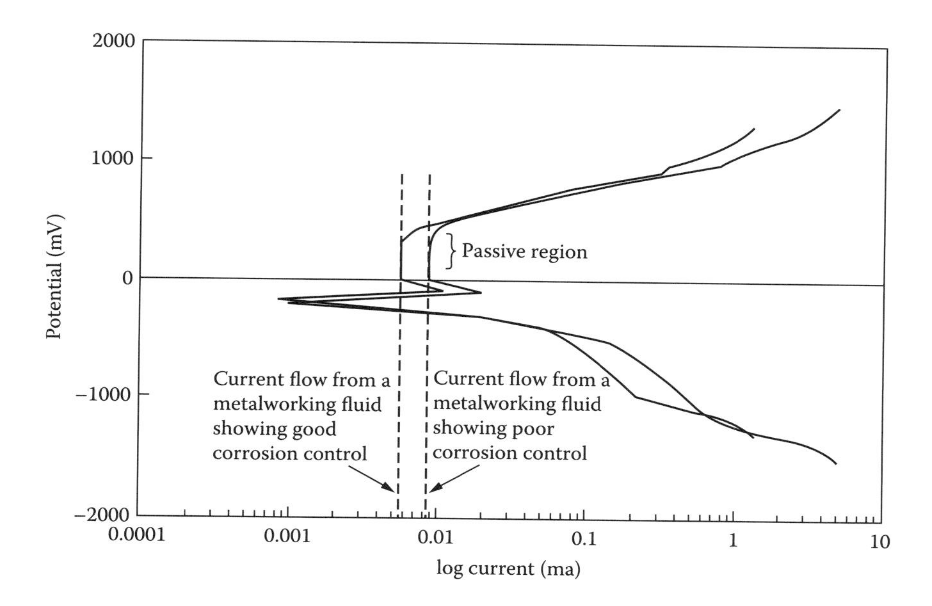

FIGURE 8.6 Potentiostatic comparison of two fluids.

A metalworking fluid that offers good corrosion inhibition will typically have a significantly lower current flow in this passive region than a fluid with poor anticorrosive properties.

AC impedance is a similar technique, but uses an alternating current (AC) whose frequency gradually changes. The real and imaginary components of the impedance of the system are determined and modeled as if they were part of an RC (resistance/capacitance) circuit. This method is preferred in systems where the metal is particularly well-inhibited, for example, if the metal is painted.

VII. PRACTICAL STEPS TO PREVENT CORROSION WHEN USING METALWORKING FLUIDS

A. Choice of Metalworking Fluid

Clearly, this is the starting point for solving (or avoiding) corrosion problems with metalworking fluids. When considering ferrous corrosion control (rust), there comes a time when further dilution of the product will no longer provide sufficient inhibitor to prevent corrosion. That, after all, is how we generally test these products; how much can we dilute them until the test pieces rust? However, there is another aspect these break point tests do not show. Most ferrous corrosion inhibitors are based on nitrogen, and this is an energy source utilized by bacteria that may grow in the diluted mix. Thus, even if a particular product shows good break point test results, if it is likely to become microbially infected quickly, the bacteria will rapidly deplete the inhibitors and leave behind corrosion by-products. A product with good break points *and* good microbial control is essential.

B. Water Quality

As stated earlier, dissolved ions can greatly increase the corrosivity of an aqueous solution by either interacting directly as a corrosive agent or by simply increasing the electrical conductivity of the fluid. Thus, if water containing high chloride or sulfate levels is used as mix water for a metalworking fluid, a degradation of the corrosion control of that product will surely occur. What actually constitutes high ionic levels depends on the nature of the fluid and the metals it comes into contact with; however, chloride levels higher than 100 ppm and sulfate levels higher than 200 ppm should, if possible, be avoided. As water is lost through evaporation from the machine tool sump or central system, these ions will naturally concentrate in the metalworking fluid mix and could then cause corrosion. Unfortunately, there is no really practical way of removing these ions from the mix, and so in areas where dissolved ions in the water are high two solutions are possible: either use dionized water for makeup or, if this is not possible, increase the mix concentration to combat their corrosive effects. Tankside additives may also be considered (as discussed in Section VII. E).

C. Dilution Control

The majority of problems experienced with metalworking fluids can be traced back to problems of not holding the mix at the recommended concentration level. A mix that is excellent at 5% may give considerable problems at 3%, and surely will at lower strengths. Not only can the corrosion inhibitors be over-diluted, but microbial damage (as discussed earlier) can set in. Holding the concentration too high will still provide good ferrous corrosion control, but it is uneconomic, it may lead to skin irritation problems, and the associated rise in alkalinity could stain nonferrous materials.

D. Treatment of Metalworking Fluids when in Use

If rust is experienced when using a water-mix cutting and grinding fluid, do not immediately blame the fluid. First investigate the situation. Where is it occurring? When is it occurring? Are the production pieces or just the machine showing rust flecks? There are numerous questions that have to be asked. If the corrosion is occurring on areas of the machines well away from where cutting fluid is being used, then it would appear to be water vapor in the air that is the problem; increasing mix concentration would not be an answer. Workpieces can suffer a similar fate if they are taken from the machine and stacked on cardboard or on any other absorbent material. The metalworking fluid (with its associated inhibitors) is mopped up by the absorbent material, which then acts like a wick giving back high water vapor levels into the surrounding air. The water vapor is devoid of inhibitors and causes the components to corrode.

If the problem is one of insufficient corrosion control because of over-dilution of the mix, then clearly the remedy is to add more concentrate. Adding more concentrate if the mix is microbially contaminated is only curing half the problem. The addition of biocide to the mix should be considered first.

E. Tankside Additives

Additives to prevent and cure problems largely take the form of just the corrosion inhibitor package from a metalworking fluid, or a biocide to keep the microbial count low. While these are useful (especially the biocide), it may often be more useful to add more concentrate than just the corrosion inhibitor. Remember, many metalworking fluid problems are concentration related and adding bits and pieces makes determining the realconcentration of the mix very difficult. Not all corrosion inhibitors will increase the apparent concentration of a cutting fluid, but some will. Biocides generally do not affect concentration measurements.

9 Microbiology of Metalworking Fluids

Frederick J. Passman

CONTENTS

I. INTRODUCTION

A. Relevance of MWF Microbiology

The microbial world is remarkably diverse, including organisms from three domains, the bacteria, archaea, and eukaryotes. Only an estimated 0.001%[1] of all microbes have been isolated and identified. A major challenge is that only a small fraction of the known species can be cultivated on growth media. Fortunately, advances in molecular biology over the past two decades have advanced our understanding of the existence and ecology of many nonculturable microbes. To date, the newer methodologies have rarely been used to advance our understanding of metalworking fluid (MWF) microbiology.

Despite the limitations of the methods that have been used to investigate microbes in the metalworking environment, the empirical evidence of successful contamination control in well-maintained systems suggests that the traditional microbiological methods have served us reasonably well.

B. Biodeterioration and Health Risks

In the metalworking environment, microbes are primarily agents of biodeterioration — they degrade the commercial value of coolants, tools, finished parts, and fluid systems. The annual net adverse economic effect of uncontrolled microbial growth can be valued in the tens of millions of dollars. This estimate reflects the costs associated with:

- Coolant disposal and replacement
- System clean-out labor
- Waste disposal costs
- Increased part rejection/failure rate
- Decreased tool life
- Lost productivity

This substantial economic exposure should provide any MWF system stakeholder with sufficient motivation to understand the fundamentals of MWF microbiology.

Over the past 15 years, there has been an accumulating body of evidence demonstrating the adverse health effects of microbes on the health of exposed workers. Although the risks of serious infectious disease (think cholera) or communicable disease transmittal remain vanishingly small, other diseases, such as allergies and toxemias may affect a significant proportion of the MWF industry worker population.

C. Waste Treatment

Notwithstanding the problems associated with uncontrolled microbial growth in MWF systems, microbes play an essential role in waste treatment. Bacteria (and perhaps archaea) are the primary agents that reduce biochemical oxygen demand (BOD) and chemical oxygen demand (COD) inside waste digesters. Moreover, bioremediation — the use of microorganisms to detoxify and clean up spills — relies on the same microbial processes as biodeterioration. Used appropriately, microbial communities enable industry stakeholders to comply with waste minimization regulations in a cost-effective manner.

D. Summary

Microbes have a tremendous economic impact on MWF system operations. As agents of biodeterioration and disease they are detrimental. As waste treatment agents they are beneficial. Learning a few fundamentals about microorganisms and their lives in the metalworking environment will help industry stakeholders to minimize the adverse effects of uncontrolled microbial growth and maximize the beneficial effects of controlled microbial growth.

This chapter provides an overview of MWF microbiology. After introducing the relevant microbes and microbiological concepts, it will survey the problems that microbes cause in MWF systems. A separate section will address health issues. The last two sections will explain condition monitoring and discuss microbial contamination control strategies. Recognizing that many readers are likely to be unfamiliar with the microbiological terms used in this chapter, definitions have been included in the glossary at the end of this book.

The typical introductory microbiology textbook exceeds 1000 pages of small print. Consequently, this chapter represents the author's priorities and opinions regarding the information most likely to be of value to nonmicrobiologist readers who have a vested interest in the safe and profitable operations of metalworking systems. As used in this chapter, the term *metalworking* refers to fluids and systems used for metal removal and forming, and parts washing.

II. MICROBIOLOGY FUNDAMENTALS

A. Types of Microbes in MWF

The two major groups of microbes that have been recovered from MWF historically are bacteria and fungi. Bacteria are single cell organisms that comprise one of the three biological domains (the other two being *Archaea* and *Eukaryota* as described below).*

Bacteria are genetically distinct from organisms of the other two domains. They are also distinguished by their unique cell structure. Muramic acid is a characteristic constituent of the cell wall of all bacteria that have a cell wall (some bacterial taxa lack a cell wall). Fungi are members of

* A detailed discussion of microbial taxonomy is beyond the scope of this chapter. Readers interested in more information about microbial taxonomy may obtain that information from any introductory microbiology text or referral to *Bergey's Manual*[2] the definitive reference on microbial taxonomy.

the domain *Eukaryota*, which also include all of the more complex life forms. Similar to all other eukaryotes, fungi have membrane-bound internal structures called *organelles*. Approximately 20 years ago, microbiologists discovered that a group of microbes that had previously seemed to belong among the bacteria were really members of a genetically unique domain. This group of single-cell microbes, originally classified as *Archaeobacteria*, is now the domain *Archaea*. First recovered from extreme environments (high temperature, high salt content, high pressure, etc.), Archaea have subsequently been found to be ubiquitous in nature. Since the Archaea are difficult to culture, most of our current information about them is based on molecular biology research. There are no reports of Archaea recovery from MWF, but this may reflect a limitation in our test methods rather than reality. For example, Zhu et al.[3] recently reported gas pipeline microbially influenced corrosion (MIC) caused by Archaea. This was the first such report in an industrial system. Extrapolating from the Zhu et al. report, it is not unreasonable that Archaea may have some as of yet unrecognized role in MWF and MWF system biodeterioration.

1. Bacteria

Unlike the eukaryotes, bacteria do not have any membrane-bound internal organs. The most recent edition of *Bergey's Manual*[2] divided bacteria into 32 groups based on classical taxonomy, but in the volume's preface, *Bergey's* editors admit that since the advent of genetic taxonomy (genomics) bacteria taxonomy has been more fluid than at any previous time in the history of microbiology. The reason for this is that a surprising number of bacteria that share common morphological and physiological characteristics have been shown to be quite dissimilar genetically. Conversely, bacteria that appear to be different in appearance and physiological properties may be very similar genetically. Recent biofilm research, using single bacterial cultures, has demonstrated that as the biofilm matures, cells from the same ancestor (single cell) take on a considerable range of taxonomic and physiological (*phenotypic*) properties.[4] This is analogous to the differentiation of stem cells into the myriad cell types of our bodies. The cells are all identical genetically (*genotypically*) but are very different phenotypically.

Outside the realm of formal taxonomy, it is useful to classify bacteria based upon their relevant activities. For example, bacteria that require oxygen are called *aerobes*. Those that cannot tolerate oxygen are called *anaerobes*. *Facultative anaerobes* (also called *facultative aerobes*) can thrive regardless of oxygen availability. Table 9.1 summarizes this type of classification scheme. Many activities such as slime and biosurfactant production depend upon a combination of environmental conditions and genotypes present.

2. Fungi

Fungi are classified by their morphology. The fungi that infect MWF are either filamentous *molds* or single-celled *yeasts*. Molds have two types of cells. Vegetative cells form hair-like filaments or *hyphae*. Spores form in special bodies that are found at the head of special aerial hyphae, which extend out from the filamentous mass. The coloration of the spores gives fungal colonies their characteristic color. In MWF systems, fungal colonies growing on surfaces typically form slimy stringers. Planktonic colonies appear as "fisheyes" (slimy balls) or fuzz-balls.

Historically, for most practical purposes, taxonomic identification of contaminant microbes, beyond the determination of whether the microbes were bacteria or fungi, provided little practical information. More detailed information had little impact on treatment options. From an operational management perspective this has not changed. However, understanding the relative abundance of health-related species is important for monitoring health risks in the plant environment. Microbes implicated as representing potentially increased health risks are discussed in Section IV.

TABLE 9.1
Classification of Bacteria Based on Their Physiology

Oxygen requirement	
Require oxygen	Obligate aerobes
Cannot tolerate oxygen	Obligate anaerobes
Grow with or without oxygen	Facultative aerobes (anaerobes)
Growth temperature	
Require <20°C	Psychrophile (psychrophilic)
Require 20 to 40°C	Mesophile (mesophilic)
Require >40°C	Thermophile (thermophilic)[a]
Salt tolerance	
Require ≥2 M NaCl	Halophile (halophilic)[b]
Tolerate 2 M NaCl but prefer <2 M	Halodure (haloduric)
Energy source	
Inorganic molecules (e.g., ammonia, sunlight)	Lithotroph
Organic molecules	Organotroph
Nutrition	
Carbon dioxide	Autotroph
Organic molecules	Chemotroph
pH requirements	
<6	Acidophile
6 to 9	Neutrophile
>9	Alkalinophile

[a] Most obligate thermophiles formerly classified as bacteria are now listed among the Archaea.
[b] All obligate halophiles are now classified among the Archaea.

B. Microbial Ecology

In the laboratory, microbes are typically studied in pure (*axenic*) culture — populations descended from a single cell. Research based on axenic cultures yields a tremendous amount of useful information about cell physiology and chemistry. However, in MWF systems, other industrial environments, and natural environments, microbes rarely exist as axenic cultures. The varied and complex interactions among different taxa present in an ecosystem affect the activities of the microbial community (*consortium*) so that the net effects of the community differ substantially from those that would be predicted based on our knowledge about the physiology of the individual taxa within that community. The following section elaborates on this critical concept.

1. Planktonic vs. Sessile Cells

Cells floating free in recirculating MWF are *planktonic*. Planktonic bacteria and fungi may be present as either individual cells or as small aggregates (*flocs*). Although planktonic bacteria and fungi may be metabolically active, they are typically transient inhabitants of the fluid. In central recirculating systems, a significant proportion of the planktonic individuals and flocs are removed during filtration. In this respect, central systems are open systems resembling *chemostats*. A chemostat is a continuous or semicontinuous flow apparatus used to maintain a constant biomass

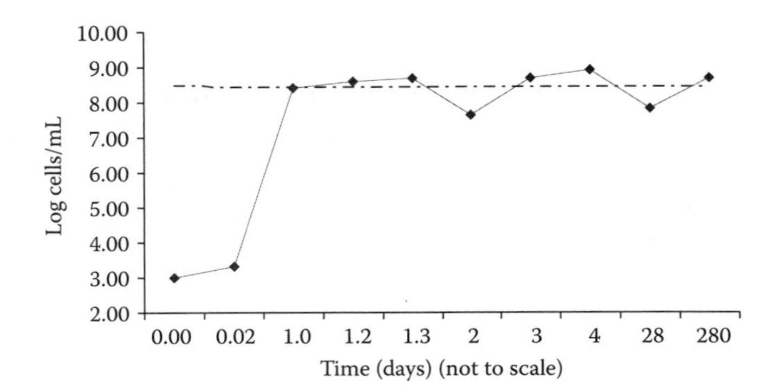

FIGURE 9.1 Microbial population density under steady-state (continuous flow) conditions. After an initial adaptation period, the population density (solid line) grows exponentially until it reaches the maximum sustainable density. Thereafter, population density varies around the average population density (dashed line) indefinitely unless the system is perturbed. *Note*: the time line is not to scale.

and nutrient concentration. Fresh growth media is introduced and spent medium plus cells are drained from the culture vessel at a predetermined rate (Figure 9.1). In contrast, small, individual sumps more closely resemble batch — or closed — culture conditions (Figure 9.2). In a closed system, microbes first acclimate to the system (lag phase). Once acclimated, the population begins to increase logarithmically with time (log phase). Reports of an organism's or population's *doubling time* (or *generation time*) are based on log phase observations. The fastest growing bacteria have generation times on the order of 20 min. Some species have generation times longer than 6 h. As will be discussed later in this chapter, the generation time can have a significant impact on viable count test interpretation.

As nutrients in a closed system become depleted, and toxic wastes from growing microbes accumulate, the generation time increases. The number of new cells produced and number of cells dying becomes approximately equal. Consequently the plot of biomass or cell numbers vs. time is

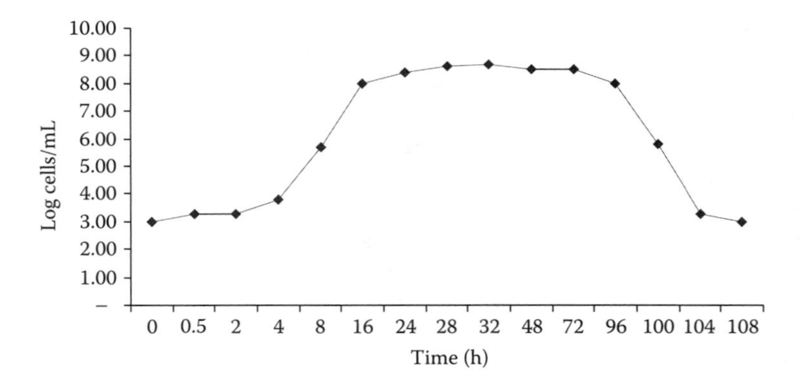

FIGURE 9.2 Microbial population density under batch culture (closed system) conditions. After an initial adaptation period (*Lag Phase*) during which cells/ml does not change with time, the population enters a period of logarithmic grow (*Log Phase*) during which the cells/ml doubles per unit time (*generation time*). As nutrients are depleted and inhibitory metabolites accumulate, the rates of cell death and production are approximately equal (*Stationary Phase*). Ultimately, the death-rate exceeds the cell production rate and the population enters into a final phase of logarithmic decline (*Death* or *Decline Phase*).

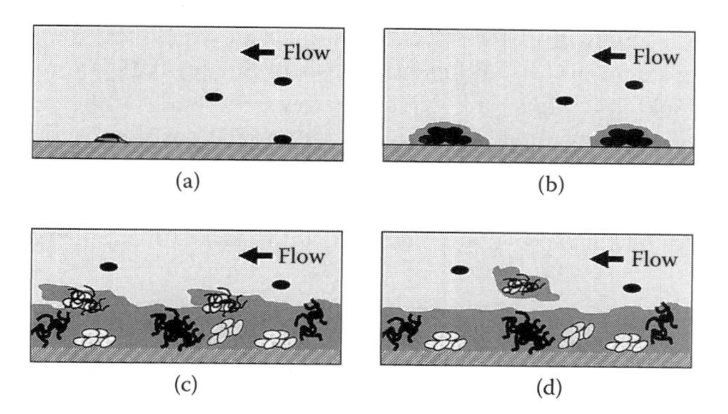

FIGURE 9.3 Development of a mature biofilm on a metal surface. (a) Planktonic microbes in MWF stream attach to pristine metal surface; secrete glycocalyx polymer. (b) Within a few hours, microcolonies develop on surface; secreted glycocalyx traps other microbes and particulates (organic and inorganic). (c) Within 1 to 2 days, mature biofilm has developed with complex ecology. (d) Fluid flow creates sheer stress on biofilm; note that the biofilm streamer in (c) has sloughed away from biofilm and is now a planktonic floc being transported downstream to either settle onto another surface or be trapped on the system's filter medium.

approximately flat. This is the *stationary phase* and resembles the dynamic steady state described above for open systems. However, in a closed system, nutrient depletion and toxic metabolite accumulation continues. In time, the production of new cells lags behind the death rate and the culture enters its terminal phase, the *death* (or decline) *phase*.

It is important to understand that the phenomena described above are based on observations of planktonic microbes. Life is much more complex. When microbes are introduced into a pristine system, some of them will adhere to system surfaces. This is the first stage of biofilm development (Figure 9.3). Microbes growing attached to surfaces are *sessile*. Some species, once they attach to a surface, produce prodigious volumes of a mucus-like (*mucilaginous)* substance (*extracellular polymer substance* — primarily sugars and amino sugars — *glycocalyx*). This glycocalyx matrix has several important features. Being sticky, it traps planktonic microbes; bringing them into the biofilm matrix. This marks the beginning of a consortium (see next section). Cells dividing (*proliferating*) within the biofilm matrix form microcolonies. Microcolony growth causes both the local depletion of nutrients and local concentration of metabolites. These chemical changes have a number of important impacts on the population and the system.

Microbial activity within the biofilm creates gradients between the bulk fluid and the surface to which the biofilm community is attached. Despite well-oxygenated (*oxic*) conditions at the biofilm-fluid interface, *anoxic* (oxygen-free) conditions are likely to exist deep within the biofilm. This creates the conditions necessary for anaerobic bacteria to grow. Similarly, pH within the biofilm decreases with distance from the bulk-fluid. It is not uncommon for deep biofilm pH to be less than 4, although the bulk fluid pH is greater than 9. Electropotential gradients between exposed and biofilm covered surfaces drive corrosion-cell currents.

Biofilm accumulation has two other features critical to fluid system management. Biofilms trap metal fines and other recirculating particles from the MWF. The glycocalyx becomes cement that glues large accumulations of swarf, tailings, and debris onto machine, pipe, and sluice surfaces.* Moreover, the glycocalyx functions as a barrier. In particular, antimicrobial pesticides are unlikely to diffuse into a biofilm sufficiently to kill microbes living deep within the biofilm. Nonspecific reactions with metallic and organic detritus, reactions with microbes near the biofilm-bulk fluid

* Some MWF also have a tendency to leave tacky residues which serve much the same detrimental function.

interface, and sequestering within the glycocalyx matrix, all reduce the effective microbicide concentration. This problem will be addressed further in Section VII.

In summary, planktonic cells are typically transient and unlikely to cause significant biodeterioration in recirculating fluid systems. Sessile populations with biofilms are more difficult to sample but more likely to cause fluid biodeterioration. These communities act as fixed-film biological reactors similar to those used for waste treatment and bioindustrial production of chemicals. Microbes growing with biofilms are much more difficult to kill than are planktonic microbes.

2. Individual Cells vs. Consortia

The difference between individual planktonic cells and cells living embedded within biofilms is considerably more complex than described in the previous section. In addition to the factors discussed above, the proximity of cells within biofilm communities enables them to function as consortia. This means that the biofilm community as a whole can carry out processes beyond the capabilities of any of its individual member taxa. A discussion of the various possible dynamics between taxa (synergy, commensalism, and competition) is beyond the scope of this chapter. In this section, two phenomena will be used to illustrate the consortium concept.

Pseudomonas aeruginosa and *Desulfovibrio desulfuricans* are two commonly recovered MWF species. A typical pseudomonas species, *P. aeruginosa* can use a diverse range of organic molecules as nutrients. Also, *P. aeruginosa* is an aerobe. Similar to all living organisms, *P. aeruginosa* produces waste by-products of its metabolism. These by-products — *secondary metabolites* — include one to three carbon fatty acids such as formic, lactic, acetic, and pyruvic acids. In contrast to *P. aeruginosa*, *D. desulfuricans* is an obligate anaerobe. Moreover, it is nutritionally fastidious — able to use a limited variety of organic molecules as food. The organic acids produced by *P. aeruginosa* are just the types of molecules *D. desulfuricans* needs. Consequently, in this two-member consortium, *P. aeruginosa* removes oxygen from the environment — creating the conditions necessary for *D. desulfuricans* to thrive — and converts complex MWF organic molecules into the simple nutrients on which *D. desulfuricans* depends. By removing *P. aeruginosa*'s waste metabolites, *D. desulfuricans* prevents them from accumulating and becoming toxic to *P. aeruginosa.* (Secondary metabolite accumulation is one of the primary factors that cause the batch-culture growth curve shift from log to stationary and then death phase — see Figure 9.2 and earlier discussion.)

In summary, *P aeruginosa* creates conditions favorable for *D. desulfuricans,* and *D. desulfuricans* creates conditions that enable *P. aeruginosa* to attack MWF molecules at maximum rate. This example is a very simplified explanation, intended to illustrate the concept of microbes acting in consortia. Recent research with *P. aeruginosa* pure culture biofilms has demonstrated that within a biofilm, cells differentiate in a manner similar to the way in which our cells differentiate from somatic (undifferentiated cells) into our various tissue cells. An individual *P. aeruginosa* cell's physiological properties seem to depend on its location within the biofilm. That is, cells originating from the same ancestor cell can play different consortium roles. It is not a great leap to think of biofilm consortia as an evolutionary link between single-cell and multicellular organisms. This also explains, in part, why biofilm communities are so much more difficult than planktonic cells to kill.

3. Proliferation, Growth, and Metabolism

The terms *proliferation*, *growth*, and *metabolism* have been used in the preceding section. Each of these terms is related but reflect different, critical attributes of a microbial community. Proliferation refers to increasing numbers of cells. The growth curve explained earlier is really a plot of population density as a function of time. However, it is important to understand that individual cells can grow without dividing. Consequently, growth reflects the net *biomass* increase. As such, it is equal to the sum of biomass increase due to proliferation plus biomass increase due to the mass per cell. Metabolism is the set of processes that are necessary for life that occur with all living cells.

Some metabolic processes convert organic or inorganic compounds into the energy needed to support life. Other metabolic processes convert food molecules into new cell material (biomass) and waste (secondary metabolites). The pace at which metabolic processes occur is called the *metabolic rate*. In general, biodeterioration rates are proportional to metabolic rates. Consequently, a relatively small population of metabolically active cells may have a higher overall biodeterioration rate than a substantially larger population of dormant or moribund cells. The importance of understanding these principles will become apparent below in context of the condition monitoring discussion.

III. MWF AND MWF SYSTEM BIODETERIORATION

Biodeterioration includes all biological processes that result in economic loss. Biodeterioration and bioremediation represent the two sides of the biodegradation coin. Biodegradation refers to the set of processes by which organisms break down large molecules into smaller molecules. The ultimate end of biodegradation is the conversion of organic molecules into carbon dioxide and waste energy. *Direct* (or *primary*) biodegradation includes the metabolic processes by which organisms convert substrates into products. *Indirect* biodegradation includes all of the other means by which organisms cause or contribute to substrate breakdown. This section will provide a brief overview of biodeterioration processes in MWF systems.

A. Selective Depletion of Components

Selective component depletion can result from primary degradation, secondary reactions or both. Most MWF constituent chemicals are biodegradable. Biodegradability is a critical requirement for waste treatability. Biodegradation rates depend on multiple factors including:

- Chemical structure
- Microbial species present
- Water availability
- Oxygen availability
- Temperature
- Surface area to volume ratio

Most MWF systems present optimal environments for MWF biodegradation. Typically used at 3 to 10%, end-use dilutions of MWF provide a balanced blend of water and organic nutrients. Fluid recirculation aerates the bulk fluid, while biofilm consortia and quiescent zones provide niches for anaerobic metabolism. Suspended metal fines (swarf, chips, and tailings) provide surface area. Plant ventilation, supply water, and employees contribute the microbial inocula.

Not all MWF components are equally biodegradable. More complex molecules tend to be recalcitrant to biodegradation. For example, naphthenic base oils are more bioresistant than paraffinic oils. Borate esters and boramides are more bioresistant than their nonborated analogues. Oxazoladines and amines, such as aminomethyl propanol, are more recalcitrant than simpler amines, such as monoethanol and triethanol amine. Among the many challenges confronting formulation chemists is the need to balance MWF biostability in application against biodegradability during waste treatment.

Primary biodegradation occurs when microbes use one or more MWF components as food. Some molecules are consumed completely; converted to biomass, carbon dioxide, and energy. Others are only partially degraded (for example phosphate may be removed selectively from phosphate esters, without further mineralization of the ester). Secondary biodegradation occurs when microbial metabolites react with MWF components. For example, as discussed above, most microbes excrete a variety of weak organic acids as metabolic waste. These acids may react with neutralizing amines directly or may react with inorganic ions, such as chloride, sulfate, and nitrate that are present in the MWF. In the latter scenario, the reactions lead to the formation of fatty acid salts plus strong inorganic acids (hydrochloric, sulfuric, and nitric acid, respectively). The resulting inorganic acids then react with any neutralizing amines that are present.

Although not considered biodegradation in a strict sense, large microbial contaminant populations place a greater *demand* than do small populations on microbicides. Since microbicide half-life ($T_{\frac{1}{2}}$) is inversely proportional to the demand, heavily contaminated systems deplete microbicides more rapidly than do uncontaminated systems.

B. Performance Characteristic Changes

As discussed in other chapters within this volume, MWF performance is determined by the types and ratios of components formulated into the product. Selective component depletion disrupts the balance of component ratios and causes the fluids performance characteristics to change. Another chapter addresses nonbiological factors that affect performance characteristics. This chapter focuses on biodeterioration.

1. Emulsion Characteristics

Direct attack of petroleum sulfonates, soaps, alkanolamides, and other MWF emulsifier additives degrades emulsion stability. Microbial activity can change emulsion characteristics either by tightening the emulsion through the production of biosurfactants, or splitting the emulsion through the production of organic acids. Biosurfactant production also increases foaming and aerosolization (misting) tendency. In synthetic fluids, biosurfactants emulsify tramp and way oils, converting synthetic MWF into unintended emulsions. In practice, both phenomena (emulsion splitting and biosurfactant production) occur simultaneously, resulting in increased mist levels and changes in micelle size distribution.

2. Lubricity and Extreme Pressure Performance

Removal of phosphate from phosphate esters was used above as an example of partial degradation. Except for having been dephosphorilated, the ester remains intact, but no longer functional as an extreme pressure (EP) additive. Sulfurized fats and paraffins and petroleum sulfonates are also biodegradable. As these components are attacked, the MWF's lubricity properties degrade. Polymers such as polyalkylene glycols, polyethylene glycol esters, and block polymers tend to be substantially more biologically recalcitrant than phosphate esters and the sulfurized products mentioned above.

3. Corrosion Inhibition

Selective depletion of neutralizing amines has been discussed above. Microbes can also contribute to metal part and system component corrosion through a number of processes known collectively as *microbially influenced corrosion* (MIC). Several critical MIC processes have been addressed in this chapter. Biofilms create electropotential gradients that accelerate electron flow from the anode (where metal dissolves) to the cathode (the electron sink). Typically, this results in pitting corrosion — rust-colored, volcano-shaped tubercles over black ooze covering a corrosion pit. Strong inorganic acids (primarily hydrochloric, but also sulfuric and nitric), formed when weak organic acids react with inorganic salts, attack metal surfaces directly. Sulfate-reducing bacteria and certain endospore-forming, Gram-positive rods (*Clostridium* sp.) have the enzyme *hydrogenase*. Hydrogenase consumes the hydrogen ions (protons) that otherwise would accumulate on the metal surface at the cathode. Unperturbed, the hydrogen ion barrier arrests (*passivates*) the galvanic corrosion process. Hydrogenase activity *depassivates* the process, thereby accelerating galvanic corrosion. Videla's monograph on biocorrosion[5] provides an excellent, detailed description of MIC processes and symptoms.

4. Aesthetics

The most common aesthetic concerns associated with the use of water-miscible MWF are odors and slime accumulation. This section provides a brief overview of odor and slime problems. Although malodorous gases are more often noxious rather than toxic, hydrogen sulfide and ammonia accumulate in areas of poor circulation. Frequently, unusual odors can be the first indication of significant microbial activity within the systems. In contrast, slime accumulation is a late symptom.

5. Odor

a. *Monday Morning Odor*

This very common phenomenon is the result of normal microbial metabolism. As discussed in Section II.B, metabolizing microbes produce a wide variety of metabolites, many of which they excrete as waste products. In particular, 2 to 4 carbon acid–alcohols, aldehydes, and dicarboxylic acids each have characteristic odors. Additionally, hydrogen sulfide, methyl mercaptan ($HS{-}CH_3$), and skatol (methyl indol: C_9H_9N) contribute to the knall gas (swamp gas) bouquet that occasionally greets MWF plant employees after a weekend shutdown.

Monday morning odor is not caused by *weekend warrior* microbes nor by unique metabolic activities that occur only during the weekend. The odiferous volatile organics that contribute to knall gas accumulation are produced by MWF microbes continuously. When the fluid is recirculating and remaining well aerated, chemical oxidation neutralizes the malodorous molecules and prevents them from escaping the fluid in their noxious form. During shutdown periods, dissolved iron and microbial activity scavenge the available oxygen and create anoxic conditions. The shutdown has a triple effect. The system becomes more like a batch culture (see description above). Since the fluid becomes anoxic, anaerobic bacteria are no longer restricted to live deep within the shelter of the biofilm. Fermentation (the use of organic molecules for both food and energy) replaces oxidative metabolism as the primary means of energy metabolism. These three factors contribute to volatile organic compounds (VOC) accumulation in the MWF.

On Monday morning, as the pumps are turned on, the initial churning of the fluid that had been quiescent through the weekend causes, in effect, a major emission of gas. Once the fluid becomes reaerated and the emitted gas flush dissipates, the plant once again becomes habitable. Good microbial contamination control and MWF aeration provide the best defense against Monday morning odors.

b. *Ammonia Flush*

Ammonia release from MWF is rarer than knall gas release. It occurs when microbes attack MWF amines oxidatively. Both bacteria and fungi have enzymes of the *amine oxidase* family. These enzymes oxidize the amine to a carboxylate, liberating ammonia:

$$RCH_2{-}NH_2 + H_2O + \tfrac{1}{2}O_2 \xrightarrow[\text{Amine oxidase}]{} RCH_2COOH + NH_3$$

The chemical equilibrium between ammonium (NH_4^+) and ammonia (NH_3) is pH dependent. At pH > 9.2, ammonia becomes the dominant species. At pH 10, NH_3 comprises 90% of the combined $NH_3 + NH_4^+$. Early reports of ammonia flush often followed shock treatment with high doses of amine-type microbicides such as triazine. Ammonium, already present due to biodegradation of monoethanolamine, diethanolamine, and triethanolamine, rapidly converts to the ammonia form, which then off-gasses into the plant atmosphere when pH 10.5 to 11 triazine is added. High concentrations of dissolved iron (typical of MWF used in ferrous metal-grinding operations) seem to exacerbate ammonia flush. As with Monday morning odor, amine metabolism by MWF bacteria and fungi occurs continuously in systems with active communities. Ammonia can accumulate in quiescent systems and be released as a flush when pumps are turned on. The most effective strategies for preventing ammonia flush include good microbial contamination control and good swarf removal.

c. *Mildew*

When MWF emit a musty, mildew, *sweat socks* odor, it is a sure sign of fungal contamination. Fungal VOC released into plant air are responsible for these odors. Almost invariably, fungal communities growing on system surfaces are the culprits. Fungi tend to populate splash zones. The underside of sluice deck-plates provides an excellent habitat for fungal colonization. Moreover, in many facilities, these surfaces are rarely inspected. Microbicides used to knock down microbial contamination in recirculating MWF and submerged surfaces will not come into contact with splash zone surfaces. Decontaminating deck-plate bottoms and other splash zone surfaces eliminates musty odors effectively.

6. Slime

Slime is the most visible evidence of uncontrolled microbial contamination. As noted above, slime is a complex amalgamation of biopolymer, microbes, and MWF particulates that become trapped within the glycocalyx matrix. Slime sloughed off from MWF system surfaces contributes to premature filter indexing and high usage of paper filter media. It can also contribute to slip hazards. Having been discussed in the section on biofilms, consideration of slimes will not be reiterated here.

IV. HEALTH EFFECTS

Passman and Rossmoore[6] recently reviewed MWF microbe health effects. This section will provide a brief overview of that article and highlight information that has become available since its publication in 2002. Microbes can cause three types of human disease: infection, toxemia, and allergy. The following paragraphs will define each of these types of diseases, and discuss their significance to workers routinely exposed to MWF.

A. Infectious Disease

Infections occur when one or more microbes gain entry into the body and proliferate. The symptoms of different infections reflect the combined effects of proliferation, tissue attack, toxin, and other metabolite production and the body's immunological response. The most common cause of infections in the MWF environment is improperly treated wounds. Relatively minor cuts and abrasions by surfaces carrying high numbers of bacteria and fungi introduce these microbes into the subcutaneous tissue. When wounds are not cleaned and treated properly, chances for opportunistic pathogens to proliferate and cause infection increase substantially. Typically, these wound infections are not easily transmitted to others. They are considered *noncommunicable*.

Although there is a finite risk of infectious disease in the metalworking environment, there is no evidence that the incidence of infectious disease at metalworking plants differs from that of the general population. Good personal hygiene practices are the best defense against infectious disease.

B. Toxemia

Toxemias are diseases caused by poisons. Microbial toxins include thousands of different chemicals produced by bacteria and fungi.[7] Although there is a variety of toxigenic bacteria and fungi routinely recovered from MWF (Table 9.2), there are few data on microbial toxin concentrations in

TABLE 9.2
Toxin-Producing Microbes Recovered Routinely from Metalworking Fluids

Bacteria	*Escherichia coli*
	Pseudomonas aeruginosa
	Staphylococcus aureus
	Streptococcus pyogenes
Fungi	*Aspergillus* sp.
	Fusarium sp.
	Penicillium sp.

either recirculating MWF or plant air. Three types of microbial toxins may be significant to MWF workers:

- Endotoxins
- Exotoxins
- Mycotoxins

1. Endotoxins

Endotoxins are the lipopolysaccharide (LPS) component of the Gram-negative bacterial outer cell membrane. Endotoxins are both toxic and allergenic. The LPS molecule's lipid portion (lipid A) confers its toxicity.[8] Symptoms can range from mild fever to death due to toxic shock. Castellan et al.[9] computed the endotoxin no observable effect level (NOEL) to be 9 ng m^{-3}. At these low concentrations the primary symptom is decreased respiratory volume. At 200 to 500 ng endotoxin m^{-3} mucus membrane irritation becomes apparent. Bronchial restriction occurs at 1000 to 2000 ng endotoxin m^{-3} and death is increasingly likely once the endotoxin exposure exceeds 10,000 ng m^{-3}.

A 1999 survey of 18 MWF facilities[10] determined that indoor endotoxin concentrations ranged from 0.3 to 2.5×10^5 ng endotoxin ml^{-1} in MWF and from <0.04 to 1.4×10^3 ng endotoxin m^{-3} in plant air.

Endotoxin can be released by growing cells and by lysing cells. The molecule is heat-stable but may be denatured with aldehyde-based microbicides such as formaldehyde donors,[11] and strong oxidizing chemicals such as peroxides, hypochlorites, and superoxides. Shock treatment of MWF systems with high ($>10^6$ cells ml^{-1}) numbers of Gram-negative bacteria can, at least temporarily, increase airborne endotoxin concentrations by several orders of magnitude. Aerosol containment and microbial contamination control are the most effective means for minimizing endotoxin exposure in metalworking facilities.

2. Exotoxins

Exotoxins are excreted by living bacteria and fungi. Fungal exotoxins are called *mycotoxins* and will be discussed below. Most exotoxins are proteins. Common* MWF isolates, *P. aeruginosa*, *Escherichia coli*, *Staphylococcus aureus*, and *S. pyogenes* produce exotoxins (Table 9.3). Exotoxin

* This list of commonly recovered taxa reflects MWF population studies performed in the 1950s and 1960s by MWF microbiology pioneers, E.O. Bennett, R.O. Hansen, E.C. Hill, and H.W. Rossmoore, respectively. Recognizing that both MWF chemistry and selective recovery growth media have evolved since then, it is possible that a taxonomic profile study performed today would yield different results.

TABLE 9.3
Exotoxins Produced by Genera of Bacteria Recovered Routinely from Metalworking Fluids

Genus	Exotoxin
Escherichia	*E. coli* LT toxin
	E. coli ST toxin
Pseudomonas	Exotoxin A
Staphylococcus	Exfoliation toxin
	Staphylococcus enterotoxins
	Toxic shock syndrome toxin
Streptococcus	Erythrogenic toxin

A, from *P. aeruginosa* causes diphtheria-like symptoms. The toxins produced by *S. aureus* and *S. pyogenes* cause toxic shock. The toxins produced by *E. coli* cause severe gastroenteritis, similar to the symptoms of cholera.

3. Mycotoxins

Mycotoxins excreted by fungi are chemically diverse and can cause symptoms ranging from hallucination to cancer. Although there is no documentation of mycotoxin concentrations in MWF or their aerosols, *Aspergillus fumigatus* and other Aspergillus species routinely recovered from MWF are known to produce aflatoxin. Species within the genera *Fusarium* and *Penicillium* also produce a range of mycotoxins. Table 9.4 lists more than 60 different mycotoxins produced by species within these three genera.[12] Whether any of these mycotoxins are present in metalworking facilities at concentrations that might affect employee health is unknown. The absence of a record of metalworking industry employees developing mycotoxin-related diseases suggests that mycotoxins may not be important health factors in the metalworking environment.

C. Allergy

Allergies develop when the body's immune system reacts to a molecule it recognizes as foreign. An *allergen* is any molecule that induces the body's release of histamine, characteristic of an allergic reaction. In contrast to toxins, any given allergen is likely to affect only a small percentage of exposed individuals. The severity of the response is more affected by an individual's sensitivity than to dose. Exposed to a particular allergen, some people may suffer minor irritation (consider mild hay fever or rashes). Particularly sensitive people may suffer *anaphylaxis*. Anaphylaxis is a general, potentially lethal, whole body condition resulting from the body's rapid release of antibodies and histamine.

Any biomolecule is likely to be allergenic to some percentage of the exposed population. Endotoxins are both toxic and allergenic. Over the past decade, hypersensitivity pneumonitis (HP) has become an increasing concern within the metalworking industry.[13] Although fewer than 300 cases of HP have been reported over a 14-year period, the cases have tended to occur in clusters.[14] Since HP is an allergenic disease, and there are multiple known agents in the metalworking environment that can cause HP, identifying a specific set of cause and effect relationships is difficult. As shown in Table 9.5, at least seven of the 19 known microbial agents linked to HP have been recovered from MWF. In at least one outbreak,[15] the acid-fast bacterium *Mycobacterium*

TABLE 9.4
Mycotoxins from Fungal Genera Recovered from MWF[12]

Genus	Mycotoxin	
Aspergillus	Aflatoxin	Cyclopiazonic acid
	Aflatrem	Fumagilin
	Altenuic acid	Maltoryzine
	Alternariol	Ochratoxin
	Austdiol	Oxalic acid
	Austamide	Patulin
	Austocystin	Penicillic acid
	Brevianamide	Sterigmatocystin
	Citrinin	Tryptoquivalene
	Citreoviridin	Viomellein
	Cytochalasin E	Viriditoxin
Fusarium	Acetoxyscirpenediol	HT-2 toxin
	Acetyldeoxynivalenol	Ipomeanine
	Acetylneosolaniol	Lateritin
	Acetyl T-2 toxin	Lycomarasmin
	Avenacein	Moniliformin
	Beauvericin	Monoacetoxyscirpenol
	Calonectrin	Neosolanio
	Deacetylcalonectrin	Nivalenol
	Deoxynivalenol diacetate	NT-1 toxin
	Deoxynivalenol monoacetate	NT-2 toxin
	Diacetoxyscirpenol	Sambucynin
	Destruxin B	Scirpentriol
	Enniatins	T-1 Toxin
	Fructigenin	T-2 Toxin
	Fumonisin B_1	Triacetoxyscirpendiol
	Fusaric acid	Yavanicin
	Fusarin	
Penicillium	Patulin	Rugulosin
	Penitrem	Sterigmatocystin
	Rubratoxin	Viopurpurin
	Rubroskyrin	Viomellein
	Rubrosulphin	

immunogenum could not be recovered from the site's MWF, nor was *M. immunogenum* precipitin detected in exposed workers. However, for a variety of reasons, not all of which appear to be technical, the industry has subsequently focused its attention on *M. immunogenum* as the putative HP agent. This, despite the following realities:

1. No consensus *M. immunogenum* detection and quantification methodology exists.[16]
2. *M. immunogenum* is known to be a nearly ubiquitous microbe.
3. *M. immunogenum* has been recovered from numerous MWF systems around which no employees have developed HP.
4. Nonacid-fast bacteria, listed in Table 9.5, have been recovered from systems proximal to workers who have developed HP.

TABLE 9.5
Microorganisms Known to Cause Hypersensitivity Pneumonitis[13,14]

Microbe	Microbe
Achromobacter sp.[a]	*Mycobacterium avium* complex
Alternaria sp.	*Mycobacterium immunogenum*[a]
Aspergillus sp.[a]	*Penicillium casei* (*P. roqueforti*)[a]
Aureobasidium pullulans	*Penicillium citreonigrum*[a]
Bacillus cereus[a]	*Penicillium frequentans*[a]
Bacillus subtilis[a]	*Pullularia* sp.
Cryptostroma corticale	*Rhizopus* sp.
Graphium sp.	*Thermoactinomycetes candidis*
Klebsiella oxytoca[a]	*Trichosporon cutaneum*
Monocillium sp.	

[a] Genera that have been recovered from MWF.

The limited epidemiological and immunological data available thus far,[17,18] are insufficient to support any hypothesis adequately. Moreover, recent work by Yadav et al.[19] challenges the conventional wisdom that *M. immunogenum* will only grow in MWF if Gram-negative bacteria such as *Pseudomonas* sp. have been eliminated through the use of certain biocides. Using polymerase chain reaction methods (see Ref. [19]), Yadav et al. found numerous MWF with substantial populations of both mycobacteria and pseudomonads. Premature focus on a single suspected agent may increase the challenge of identifying the factors contributing to HP correctly. Misidentifying the causative factors will undoubtedly confound efforts to reduce HP risk.

V. WASTE TREATMENT

Most often, MWF waste treatment is a combination of physical, chemical, and biological processes. At many plants, only physical and chemical processes are used. The on-site objective is to separate spent MWF concentrate from water so as to minimize the total volume of MWF waste that must be hauled off for further treatment. In order to comply with local-specific water discharge regulations, plants may also need to operate a wastewater treatment system. This system may rely on filtration, biological oxidation or both.

Microbes associated with MWF can facilitate or disrupt waste treatment. Microbes disrupt waste treatment by fouling filtration systems or by producing biosurfactants that can tighten emulsions and exacerbate foaming problems. Biofilms developing on the walls of ultrafilter or nanofilter tubes will blind-off the tubes and arrest filtration.

Biochemical oxidation of organic molecules in wastewater is a microbiological process. The microbial processes discussed in the section on MWF biodeterioration are also relevant to the discussion of MWF wastewater treatment. Aerobic and anaerobic digesters are designed to maximize the rate at which microbes inside the digester mineralize organic matter. Chapter 13 provides more detail on waste treatment processes.

VI. CONDITION MONITORING FOR MICROBIAL CONTAMINATION CONTROL

The general tests for MWF condition monitoring are treated elsewhere in this volume. Standard protocols for many MWF tests are provided in ASTM compilation of metalworking industry standards.[20] In this chapter we will focus on using MWF condition monitoring parameters to monitor microbial contamination and its symptoms. The tests sort into four categories: gross observations, physical tests, chemical tests, and microbiological tests. Except for the microbiology, most of the variables monitored routinely may change due to biological, nonbiological, or a combination of both classes of factors. Consequently, no single parameter provides adequate microbial contamination control condition monitoring data. Also, increasing awareness of noncommunicable disease health risks associated with bioaerosol exposure has dictated a strategic shift in microbiological testing. Historically, the primary — if not exclusive — focus was system biodeterioration (fluid degradation and system fouling). Monitoring strategies that are adequate for that purpose do not provide information necessary to assess bioaerosol exposures.

As for all other condition monitoring efforts, each parameter should have either a predetermined upper control limit (UCL — for example, viable counts not to exceed a predetermined criterion level), lower control limit (LCL — for example, bulk fluid microbicide concentration should not fall below a specified minimum level), or both (for example, pH should be within an acceptable range with designated UCL and LCL). Moreover, the fluid manager should have a written, specified set of actions to be taken when a parameter falls outside its UCL or LCL.

A. Gross Observations

Gross observations include three senses: sight, touch, and smell. A well-maintained system will not have visible accumulations of slime on machine or sluice surfaces. Splash zones, particularly the underside of sluice deck-plates (Figure 9.4), are primary zones for biomass accumulation. Splash zones are particularly troublesome because microbicides added to recirculating MWF do not contact biomass growing on surfaces that are not in constant contact with the fluid. Slime accumulation should be noted during daily plant tours. Deck-plate underside visual inspections should be scheduled as routine checks. Filtration systems should be inspected for the appearance of slime stringers on supports and chip-drag flights.

FIGURE 9.4 Microbial slime (biofilm) coating on underside of a metalworking fluid sluice cover plate (deck-plate). (Photo courtesy of BCA, Inc.)

Tramp oil pooling indicates zones where conditions may become anoxic and promote the proliferation of anaerobic bacteria. It is necessary to ensure that skimmer or alternative tramp oil removal systems are functioning properly and have adequate capacity to remove tramp oil from the system before it accumulates.

Visual inspections also detect evidence of poor hygiene and housekeeping, which often exacerbate microbial contamination. Accumulations of garbage, refuse, and other nonsystem substances in chip hoppers, MWF sluices, and on the shop floor stimulate microbial activity and indicate the need for improved industrial hygiene education and training.

The malodorous biochemicals discussed in Section III.B are known as microbial volatile organic compounds (MVOC). In some cases, these MVOC can be detected before there are obvious signs of slime accumulation. Characteristic knall gas, rotten egg, ammonia, or sweat sock odors are unequivocal signals of the need for prompt corrective action. During plant inspection tours, be particularly mindful of areas of poor circulation. Where MWF appears to be stagnant (systems not operating, eddies in fluid flow, pools on transfer lines of floor, etc.) stir the coolant and sniff the air. If any of the aforementioned telltale odors are detected, corrective action must be taken.

Tactile testing is useful when it is not certain whether residue accumulations are biological or not. For example, MWF mist coats machine and other surfaces with which it comes into contact. As water evaporates from the deposited mist, the residual MWF typically becomes tacky. Swarf entrapped in tacky MWF residue can build up on machine and sluice surfaces. This is a nonbiological process. In contrast to the slippery feel of biofilm residue, this swarf-MWF residue aggregate feels dry and gritty. Apparent slime accumulations on filtration media may be biomass, or globs of lubricant or tramp oil. Again, each type of accumulation will have a characteristic feel. Most of the tactile tests can be performed while wearing surgical gloves.

B. Physical Tests

There are no physical tests unique to microbial condition monitoring. Using physical tests to diagnose microbial contamination is particularly challenging because, except for their consistently adverse impact on MWF filterability, microbes can have diametrically opposite effects on physical test results. Biopolymer production and biomass accumulation affect MWF filterability adversely in both recirculation and waste treatment systems.

Microbial activity can cause tramp oil emulsification into all classes of MWF (although the symptoms are most obvious in synthetics). Conversely, microbes can also split emulsions. This is particularly problematic in applications such as nonferrous metal rolling where surface finish is critical and dependent on uniform emulsion droplet size distribution. Microbial activity may also affect foaming tendency test results. Biosurfactants will tend to increase foaming tendency. Acidic metabolites will tend to decrease foaming tendency.

C. Chemical Tests

This section addresses only chemical parameters indicative of MWF biodeterioration. Chemical tests used to characterize microbial contamination will be discussed in the next section. Other chemical tests are described in Chapter 7 (Laboratory Evaluation of MWF) and Chapter 11 (MWF Management and Troubleshooting).

Microbial depletion of specific MWF components was addressed in Section III.A. Analytic methods including gas chromatography (GC) and high performance liquid chromatography (HPLC) can be used to track this selective depletion process.[21] More basic tests can be run routinely to determine the concentration of specific MWF components. Augmenting pH measurements with reserve alkalinity analysis facilitates detection of buffering capacity loss (neutralizing amine depletion) before fluid performance degrades. Biodeterioration tends to contribute to increased conductivity (total dissolved solids). High bioburdens deplete microbicides at abnormally high rates. Understanding the normal depletion rate for each microbicide used at a facility, and monitoring microbicide depletion rates after tankside treatment, is a powerful but generally neglected tool for detecting high bioburdens before they affect operations. Most MWF microbicide manufacturers can provide protocols for monitoring the concentrations of their products in application.

Although dissolved oxygen is a chemical parameter, this author recommends against its use as a chemical test. Oxygen depletion in a recirculating system is a relatively late symptom of substantial microbial activity. The next section will discuss using dissolved oxygen measurements as part of a 2-h oxygen demand test.

D. Microbiological Tests

Ultimately, the only means for quantifying MWF microbiological contamination is by running microbiological tests. No single microbiological test provides sufficient information about bioburden, biological activity, and the presence of particular species of biomolecules of interest. Consequently, both the methods used and test frequency should be defined by carefully considered objectives rather than tradition.

There are three primary categories of microbiological test methods:

- Microscopy
- Culture
- Chemistry

Each provides useful information and has significant limitations. The following paragraphs will summarize the general approaches.

1. Microscopy

Microscopy includes all methods that involve the use of a light or electron microscope to observe cells directly.[22] The most basic microscopic procedure is to place a drop of fluid onto a watch-glass and observe it through a stereoscope. The relatively low magnification afforded by binocular scopes permits visualization of fungal filaments, yeast cells, and flocs of bacterial biomass. Higher magnifications ($>1000\times$) are needed to visualize individual bacterial cells. Light microscopes are used for this purpose. Various technologies for directing the light path through a microscope's lenses provide means for seeing microbes in the unprocessed sample, however, enumeration (cell count) methods generally require preliminary sample processing steps. In general, sample processing includes staining.

There are numerous staining procedures used to facilitate differentiation between living and dead cells, endospores from vegetative cells, microbial species, and specific cell constituents. Two methods are particularly relevant for MWF microbiological testing.

Historically, the Gram stain has been the most common procedure by which samples were prepared for microscopic observation. A small (approximately, 10 μl) sample is smeared onto a microscope slide over an approximately 1.5-cm diameter circular area. The resulting droplet is then either heat fixed by gentle heating, or permitted to air dry. Through a series of staining and rinse steps, the smear is stained with an iodine preparation, decolorized, and then stained with a nonspecific safranin stain. Since iodine turns purple when it reacts with carbohydrates, Gram-positive cells appear violet to purple when observed through a microscope. Gram-negative cells are coated with their LPS outer-envelope (see Section IV.B.1) and do not retain the iodine stain. Instead, they retain the pink–red color imparted by the nonspecific stain (safranin).

Increasing concern about the possible relationship between *M. immunogenum* and HP (Section IV.C) has made the acid-fast stain relevant to MWF system stakeholders. Mycobacteria and actinomycetes cell envelopes have characteristically high lipid contents. Consequently, after they are stained with a carbolfuchsin dye, they appear red when viewed through a light microscope. Rossmoore et al. have recently proposed a method for enumerating acid-fast bacteria (AFB) in MWF.[23]

Quantitative direct count methods make use of specialized counting chambers that compress fluid samples into a thin film spread over a calibrated grid. Alternatively, cells may be concentrated onto a filter membrane. Cells may either be prestained (as is done for AFB enumeration) or visualized using phase contrast optics. The analyst counts the number of cells per microscope field (circular area visible when looking through the microscope), computes the average number of cells per field and then, based on the predetermined field dimensions, calculates the number of cells per milliliter of sample.

Direct count methods have the advantage of not requiring microbes to proliferate in order to be detected. However, most direct count techniques do not differentiate between cells that were alive and those that were dead at the time of sampling. Given the very small sample volume and surface area visible in a microscope field, direct count detection lower detection limits are approximately 2×10^6 cells/ml. Direct counting is labor-intensive and requires a moderate amount of technical expertise.

2. Culture Methods

Culture methods depend on the ability of cells to proliferate in or on a particular growth medium under a predetermined set of conditions. The two most common culture strategies are most probable number (MPN) or plate count. For MPN determination, a sample is diluted serially into culture tubes containing a liquid growth medium. The sample is diluted through three to five 10-fold dilutions. Order of magnitude population density estimates can be obtained using a single dilution series. However, the standard MPN protocol recommends that at least three (preferably five) replicate culture tubes be inoculated at each dilution.[24] After incubation, the tubes are scored as positive (turbid — implying growth occurred) or negative (clear — no evidence of growth). The pattern of positive tubes per dilution is compared against a statistically derived reference table and corrected for dilution factor to determine MPN per milliliter. Setting up an MPN array is labor-intensive, but has advantages. Through careful selection of growth media, the MPN method can be used to quantify microbes with specific metabolic capabilities. Microbes that are acclimated to growth in a fluid may not form colonies on the solid media used for plate counts. Using broth media can improve culturable count recoveries.

Plate counts are performed by diluting a known sample portion (this may be milligrams of solid or milliliters of liquid) serially into a buffered dilution solution. For spread plates, a 0.1-ml portion of diluted sample is transferred to the surface of a prepoured petri plate containing the required solid growth medium. The transferred droplet is then spread uniformly across the plate's surface using a sterile glass rod bent into the shape of a hockey stick. For pour plates, either 0.1 ml or 1.0 ml of the diluted sample are transferred to an unused petri plate. Molten (approximately, 45°C) growth medium is then poured into the plate. The plate is swirled gently to disperse the sample throughout the growth medium, and the medium is allowed to solidify as it cools. The remaining steps are the same for both pour and spread plates. The plates are inverted, incubated for a specified interval, and then observed for the presence of colonies. The number of organisms or colony-forming units (CFU) in the original sample (CFU/ml, CFU/g, or CFU/cm^2) is determined by multiplying the number of colonies by the dilution factor (Figure 9.5). There are a number of commercially available dip-slides (paddles) available that serve as spread plate surrogates. Dip-slide manufacturers provide comparator charts to correlate the number of colonies on a paddle to CFU per milliliter.

Plate counts are easy to perform, particularly when using dip-slides. Also, the cells in each individual colony can be isolated easily for further testing. However, culturable microbe

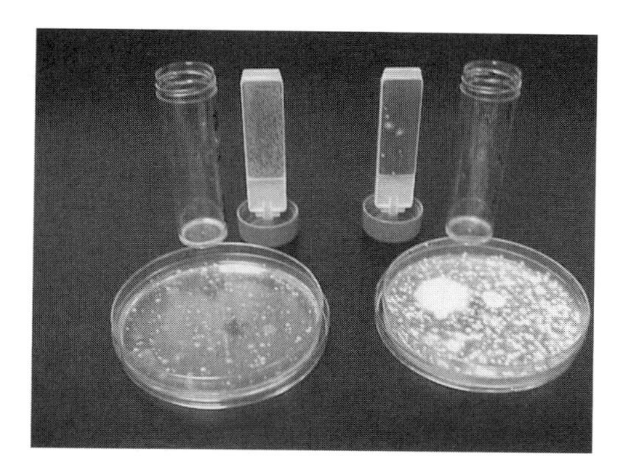

FIGURE 9.5 Culturable bacteria and fungi. Clockwise from left rear: bacterial enumeration side of dip-slide; fungal enumeration side of dip-slide; fungal colonies on culture plate; and bacterial colonies on culture plate. (Photo courtesy of Milacron Inc.)

recoveries are dependent on many variables, few of which are controllable by the analyst. Viable cells in the sample are only detected if they proliferate in or on the growth medium. There are thousands of different growth media available commercially. Each one does a better job than the others at recovering specific microbial species. No single growth medium supports the growth of more than a small fraction of the microbes present in a sample ($<10\%$ for transfer of a pure culture from a broth to a solid medium with the same nutrient composition; typically $<0.01\%$ for natural mixed populations). Microbes with longer doubling times will take longer to reach detectable population densities in broth ($>10^6$ cells/ml) or on solid media ($>10^9$ cells to form a visible colony). Consequently, routine incubation periods of 48 to 72 h will be insufficient for mycobacteria (10 to 14 days incubation required) or sulfate-reducing bacteria (up to 30 days incubation required). Sample handling, the environment of the system from which the sample was collected, and the incubation conditions all have substantial effects on culturable counts. For example, anaerobic bacteria will not form colonies on plates unless they are in an anoxic environment.

Some of the early speculations regarding the relationships between mycobacteria and Gram-negative bacteria prevalence in MWF resulted from misinterpretation of culturable count data. Since they grew slowly, mycobacteria were never detected on growth media covered with colonies of recovered Gram-negative bacteria. This led to speculation that mycobacteria only grew when Gram-negative populations had been suppressed. Further consideration of the issue indicated that it was the fact that the colonies of fast growing microbes would generally be confluent (run together) before mycobacterial colonies were large enough to be visible. Any mycobacterial colonies on the medium would be obscured by the other colonies. Only when the faster growing microbes have been killed off do plates remain colony free long enough to permit mycobacterial colonies to become visible. Once this artifact was understood, antibiotic-containing growth media (designed to suppress nonAFB) were developed. It is now apparent that the presence or absence of mycobacteria is independent of the presence or absence of Gram-negative bacteria.[19]

In the course of 48 h, a single cell can proliferate to the billions of individuals needed to create a visible colony. Similar population explosions are also possible within MWF systems. The time delay between test initiation and data availability represents a significant culture test limitation. Moreover, as interest in specific cell constituents and microbial community biodeteriogenic activities increases, the need for chemical tests has become more apparent.

3. Chemical Tests

Chemical tests are used to either measure a microbial activity or detect a chemical component of the population. Oxygen demand illustrates the first class of chemical tests. Catalase activity, fatty acid methyl ester (FAME), adenosine triphosphate (ATP), LPS, and polymerase chain reaction (PCR) tests are examples of the second class of methods.

1. *Oxygen demand.* As discussed earlier in this chapter, active MWF microbial communities invariably include obligate aerobes. By definition, these microbes consume oxygen. To run a simple oxygen demand test, an analyst aerates a sample, tests the dissolved oxygen concentration, lets the sample sit for 2 h and retests the dissolved oxygen concentration. Active microbial populations will consume $>50\%$ of the available oxygen within 2 h. The pre- and poststanding period dissolved oxygen concentrations for uncontaminated samples will be different by $<10\%$. Intermediate consumption rates typically reflect a combination of chemical oxygen scavenging (ferrous metal particles react with oxygen — this is the reaction that produces rust) and microbial activity.

2. *Catalase activity.* Catalase is an enzyme found in all fungi and most aerobic bacteria. Although the catalase concentration per cell (specific catalase activity) depends on both species and the cell's physiological state, the total catalase concentration tends to be proportional to the population's metabolic activity. A simple method was developed by Gannon and Bennett in the early 1980s, which relies on the reaction of catalase with hydrogen peroxide to liberate oxygen gas.[25] The amount of oxygen generated from a fixed-volume sample in a sealed reaction tube is proportional to the catalase concentration. Since the volume of the sample and total volume of the reaction tube are constant, Boyle's law dictates that the oxygen generated by the catalase–hydrogen peroxide reaction will cause a pressure increase in the reaction tube's headspace. The 15-min test developed by Gannon and Bennett uses the pressure as an indirect but accurate measurement of catalase concentration.

3. *FAME analysis.* The FAME profile of each microbial species is unique. Consequently, FAME analysis can be used to analyze the taxonomic diversity of microbes isolated from a sample. Isolates from the original sample are cultured and their fatty acid methyl esters (FAME) are extracted for gas chromatographic (GC) analysis. Speciation is accomplished by comparing spectra from test isolates against spectra from known species. Currently, FAME has two major drawbacks. Since the analysis is performed on cultures of isolates obtained from plate count colonies, FAME cannot be used to detect or characterize nonculturable microbes. Additionally, since FAME analysis depends on GC spectrum matching, FAME-based identifications are subject to library limitations. Many spectra from environmental samples do not match catalogued spectra or may be classified erroneously (for example, an 80% match may result in the incorrect assignment to a species or genus). As FAME methodology improves, so will its utility.

4. *ATP.* ATP is the primary energy molecule in all cells. Unlike catalase, ATP is present in all bacteria and fungi. Similar to catalase, the specific ATP concentration depends on both species and physiological state. More dynamic populations will have more ATP per milliliter than will moribund populations. Although the ATP test has been used for more than 50 years, it has only recently been applied successfully to complex fluids containing organic molecules.[26] To determine ATP in MWF, using the method described by Passman et al.,[26] a 50-μl sample is treated with a surfactant reagent to remove interferences. Next, a strong surfactant is used to lyse the bacteria and fungal cells in the sample. This releases ATP into the lysing reagent. The ATP–reagent mixture is then applied to a luciferin–luciferase impregnated ticket which is then placed into a bioluminometer (Figure 9.6). When ATP reacts with the luciferin reactant–luciferase enzyme couple, it splits into adenosine monophosphate (AMP) and diphosphate $(PO_4)_2$, liberating a photon of light. The bioluminometer records these photon emissions as relative light units (RLU). The ATP test is a field test that takes approximately 2 min to complete.

5. *Endotoxin.* Endotoxin has been discussed in Section IV.B.1. ASTM method E2250[27] provides detailed instructions on how to perform endotoxin tests on MWF samples.

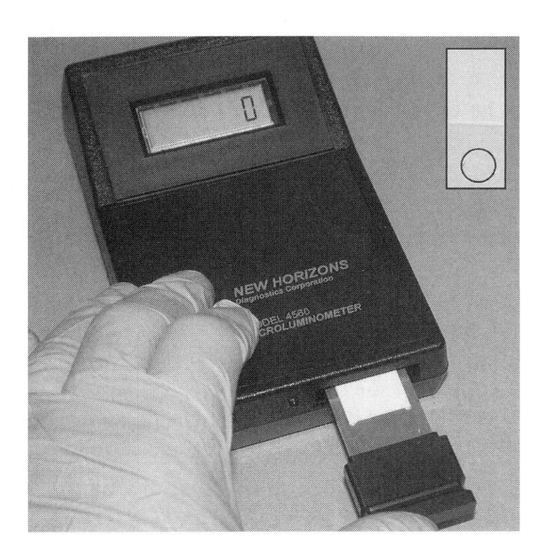

FIGURE 9.6 ATP bioluminometer. A 10-μl solution of ATP extracted from MWF sample is placed onto luciferin-luciferase impregnated filter pad (inside circle in ticket — photo insert). The ticket is folded closed, placed into bioluminometer drawer (lower right), which is then closed. Bioluminometer detects light emitted by ATP reaction with the luciferin–luciferase enzyme-substrate pair and converts data into display readout (top center) as relative light units (RLU). 1 RLU ≈ 1 ng ATP. (Photo courtesy of BCA, Inc.)

6. *PCR*. PCR testing was mentioned in Section II.A. PCR is particularly useful for characterizing microbial communities both qualitatively and quantitatively.[28] The enzyme *DNA polymerase* is used to amplify deoxyribonucleic acid (DNA) extracted and purified from a sample. Special *primers* derived either from ribosomal ribonucleic acid (r-RNA) or from known DNA gene fragments are used to tag DNA strands through repeated heat-denaturation and annealing cycles. This process, which can be repeated several dozen times within a 1- to 2-h period, creates measurable concentrations of each type of DNA in the sample. After the DNA concentrations have been amplified, they are separated by gel electrophoresis. This process concentrates each type of DNA into discrete bands. The DNA in each of these bands is then analyzed to determine the portion of the total biomass represented by the taxon and the identity of the taxon. As noted earlier[3] PCR analysis is independent of culturability. It promises to be a valuable tool for helping industry stakeholders to understand MWF microbiology better. ASTM Subcommittee E 34.50 on health and safety of MWF is working on developing a consensus method for quantifying *M. immunogenum* by PCR. A number of commercial testing laboratories currently run PCR analysis for a variety of environmental bacteria and fungi of interest to the MWF industry.

7. *Other methods*. At present there are no consensus methods for quantifying the concentrations of the various exotoxins and mycotoxins discussed in Section IV.B. Until the available methods are standardized, it will remain challenging to interpret data developed by different investigators.

4. Sampling and Sample Handling

Many of the methods described in this section are applicable to bulk fluid, surface swab/scraping, and aerosol samples. Each type of sample provides important information about the microbial state of the system. Short of immunological testing, analysis of aerosol samples provides the most direct assessment of employee exposure to MWF microbes and biomolecules. Consequently, aerosol samples are essential to health risk assessments. ASTM Guide E 1370[29] offers strategies for plant air quality monitoring. ASTM Practice E 2144[30] provides guidance for the collection of aerosol samples to be used for endotoxin analysis. Much of the information provided in Practice E 2144 is generally applicable to bioaerosol sampling.

Bulk fluid samples are the easiest to collect and test. Consequently, they are most useful for routine MWF system condition monitoring. Bulk samples should be collected in clean, sterile containers. If sterile containers are not available, unused sample bottles are acceptable. Bulk fluid samples may be drawn from a sampling or drain valve, or collected by dipping the bottle into the MWF sump. Before sampling from a valve, clean the valve and disinfect the surface with alcohol. Let the drain dispense MWF into a bucket for ≥ 30 sec in order to purge the line. Once the line is purged, place the open sample bottle into the stream, fill it, and recap it. Bulk fluid dip samples should be taken from ≥ 5 cm below the fluid surface. Before submerging the sample bottle, use a paddle to sweep away tramp oil from the sampling area. Invert the sample bottle before submerging it, then turn it right-side up to fill. To capture a surface sample, place the open bottle horizontally so that it is approximately two thirds submerged below the MWF surface.

Since most biomass is concentrated on system surfaces, periodic testing of swab or scrape samples provides the only means of assessing whether contamination control measures are fully effective. Commercially available biomedical swab kits are useful for MWF system testing. To collect a sample, sweep a sterile swab back and forth across a premeasured surface area (a 2×2 cm^2 generally works well). After swabbing, return the sampling device to its container. Scrape samples may be collected from a known area or by transfer to a sterile tared vial. Before sampling, clean the sampling tool (a spatula) and rinse it in alcohol. Allow the alcohol to evaporate before collecting the sample.

ASTM guide E1370[29] and Chapter 18 address aerosol sampling issues. The reader is referred to those documents for guidance on bioaerosol monitoring.

Samples for biological analysis age rapidly. The total and relative abundance of different microbes in a sample changes over time. Culturability may also be affected. Optimally, microbiological testing should be initiated on-site, within an hour after sample collection. If this is not possible, samples should be either refrigerated or stored on ice and tested within 18 h. Precautions should be taken to prevent sample freezing. All samples should be labeled properly. An informative label contains the following information:

- Time and date sample collected
- Identity of person who collected sample
- Sample source: sump number or shop floor grid coordinates
- Sample type: sample valve; subsurface dip; air sample, etc.
- Source fluid identity: fluid classification (emuslifiable oil, etc.); manufacturer's name and product name or code

If each sample is given a unique identification number, the sample information can be entered onto a chain of custody or sample log sheet. This sheet should also be used to track the sample's handling history:

- Date and time delivered to laboratory
- Date and time testing started
- Tests performed and analyst(s)
- Date and time testing completed

E. Data Interpretation

Typically, microbial contamination monitoring is integrated into the fluid management program as described in Chapter 11. Routine completion of all of the tests described in the preceding paragraphs would be prohibitively expensive and labor-intensive. Consequently, microbiological testing is best accomplished using an echelon approach. This means that one or two simple procedures should be run routinely (for example, bulk fluid, dip-slide colony counts, and oxygen demand). When data approach the control limits, or begin to trend away from normal variation, both the number of tests and number of samples should be increased to facilitate root cause analysis. The supplemental tests should differentiate between alternative likely causes for the reported test results. For example, consider a system with a UCL of $\log_{10}$ CFU bacteria/ml = 5 and a 2-h oxygen demand $\leq 30\%$ UCL. Testing yields the following results:

(a) Log_{10} CFU bacteria/ml = 6
(b) 2-h oxygen demand = 10%

The high culturable bacterial count could be due to several causes:

- Dormant bacteria in the system were revitalized when transferred to the nutrient medium.
- The sample contained a single floc of bacteria that dispersed during plating, but did not create a significant oxygen demand.
- The growth medium was contaminated during the test.
- The dissolved oxygen meter malfunctioned.

Collecting several samples from the system and running both culturable counts and oxygen-demand tests (after recalibrating the dissolved oxygen meter) on each sample will provide the data necessary for distinguishing among the possibilities listed above.

The critical issue behind data interpretation is to have a clear definition of normal for each parameter. Fluid managers must know both average and acceptable ranges (data variability) for each measured parameter. Personnel responsible for routine condition monitoring or data interpretation should have at least basic training in statistical process control (SPC). In particular, personnel interpreting test results must recognize the difference between special causes of variation and normal causes of variation.

VII. MICROBIAL CONTAMINATION CONTROL

Although microbial contamination control begins with good system design, and depends on well-conceived and -executed industrial hygiene programs, except for the following two caveats, this chapter will focus on the more direct aspects of microbiological contamination control. The two caveats are these. Caveat 1: Recognizing that ease of microbial contamination control is only one of many — often conflicting — considerations for system design. Design features that facilitate microbial contamination control should be incorporated to the extent that they do not degrade system performance. For example, sluices should be designed to minimize eddy formation. Eddies are effectively stagnant zones conducive to biomass accumulation. Machine housings should include easy access to sumps to facilitate sampling and cleaning. Piping dead-legs should be removed from the piping network. Numerous other relatively simple design features can simplify microbial contamination control efforts substantially. Caveat 2: Increased system insult (trash, tramp oil, shop debris, human waste, etc.) results in increased susceptibility to uncontrolled microbial contamination. Most attempts to control microbial contamination in the absence of effective housekeeping and hygiene programs fail.

A. Strategic Planning

Effective microbial contamination programs meet their objectives. This statement implies that management has defined the program's objectives and the criteria by which their successful attainment will be measured. Common microbial contamination control objectives include:

- MWF functional life extension
- Biogenic odor prevention

- Waste reduction
- Bioaerosol control/elimination
- Slime accumulation prevention

Although each of these objectives shares the common goal of reducing bioburden, each requires somewhat different monitoring efforts (sample collection — sample type, source, and collection frequency; parameters tested), and may dictate objective-specific parameter control limits. As discussed earlier under *Condition Monitoring*, data collection is only useful to the extent that it guides management decisions. ASTM E 2169[31] addresses the strategic considerations relevant to antimicrobial pesticide use. Similar strategic thinking will facilitate overall microbial contamination control efforts.

B. Physical Treatment

All physical treatment technologies share two characteristics. First, they are point source treatments. This means that MWF flows through the treatment device. Surviving microbes that settle onto downstream system surfaces are never again exposed to the treatment. Second, they function by imparting energy into the MWF. This energy is intended to kill exposed microbes. Efficacy depends on the energy's intensity and exposure time. Many physical treatment systems have a large footprint to capacity ratio, are energy intensive, and require substantial maintenance. When considering the installation of physical treatment systems, annualized capital, energy, labor, and operational costs should be compared against the costs associated with chemical treatment or disposal. Physical treatment systems use one or more of the following five types of energy to kill microbes:

- Sound
- Light
- Heat
- Radiation
- Filtration

Although physical treatments have been demonstrated successfully in bench scale and pilot studies, they have not received general acceptance by the metalworking industry. Overall economics, efficacy limitations, and scale-up challenges appear to be the primary barriers to wider acceptance.

1. Sound

Sonication systems (*sonicators*) transform line voltage (voltage) into high frequency (kHz) high energy (kJ), which in turn is transmitted to a vibrating probe or horn. For example, a sonicator can transform a 120 V current into 20 kHz frequency at 120 to 960 kJ/min energy. The probe's (or horn's) vibrations convert the electrical energy into mechanical energy. In a fluid, the result is *cavitation* — the production of small bubbles. The bubbles subsequently implode, creating mini shock waves. The force imparted by these shock waves depends on the sonication energy. At the lower end of the energy range, sonication is used to disaggregate flocs of bacteria in order to improve plate count recoveries. At the higher end of the range the energy will disrupt cell envelopes, thereby causing the cells to lyse. Sonication is used routinely to lyse cells for DNA extraction — the first step in PCR analysis.

Since cavitation decreases with distance from the vibrating device, sonication effectiveness also diminishes with distance from the device. Commercial scale, flow through sonicators use probe arrays to improve the uniformity of cavitation and duration of exposure.

Sonicators are best used to reduce the planktonic bioburden, either as components of batch treatment systems or downstream of central system filtration units.

2. Light

Ultraviolet (UV) radiation is the electromagnetic radiation of light with wavelengths (w) between 10 nm and 0.3 μm ($w_{\text{visible light}} = 0.3$ to 0.9 μm). At $w = 10$ to 280 nm, UV radiation is microbicidal. Antimicrobial performance depends on wavelength, energy (kJ/min) and exposure time. Since UV energy dissipates rapidly as it passes through water, UV systems typically are designed to have thin (<1 cm) films of water pass through a glass enclosed path (sandwiched sheets or cylinders — the *lens*) surrounded by fluorescent UV lamps. This form of disinfection is most effective for clear water treatment. Particulates, emulsion droplets, and other fluid constituents that contribute to turbidity, decrease UV light penetration into the fluid. Moreover, since radiation is not 100% effective, biofilms can develop on UV system lenses, further attenuating microbicidal performance.

In practical terms, UV irradiation can be used effectively to reduce bioburdens in MWF system makeup water. There may also be application for disinfecting mist collector exhaust air; however, UV irradiation will not neutralize antigenic biomolecules.

3. Heat

Pasteurization is the most commonly used heat treatment-based fluid disinfection process. Although its applicability for MWF treatment may be limited, pasteurization is used broadly in the food industry to disinfect both food products and fluid transfer systems. Pasteurization may be accomplished by either batch or continuous flow processes. Portable systems used to recondition MWF from individual sumps are typically batch units. Continuous flow pasteurizers are more appropriate for central systems. True pasteurization is a three-step process. The incoming fluid is heated to ≥60°C for 30 min (the specified temperature may vary among manufacturers or may be adjusted based on culturable count data from freshly pasteurized MWF). The fluid is then heated briefly to a higher temperature (16 sec at 72°C for milk). The hot fluid is then cooled to ambient temperature. Since pasteurization was developed initially to prevent food spoilage, the historical treatment parameters may be inappropriate for MWF pasteurization. Pasteurization was never intended to function as a sterilization process; consequently, determination of the desired postpasteurization culturable count is a management decision. Pasteurization systems are most appropriately used in the same applications as sonication. There is some concern that heat may destabilize emulsions and cause the breakdown of certain biocidal molecules.

4. Radiation

There are two general types of radiation equipment available. One class of systems uses high-energy electron (HEE) irradiation. The other uses gamma radiation — typically using spent nuclear reactor fuel (uranium) as the gamma radiation source. The dynamics of gamma and HEE radiation treatment are very similar to those described above for UV irradiation. Passage through fluids attenuates radiation energy rapidly; consequently, the fluid must pass through the energy beam as a thin film. Both gamma and HEE radiation are forms of ionizing radiation. Their energy is sufficient to knock electrons off of biomolecules and water. The former effect denatures biomolecules directly. The latter effect drives the reaction:

$$H_2O \xrightarrow{\text{Energy}} H_2O^+ + e^-$$

The ionized water molecule, H_2O^+, is a very reactive oxidizing agent and will denature biomolecules through a mechanism similar to that of sodium hypochlorite (bleach). Both HEE and gamma radiation have been used to disinfect wastewater in demonstration projects. Neither has been used broadly in commercial applications.

5. Filtration

Filter sterilization is accomplished by passing fluid through a physical barrier that traps microbes and nonbiological particles larger than the barrier's pore size. Used routinely in the semiconductor and biomedical industries for water sterilization, its utility for MWF disinfection is limited. Although ultrafiltration and nanofiltration have been used successfully for waste treatment (see Chapter 13), the >2-μm diameter particle loads in most recirculating MWF tend to plug the filters' pores rapidly. The resulting maintenance and consumable costs represent considerable barriers to the use of filtration for recirculating MWF sterilization.*

High efficiency particulate air (HEPA) filters used on mist collector exhaust vents remove most of the airborne bacteria and fungi from mist collector exhaust air. However, since traces of oil mist and moisture can condense on these filters, they can also become sites for microbial colonization. This is rarely a problem when filters are replaced as required. If HEPA filters are not replaced routinely, microbes colonizing the filters can become a major source of airborne inocula. Mist collector HEPA filters will retain a percentage of biomolecules, but do not capture 100% of endotoxin or other antigenic cell constituents.

C. Chemical Treatment

Two fundamental, but not necessarily mutually exclusive, strategies fall into the chemical treatment category. The first is to formulate with bioresistant (*recalcitrant*) molecules. The second is to use *antimicrobial pesticides* (*microbicides*). Chapter 6 has addressed MWF formulation thoroughly. Recalcitrant molecules do not have any intrinsic antimicrobial activity. Rather, they resist microbial attack thereby depriving contaminant microbes of a food source. It was noted earlier (Section III.A) that, as a general rule, more complex molecules are less biodegradable than simpler molecules. Distinguishing between recalcitrance and microbicidal activity has important legal implications, particularly in the U.S.

In the U.S., microbicide registration and usage is controlled under the provisions of the Federal Insecticide, Fungicide and Rodenticide Act[32] and regulated under 40 CFR 152 *et seq.*[33] Chemicals used to control pests (including microbes) in industrial, commercial, domestic, or agricultural applications must be registered in accordance with the cited regulations. The *intent* stipulation creates room for confusion. For example, 0.5% sodium hypochlorite used as household bleach (nonpesticide usage) is not registered. That same 0.5% solution sold as a swimming pool algaecide is registered under 40 CFR 152.

In MWF formulations, oxazoladines and triazines used as neutralizing amines are marketed as technical-grade chemicals without pesticide registration. A number of the same chemicals are also sold as registered microbicides. Others, with demonstrated microbicidal activity are marketed as bioresistant additives. There is a balance between ethics and application. Current costs for the complete package of toxicological tests needed for microbicide registration may exceed $250,000. The testing and application process may span several years. Some companies choose to bypass these monetary and time investments by invoking (incorrectly) the term *bioresistant*. If the manufacturer has data to support their nonmicrobicidal performance claims (corrosion inhibition, emulsion stabilization, etc.), and they fully intend to compete against comparable, nonmicrobicidal performance additives, they may do so legitimately.

* Regarding filtration, *sterilization* rather than *disinfection* is the appropriate term because the objective is 100% removal of planktonic microbes.

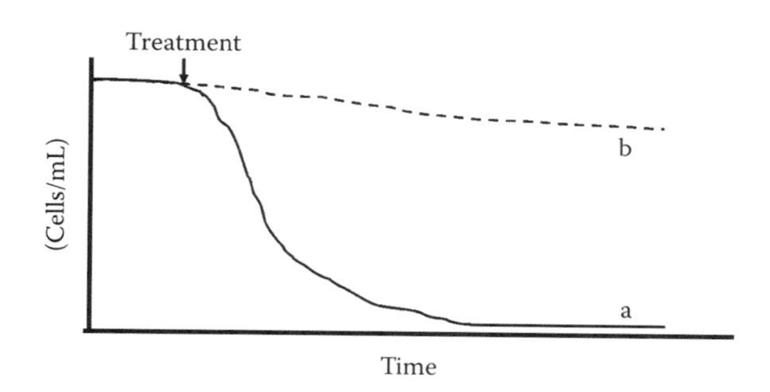

FIGURE 9.7 Comparison between effects of microbicide treatment and bioresistant additive treatment. Curve (a) After microbicide addition, population density decreases precipitously. Curve (b) Bioresistant additive has negligible short-term effect on microbial population density, although population may die off slowly.

If an additive is introduced into a heavily contaminated MWF at its normal used concentration, and the microbial population plummets (Figure 9.7, curve a) then the product is a microbicide, regardless of claims. A bioresistant MWF will not support the growth of a contaminant population. If added to a system with a high bioburden, the population will die off slowly (Figure 9.7, curve b).

As detailed in ASTM E 2169,[31] microbicides may be used to disinfect formulation equipment, preserve MWF concentrate in storage, protect dilute MWF in application, disinfect MWF systems or, most commonly, combinations of the above. At present, there are at least 70 active ingredients approved for use in MWF in the U.S. (see Ref. [34], Table 2). Product selection should be based on intended application, target microbes, and chemical compatibility with other formulation chemicals. For example, fast-acting (>2 log CFU ml^{-1} reduction in ≤ 30 min), short half-life ($T_{\frac{1}{2}} < 96$ h) products are useful for tankside addition and for equipment disinfection. They should not be built into formulations if the compounder expects to have a residual active ingredient available to protect the formulation in end-use application.

Glutaraldehyde, bromo-nitro-propanediol (BNPD) and 2,2-dibromo-3-nitrilopropionamide (DBNPA) and the blend, 5-chloro-2-methyl-3(2H)-isothiazaolin-3-one + 2-methyl-3(2H)-isothiazolin-3-one (CIT/MIT) are examples of short $T_{\frac{1}{2}}$ microbicides. In particular, glutaraldehyde is used as a surface contact disinfectant in the biomedical industry. Phenolic microbicides, such as orthophenylphenol (OPP) are also used as surface contact disinfectants. In contrast to glutaraldehyde, BNPD, DBNPA, and CIT/MIT, OPP is a persistent product. Its $T_{\frac{1}{2}}$ depends primarily on MWF turnover rate and consumption through reaction with microbes than with the chemical degradation processes that decrease the $T_{\frac{1}{2}}$ of the nonpersistent products. Primarily soluble in nonpolar solvents, phenolic microbicides are appropriate for use in emuslifiable oil and semisynthetic MWF formulations. Phenolic microbicides are generally stable in formulation concentrates. Consequently, they can be formulated into MWF concentrates or used tankside.

The most commonly used microbicides (hexahydro-1,3,5-tris(2-hydroxyethyl)-s-triazine — *triazine*; various oxazoladines and 4-(2-nitrobutyl)morpholine — NMEND) are suitable for use both in formulations and as tankside additives.

The European Economic Union and several nations also maintain lists of approved microbicides. Products approved for use in one nation are not necessarily approved for use elsewhere. Within the U.S., most states have product registration requirements. Confirming legality of use is a critical early step in microbicide selection.

Microbicides should not be used in MWF applications until their performance has been evaluated in bench and pilot-scale studies. Product testing protocol should be guided by the application specifics.[34] For example, microbicides used for surface disinfection do not need to be persistent. Conversely, persistence is more important than speed of kill for products used to preserve MWF concentrate during in-drum storage. *Bactericides* are microbicides that are specifically effective against bacteria. Some bactericides such as formaldehyde-condensates (oxazoladines and triazines) are primarily effective against Gram-negative bacteria. Although it is also a formaldehyde condensate, NMEND is a true broad-spectrum antimicrobial agent. Lipid soluble microbicides (*ortho*-phenylphenol, *para*-chloro-*meta*-cresol and *para*-chloro-*meta*-xylol) are particularly effective against acid-fast bacteria.

Microbicides that target fungi are called *Fungicides*. Only four of the approximately 70 U.S. EPA-registered microbicides referred to above are fungicides: iodo propynyl butyl carbamate (IPBC), and *n*-octylisothiazolinone (NOIT), sodium pyridinethione (NaP), and 2-(thiocyano-methylthio) benzothiazole (TCMTB). The antifungal activity spectra of these microbicides are similar, but their handling characteristics differ. For example, IPBC has limited water or oil solubility and is therefore used most often in semisynthetic MWF. Dissolved iron will react with NaP, forming insoluble, but fungicidally active iron pyridinethione.

Products that are effective against both bacteria and fungi are considered broad-spectrum microbicides. NMEND and several two-ingredient blends (in particular, NaP + triazine, and methylene-*bis*-thiocyanate + TCMTB) function as broad-spectrum microbicides. A number of manufacturers claim broad-spectrum performance for their products, but the doses needed to kill the secondary target* are at, or above the maximum permissible treatment level listed on the U.S. EPA-approved product label.

Blending microbicides into MWF formulations is a more complex undertaking than generally appreciated in the industry. The end-use microbicide concentration will depend on the MWF dilution in-application. Unless the formulation is to be used at the same end-strength in all applications, the microbicide concentration may be too low when the MWF is used lean or too high when the MWF is used rich. For example, consider a MWF formulated with 3% of a microbicide designed to work best at 1000 ppm (0.1%). Assume that for this product, the permissible end-use concentration is 2000 ppm. A 1:20 dilution of the MWF will dilute the microbicide to 1500 ppm; well within the desired use range. However, if the MWF is used at 3% instead of 5%, it will deliver only 900 ppm of the microbicide (see the discussion below on *hormonesis*). At 7%, the formulation will deliver 2100 ppm of the microbicide — 100 ppm in excess of the maximum permissible dose.

If microbicide delivery via the MWF concentrate is not augmented with system bioburden monitoring, there is a considerable risk that microbial demand will deplete the microbicide package selectively, relative to the other MWF components. Conversely, in a low-bioburden system, the concentration of a persistent microbicide may build up so that it exceeds the maximum permissible concentration. Many effective antimicrobial pesticides that are safe to use within the recommended concentration range, can become significant dermal and respiratory irritants at excessive concentrations. Complete reliance on in-formulation microbicides is likely to: (a) reduce overall microbial contamination control efficacy; and (b) increase the risk of worker over-exposure.

The universal microbicide has yet to be invented. The art lies in matching appropriate actives with their applications. Much of the competitive marketing literature is similar to arguing over the relative merits of one size crescent wrench vs. another. The appropriate wrench is the one that fits the hexagonal nut to be turned. Arguments regarding the relative toxicities of microbicides are similarly misleading. Products may differ in their relative oral, dermal, or inhalation toxicities, but overall, none of the commercially available products is substantially more toxic than the others.[35] When products are used in accordance with label, material safety data sheet, and manufacturers' technical application instructions, they create no significant incremental toxicological risk (see Chapter 15). Some of the microbicides used most commonly in MWF are also used in personal care products (for example a blend of chloro-methyl isothiazolinone and methyl

* For a bactericide such as triazine, fungi would be the secondary target. For a fungicide such as NaP, bacteria would be the secondary target.

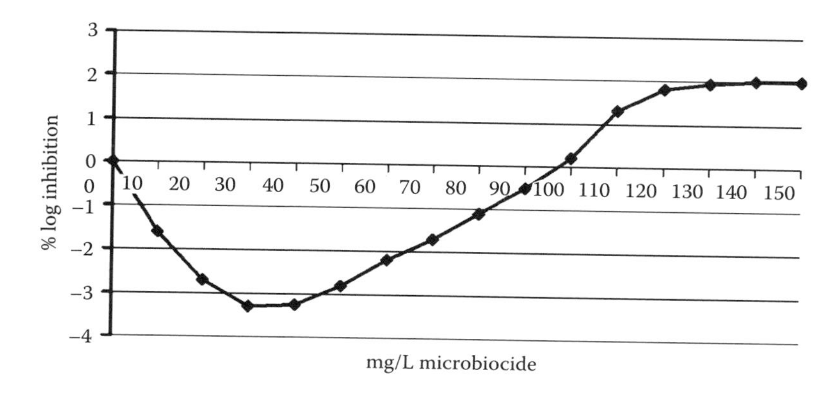

FIGURE 9.8 Hormonesis. At 40 mg microbicide/l, population is >3 orders of magnitude more dense than in the untreated control. At the 100 mg/l dose, the population density is not significantly different from that of the untreated control. Inhibition approaches 100% at doses ≥120 mg microbicide/l. *Hormonesis* is the apparent stimulation effect of low dosage treatments.

isothiazolinone). Substantially higher microbicide concentrations are used in the latter application than in MWF.

Chemical treatment is most effective when used as a preventive measure. Microbicide demand is greater when a product is used to treat a heavily contaminated system. Frequently, once a heavy bioburden has developed, microbicide concentration is depleted to below the minimum effective concentration before the target population is eradicated. Unless microbicide concentration is monitored, systems are likely to be either undertreated or overtreated. Underdosing may select for resistant microbes or cause *hormonesis*. Hormonesis is the phenomenon of apparent stimulation at sublethal doses (Figure 9.8). The treated population can respond to sublethal doses of a toxic agent by increasing slime production, producing metabolites that react with and neutralize the agent, or changing cell envelope chemistry to reduce the cell's susceptibility to the agent. As depicted in Figure 9.8, one obvious symptom of this response may be dramatic increases in cell proliferation.

Figure 9.9 illustrates the advantages of data-driven microbicide dosing. This approach reduces the risk of developing high bioburdens or selection for treatment-resistant populations. The

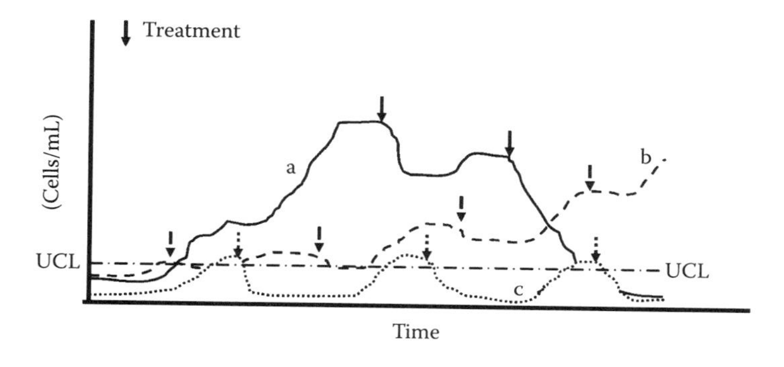

FIGURE 9.9 Comparison of relative impacts of three microbicide treatment strategies. (a) Corrective shock treatment — may require multiple doses to reduce population density to below UCL. (b) Time-based preventive treatment — may result in overtreatment or undertreatment. Undertreatment will select for resistant microbes. (c) Data-driven treatment — timing, dosage, and selection of microbicides are dictated by microbiological data.

increased level of effort required for data-driven treatment is balanced by the overall benefits of improved microbial contamination control.

D. Disposal

Once a system has developed visible slime and noticeable odors, the aforementioned contamination control strategies are unlikely to be effective. As noted earlier, single-point treatment systems will have no impact on downstream biofilm communities. Gross symptoms of microbial contamination provide ample evidence that microbicides and bioresistant additives built into MWF formulations have been overwhelmed. Additional tankside microbicide additions may provide transient relief of some symptoms, but they are unlikely to return the system to a satisfactory level of control. It is now time to drain and clean the system.

Draining and recharging MWF systems without cleaning them adequately also provides only a short-term fix. Biofilm growth is unaffected by simply draining and recharging a system. Each fluid management service provider has their own preferred process. However, all effective system cleaning processes share the following common elements.

High pressure wash. Splash zone surfaces accumulate MWF residue and mist vapors, which in turn provide an excellent habitat for colonization and biofilm development. All splash zone surfaces should be washed down prior to system draining. Surfaces, such as sluice-walls and deck-plate undersides may need to be treated with a dilute microbicide before being washed.

Machine cleaner recirculation. Whether incorporated into the recirculating MWF before draining or used as a postdraining step, recirculating a good machine cleaner helps to disaggregate and flush residues off system surfaces. The duration of the recirculation process will depend on the system size and cleaner chemistry. Fluid managers should have the technical expertise to determine the appropriate exposure period for effective cleaning.

Drain and clean. These two previous steps will generate a substantial volume of solid waste. Once the machine cleaner has been recirculated and drained, sludge, swarf, chips, and other debris accumulated in the system sump and in sluiceways should be removed physically.

Dilute microbicide recirculation. The next step is to recirculate a disinfection rinse. This rinse will normally contain water, one or more microbicides, and a corrosion inhibitor. It may also include machine cleaner. As for the original machine cleaner recirculation process, the duration of the microbicide treatment will depend on parameters best specified by the fluid management service provider.

System flush. The microbicide-treated system should be flushed with clean water, before fresh MWF is added. The flush solution may contain a corrosion inhibitor or dilute MWF, to prevent flash corrosion. If the flush fluid is clean after it has run through the MWF system, the system is ready for recharging. If not, the system should be treated again with a microbicide-machine cleaner solution. The cycle of cleaning and flushing should be repeated until the flush fluid is clean after it has run through the system.

Other considerations. Thorough system cleaning is often perceived as providing an inadequate return on investment. Cleaning may require a day or more of downtime. Personal protective gear and portable ventilation equipment will be needed in areas where workers are exposed to MWF residues and aerosols generated during the cleaning process. Obviously, the process generates substantial volumes of waste. Without a doubt, the cleaning process is labor-intensive. Notwithstanding the short-term productivity losses and direct costs associated with thorough system cleaning, the return on investment is often substantial. Clean systems translate into increased productivity as measured in parts per unit time, parts per tool, decreased MWF and tankside additive consumption, and increased filter indexing rates. Moreover, clean systems create a healthier work environment.

To provide the greatest benefit, system cleaning should be coordinated with facility cleaning. Ventilation systems and mist collectors, both major sources of microbial contamination, should be cleaned and disinfected. MWF piping dead-legs should be identified and eliminated from the system. Stagnant MWF accumulates in dead-leg pipes and supports heavy biomass accumulation. Recirculating MWF passing by the intersection of a dead-leg creates a vacuum (Venturi) effect that draws contaminated MWF from the dead-leg into the recirculating fluid, thereby reinoculating the system as soon as its been recharged.

System cleaning inevitably involves worker exposure to hazardous chemicals. Personnel involved in cleaning operations should be trained in the relevant hazardous material and confined space entry considerations, should apply the appropriate health and safety risk minimization practices, and be supervised appropriately.

VIII. SUMMARY AND CONCLUSIONS

Uncontrolled microbial contamination of MWF systems represents both economic and health risks. Effective microbial contamination control depends on a fundamental understanding of MWF system microbial ecology. In particular, it is important to recognize that the percentage of microbes recirculating with the fluid is a fraction of that growing on system surfaces.

Bacteria and fungi enter MWF systems with the make-up water, are carried along by ventilation system air, and are tracked in with vehicle tires and personnel footgear. Although it is impracticable to prevent system inoculation completely, it is cost-effective to mitigate inoculation substantially through good housekeeping and industrial hygiene practices. Effective condition monitoring, coupled with timely and effective microbial contamination control measures can minimize the adverse operational problems caused by uncontrolled microbial contamination. Moreover, effective contamination control can reduce bioaerosols in the plant environment.

The definition of effective microbial contamination control is in flux. Ultimately defined by management, adequate control needs to encompass both operational and health considerations. Strategies for providing adequate protection against performance-related biodeterioration are informed by over 50 years of MWF microbiology literature, including several consensus documents. In contrast, effective strategies for minimizing bioaerosol exposure risks are still being debated. There are few studies that correlate specific microbial contamination control measures with their impact on bioaerosol concentrations. Moreover, the relationship between bioaerosol exposure and specific health effects still requires considerable research. In the absence of a complete understanding, the prudent manufacturer will err on the side of caution, rather than assume that an unproven risk is the same as no risk.

REFERENCES

1. Rozack, D. B. and Colwell, R. R., Survival strategies of bacteria in the natural environment, *Microbiol. Rev.*, 51(3), 365–379, 1987.
2. Holt, J. G., Krieg, N. R., Sneath, P. H. A., Staley, J. T., Williams, S. T., Eds., *Bergey's Manual of Determinative Bacteriology*, 9th ed., Williams & Wilkins, Baltimore, MD, p. 787, 1994.
3. Zhu, X. Y., Lubeck, J., and Kilbane, J. J. II, Characterization of microbial communities in gas industry pipelines, *Appl. Environ. Microbiol.*, 69(9), 5354–5363, 2003.
4. Xu, K. D., McFeters, G. A. and Stewart, P. S., Biofilm resistance to antimicrobial agents, *Microbiology*, 146, 547–549, 2000.
5. Videla, H. A., *Manual of Biocorrosion*, Lewis Publishers, New York, p. 273, 1996.
6. Passman, F. J. and Rossmoore, H. W. R., Reassessing the health risks associated with employee exposure to metalworking fluid microbes, *Lubr. Eng.*, 58(7), 30–38, 2002.
7. Proft, T., Ed., *Microbial Toxins*, BIOS Scientific, Oxford, p. 600, 2005.
8. Frecer, V., Ho, B. and Ding, J. L., Interpretation of biological activity data of bacterial endotoxins by simple molecular models of mechanism of action, *Eur. J. Biochem.*, 267, 837, 2000.

9. Castellan, R. M., Olenchock, S. A., Kinsley, K. B. and Hankinson, J. L., Inhaled endotoxin and decreased spirometric values, *N. Eng. J. Med.*, 317, 605–610, 1987.
10. Laitinen, S., Linnaimaa, M., Laitinen, J., Kiviranta, H., Reiman, M. and Liesivouri, J., Endotoxins and IgG antibodies as indicators of occupational exposure to the microbial contaminants of metal-working fluids, *Int. Arch. Occup. Environ. Health*, 72, 443–450, 1999.
11. Douglas, H., Rossmoore, H. W., Passman, F. J. and Rossmoore, L. A., Evaluation of endotoxin-biocides interaction by the *Limulus* amoebocyte assay, *Dev. Ind. Microbiol.*, 31, 221–224, 1990.
12. Anonymous. *Some Common Mycotoxins and the Organisms that Produce Them.* http://www.mold-help.org/fungi.mycotoxins.currentresearch.htm.
13. Cormier, Y., Hypersensitivity pneumonitis, In *Environmental and Occupational Medicine*, 3rd ed., Rom, W. N., Ed., Lippincott-Raven, Philadelphia, PA, p. 457–465, 1998.
14. Rose, C., Hypersensitivity pneumonitis, In *Occupational and environmental respiratory disease*, Harber, P., Schenker, M. B., and Balmes, J. R., Eds., Mosby-Year Book, St Louis, pp. 293–329, 1996.
15. Bernstein, D. I., Lummis, Z. L., Santili, G., Siskosky, J., and Bernstein, I. L., Machine operator's lung. A hypersensitivity pneumonitis disorder associated with exposure to metalworking fluid aerosols, *Chest*, 108(3), 593–594, 1995.
16. Passman, F. J., ASTM symposium on the recovery and enumeration of mycobacteria from the metalworking fluid environment, *J. Test. Eval.*, 332005, paper ID 12835, www.astm.org.
17. Kreiss, K. and Cox-Ganser, J., Metalworking fluid-associated hypersensitivity pneumonitis: a workshop summary, *Am. J. Ind. Med.*, 32(4), 423–432, 1997.
18. Wallace, R. J. Jr., Zhang, Y., Wilson, R. W., Mann, L., and Rossmoore, H. W., Presence of a single genotype of the newly described species *Mycobacterium immunogenum* in industrial metalworking fluids associated with hypersensitivity pneumonitis, *Appl. Environ. Microbiol.*, 68(11), 5580–5584, 2002.
19. Yadav, J. S., Izhar, U. H., Khan, F. F., and Soellner, M. B., DNA-based methodologies for rapid detection, quantification, and species- or strain-level identification of respiratory pathogens (Mycobacteria and Pseudomonads) in metalworking fluids, *Appl. Occup. Environ. Hyg.*, 18, 966–975, 2003.
20. Anonymous, *Metalworking Industry Standards: Environmental Quality and Safety, Fluid Performance and Condition Monitoring Tests*, ASTM International, West Conshohocken, Standards on CD-ROM: FLUIDSCD, 2003.
21. Rossmoore, L. A. and Rossmoore, H. W., Metalworking fluid microbiology, In *Metalworking Fluids*, Byers, J., Ed., Marcel Dekker, New York, pp. 247–271, 1994.
22. Lawrence, J. R., Korber, D. R., Wolfaardt, G. M., and Caldwell, D. E., Analytical imaging and microscopy techniques, In *Manual of Environmental Microbiology*, Hurst, C. J., Knudsen, G. R., McInerney, M. J., Stetzenbach, L. D., and Walter, M. V., Eds., ASM Press, Washington, DC, pp. 29–51, 1997.
23. Rossmoore, L. A., Rossmoore, K., Cuthbert, C., and Cribbs, C., Direct microscopic count of mycobacteria from metalworking fluids, *J. Test. Eval.*, 332005, paper ID 12836, www.astm.org.
24. Anonymous. Estimation of bacterial density, In *Standard Methods for the Examination of Water and Wastewater*, Eaton, A. D., Clesceri, L. S., and Rice, E. W., Eds., American Public Health Association, Washington, DC, pp. 9–49, see also 9–51, 2005.
25. Gannon, J. D. and Bennett, E. O., A rapid technique for determining microbial loads in metalworking fluids, *Tribol. Int.*, 14(1), 3–6, 1981.
26. Passman, F. J., Loomis, L., and Tartal, J., Non-conventional methods for evaluating fuel system bioburdens rapidly, In *Proceedings of the Sixth International Filtration Conference*, Bessee, G. B., Ed., Southwest Research Institute, San Antonio, 2004, CD-ROM.
27. Anonymous, E2250 *Standard Method for Determination of Endotoxin Concentration in Water Miscible Metal Working Fluids*, ASTM International, West Conshohocken, www.astm.org.
28. Kahn, I. U. and Yadav, J. S., Real-time PCR assays for genus-specific detection and quantification of culturable and non-culturable mycobacteria and pseudomonads in metalworking fluids, *Mol. Cell. Probes*, 18(1), 67–73, 2004.
29. Anonymous, E1370 *Standard Guide for Air Sampling Strategies for Worker and Workplace Protection*, ASTM International, West Conshohocken, www.astm.org.

30. Anonymous, *E 2144 Standard Practice for Personal Sampling and Analysis of Endotoxin in Metalworking Fluid Aerosols in Workplace Atmospheres*, ASTM International, West Conshohocken, www.astm.org.
31. Anonymous, E2169 *Standard Practice for Selecting Antimicrobial Pesticides for Use in Water-Miscible Metalworking Fluids*, ASTM International, West Conshohocken, www.astm.org.
32. Anonymous, Federal Insecticide, Fungicide and Rodenticide Act. 7 U.S. Code 136 *et seq.*, 1996.
33. Anonymous. *Pesticide Classification and Registration Procedures. 40 Code of Federal Regulations 152*, US Government Printing Office, Washington, DC, 2003.
34. Anonymous, E2275 *Practice for Evaluating Water-Miscible Metalworking Fluid Bioresistance and Antimicrobial Pesticide Performance*, ASTM International, West Conshohocken, www.astm.org.
35. Passman, F. J., Formaldehyde risk in perspective: a toxicological comparison of twelve biocides, *Lubr. Eng.*, 52(1), 68–80, 1996.

10 Filtration Systems for Metalworking Fluids

Robert H. Brandt

CONTENTS

I. INTRODUCTION

Metalworking fluid chemistry and its uses have gone through many changes over the years. With the greater performance requirements for both direct metal removal attributes and indirect functional attributes, upsets in metalworking fluid integrity will impact performance. If a once-through use was practiced, contamination from the operation itself would not cause difficulty. Even once-through use could be compromised by the water used to dilute the concentrate and the containment or delivery methods chosen; but for the most part, the metalworking fluid would contain all the active ingredients and performance packages required.

In the real world, the practice of once-through fluid use is not acceptable except in some specific processes. Therefore, the metalworking fluid is reused for months or years. The reuse consists of collecting the used fluid in some sort of container and then recirculating the fluid back to the tool–workpiece interface. Units as simple as a tank and pump may constitute a recirculation system.

When reuse is instituted, a variety of interactions from an assorted number of contaminants begin to occur. The metalworking fluid is subjected to the metal chips and fines of the process, airborne contamination from cascading fluid over a part and the machine, machine leakages, residues left on the part from previous operations, water, operators, etc.[1] This list can be quite long.

In order to provide metalworking fluid in an acceptable operating condition after it has been subjected to these and other degradation contaminants, the impact from these contaminants needs to be minimized. This can be accomplished chemically by adding new concentrate or additives to the working solution. However, some contaminants may not be appreciably affected by additives or concentrate additions. These include metal chips, fines, and free oil.

Whenever possible, contaminants need to be removed from the metalworking fluid and the system. The removal process is generally some means of separation or filtration. Many aspects of the process need to be discussed before a final approach or system can be selected. It is not just a matter of saying we want to remove the metal chips. Various criteria for design need to be addressed and a number of questions need to be answered. These criteria include: material being worked, type of machine processing, chip shapes produced, amount of material removed, production rates, machine horsepower, metalworking fluid type, amount of fluid required, and a floor plan layout. As we begin to develop a system, typical components will be addressed and some answers will be provided. Typical system components include: return troughs, chip conveyors, filters, supply line pumps, makeup systems, and electrical and pneumatic controls.

Application of ultrafiltration, nanofiltration, or reverse osmosis is generally not applied to in-plant metalworking systems. These filtration regimes are used in selected processes, such as incoming water preparation and metalworking fluid waste treatment. These membrane systems, if used as the metalworking fluid filtration on-site, would have a deleterious effect on the fluid by selectively removing certain ingredients.

II. PARTICULATE

Before moving to a discussion of various particulates, let us first address key issues: How free of particulate should the fluid be kept? How clean should the fluid be? Various answers are given and various approaches taken to these questions. A piece of equipment may be purchased with an implied guarantee of particulate cleanliness. However, each piece of equipment will provide cleanliness to a certain equilibrium level. We can always expect some residual, equilibrium level of metal particulate in the metalworking fluid. There is no absolute filtration in the metalworking fluid industry such that all metallic particulate will be removed. Given this fact, the best that can be attained is to minimize the metallic contamination equilibrium level being maintained by one or a combination of separation or filtration devices. Considering only the usual filtration and separation processes then, let us look at the equilibrium level of the metal fines in the metalworking fluid.

The equilibrium level occurs because the filtration device will not remove 100% of the metal removed in the metalworking process each time the fluid passes through the filter.[2] This residual quantity stays in the system and builds until the filter reaches the equilibrium level. This level is different for each process and filter system. The level is related to the fluid used, the maintenance of the fluid and machines, the filter, and a host of other outside influences. To determine the equilibrium level which is reasonable and necessary, tests can be run on existing filtration systems and then correlated.[3] These tests will provide a direction toward the particulate levels which should be maintained. Table 10.1 provides some information on two aspects of the metal particulate

TABLE 10.1
Equilibrium Average for Suspended Solids from Various Equipment

	Quantity[a] (ppm)	Quality[b] (μm)
Cast iron		
Machining	20	15
Grinding	30	30
Steel		
Machining	25	20
Grinding	12	16
Aluminum		
Machining	10	15
Grinding	10	15
Glass		
Grinding	100	<5

[a] Metallic particulate only.
[b] Most probable size: measurement by microscope and electrical sensing zone instrument.

to consider at the equilibrium level. The first is the amount of metal fines that would be tolerable, i.e., the quantity of the dirt in the system. It is not sufficient, however, to describe only a quantity of dirt. This quantity of dirt, expressed in mg/l or parts per million (ppm), gives only an indication of the amount by weight of dirt at equilibrium. For example, a quantity represented by 10 mg/l (ppm) would seem small. Extrapolated to its meaning in a large volume system, however, it could be quite significant. In a 10,000 gal (38,000 l) central filtration system, the weight of metalworking fluid may approach 83,000 lb (38,000 kg). This would mean approximately 0.83 lb (0.38 kg) of metal fines circulating in the system. If a system contained 100,000 gal (380,000 l), the quantity would be 8.3 lb (3.8 kg). Increasing the parts per million would mean higher levels of recirculating metal fines.

The other necessary number to deal with at the equilibrium point is the quality of the recirculated metal fines. The quality refers to the size of the fines being circulated. How large are the particles reaching the tool–workpiece interface? It has been suggested that particulates in the 3 to 8 μm range have more effect than had been previously suspected.[4] The size of the particle, however, is open to some discussion because there are a number of ways to describe and also determine the particle size. Typically, particle size is considered in terms of spheres. Therefore, the particle size number may be referred to as the spherical diameter. This, however, is generally different from the real world. Chips and particulate come in a variety of shapes including flat platelet, cylindrical, parts of a broken helix, etc. Rarely is the particle a sphere. However, a guide or common reference is necessary and therefore we talk in terms of only a single micron size — a linear dimension — to describe the particulate. Figure 10.1 shows some linear dimension comparisons. Agreement on an acceptable level of fluid cleanliness is one of the first requirements of system design. There should be two numbers presented: a quantity of recirculated dirt at equilibrium and the average particle size.

The particulate being recirculated at equilibrium is only the resulting end of the particulate produced by the metal removal operation. The particulate produced consists of a large number of different shapes, sizes (up to feet in length), and volume. Volume will be discussed in a subsequent section. The shapes and sizes vary according to the metal being worked and the tooling on the machine. If the metal is steel and it is being machined, the chips can be long and stringy or curled and small (see Figure 10.2). If the steel is being processed by grinding, the chips produced are generally referred to as "fish hooks" because they lock together in a steel wool pad arrangement (see Figure 10.3). Aluminum forms a variety of shapes. The shape of aluminum

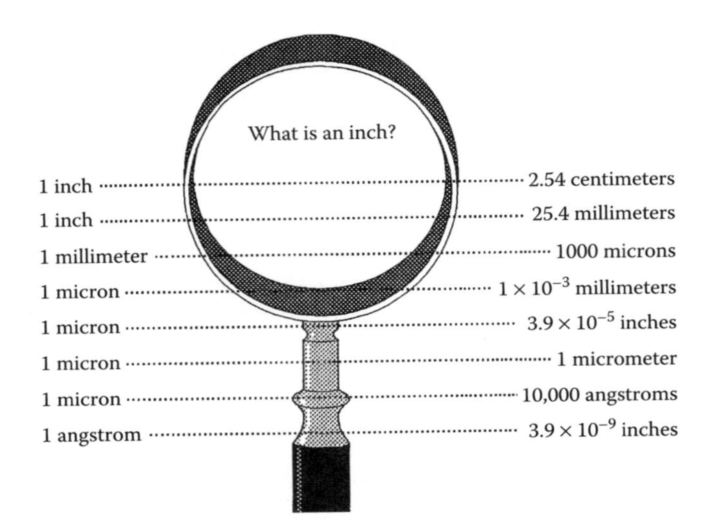

FIGURE 10.1 Linear dimension comparisons.

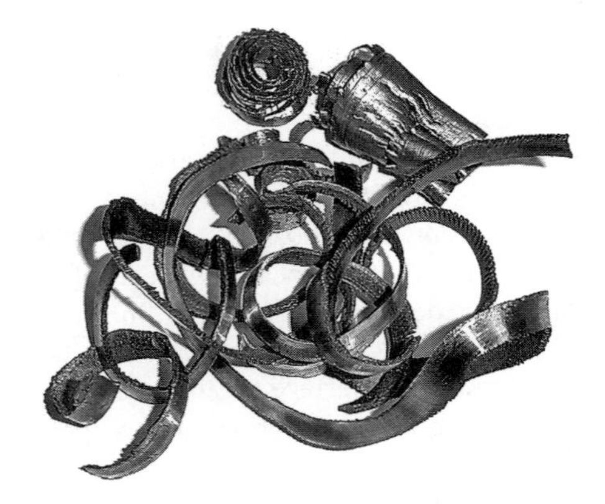

FIGURE 10.2 Varieties of steel-machining chips.

FIGURE 10.3 Varieties of steel-grinding chips.

FIGURE 10.4 Varieties of aluminum-machining chips.

machining chips are in general the same as aluminum grinding chips (see Figure 10.4). Cast iron produces a much different chip than either steel or aluminum, more of a flat type and the chips form a more dense mass (see Figure 10.5). Under usual machine operations, the metals produce a variety of large and small chips. In certain operations, however, such as honing, finish grinding, lapping, and others, the average size of the chip is small and presents a different filtration requirement. With different alloys, different machines and processes, and different tooling, it is necessary to explore the chip configuration closely in order to apply the best filtration system available to reach the desired effect.

III. TRANSPORT SYSTEMS

While the metalworking fluid is being delivered to the machine at the tool–workpiece interface, work is being done in the form of metal removal. The fluid flushes chips produced in the operation and carries them as part of the fluid flow. This fluid mixture flows off the machine and into a variety of devices, which will transport the fluid and chips back to a point of separation or filtration. These methods include H-chain, chain and flight, push bar (harpoon), metalworking fluid, and in some cases, overhead troughs and sumps.

FIGURE 10.5 Varieties of cast iron-machining chips.

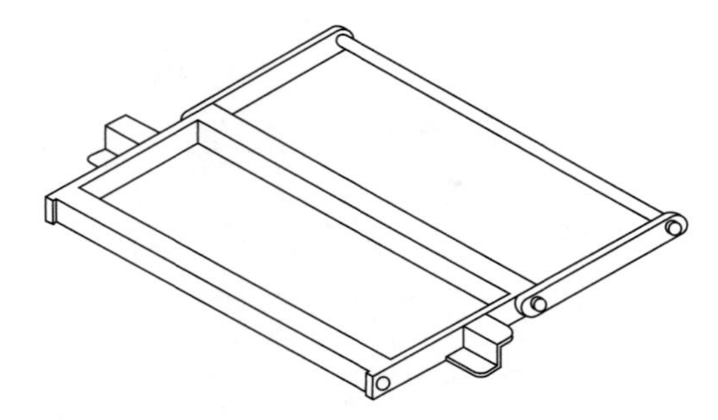

FIGURE 10.6 Typical H-chain section. (Courtesy of Brandt and Associates, with permission.)

A. H-Chain

The H-chain or rubbish chain is not commonly used because of potential repair difficulty when the chain breaks under machines (see Figure 10.6). In use, the H-chain is applied to cast iron-type materials which have been machined. This system does provide one advantage; it can deliver the chips to a tote-box prior to the separation or filtration equipment. A problem with this method is the need for additional fluid-holding capacity in the reservoir of the filter in order to accommodate the "draw down" resulting from the retention of fluid in the conveyor trench.

B. Push Bar (Harpoon)

The push bar conveyor or oscillating system is typically applied to steel-machining systems (see Figure 10.7). Steel chips can be delivered to a tote-box before the fluid is separated or filtered. The typical installation requires a large amount of mechanical apparatus under the floor, usually in troughs. This system requires maintenance of the system's hydraulic components as well as the push bar itself. In typical setups, the fines and metalworking fluid are allowed to exit the system through panels of perforated plate into a filtration system. Although these conveyors can be used for other metals, steel machining is the most common application. The same draw-down considerations are needed with this system as with the H-chain because of the retention of metalworking fluid in the troughs.

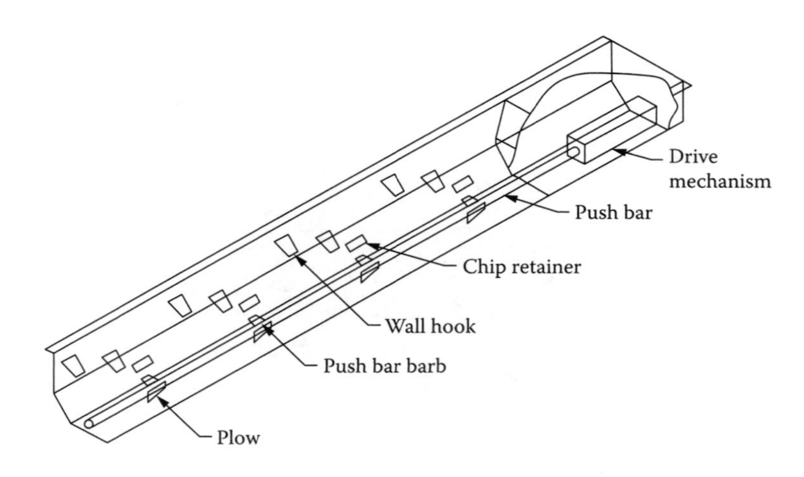

FIGURE 10.7 Typical push bar (harpoon) section. (Courtesy of Brandt and Associates, with permission.)

FIGURE 10.8 Chain and flight conveyor.

C. Chain and Flight

Another type of conveyor system applied in special cases is referred to as the chain and flight system (see Figure 10.8). This system is composed of two continuous loops of chain, between which has been bolted or welded iron bar stock or angle iron. This chain and flight system is put into a trough and the movement of the conveyor removes heavy settled solids up a ramp to a discharge point. The trough usually has an overflow opening so the liquid can flow out of the trough, along with fine metal particles, and into a filter system. The advantage of this system is that it removes bulk solids before the filtration process system. The metal usually machined or ground when this type of system is employed is cast iron or nodular iron. Draw down is also a concern with this system as with the two mentioned previously.

D. Flume System

The most commonly used method for moving chips and fines generated during the machining or grinding process is the metalworking fluid itself. This is referred to as a velocity flume system (see Figure 10.9). The fluid and the metal particles fall from the machine into a trough in the floor under or alongside the machine. This trough contains nozzles, which deliver metalworking fluid under pressure to the flume or trough system. The momentum of the fluid discharged from the nozzles is transferred to the cascading machine fluid and metal particles. This means the fluid and particles are moved down the trough and into the filtration system. Typically, the velocity of the fluid in the

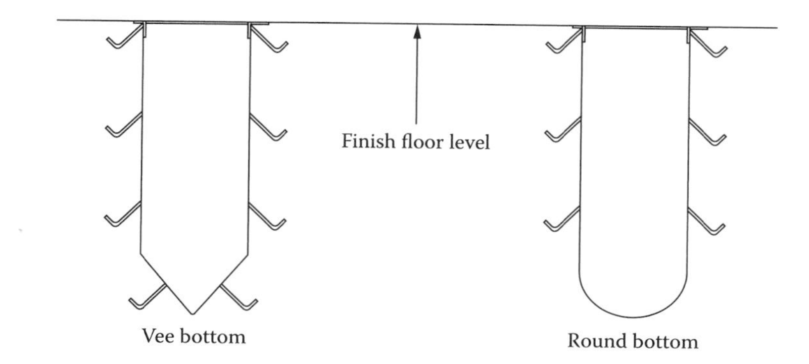

FIGURE 10.9 Troughing advantages and disadvantages. Primary applications include: (a) cast iron machining, grinding, honing; (b) aluminum machining, grinding; (c) steel machining, grinding (most universal return system). Advantages include: (a) flexibility in layout, (b) readily adapts to machine wet decks and foundations, (c) adapts to changes to material machined, (d) cost. Disadvantages include: (a) requires additional filtration capacity to supply flushing capacities required, (b) velocity flushing may extenuate foaming tendency of any given metalworking fluid, (c) misdirected, plugged, or incorrect flow may produce "dead spots" and plugging may occur. (Courtesy of Brandt and Associates, with permission.)

TABLE 10.2
Metalworking Fluid Velocities in Fluming Systems[a]

	Machining	Grinding
Cast iron	8 (2.4)	8 (2.4)
Aluminum	6 (1.8)	6–7 (2.0)
Steel	12 (3.6)	10 (3.0)

[a] In units of ft/sec (m/sec).

flume system varies between 6 and 12 ft (1.8 and 3.6 m)/sec. Table 10.2 shows the different velocities needed to maintain chip and fluid movement down a trough with a slope of 1/4 in. (6.4 mm)/ft (0.3 m).

The flume system is provided to transport material to a central process point. It should not retain particles or fluid when the system is turned off. There are different opinions as to the slope of the trough and number of nozzles used that directly impacts the gallons of fluid needed for the flushing process. A slope of 1/2 in. (12.8 mm)/ft (0.3 m) would require less gallons and fewer nozzles. If this steep slope is used, the usual practice is to place a nozzle at the end of each trough run, with a minimum number or no additional nozzles in the trough. The difficulty with this system may be that for long trough runs, the invert or depth of the trough at its discharge point into the filter is twice as deep as the usual 1/4 in. (6.4 mm)/ft (0.3 m) slope. This will mean a deeper pit and more steel for the flume. An advantage of the 1/2 in. (12.8 mm)/ft (0.3 m) slope system is the reduced requirement of flushing gallons from the filtration system.

Typically, a transfer line requiring 1500 gal (5700 l)/min of metalworking fluid on the machine tool may require an additional 1500 gal (5700 l)/min for flushing the fluid down the trough to the filter system. Usually, the filter supplies both the machine and flushing requirements. This can substantially increase the size of the filter. However, a compromise could be used. It is possible to conceive a system with separate pumping and piping systems for the machine and flushing system. This would allow complete flexibility of the gallons used for flushing and yet ensure clean filtered metalworking fluid at the tool–workpiece interface where the best filtered liquid is needed. This type of system would use the initially received fluid after some settling or pre-separation for the flushing. The flushing system usually consists of stream-directing nozzles such as fire hose nozzles, which have 3/8- to 5/8-in. openings. It is not necessary to filter fluid finely which will be delivered to these size nozzles. The typical size used is 1/2 in. The other liquid would then be further processed through a positive filter to remove fines and be delivered to the machine tool.

The pressure of the fluid at the nozzles will vary based upon the pumping volume of the pumps and the amount of fluid allowed to flow. In systems where the pump supplies both the machine and flushing system, the velocity can vary appreciably. This variation can contribute to higher than designed pressures at the nozzles, resulting in higher velocities of exiting fluid. When this happens, the nozzle discharge liquid becomes a venturi-type device drawing air into the stream and causing or enhancing foam conditions. A dual system would alleviate this fluctuation in pressure. At too low a pressure, the velocity may be decreased enough to cause inadequate flushing, leaving chips in the trough system. Although the fluid flowing over these chips may appear aerated and turbulent, the deposited chips may become stagnant and contribute to metalworking fluid microbiological control problems.

IV. BULK CHIP SEPARATION SYSTEMS

Some metalworking operations produce a volume of chips that would interfere with the normal cycling or indexing of a filter system. These operations are steel machining, aluminum grinding,

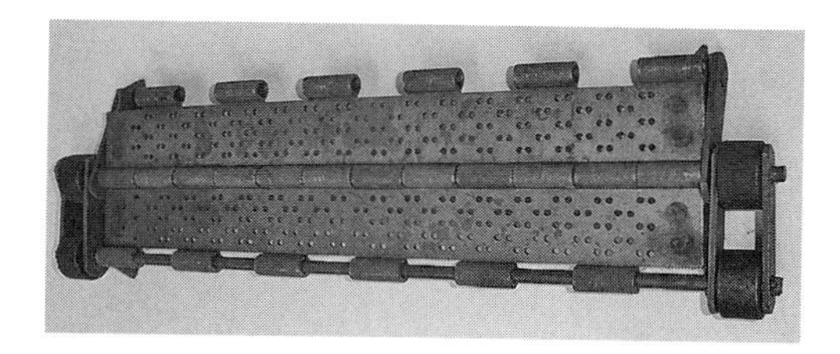

FIGURE 10.10 Hinged belt conveyor.

and machining. When these operations are performed, consideration should be given to removing the bulk of the chips by a distinct separation means. Some of the bulk separation can be accomplished by using the push bar or conveyor-type transport systems. However, because velocity flume flushing is more commonly used, another separation device needs to be placed at the end of the trough. These devices are usually referred to as primary separators. These units are tanks that are equipped with a perforated plate hinged belt conveyor or a chain and flight conveyor. These conveyors travel in an inlet trough or over a stainless steel wedge wire panel, respectively. The perforated plate hinged belt conveyor system allows the passage of fine particles and fluid into the filter for further processing (see Figure 10.10). Most of the large particles of stringy steel machining chips are retained on the conveyor belt and deposited into a tote-bin. The same process occurs for aluminum grinding and machining chips moving into and through wedge wire panels. The bulk of the aluminum chips are removed as the chain and flight conveyor moves over the wedge wire screen. The openings in the wedge wire can vary but are usually 1/8 in. (3.2 mm) (see Figure 10.11). This provides for large particle and large volume removal of the aluminum. These types of primary separation systems should be applied on most operations producing large chips or large volumes of chips.

V. RECIRCULATION SYSTEMS

After the transport of the chips and fluid has been accomplished along with primary separation, if necessary, the fluid needs to be further processed. This takes place by a process of clarification. Usually the word filtration is loosely applied to metalworking fluid and particle separation processes, even if the process relies on a physical characteristic and does not involve a filter. Recirculating systems are just that; they receive dirty fluid and, after processing, send it continuously back to the machines for further use. The time taken for this to occur may be a minimum of 3 min or as long

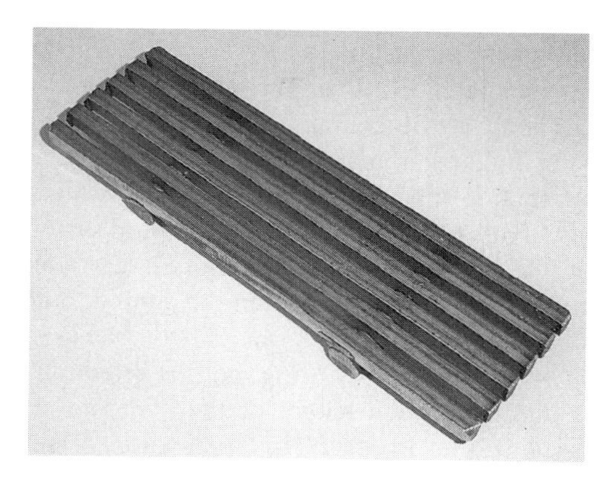

FIGURE 10.11 Wedge wire for primary separation.

TABLE 10.3
Clarification Chart

Separation	Simple Settling
	Flotation
	Centrifugal
	Cyclone
	Centrifuge
	Magnetic
Filtration	Disposable Media
	Bags
	Cartridges
	Rolled media
	Precoats
	Fiber
	Diatomaceous
	Permanent media
	Metal mesh
	Fabric belts
	Wedge wire screens

as 1 h. Whatever the length of time, it is a circular flow of metalworking fluid and a continuous removal of particles.

These recirculation systems can be divided into two groups, and are discussed here in two main categories: separation systems and filtration systems (see Table 10.3). These categories contain a variety of equipment developed to accomplish the same thing: cleaner fluid. The driving force found in these categories is either gravity, vacuum, or pressure.

A. Separation Systems

The physical characteristics used in separation processes are specific gravity differential, foam bubble inclusion, or ability to be magnetized. These various characteristics are put to use in a variety of equipment.

1. Settling Tanks

The basic separation device for the individual machine is the settling tank (see Figure 10.12). This unit is generally placed alongside a machine and receives liquid from the process. The fluid capacity of the tank provides for a retention time. The retention time is an indication of settling that will take place in the tank. The type of metal and size of chip or particulate formed in the metal removal process, as well as the fluid and retention time, will determine what equilibrium point will be reached. Typically, the retention time for the average tank is 5 min. Less retention time will usually mean a large quantity of fine recirculation. In large central systems where settling is the only method of particle removal, the retention time should not be less than 10 min and ranging up to 15 min. These settling devices are applied to cast iron machining, with some application to other noncake-forming particles, such as glass grinding and silicon sawing. The particles produced by grinding cast iron may also be separated by settling, but the retention time is customarily double that of machining. This is due to the fine nature of the particles produced and their lightness compared with the fluid used in the process. The application of settling systems to cast iron machining and grinding does not produce metalworking fluid clean enough for the requirements of most critical metal-removing processes.

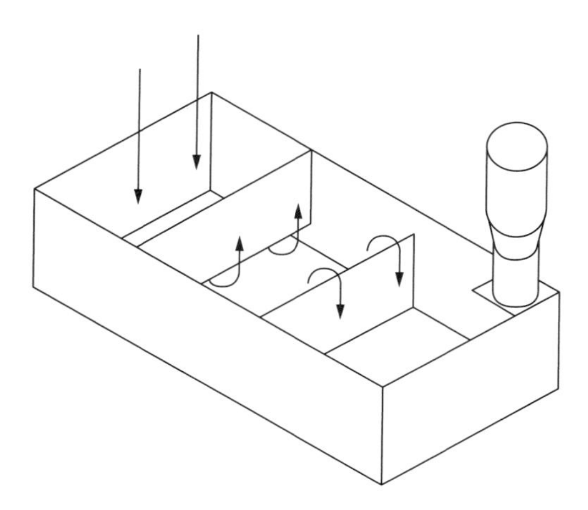

FIGURE 10.12 Setting tank. (Courtesy of Brandt and Associates, with permission.)

The settling system can provide primary separation of the larger chips and particles. However, separation processes can be added to the settling system in the form of centrifugal and magnetic devices.

2. Foam Separators

In some separation systems, particle removal has been reported as being enhanced by the occurrence of foam. The generation of foam or small air bubbles tends to entrap the fine particles causing them to float with the foam. The fines brought to the surface of the tank can be removed along with the foam by a conveyor or bar mechanism. The particle-filled foam is generally moved into a tank, which settles the fines when the foam breaks. This method has been used but is limited in application to metalworking fluids that can sustain a foam condition. For fines removal, this method has proved difficult to control in a consistent manner.

3. Centrifugal Separators

The addition of centrifugal devices could be in the form of a hydrocyclone and/or centrifuge. Hydrocyclones are devices which, when supplied tangentially with metalworking fluid, move the fine particulate to the outside wall of the hydrocyclone device (see Figure 10.13). The particulate

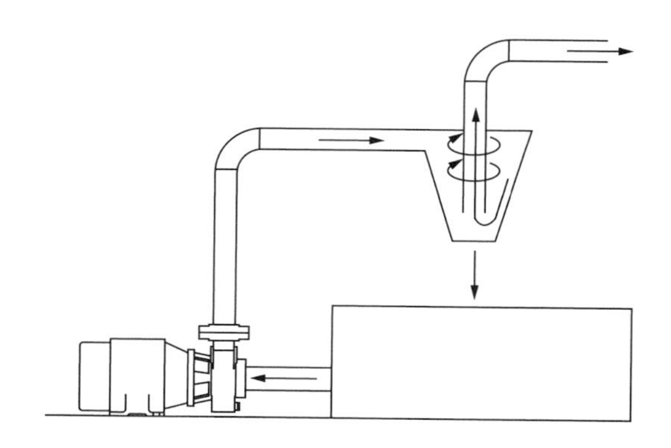

FIGURE 10.13 Hydrocyclone system. (Courtesy of Brandt and Associates, with permission.)

concentrates on the outside wall of the device and moves down the cone wall. A proportion of the liquid and concentrated particulate flows out of the bottom of the cone unit. The rest of the liquid moves back towards the top of the hydrocyclone through a vortex-finding device. The liquid processed in this manner reaches an equilibrium based on the differential pressure across the hydrocyclone and the specific gravity differences between the fluid and the particles. Under the same conditions, a smaller diameter hydrocyclone can remove finer particulate than a larger diameter hydrocyclone. Larger diameter hydrocyclones are used because they tend to produce more gallons per minute of cleaned fluid. Smaller diameter hydrocyclones are usually manifolded together to produce a common inlet and outlet for a cluster of units. The best performance will be obtained by maintaining proper differential pressure and applying the units to particulate where specific gravities are highly divergent from the liquid. An example of a system which may not reach a satisfactory equilibrium level would be cast iron grinding fines suspended in a heavy opaque soluble emulsion product. Better separation may occur using a solution-type water-miscible product containing particulate from cast iron grinding. Modification of the typical hydrocyclone has been carried out and has resulted in a dumbbell-shaped unit instead of the typical conical shape. Whatever the outside configuration, the separation parameters are the same.

Another device which can be added to primary separation tanks is the solids-separating centrifuge. This unit is designed to remove particles by centrifugal force. A spinning bowl receives the particle-laden metalworking fluid. The force against the fluid and particles separates the particles by moving them to the side of the bowl. The bowl fills with particulate and eventually requires cleaning. Some centrifugal solid separators have liners to facilitate easier cleaning. It is customary to use one of these units on small flows and small dirt load systems. These conditions are usually found in individual machines and not large central systems. There is another type of centrifugal separator which will remove particles; however, it is generally applied to systems to solve another separation problem. These will be addressed later.

4. Magnetic Separators

The metalworking industry works a variety of metals, the properties of which are different. Some of the metals interact with a magnetic field. When such a metal is worked, a magnet can be used to remove fine particulate (see Figure 10.14). The metalworking fluid is passed in close proximity to the permanent magnet system. As the fluid passes by the magnet, the particles which are magnetizable stick to the magnet. The magnet is generally rotating and brings the particles up and out of the liquid. As the magnet continues to move, a nonmagnetic blade removes the accumulated particles from the magnet. If operational changes are made to a system, such as intermittent magnet movement, the particles accumulated can form a "cake" on the magnet. Nonmagnetic particulate can be trapped in this cake as the metalworking fluid passes through it. In this manner, the fluid can

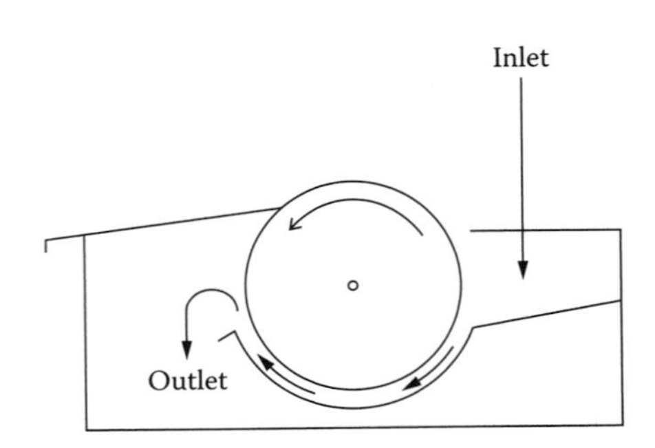

FIGURE 10.14 Magnetic system. (Courtesy of Brandt and Associates, with permission.)

be cleaned of both magnetic and nonmagnetic particulate. The difficulties with this system are the flow rates attainable and the cleanliness obtainable. The application of magnetic separation systems is generally used on systems requiring less than 250 gal (946 l)/min. Another area of use for magnetic systems is in the primary separation of metal fines prior to positive filtration. This application allows for bulk separation of solids to take place before finer filtration. The magnetic system is satisfactory for magnetizable particle removal, but has its drawbacks in gallons per minute provided and consistency of fines removed.

B. Filtration Systems

The actual separation of particulate by introducing a media or filter into the fluid stream constitutes filtration of the fluid. This filter needs to be supported in some sort of system hardware. A force is applied and the filter system is activated.

There are three driving forces for filtration systems: gravity, vacuum, and pressure. There are a number of different filter materials that can be selected. It is not within the scope of this chapter to give a presentation of selection criteria, but it can generally be said that the more dense the fabric in ounces per square yard, the finer the attainable filtration. This is valid in comparing fabric densities within the same family. There are a number of criteria used for the selection of filter materials.[5]

As with the selection of fabric, the selection of hardware also varies and can be discussed in two categories: those units that use disposable media and those using permanent media.

1. Disposable Media

Systems are produced in a number of configurations. Disposable media types include bags, cartridges, rolled goods, chopped paper, and a host of material referred to as precoats (see Figure 10.15). The operation can be manual or automatic. Bag filters are filter media in a bag form. The bag(s) may be suspended at the end of a pipe with a special fitting or enclosed in a housing. The driving force is pressure: less than 15 lb/in.^{2} (103 kPa) for the end-of-pipe assembly and up to 150 lb/in.^{2} (1034 kPa) for housing units. The bag is selected to retain chips and particulate but allow a flow of metalworking fluid through the system. When a pressure drop is noted across the bag, indicating the bag is plugged or there is a significant reduction in flow, the bag filter is manually changed. The bag filters can be purchased in various sizes. This provides some flexibility in the square foot of filter area available. A wide range of micron retention bags are available providing additional flexibility. Typically, a flow rate of 25 to 50 gal/min/ft^2 (1000 to 2000 l/min/m^2) can be obtained.

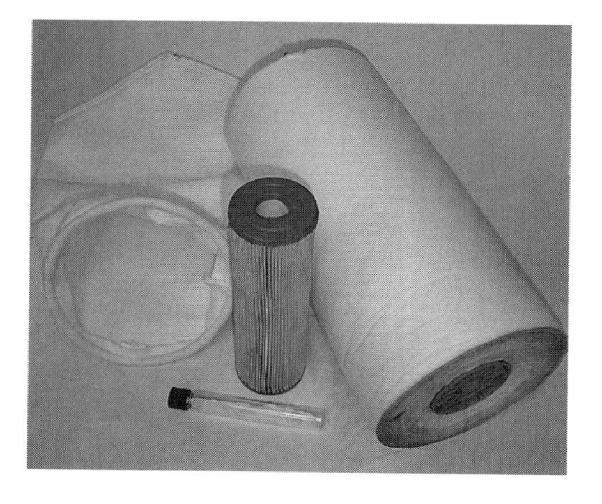

FIGURE 10.15 Forms of disposable media.

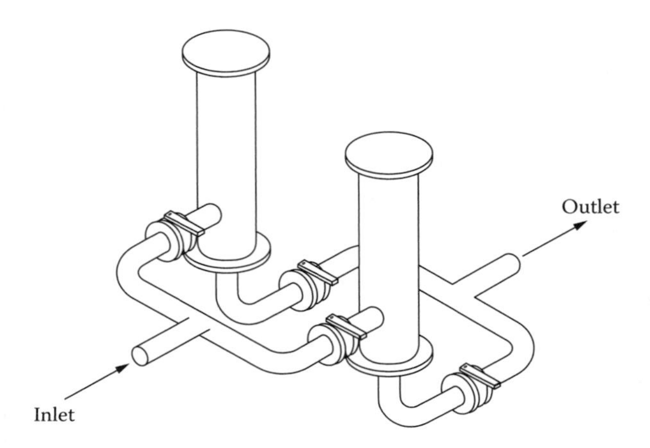

FIGURE 10.16 Cartridge filter system. (Courtesy of Brandt and Associates, with permission.)

However, type of dirt and viscosity of the fluid may modify those numbers downward. For selected applications: washers, individual machines, and recycling equipment, the bag filter has proven effective.

Cartridge filters are used to provide a positive after-filter for other filtration systems. These units consist of individual cartridges or clusters contained in multiple cartridge housings (see Figure 10.16). The driving force for cartridge filtration is also pressure. Therefore, the housings for the cartridges are made for low (75 lb/in.^{2} [517 kPa]) to high (300 lb/in.^{2} [2068 kPa]) pressure. The flow rate for cartridge filtration is less in gallons per minute per square foot than bag filters. The application of these units is not generally on large flow rate systems but as guard filters to prevent spurious particulate from entering a close tolerance area. This close tolerance area could be a through-the-drill supply of coolant, a tapping operation, or application to a fine finish machine.

Disposable media on a roll provides the maximum flexibility because it can be used for gravity, vacuum, or pressure separations. The basic use of the rolled goods is in the gravity-driven pieces of equipment. In a gravity separation device, the metalworking fluid containing particulate flows or is pumped from the machine into an open container (see Figure 10.17). In the bottom of the container is a perforated plate which supports the media (see Figure 10.18). As the liquid level deepens, the head of liquid forces metalworking fluid through the media. The particulate is trapped on the filter media. As more and more particulate is collected on the filter media, a greater resistance to flow is encountered. The metalworking fluid becomes deeper and a cycle-inducing device is activated. This may be a float ball, limit switch, conductivity probe, or other device. When the device is activated the filter media is indexed through the tank by means of a conveyor. This provides new filter media in the tank and results in improved flow rate. The liquid level drops and filtration continues until the cycle repeats itself. The indexing and movement of the media could be done manually, replacing the filter media or pulling the media through the system.

Vacuum systems are similar to gravity systems except there is a pump that is used to create a negative pressure under the media and its support structure (see Figure 10.19). This pump can be an air pump used to evacuate a large volume of air from under a media and support system, or more typically, a pump to draw liquid through the filter media faster than gravity would normally allow. The use of pumps generally increases the time between indexes of the media. The liquid-drawing filter pump had been selected on its merit of being able to provide good vacuum characteristics. However, the concept of a single pump being used as the filter vacuum pump and machine supply pump has meant a compromise of characteristics. These characteristics are pressure and volume. As the vacuum increases, the flow and pressure decrease. It is, therefore, important to select a pump with a good net positive suction head. This system pump concept is a compromise because it

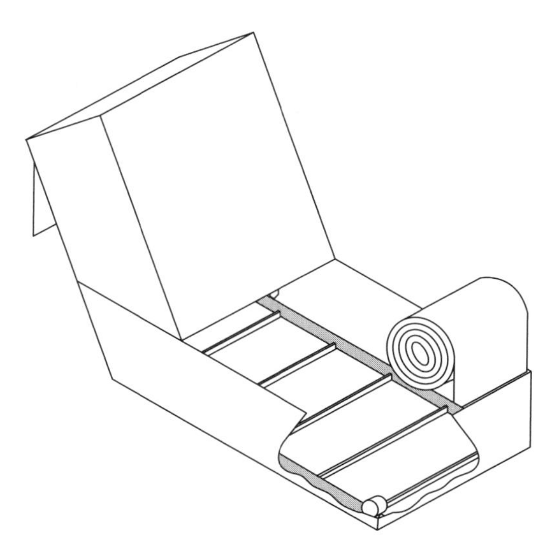

FIGURE 10.17 Gravity media system. (Courtesy of Brandt and Associates, with permission.)

requires good vacuum character as well as providing adequate pressure and volume for the machine and flushing systems. Typical vacuum conditions can reach 12 to 15 in. of mercury (41 to 51 kPa). Beyond 15 in. of mercury (51 kPa), justified benefit is low. Typical indexing methods for a vacuum media system include a mercury vacuum switch, differential pressure switch, or a timer. The first two index relative to the reduction in flow caused by the buildup of particulate on the filter media. This causes the vacuum pump to draw more vacuum and this is subsequently sensed by the switch. The timer is used on materials that do not cause a substantial decrease in flow rate or buildup in vacuum, such as steel machining, aluminum machining, or grinding. Since these chips and particles form a porous pile or cake, indexing on vacuum may not be adequate to keep ahead of the chip loading in the system.

To ascertain the timer setting, a calculation is performed to determine the quantity of solid material removed. Based on an expansion factor, a volume of chips can be approximated from the solid stock removed (see Table 10.4). Knowing the capacity of the conveyor in the filtration system,

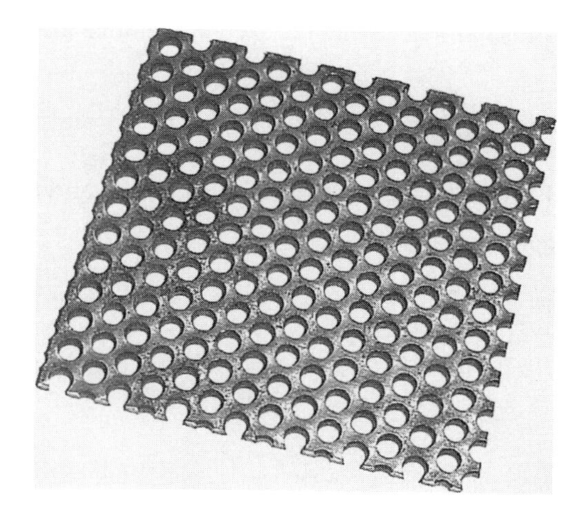

FIGURE 10.18 Perforated plate support.

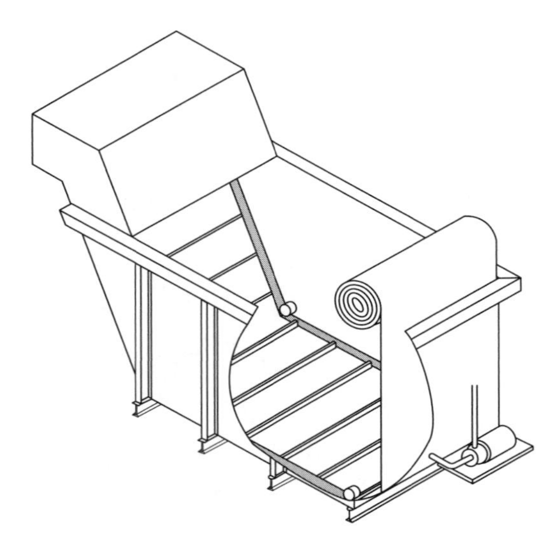

FIGURE 10.19 Vacuum media system. (Courtesy of Brandt and Associates, with permission.)

the timer can be set for an interval which will prevent overloading of the system. This setting has to be done for the worst condition, i.e., the most chips produced at maximum production. At any other condition which would make less chips, the indexing will be more frequent than necessary.

When the vacuum system indexes, the filter media moves a short distance. Therefore the media is never fully replaced. This changes the gallons per square foot passing through the new area. If a system is purchased to provide 10 gal/min/ft^2, this condition only occurs at start up. At any other time during operation, the filter's moderate indexing minimizes the new area and causes the gallons per minute per square foot to increase slightly. In fact, the only reason for large tanks with large filter areas is to provide physical flow volume of metalworking fluid.

Another system is the pressure-driven unit in which the filter media is supported on a metal or cloth belt, slotted, or perforated plate. The support and filter media are enclosed between two shells. The metalworking fluid and particulate are introduced into the cavity above the filter media and forced through the media. This process continues until the particulate builds a cake and the resistance causes an increase in pressure. When a preset pressure on a switch is reached, the filter supply pump is turned off, air is introduced into the cavity to remove the metalworking fluid and dry

TABLE 10.4
Approximate Expansion Factors Based on Densities of Solids and Chips

	Density of Solids[a]		Volume Expansion Factor
Steel	490 (7850)	Grinding	5
		Machining	7
Cast iron	480 (7690)	Grinding	5
		Machining	5
Aluminum	170 (2720)	Grinding	5
		Machining	10
Brass	560 (8970)	Turnings	3

[a] In units of lb/ft^3 (kg/m^3).

the particulate cake. After this process is complete, the two shells separate and the filter media is advanced entirely out of the system. The two shells close and the process begins again. The usual configuration is for the two shells to be in the horizontal position. Multiple horizontal shells can be used with multiple rolls of filter media. The pressure settings vary for each type. Indexing pressures for two-shelled units are usually 7 to 10 lb/in.^{2} (50 to 70 kPa) while multiple-shelled units can run to 35 lb/in.^{2} (240 kPa). The limitation for any of these systems is the mechanism which keeps the shells closed. The closing mechanism used is air pressure in conjunction with air bags or cylinders. The advantage of the pressure-driven unit is that a reasonably dry cake can be discharged from the system. With pressure filtration, each time the filter indexes, a new complete filter media is introduced into the system.

The disposable media filter material comes in a variety of retention capabilities. As the filtration process proceeds through its cycle, the particulate being deposited on the media is building up. This buildup in particulate forms a cake. In some cases, this cake is actually necessary in order to attain fine particle filtration. The filter media acts only as an initial barrier upon which the cake can build. This process is referred to as depth filtration.

2. Permanent Media

Permanent media filters use the same driving forces for the filtration process as disposable media filters. However, the indexing cycle moves or removes only the cake from the media. This leaves the same media to be the support and initial barrier for the particulate in the next cycle. Permanent media are made of wire mesh, woven fabric belts, or wedge wire screen (see Figure 10.20 to Figure 10.24). The primary difference is the backwash or blow down of the cake of particulate formed on the media. In the disposable media system, the cake is carried out of the system with the filter media. In permanent media systems, the cake accumulated on the media can be blown off with air or metalworking fluid. The edge of a piece of metal, called a "doctor" blade, can also be used to remove the cake from the media. This essentially removes the particulate cake and provides a renewed area for reestablishing the cake. Generally, permanent media have a very open character. The percent open area and micron size of the openings is large in comparison to the small particles to be removed. It is necessary and a requirement for fine filtration that a cake be established and maintained for as long a period as possible. After index and before a new cake is established, migration of large particulate can occur. After the cake has been established there is an increasing improvement in the cleanliness of the filtrate. As with disposable media systems, the cake carries out the filtering in a permanent media system, rather than the media itself.

FIGURE 10.20 Wedge wire permanent media panels or drum.

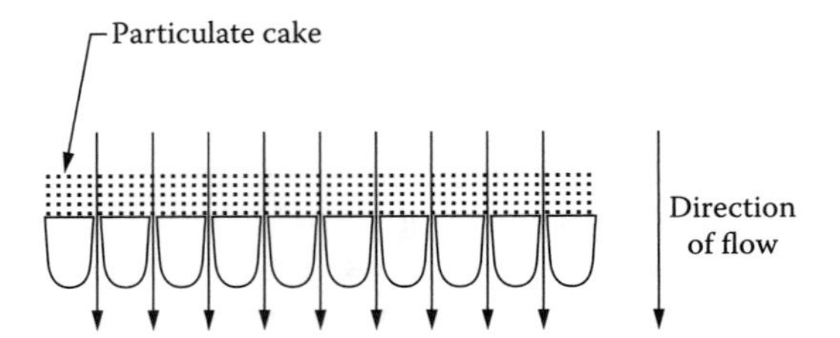

FIGURE 10.21 Wedge wire screen cross-section. (Courtesy of Brandt and Associates, with permission.)

VI. ANCILLARY SYSTEMS

A. Extraneous Oil Removal Units

The metalworking fluid being recirculated in the filtration system is subject to varieties of outside contaminations. One of these is oil, which is introduced from the machine tool hydraulic, way, and lubricating systems. This oil leaks into the fluid and becomes in varying degrees part of the recirculating metalworking fluid. Most units incorporated into a system for the removal of this oil act best on free oil, which will separate from the recirculating metalworking fluid given the time. These units pick up this free oil by a wetting of oil-loving material in the form of belts, ropes, and coalescer media. These removal vehicles utilize the affinity of polypropylene or stainless steel to surface coat with free oil. As these vehicles become wetted with oil they are constantly having oil removed from them by the natural separation process of gravity (coalescer) or a blade squeegee device. The free oil removed travels down a chute into a container for disposal or reuse. One difficulty with these devices is the need for cleaning because of the oil-wetted fine particulate which floats with and is removed by the vehicles. The coalescer media and squeegee devices need cleaning or they will become plugged. The coalescer-type units have an advantage over the ropes and belts because they can be set up to remove free oil and not a large quantity of metalworking fluid.

Other removal units available work on the difference in specific gravity between the oil and metalworking fluid. These devices are centrifugal and were originally designed for the removal of one fluid from another. The separation process takes place under varying amounts of relative

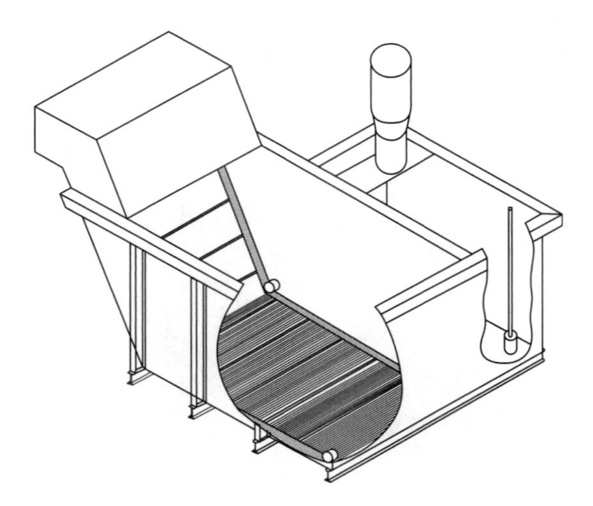

FIGURE 10.22 Wedge wire system. (Courtesy of Brandt and Associates, with permission.)

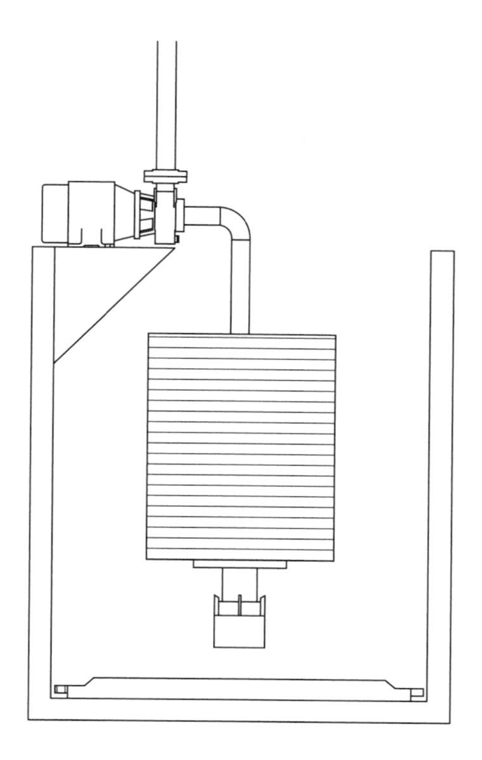

FIGURE 10.23 Vertical wedge wire drum system. (Courtesy of Brandt and Associates, with permission.)

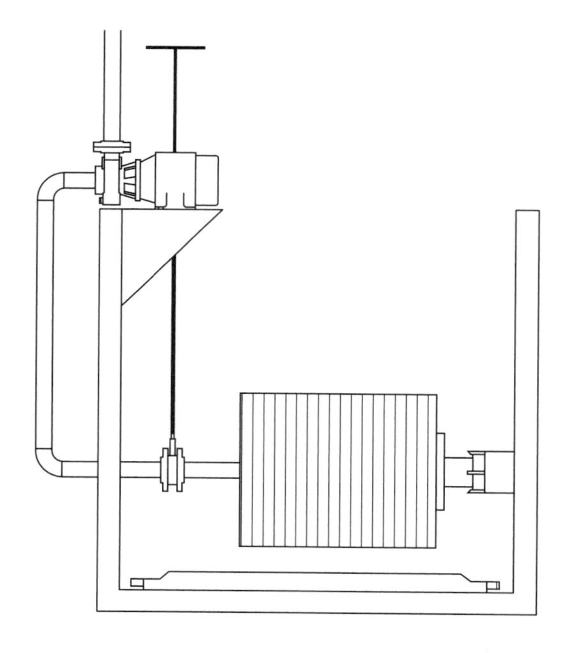

FIGURE 10.24 Horizontal wedge wire drum system. (Courtesy of Brandt and Associates, with permission.)

centrifugal gravities. These can be low or high depending on the equipment purchased. The low-speed units do not subject the metalworking fluid to high gravity forces and therefore separate the free oil with less retention time than would be required under normal gravity conditions. The higher-speed units can subject the metalworking fluid to very high forces. These forces can be high enough to separate not only the free oil but also the metalworking product itself. Besides the possible deleterious effect on the fluid, these units tend to require more maintenance and a higher degree of technical support. All of the units will tend to remove some material that can form in the metalworking fluid other than free oil. These consist of invert emulsions, soap scum, hard-water precipitates, and fine particulate.

B. Metalworking Fluid Makeup Units

The very use of the metalworking fluid will require an addition of either water to compensate for evaporative loss, or metalworking fluid concentrate to replenish various ingredients lost. The additions to meet these requirements can be done manually by adding water or concentrate as needed based on an analytical determination. This has proved an effective method for smaller clarification systems servicing one or a few machines. In systems of large capacity, the need for addition can be high, and alternative addition techniques are used. These techniques rely on a control means to determine whether water or premixed water and metalworking fluid should be added. Additions are not made because of on-line analytical tests but rather the indication of reduction in overall system volume. Owing to this method of replenishing the system, only water can be added with adjustments made later in the concentration, or a premixed metalworking fluid concentrate and water can be added. The latter is the preferred method because it adds a new amount of concentrate each time volume replenishment is required. This type of addition also means that there is less fluctuation in the metalworking fluid concentration from day to day. One unit available works on a venturi principle. As water passes over a fixed orifice plate a certain amount of fluid concentrate is drawn into the water stream. This mix is discharged into the system. Owing to changes in water pressure and flow, clogged orifices, and different concentrations required at different seasons of the year, this unit may not provide uniform and consistent concentration deliveries as required.

A few units use pumps to pump the concentrate into a water stream. One unit employs a water-actuated proportioning pump, which when supplied with water, draws up a quantity of concentrate into a separate chamber. When the pump continues to function, the concentrate is mixed with the water stream. This unit has a variable screw adjustment, which can be changed to give different concentrate deliveries to the water stream. The concentration in this type of unit varies depending on the stroke of the water-actuated pump. For more truly premixed metalworking fluid, a mixing chamber is needed on the downstream side of the pump. Another pump system includes an electrically operated pump — either centrifugal, piston, or tube — to deliver concentrate into a flowing stream of water. Concentration adjustments are made by either changing the feed or stroke on the pump, or opening or closing a valve limiting or increasing the flow of concentrate. These systems also require a mix chamber in order to provide more uniform premix additions. A variety of makeup units are also available which measure the amount of water added to a system and add a preset amount of concentrate. Usually these additions are not premixed but added as two separate streams of liquid. This unit relies on the filter system turbulence to mix the metalworking fluid. In a few cases pumps have been placed on systems, and set to add an amount of metalworking fluid concentrate to the system each and every day with no regard to variations in the requirements of water additions. This addition technique is used based on the assumption that concentrate replenishment is required uniformly each day and evaporative loss of water is consistent each day.

C. Temperature Control

Certain applications require that the machine, tooling, and workpiece maintain a relatively uniform temperature. These applications include fine tolerance work, such as honing and mirror finish grinding. However, where dimensional stability is important, some machining systems are being considered for temperature control. The temperature of the fluid is selected based on the ambient room temperature plus or minus two or three degrees, or a compromise which is economically justifiable. The heat input is determined by the horsepower of the machines, peripheral equipment, and the work done. Other considerations are the evaporative loss of water, room temperature, water added to the system, and temperature of the parts. Once a temperature has been determined and the necessary calculations made, the selection of a cooling system can be pursued. Two different types of cooling can be used: evaporative cooling from an outside water-cooling tower or a mechanical chiller. Each of these choices has a number of different operational parameters that contribute to their particular advantages and disadvantages. Selection is made by working closely with those trained in this field.

D. Alarms and Controls

The controls available on most systems have changed as the technology and electronic gear have advanced. The changes have occurred in the electronic hardware and the information retrieval which can be interfaced with other systems plant-wide, but this has not altered the functional requirements of the system. This function is to produce consistently clean metalworking fluid for use at the machine tool at adequate volume and pressure. Indications of trouble in the operation of the filtration system have always required observation or physical interaction. It is necessary to provide sensors and alarms on filtration systems to monitor their continued mechanical performance.

VII. RECENT DEVELOPMENTS

In the past 10 years, equipment used in the separation and filtration of metalworking fluid has undergone very little change. Some of the electronics in the control panel have changed with the addition of touch screens and computer software improvements. However, these are only convenience changes based on newer available electronic technologies. The basic method of removing particulate from the metalworking fluid has not changed because of the new convenience items.

People have come to the realization that systems can be more focused on the primary goal of providing metalworking fluid as clean as economically justifiable to the tool-workpiece interface by combining several different types of units. Engineers are now incorporating into central systems features that take into account the type and quantity of chips to be removed, the flow rate required at the machine, and the fact that the cleanliness of the metalworking fluid for various aspects of the machine operation may be different. At present, more systems are using primary separation techniques. Primary separation tanks are being equipped with devices such as magnetic conveyors and drums, screen drums, perforated belt conveyors, or continuous chain and flight conveyors to remove the bulk of the metalworking chips. These separation methods take advantage of the ever present natural settling caused by gravity, the physical nature of the chips and bulk separation by straining. Primary separation processes handle the chips in such a way as to allow for recycling or direct chip disposal, without added media contamination. However, the primary processing that takes place often leaves the metalworking fluid too dirty for certain operations at the machine tool. Even with the addition of wedge wire screen drums or microscreens in the primary separation units, the cleanliness of the metalworking fluid may still be inadequate to meet the needs of specific metal removal operations. The settled and/or screen-cleaned metalworking fluid is generally clean enough to be used for trough flushing and for washing of the machine surfaces.

To provide better filtration to selected machine operations, filters with positive media are being added to the settling, wedge wire drum, and microscreen units. This filter addition provides cleaner processed metalworking fluid to the point of cut on the machine tool. Only a proportion of the entire volume of the metalworking fluid is being processed through the filter on each pass. Eventually, all of the fluid will be filtered by the positive filtration system, and thus it becomes the unit providing true filtration. Owing to the primary separation method of the particulate, smaller suspended particles will be delivered to the added positive media filter. These smaller particles can plug the filter media more quickly and cause more filter media to be used.

The advantage of using the added positive media filter is to minimize the size and cost compared with a full flow disposable media filter. This unit will be much smaller than a full flow media filter, and is usually sized for the flow rate needed at one specific or multiple functions on the machine, where cleaner metalworking fluid is required. As all the fluid volume is eventually filtered, the entire system will achieve a particulate equilibrium based on the density of the filter media, the type of metal being worked, and the size of the fines being removed.

Systems that appear to be new are often simply a repackaging of that which has been proved over time to be reasonable and acceptable for the filtration of metalworking fluid used for machining and grinding metal. Older systems that relied solely on settling or basic separation processes have been shown to be less than efficient, and unable to meet more stringent requirements at the workpiece. This realization has driven engineers and other personnel in manufacturing facilities to explore and purchase equipment designed to meet the cleanliness required for today's operations. The justification for the added expense to provide this equipment is a more functional system that supplies cleaner metalworking fluid to the metal removal process, and better part production with fewer rejections.

VIII. CONCLUSION

Of all the alternatives available in the filtration of metalworking fluid, the best option and primary goal is to provide a mechanically sound, continuously functioning system which will deliver acceptably clean fluid at the tool-workpiece interface. This can be achieved by reviewing the particulate produced in the metalworking operation, setting standards to be met, examining methodologies for accomplishing these standards, and paying attention to the primary goal. Filtration systems can contain a plethora of devices and controls. If they do not add to the primary goal, they are a nicety and not a necessity. The overall economics and cost of the system should be discerned and the emphasis placed on clean metalworking fluid.

ACKNOWLEDGMENTS

I wish to thank my Chief Engineer, Carl H. Brandt for his continued support and understanding, and for drawing the representations of filter units, and Addie Brandt for her support and timely help.

REFERENCES

1. Nehls, B. L., Particulate contamination in metalworking fluid, *J. Am. Soc. Lubr. Eng.*, 179–183, 1976.
2. Joseph, J. J., *Coolant Filtration*, Joseph Marketing, East Syracuse, NY, pp. 15–18, 1987.
3. Brandt, R. H., The analysis of particulate in "filtered coolant", *Lubr. Eng.*, 254–257, 1972.
4. Marano, R. S., Cole, G. S., and Carduner, K. R., Particulate in cutting fluids: analysis and implications in machining performance, *Lubr. Eng.*, 376–382, 1991.
5. Chrys, P. Z., Selecting filter media for coolants, *Man*, 42–48, 1991.

11 Metalworking Fluid Management and Troubleshooting

Gregory J. Foltz

CONTENTS

I. INTRODUCTION

The metal removal process consists of four variables: (1) the machine tool, (2) the cutting tool or the grinding wheel, (3) the part, and (4) the metalworking fluid. Each of these is significant and important in the production of any metal part. There are a number of aspects pertinent to each variable and their interaction that must be understood in order to make this metal removal process occur.[1,2] While the purpose of this chapter is to discuss the control and management of the metalworking fluid variable, it is also important to understand the others.

The variable aspects of the machine tool include its setup, the feed and speed rates, the metal removal rate, its alignment, drive systems, and vibration. The age of the machine and how it has been maintained will also influence its performance. The type of cutting tool (HSS, carbide, ceramic, etc.) or grinding wheel (silicon carbide, aluminum oxide, CBN, diamond, resin bond, vitrified bond, etc.) is very important. The sharpening or regrinding of the tools and the truing and dressing of the grinding wheels can affect their performance. The composition of the part (cast iron, steel, aluminum, copper, glass, ceramic, etc.), as well as the required shape and tolerances, are also very important to the manufacturing process.

All of these variables can be optimized, but if the proper fluid is not used, poor performance can result. It is also important to remember that when troubleshooting a system, these variables, as well as the fluid, should be considered. In most cases, a well-organized maintenance/service plan exists for each machine tool, part geometries are well defined, and plant engineers can precisely define the number of parts per tool or pieces per wheel dress. Yet selection and maintenance of the metalworking fluid, the fourth important variable, is usually not well understood.

The main functions of a metalworking fluid are to control heat and provide lubricity.[3–5] It must also flush away the chips and protect the machines and workpieces from corrosion. When these functions are integrated into the machine tool and cutting tool/grinding wheel framework, an efficient metal removal system will result. It is, therefore, very important that this fluid variable be properly selected, controlled, and maintained, in order to achieve top performance. This fluid management program then becomes an integral part of a plant's operation, as the importance of the metalworking fluid to the entire process becomes understood.

When a metalworking fluid management program[6] is in place, the fluctuation in this variable (correct fluid, concentration, pH, dirt volume, tramp oil, etc.) is reduced and more consistent quality parts can be produced. Finish, size, and geometry problems are eliminated. Productivity can be increased as machine and tool/wheel performance are optimized with the proper fluid. The plant's working environment is improved as offensive odors, irritating mists, skin problems, and dirty machines are controlled. The bottom line is improved costs.

Properly controlled fluids do not need to be dumped as often.[7] This eliminates costs associated with machine downtime, disposal, and new fluid purchase. As efficiency of operations is improved, the cost of producing each part will drop. More parts can be produced and less wheels or cutting tools are required. The advantages of a metalworking fluid management program and the ability to troubleshoot any problems that may occur are, therefore, key elements in a plant's operation.

Concerns with the safety and health of workers exposed to metalworking fluids led to the creation of two important documents in 1999, dealing in great detail with all aspects of metalworking fluid management. Both documents are only found on-line, because they are regularly updated with new information:

- *Management of the Metal Removal Environment*, published by the Organization Resources Counselors (ORC), www.aware-services.com/orc
- *Metalworking Fluids: Safety and Health Best Practices Manual*, published by the U.S. Department of Labor, Occupational Safety and Health Administration, www.osha.gov/SLTC/metalworkingfluids/metalworkingfluids_manual.html

Another source of information is ASTM E-1497-00, *Standard Practice for the Selection and Safe Use of Water-Miscible and Straight Oil Metal Removal Fluids.*

II. FLUID SELECTION PROCESS

A fluid management program begins with the selection of the proper fluid for the job. There are four categories of fluids[8–10]: straight oils, soluble oils, semisynthetics, and synthetics. The performance of these products can range from light duty to very heavy-duty operations. Metalworking fluids will have different performance properties depending on their chemical composition. This can be affected by the oil levels, the amount of chemical lubricants and extreme pressure additives, cleanliness properties, biocide levels, and a variety of other factors. The selection criteria that follow are designed to define the requirements for a particular job.

A. Selection Criteria

In order to achieve optimum performance, the correct fluid must be selected, based on a review of the variables of the entire operation.[11] These include the following.

1. Size of Shop

For a small shop with a few machines, doing a variety of work on a variety of metals, a very general-purpose product is selected to minimize the number of products required. For a large plant producing large quantities of the same part, a product specific to the needs of that operation can be selected.

2. Type of Machines

It is important to consider the age and design of a machine tool before selecting a product. Some machines, especially older models, were designed so that the metalworking fluid also serves as the lubricating fluid for the moving parts and gears. In that case, a fluid with a high degree of physical lubricity will be required. The seals on the machines must also be inspected to insure they are designed to be used in a water environment. If not, it may be necessary to use a straight oil type product.[12,13]

3. Severity of Operations[14]

The severity of the operation will dictate the lubricity requirements of the fluid. Two types of fluid lubricity exist,[15] chemical and physical, so that it is not always necessary to use an oil-containing product to achieve good machining/grinding characteristics. Stock removal rates, feeds and speeds, together with finish requirements must be considered. Metalworking operations can be divided according to their severity, light duty (surface grinding cast iron), moderate duty (turning, milling steels), heavy duty (centerless grinding, sawing steels), and extremely heavy duty (form and thread grinding, broaching). If a series of operations are to be performed with one fluid, it is necessary to select the most critical operation, because in most cases it will dictate the fluid selection.

4. Materials

The type of material being worked (cast iron, steel, aluminum, titanium, copper, glass, carbide, plastics, etc.) is very important in fluid selection.[16,17] The corrosion control and/or staining properties of some fluids may not be compatible with all materials. Some fluids are formulated specifically for certain metals. The hardness and machinability of the material must also be considered.

5. Quality of Water

Since water is the main component (90 to 95%) of any water-based metalworking fluid mix, its quality can be an important factor in performance of the fluid.[18] Water quality is covered in greater detail later in the chapter. Water hardness greater than 200 ppm can produce mix stability problems with many emulsion type products. Water with a high chloride or sulfate level (greater than 150 ppm) can promote corrosion and/or rancidity. On the other hand, soft water (less than 50 ppm hardness) can lead to foam with many products. It is important to know the water quality before selecting a product. This will be discussed in more detail later.

6. Type of Filtration[19–21]

Individual machine sumps or central systems each make different demands on a fluid. Also, the type of filtration used, settling, or some type of positive filtration using media (paper, cloth, or wire screens) or a separator such as a centrifuge or cyclone, will affect the fluid selection process. Settling systems obviously require fluids with good settling characteristics. Media filters require fluids capable of passing through the media without clogging. Separators require products that are sufficiently stable to undergo the demands of this process. Filtration is described in greater detail in a specific chapter on that subject.

7. Contamination[22]

Contamination has a drastic effect on the life and performance of a metalworking fluid. Lubricating oils, way lubes, hydraulic oils, rust inhibitors, floor cleaners, and heat treat solutions are some of the contaminants often found. Different fluids have different mechanisms for handling these contaminants, especially the oils. Some may be emulsified and others rejected. While most cutting fluids can handle some contamination, greater amounts of contamination will shorten the fluid life and cause more erratic performance.

8. Storage and Control Conditions

Where and how a fluid is stored prior to use can affect its performance. Many products will freeze and eventually separate if stored outside or in unheated warehouses during winter conditions. Other products, if stored outside under the hot sun, will be degraded. The compatibility of the fluid with the plant mixing conditions must be considered. For water-based fluids, the concentration control procedures must be considered. If they are very lax or nonexistent, then a product with a very wide operating range should be selected.

9. Freedom from "Side Effects"

In some grinding operations, the use of a very transparent fluid is desirable. At some plants, a particular product color or odor may be requested. Certainly fluids should be free from misting and dermatitis problems. They should all be safe and pleasant to use. The fluid should not leave an objectionable residue or cause problems with the paint on machine tools.

10. Ease of Disposal/Recycling[23–25]

At many factories, the most critical element in the selection of a metalworking fluid is its compatibility with the waste treatment process. If the fluid cannot be effectively and economically treated, then any performance advantages are negated. Plants will typically have a very specific waste treatment test that a product must pass before it can even be considered for testing. With the advent of more in-plant recycling systems, products are also being judged on their ability to be effectively recycled through the plant's existing or planned treatment system. Frequently in recycling operations, it is necessary to standardize an entire plant on just one product, in order to have the recycling system work. This must be considered in fluid selection. Refer to the specific chapters on Recycling and Waste Treatment.

11. Chemical Restrictions

Due to concerns over the health and safety aspects of a particular chemical or some environmental or disposal issue, certain plants may restrict the use of certain chemicals that may be found in some formulations. Certain industries, such as aerospace and nuclear power plant component manufacturers, have restrictions on halogen compounds. It is necessary to obtain not only a list of the restricted chemicals, but also the allowable limits for them. In some cases, trace amounts of these materials may appear as an impurity in a formulation, and may not be present in a sufficient quantity to restrict the use of the product.

12. Performance vs. Cost

The objective with any fluid is to achieve maximum performance at minimum cost. In calculating cost it is necessary to consider all the factors and not just the cost of the product. Considerations include used fluid disposal costs, downtime for cleaning, lost production, machine cleaning, recharging costs, tool life, tool resharpening costs, etc. Small improvements in tool life may be difficult to measure, but consider the fluid's sump life, additive costs, shop cleanliness, operator acceptance, and other related factors that contribute to the cost of the fluid in use. It is also important when considering cost to use the common denominator of "mix gallon cost," the cost of the product per gallon multiplied by its recommended dilution ratio. If a product costs less ($6.00 per gallon vs. $8.00 per gallon), but is used at a stronger concentration (10 vs. 5%), then the actual cost comparison for a mix gallon is $0.60 vs. $0.40. The product that costs more is actually less expensive to use, without considering the other factors noted above.

B. Supplier Evaluation

Using these selection criteria, the requirements for any metalworking job can be very well defined. However, before a particular product can be selected, many other parameters must be considered. There are many suppliers of metalworking fluids in the U.S. offering a wide range of products. It is necessary to evaluate the suppliers and determine how their products, business practices, and philosophies compare with the needs of the plant. In this analysis, we assume that the major goal is not necessarily to find the lowest cost product, but to find the product that is most cost effective in terms of performance.

1. Quality

Dr. Deming[26] stresses that we should not be dependent on mass inspection of incoming goods or finished materials. Statistical evidence that quality is built in to the finished product should be required of all suppliers. Many users of metalworking fluids have their own quality standards that must be met and these are well defined for the supplier. Suppliers may have their own programs of quality control and quality assurance. The ability to produce consistent lots of quality material is essential when the production process is so dependent on the metalworking fluid. A visit to a supplier, with a review of processes and procedures for quality may be beneficial.

2. Delivery

The ability of a supplier to quickly deliver material, and the ability of the user to maintain a minimal inventory, are becoming more critical in the metalworking fluid industry. Supplier location, involving both production and warehouse facilities, should be evaluated. For large fluid users, there is a growing trend to be less dependent on material supplied in drums and more dependent on material supplied in refillable totes or in bulk.

3. Health and Safety Testing[27,28]

The users of any chemicals need to be assured that the products are reasonably safe for use. This can be accomplished in a number of ways. A review of the Material Safety Data Sheet (MSDS) supplied by the manufacturer is the primary method. This should include information on any hazards associated with the product, as well as information on the safe use of the product. Some users may also require product composition information in order to make their own evaluation of the product's safety. Testing procedures are available for assessing the acute toxicity of metalworking fluids formulations.[29]

4. Service

Metalworking fluids are used in a very dynamic working environment. The selection of the proper fluid may require some assistance. The methods to control the fluid and care for it in use must be explained to the user. Questions relating to safe use of the product must be addressed. When a problem arises, laboratory evaluation may be needed to resolve any questions. All of these issues can be addressed if the supplier has a good service program. This user should investigate the supplier's capabilities in these areas before making a product selection.

5. Performance Data/Laboratory Testing

When a metalworking fluid is developed, the supplier will have run many laboratory screening tests in areas such as corrosion control, lubricity, oil emulsification, rancidity control, and foam. Some of this testing may be carried out based on the supplier's procedures and other testing according to industry standards or guidelines (ASTM).[30] This data should be reviewed in terms of the customer's desired performance and selection criteria.

6. Case Histories

In judging how a product will perform in a particular application, it is frequently very helpful to review any case history data that may be available from the supplier. It is important to compare these applications in terms of setup, work material, water quality, filtration, flow rates, etc. If a particular product is successfully being used in a certain application, there is a higher level of confidence in the new application.

C. Product Evaluation

A metalworking fluid user can define an answer to the various selection criteria and most suppliers can furnish the information requested in an evaluation. If additional information is needed in the selection process, it is typically obtained through a laboratory evaluation and/or an in-plant testing program.

1. Laboratory Testing

Many plants will have a set of laboratory screening criteria[31–35] that a product must pass before it can move any further into the plant. Tests such as lubricity,[36] corrosion control, and rancidity control are some of the many performance procedures used to screen metalworking fluids. Also, chemical tests may be run to develop a product profile, determine compatibility with waste treatment processes, and generate background information on product quality. Several tests are typically chosen that are known to be key to the success of the product in a particular operation. For example, on cast iron machining applications, a corrosion test using cast iron chips is a typical laboratory evaluation.

2. In-Plant Testing[37,38]

The true measure of any product's performance is a test on the actual application. An individual machine or a small central system with several isolated machines may be used. This is the best method to simulate all of the variables that will be encountered in a normal use situation. It is important in this type of testing to "qualify" the entire process with the existing product or standard. Define a measurable set of performance criteria that are to be evaluated, i.e., tool life, parts per dress of the wheel, machine cleanliness, sump life, product odor, etc. Then set up a system for measuring this data, along with key product specifications, such as concentration, pH, bacteria counts, etc. Typically it will take at least 8 weeks, maybe more, to develop a sufficient database by which products and their performance can be judged. Work with the supplier on these tests, because even though an objective is to hold certain variables constant, improved product performance may be achieved by altering some variables in combination with the test fluid. For example, modifying the coolant application to obtain a better flow to the cut zone with a synthetic may show improved performance over a soluble oil type fluid. The most important item in in-plant testing is to establish measurable parameters before the testing begins. In this way the actual performance of various products can accurately be compared and judgments made on product selection. With this information on selection criteria, supplier evaluation, and product evaluation, the proper fluid can be selected for any application. This is the first step in fluid management. It is now necessary to control and maintain that fluid in the work environment to achieve optimum long-term performance.

III. WATER

Water is the major ingredient in a water-soluble metalworking fluid mix. It may amount to as much as 90 to 99% of the fluid as used. Therefore, the importance of water quality to product performance cannot be ignored.[39,40]

Corrosion, residue, scum, rancidity, foam, excess concentrate usage, or almost any metalworking fluid performance problem can be caused by the quality of the water used in making the mix. Untreated water always contains impurities. Even rainwater is not pure. Some impurities have no apparent effect on a metalworking fluid. Others may affect it drastically. By reacting or combining with metalworking fluid ingredients, impurities can change performance characteristics. Therefore, water treatment is sometimes necessary to obtain the full benefits of water-soluble metalworking fluids.

A. Water Quality

Water quality varies with the source. It may or may not contain dissolved minerals, dissolved gases, organic matter, microorganisms, or combinations of these impurities that cause deterioration of metalworking fluid performance. The amount of dissolved minerals, for example, in lake or river water (surface water), depends on whether the source is near mineral deposits. Typically, lake water is of a consistent quality, while river water varies with weather conditions. Well water (ground water), since it seeps through minerals in the earth, tends to contain more dissolved minerals than either lake or river water. Surface water, however, is likely to contain a higher number of microorganisms (bacteria and mold) and thus need treatment. Typical water hardness throughout the U.S. is shown in Figure 11.1.

Some metalworking plants use well water and have detailed information on its composition. Most, however, use water supplied by a municipal water works, which maintains daily or weekly analyses of the water. To estimate the effect of water on a metalworking fluid mix, measurement of the following will provide sufficient data in most cases:

- Total hardness as calcium carbonate
- Alkalinity "P" as calcium carbonate
- Alkalinity "M" as calcium carbonate
- pH
- Chlorides
- Sulfates
- Phosphate

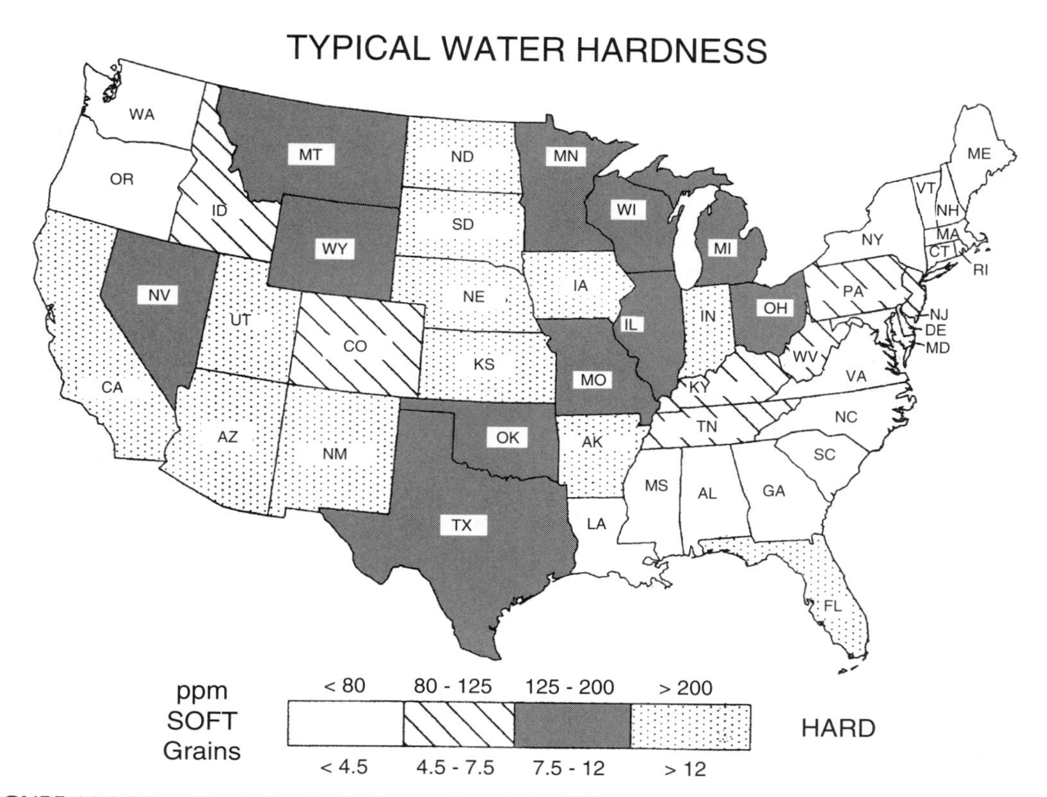

FIGURE 11.1 Map of typical water hardness values in the U.S.

1. Total Hardness

Of the water analysis results, total hardness has perhaps the greatest effect on the metalworking fluid mix. Hardness comes from dissolved minerals, usually calcium and magnesium ions reported in parts per million (ppm) and expressed as an equivalent amount of calcium carbonate ($CaCO_3$). Hardness may also be expressed in terms of "grains," with one grain equal to 17 ppm hardness. The ideal hardness of water for making a metalworking fluid mix ranges from 80 to 125 ppm. The term "soft" is used for water if it has a total hardness of less than 100 ppm or the term "hard" if total hardness exceeds 200 ppm. Test kits and test strips are available from many manufacturers for testing water quality.

a. *Soft Water*

When the water has a total hardness of less than 50 ppm, the metalworking fluid may foam — especially in applications where there is agitation. Foam causes problems when it overflows the reservoir, the machine, the return trenches, etc. Foam may also interfere with settling type separators (since it suspends metal swarf and prevents settling), obscure the workpiece, and diminish the cooling capacity of a water-based metalworking fluid. Soluble oil and semisynthetic products, typically, foam more readily in soft water than synthetics.

After exposing a metalworking fluid to chips, dirt, and tramp oil for a few days, foam tends to dissipate. If it must be eliminated immediately, inspect the system for physical conditions that contribute to excessive foam. Sharp turns or drops in fluid flow, high-pressure nozzles, or malfunctioning pumps could be responsible. If not, foam depressants, chemical water hardeners, antifoam, or oil may be used to decrease foam.

b. Hard Water

Hard water, when combined with some water-soluble metalworking fluids, promotes the formation of insoluble soaps. The dissolved minerals in the water combine with anionic emulsifiers in the metalworking fluid concentrate to form these insoluble compounds that appear as a scum in the mix. Such scum coats the sides of the reservoir, clogs the pipes and filters, covers machines with a sticky residue, and may cause sticking gauges.

Because soluble oils typically have the least hard water stability, hard water has a more obviously detrimental effect on them. Separation of the mix is apparent in severe cases, and is characterized by an oil layer rising to the top of a fresh mix.

Semisynthetics and synthetic metalworking fluids may not be visibly affected by water hardness. Some are formulated with good hard water tolerance. However, dissolved minerals may react with ingredients other than emulsifiers. In these reactions, the metalworking fluid ingredients change or are tied up and, consequently, the product never attains peak performance.

Dissolved mineral content increases in a metalworking fluid mix with use. After a 30-day period, the amount in the mix can increase three to five times the original amount. This results from the "boiler effect" that exists in a metalworking fluid reservoir. That is, water evaporates and leaves dissolved minerals behind. Then, makeup (usually 3 to 10% per day) introduces more with each addition, and they continue to accumulate. Therefore, even with water that has very low dissolved mineral content initially, dissolved minerals can build up rapidly and cause problems.

2. pH

pH is an expression that is used to indicate whether a substance is acidic, neutral, or alkaline. A pH of 7 is neutral, from zero up to 7 is acidic, and from above 7 up to 14 is alkaline (basic). Water in the U.S. normally varies from 6.4 to 8.9 in pH, depending on the area and source of water. The buffering ability of a metalworking fluid is far greater than that of any clean water supply. Adjustments to the pH of the water supply are rarely needed.

3. Alkalinity

Two kinds of alkalinity exist in water: "P" alkalinity and "M" alkalinity. "P" alkalinity is the measure of the carbonate ion (CO_3^{-2}) content and is expressed in ppm calculated as calcium carbonate. This is sometimes referred to as permanent alkalinity and, as such, is not changed by boiling as is the "M" alkalinity.

"M" alkalinity is the measure of both the carbonate ion content ("P" alkalinity) and the bicarbonate ion (HCO_3^{-1}) content. This value is also expressed in ppm, calculated as calcium carbonate. It is referred to as total alkalinity and temporary alkalinity. This is because its value can be lowered to that of "P" alkalinity by boiling.

Metalworking fluids typically perform best when the pH is between 8.8 and 9.5. They require a certain amount of alkalinity for good cleaning action, and corrosion and rancidity control. If pH and total alkalinity become too high, however, pitting and staining of nonferrous metals may occur. Skin irritation is another possible problem. Currently, there appears to be no satisfactory treatment for alkaline water, so careful product selection is critical.

4. Chloride

When chloride (Cl^-) ion content is high (above 50 ppm) in the water used in making metalworking fluid mixes, it is more difficult for the product to prevent rust. Richer concentrations of the metalworking fluid mix may sometimes counteract the effect of chlorides. In other cases, excessive chloride ions must be removed from the water prior to use by demineralization.

5. Sulfate

Sulfate (SO_4^{-2}) ions also affect the ability of a metalworking fluid to prevent rust, though not as much as chloride ions. In addition, they can promote the growth of bacteria. If sulfate ion content exceeds 100 ppm, richer concentrations of the metalworking fluid mix may improve corrosion and rancidity control.

6. Phosphate

Phosphate (PO_4^{-3} and others) ions contribute to total alkalinity and stimulate bacterial growth, leading to potential problems of skin irritation and rancidity, respectively. If phosphate ions are found in the mix water, they should be removed by demineralization to prevent these problems.

A. Water Treatment

There are two processes that are commonly used in treating hard water: water softening and demineralization.

1. Water Softening

In this process, the water passes through a zeolite softener. The softener exchanges calcium and magnesium ions (positively charged ions that are largely responsible for hardness) for sodium ions. In effect, water that was rich in calcium and magnesium ions becomes rich in sodium ions. The total amount of dissolved minerals has not decreased, but sodium ions do not promote the formation of hard water soaps. Corrosive, aggressive negative ions are not removed by the zeolite and can continue to build up in the metalworking fluid mix, leading to corrosion problems or salty deposits. Thus, the use of softened water is not recommended with water-soluble metalworking fluids.

2. Demineralization

Deionizers or reverse osmosis units are used to demineralize water. Deionizers remove dissolved minerals. This is carried out selectively or completely, depending on the type and number of resin beds through which the water passes.

It is not necessary to obtain pure water for metalworking fluid mixes. A hardness level of 80 to 125 ppm is suitable. Usually a two-bed resin deionizer produces water of sufficiently high quality, as opposed to a more expensive mixed-bed deionizer needed to obtain pure water.

Reverse osmosis removes dissolved minerals by forcing water through a semipermeable membrane under high pressure. Typically, this process removes 90 to 95% of the dissolved minerals.

3. Choice of Water Treatment

The chemistry of the water as determined by a water analysis, water quantity needs, water quality requirements, and economics (capital and operating costs) are considerations in selecting suitable water treatment. Softening of hard water eliminates the scum that forms in some metalworking fluid mixes, but increases the possibility of rust problems. Deionizers, typically, are lower in capital costs than reverse osmosis units, but higher in operating costs. Deionizers can provide higher quality water; however, resin beds must be regenerated frequently. If not regenerated frequently, water quality deteriorates and the resin beds also serve as an excellent environment for massive growth of bacteria. Reverse osmosis units do not require regeneration, but do require membrane replacement in time, depending on the water quality fed into the units. Pretreatment systems, prior to either the deionizing or reverse osmosis unit usually lengthen resin or membrane life.

With either method of demineralization, foam can be a problem when initially charging a metalworking fluid system. To avoid foam, the initial charge could be made with untreated water (except in cases where dissolved mineral content is excessive) and subsequent makeup could be mixed with the demineralized water. Chips, grinding grit, and debris eventually will add impurities to the initial charge, but the amount is not significant when compared to using untreated water daily for makeup.

Many metalworking fluid users treat poor quality water before using it in fluid mixes. The benefits vary, depending on the water quality before treatment and the type of metalworking fluid that is used. In one case, composition of a fluid user's city water varied widely in dissolved minerals content because of frequent changes in processing by the municipal water works. After passing this water through a mixed-bed deionizer, consistent quality water with zero hardness was obtained. The cost of demineralization roughly equaled the amount saved in reduced usage of soluble oil concentrate. In addition, filter media consumption was reduced, while fluid filtration improved significantly. Demineralized water has also decreased additive usage and a corresponding incidence of skin irritation. Likewise, the amount of residue on machines was less, and, what was present was more fluid in nature. This user concluded that the benefits of using demineralized water were well worth the investment. Also, the water is now of consistent quality, which eliminates one major variable when looking for the source of any metalworking fluid performance problem.

IV. METALWORKING FLUID CONTROLS

Metalworking fluids, specifically the water-soluble types, are all formulated to operate within a certain range of conditions in areas such as concentration, pH, dirt levels, tramp oil, bacteria, and mold. When fluid conditions fall out of this range, in one or more of these areas, performance problems can develop. It is, therefore, necessary to have a set of tests, to be run on some regular basis on the fluid mix to keep it within these operating conditions. Some general-purpose type products may have very wide operating ranges in several areas so that they can withstand the abuse of being used in an environment with limited control. The performance of other products may require that they be controlled in certain areas (i.e., concentration) very close to the listed specification.

The importance of controlling a metalworking fluid has been understood for many years and continues to grow.[41–44] Working with the supplier and understanding the needs of a particular operation will usually dictate the frequency and degree of the control required. In some plants, where the fluid is very critical to the operation, such as aluminum can production, checks on concentration and pH are made every 4 h. In other manufacturing plants, such as automotive components or bearings, large central systems are used and checks are typically made once a day. It is much more difficult to control a plant where many individual tanks are utilized. They may be checked once a week, or the fluid may be controlled by simply monitoring the output of a premix unit or a recycling system.

In this section, some of the typical tests used to control water-soluble metalworking fluid mixes, in use, will be explained.

A. CONCENTRATION

Water-soluble metalworking fluids are typically formulated to operate in a concentration range of 3 to 6%, although concentrations of 10% or higher are not uncommon for heavy-duty applications. Concentration is the most important variable to control. Concentration is not an absolute value, but rather a determination of a value for an unknown mix based on values obtained from a known or "control" mix. There are certain inaccuracies, variables, and interferences in any method. This must be considered when evaluating the data.

On some products, several methods for measuring the concentration may exist, based on different components of the product. One method may measure alkalinity, another one the anionic components, and another the nonionics. Initially, all the methods may agree, but as the fluid mix

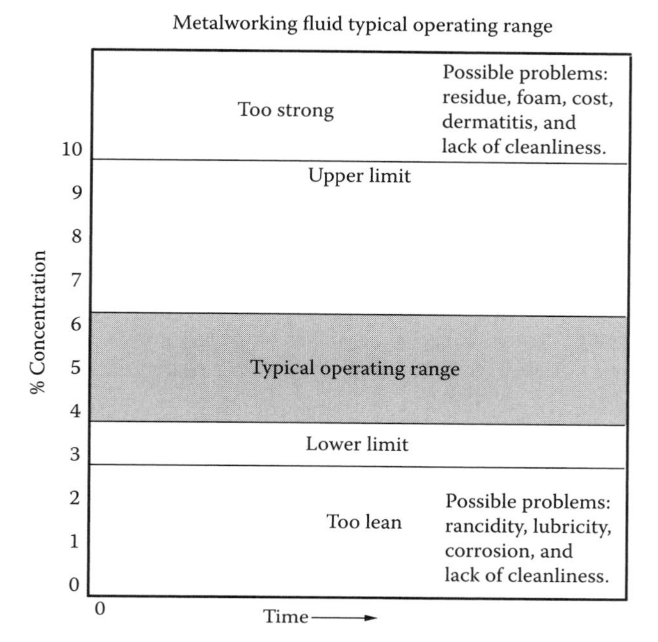

FIGURE 11.2 Metalworking fluid typical operating dilution ranges.

ages and contaminants are introduced, different values may be obtained. While this does not necessarily imply a problem, it may certainly lead to some concern. In some cases, an additive package may be necessary to rebalance the product. With other products, this may be considered a normal aging process.

Metalworking fluid suppliers formulate their products so that, at the correct operating range (see Figure 11.2) the proper levels of the chemicals needed for performance (lubricants, rust inhibitors, biocides, etc.) are present. To achieve this performance, the fluid must therefore be maintained within this range. This is carried out by means of some concentration control procedure. Various methods are available.

1. Refractometer

This is an optical instrument that measures the refractive index of a metalworking fluid. This refractometer reading, obtained from a scale of numbers in the unit, is then converted into a concentration value via a factor or graph made from taking readings on known mixes of various concentrations. A 0% concentration would have a zero refractometer reading. Depending on the model, the refractometer should always be zeroed with water before taking a reading. Generally, synthetic fluids have very small values for refractometer readings in the 1 to 3 range. Small differences in scale readings become rather large differences in concentration. Soluble oils have rather high refractometer readings and, in many cases, the reading will correspond directly to the concentration.

On a refractometer, the reading is taken where there is a distinction between two colors on the scale, typically black or dark blue and white. Since metalworking fluids during use will pick up more contamination, especially tramp oil; the distinction between these two colors becomes much less clearly defined. The ability to obtain an accurate reading becomes much more difficult.

Overall, refractometers are a fairly quick method to check concentration and are certainly sufficient for many operations, however, the inherent inaccuracies must always be kept in mind.

2. Titrations

Chemical titration methods can be established to measure certain components or groups of components in any fluid mix. These would include measuring the alkalinity, the anionic content, the nonionic level, or sulfonates. It is important to establish known control values for any of these methods. The interferences from contaminants or the change in a titration value due to the aging of the mix should also be established in order to give more accurate results. It is also possible to utilize specific ion electrodes and automatic titrators to run these types of concentration checks.

3. Instrumental

Using instrumental methods typically allows for a very specific measurement of one compound. Instruments used include gas chromatography (GC), atomic absorption (AA), high-pressure liquid chromatography (HPLC), and Fourier transform infrared (FTIR).[45] These can be quite sophisticated and involved methods, requiring expensive instrumentation and lengthy sample preparation. To that extent, they are more frequently found in the laboratories of the metalworking fluid supplier and not the customer. One or more key components are chosen to track, sometimes with the assumption that other ingredients will stay in relative balance. In other cases, a particular component may be measured to detect depletion or buildup. In cases of depletion, an additive may be used to restore this component.

B. pH

A pH measurement is used to determine the degree of acidity or alkalinity of a metalworking fluid mix. Metalworking fluids are typically formulated and buffered to operate in a pH range of approximately 8.5 to 9.5. This is somewhat of a compromise. If the pH ran higher, the fluid would provide excellent ferrous corrosion control, but could have problems in the areas of skin mildness and nonferrous corrosion protection. A lower pH would be good for mildness and nonferrous corrosion control but may cause problems with rancidity control and ferrous corrosion protection. It should be noted that some fluids, especially those used in certain aluminum applications, are formulated with a mix pH in the 7 to 8 range. These fluids are designed to operate at the lower pH values and should be controlled and maintained based upon the manufacturer's recommendations.

The pH is also a good, quick indicator of the condition of the fluid. A pH below 8.5 is typically the result of bacterial activity. This can affect mix stability, ferrous corrosion control, and microbial control. Additives can be used to increase the pH of a mix. A high pH, greater than 9.5, is generally the result of some form of alkaline contamination, and will affect the mildness of the fluid. It is very difficult to correct a high pH situation, short of dumping the system.

The pH values on a fluid mix can be obtained by using pH paper, which is dipped into the mix and observed for a color change, or by using a pH meter. Many models are available, from inexpensive hand-held units to rather elaborate laboratory bench models. When using any meter to check the pH of a metalworking fluid, it is important to ensure that the meter has been standardized with the appropriate buffers and also that the electrode(s) is clean. The oils found in most used fluid samples can quickly foul pH electrodes causing inaccurate readings. Simple cleaning with isopropyl alcohol can eliminate this problem.

C. Dirt Level

Dirt or total suspended solids (TSS) in a metalworking fluid mix include metal chips and grinding wheel grit. Recirculating dirt, whether it is a large quantity of small particles or just one or two large particles, can affect part finishes,[46] lead to dirty machines, and clog coolant supply lines. Recirculating metal fines can also lead to rust problems if they deposit on parts. Dirt or TSS measurement is typically an indication of the effectiveness of the filtration system and/or the settling and chip agglomeration properties of the fluid. To obtain a representative sample for testing, it is important to sample the fluid from the clean coolant nozzle and be aware of the current indexing cycle of the filter system.

Many procedures exist for determining dirt load in a fluid. A simple method is to centrifuge a sample in a calibrated centrifuge cone. Other methods involve filtering a sample through a specific size filter paper, drying, perhaps ashing, and then weighing. Typically, dirt volumes in excess of 500 ppm or 20 μm in size can lead to problems. Each operation will dictate the type of filtration required. Working with suppliers of both the metalworking fluid and the filtration system is the best method to address concerns in this area, especially if an adjustment to the filtration system is required. There are chemical additives available to assist in chip settling and filtration.

D. Oil Level

Almost every metalworking fluid contains oil. It is either an ingredient, the oil formulated into a soluble oil ("product oil"), or one of the major contaminants in the form of tramp oil. In both cases, it is very useful to know the amount of oil present. Product oil can give an indication of the fluid concentration, while tramp oil levels indicate the amount of contamination.

Tramp oil can be in two forms, free, or emulsified. Free oil is that oil which is not emulsified and basically floats on the top of the mix. Emulsified tramp oil is nonproduct oil, which is either chemically or mechanically emulsified into the product. Free oil can generally be removed by skimmers or belts, while emulsified tramp is much more difficult to remove, even with a centrifuge or a coalescer.

The sources of the tramp oil can be hydraulic leaks, way or gear lube leaks, or from lubrication systems that are found on many machines. The type of oil used can also make a difference in its emulsification or rejection properties with a particular metalworking fluid. Some oils are formulated to emulsify themselves into any water systems, while others have better rejection capabilities. The same goes for metalworking fluids. Depending on the formulation, fluids may be designed either to emulsify a certain level of oil or to completely reject it. Additive packages found in some lubricating oils can be problematic for the metalworking fluid.

"Free oil" is measured by centrifuging a sample of mix, or simply by allowing it to stand for several hours, and then reading the amount of oil that is floating on the surface. "Total oil" is determined by completely splitting the emulsion with sulfuric acid, and then reading the oil content. If a concentration has been determined by some alternate method, or if a "clean" sample at the same concentration has been subjected to this same break, then the product oil can be calculated. Subtracting this product oil from the total oil will give the total tramp. Subtracting the free oil from the total tramp, will indicate the amount of emulsified tramp oil in the system.

Generally, high tramp oil levels will affect a product's cleanliness, filterability, mildness, corrosion and rancidity control. Many mechanical methods exist for minimizing the leaks and removing the oil.[47]

E. Bacteria and Mold Levels

Metalworking fluids do not exist in a sterile environment and can develop certain levels of microbial growth.[48,49] The water environment of most fluids is conducive to biological activity. Fluids are formulated to handle bacteria and mold in different ways. Some products contain bactericides and fungicides to kill any organisms that may be present. Other products are formulated with ingredients that will not support biological growth. These are sometimes referred to as "bioresistant" fluids. Still others are formulated so that, while bacteria and mold may grow in them, no offensive odors or performance problems develop. Regardless of the type of product, knowing the level of microbial growth and, in most cases, being able to control it, is a very useful tool.[50] Metalworking fluid microbiology is covered in much greater detail in a separate chapter within this book. Bacteria and mold levels can be determined in several ways:

1. Plate Counts

Specific agars for bacteria and mold are prepared. The fluid mix is appropriately diluted and then introduced to the plates along with the agar. The plates are incubated for 48 h at approximately 36°C. After that time the colonies are counted.

2. Sticks

Several companies have developed very specific test systems for determining bacteria and mold counts in metalworking fluids, using "sticks," "paddles," or "dip-slides" that have already been prepared with various agars. These sticks are immersed in a fluid sample for 2 to 5 sec, placed into a plastic container and incubated at 25 to 30°C for 24 to 48 h. They are then "read" by comparing their appearance to a chart that relates appearance to a specific number of organisms.

Most suppliers set limits of about 10^5 as the maximum bacteria levels and 0 as the maximum mold level. If counts exceed these levels, some form of treatment may be needed.[51]

3. Dissolved Oxygen

Another method to obtain an indication of biological activity, although no specific counts, is with a dissolved oxygen (DO) check. When a metalworking fluid mix is exposed to the air or is pumped about in air, it will absorb a certain amount of oxygen. At 68°F, a circulated fluid mix will dissolve approximately 9 ppm oxygen. When aerobic bacteria grow, they use some of the oxygen and also excrete certain gases that drive some of the remaining oxygen out of the mix. Using this phenomenon, the DO of a mix can be measured to obtain a relative indication of biological problems. This determination can be based on one DO reading. A value of less than 3 ppm would indicate a problem. Alternatively, an initial reading can be taken, followed by a reading after the mix has been allowed to sit for 2 h. A difference of 2 to 3 ppm usually indicates a problem. DO is a good method for a quick, on-site determination of any biological problems.

F. Conductivity

Another metalworking fluid parameter frequently measured for an indication of the fluid's condition is conductivity. The unit of conductivity is the microSieman (μS). Conductivity can be measured on any type of commercially available conductivity meters. A typical 5% metalworking fluid mix prepared with tap water will have a conductivity of approximately 1500 μS. Conductivity can be altered by the mix concentration, buildup of water hardness, buildup of chloride or sulfate from the water, mix temperature, dissolved metals, and just about any other contaminant. Since so many ever-changing variables can affect conductivity, a single reading is of little value. Observing any trends in these conductivity readings over a period of time may be useful in assessing mix condition and aging, as well as helping in problem solving for residues or unstable mixes. Relate the conductivity values to the concentration values to look for any indications of contamination.

V. CARE AND MAINTENANCE OF THE FLUID

Prolonging the life of the metalworking fluid and optimizing its performance are very dependent on the control of the metalworking fluid system.[41–43,52,53] This control is as important as the selection of the proper fluid and includes maintenance of the mechanical components as well as the metalworking fluid. The problems that beset metalworking fluids in central system applications are the same as those in individual machines, only the magnitude is greater. A program to accomplish this control should include the following steps.

1. *Assign the responsibility for control.* If a coordinated program is not established to control the system, it will result in no control. One department or one individual should be responsible for checking fluid concentration and other specified parameters and for making any additions of water, concentrate, or additives to the system. These additions should be recorded for future reference. This person or department will be more mindful of additions, know the reason for making them, and not use concentrate or additive additions as the only means to resolve a production problem. When a control program is not utilized, frequent system dumps and excess usage, resulting in increased costs, can easily occur since no one really knows the status of the system.

2. *Clean the system thoroughly before charging with a fresh mix.* Dirt and oil can accumulate in relatively stagnant pockets or quiet areas in the central systems or individual machine. If not removed, such accumulations not only cause dirt recirculation in a fresh charge, but also provide an instant inoculation of bacteria to the fresh mix.

3. *Maintain the concentration of the metalworking fluid at the dilution recommended for the particular operation.* Recommended dilutions are indicated on the label and in the product literature. Many plants run daily concentration checks on central systems. Individual machines are usually checked on a less frequent basis. As mentioned earlier, the fluid concentration can be checked with a refractometer, a laboratory titration procedure, or an instrumental method. Concentration can be controlled by the use of premixed fluid or a proportioning system. Reviewing this concentration information can indicate trends and possible problems long before they show up on the production line. Lean concentrations can lead to rust, rancidity, poor tool life, lack of lubricity, and other problems. Maintaining a stronger than recommended concentration can result in foam, skin irritation, residue, increased costs, and other problems.The fluid mix is lost from the system by both evaporation and carry-off or splashing. Depending on the type of operation, type of fluid, and part configuration and handling, the amount of mix lost by either of these means can vary. Through evaporation, only water is lost. Splashing or carry-off will cause loss of both water and fluid concentrate. Therefore, each time water is added to the system, metalworking fluid concentrate should also be added at a ratio that has been selected to maintain the proper dilution in the system. This will keep product components in their proper balance and minimize any selective depletion of these components. For grinding operations, the premixed fluid is usually a leaner dilution than for machining. This is because grinding typically loses more water to evaporation.

4. *Keep the metalworking fluid free of chips and grit.* This is a major factor in fluid life. Recirculating dirt can lead to unsightly buildup on the machines, plugged coolant lines, poor finish in grinding,[54] and tool wear in machining. Chip buildup in reservoirs can drastically reduce the volume of the system and deplete product ingredients. Positive filters with some type of disposable media do a better job of removing small fines than settling tanks. On individual machines, regular cleanouts of the reservoir or sump should be utilized to keep this buildup under control. The use of fluid recycling could be a cost efficient option.

5. As mentioned earlier in this chapter, the quality of water to make a metalworking fluid mix is a very important factor in performance. Remember, the ideal hardness of water for making a metalworking fluid mix ranges from 80 to 125 ppm.

6. *Aerate the metalworking fluid mix by keeping it circulated.* This circulation prevents the growth of anaerobic bacteria that cause offensive odors. Many central systems continually circulate even when production is not running, others utilize timers to circulate the fluid for a short time on

a set schedule during any nonproduction hours or days. In individual machines, an air hose can be used to bubble air through the mix while the machine is not operating. Atmospheric oxygen is detrimental to the growth of odor producing anaerobes. During circulation, oxygen enters the metalworking fluid at a maximum rate, but at a much lower rate when the system is shut down.

7. *Provide good chip flushing at the machines and in the trenches.* If chips do not reach the filter, they deplete certain constituents of the metalworking fluid and furnish an excellent breeding ground for bacteria. It is essential that the chips reach the filter in order that they might effectively be removed. Trenches, return lines, system capacity, retention time, flow rates, and other design parameters must all be adequately sized to provide this good filtration. Wash-down nozzles may need to be installed on the machines or in the trenches to keep the metalworking fluid moving back to the sump or filter. Check that these nozzles are set at flow rates sufficient to keep the chips moving but not so excessive as to result in foaming.

8. *Employ good housekeeping practices.* Foreign matter that is allowed to accumulate in a metalworking fluid has a drastic effect on its life and performance. While a good high-quality metalworking fluid is formulated to cope with a certain amount of contamination, the greater the amount of contamination, the shorter the fluid life, and the more erratic its performance becomes. Avoid using reservoirs as a "garbage" disposal. Cigarette butts, food scraps, sputum, and candy wrappers, for example, inoculate the metalworking fluid with bacteria and furnish food for their growth. Do not dump floor-cleaning solutions into the reservoir. Many contain chemicals, such as phosphates, which may contribute to skin irritation, promote the growth of odor producing microorganisms, or cause the product to foam.

9. *Remove extraneous tramp oils.* Minimize the leakage of oils into the system through proper maintenance of seals and lubrication systems. If excess quantities of oils leak into the system, the metalworking fluid performance can be reduced. High oil levels can extract oil soluble lubricants, emulsifiers, and microbicides from the fluid. Lubricating and hydraulic oils contain food for bacteria. They may also blanket the surface of the fluid, excluding air, and thereby provide ideal conditions for the growth of odor producing bacteria. If allowed to build up, extraneous oil causes smoking, reduces the cooling action of the fluid, increases residue around the machine area, and makes the machines look dirty. Oil removing devices such as skimmers, coalescers, oil wheels, or centrifuges can be used to prevent oil buildup.

By following this program, it is possible to achieve improved productivity and long, trouble free metalworking fluid life in central systems and individual machines.

VI. METALWORKING FLUID TROUBLESHOOTING

Problems with metalworking fluids can be related to a number of causes, many of which are not inherently fluid problems. Improper machine set up, coolant application, or product selection can all lead to problems. It is certainly more complex to solve the problems in central systems compared to individual machines. A much larger volume of fluid, more production, and many more operators are involved. In some cases, there may be a combination of many factors causing the problem.

Logical thinking is the first step in any problem solving effort. Obtain the facts, analyze them, plan a course of action, and implement the plan. For an individual machine, the solution may be a simple dump and recharge. For a central system, the problem and the resolution will usually be much more complicated. In these situations, service help from the supplier via a phone call or personal visit is a likely course of action.

The problems most commonly attributed to metalworking fluids include: corrosion, rancidity or objectionable odors, excessive foam, insufficient lubricity resulting in poor tool life or an unsatisfactory finish on the part, objectionable residue or a dirt buildup on the machine, and safety concerns such as dermatitis or irritation of the eye, nose, or throat. In this section, many of the possible causes and the corresponding remedies have been listed.

A. Corrosion of the Work or the Machine

Possible Causes	Corresponding Remedies
The concentration of the metalworking fluid mix may be too low.	Make a concentration analysis by one of the described methods and adjust the mix to the recommended dilution. Determine and correct the cause of the low concentration (mixing errors, water leaks, hard water, etc.). For a central system mix, it may also be necessary to check and adjust the pH to the recommended standard.
The recommended concentration range of the fluid may be too low for this application.	Increase the concentration of the mix by 0.5 to 2% increments, depending on the product being used, to find the optimum concentration range.
The buildup of ions from the water supply (total hardness, chloride, or sulfate) may be too high for the product or the current concentration.	If possible, conduct laboratory testing to determine water quality and any buildup in the fluid. In some cases, increasing the concentration by 0.5 to 2% may provide control for a while. If the buildup is excessive, a system dump may be required. In areas of poor quality water, consider a water treatment system or products with optimum corrosion control.
The metalworking fluid tank may be full of chips or swarf, contaminated by tramp oil or other contaminants.	Clean the fluid, if possible, using filtration or recycling equipment. If it is not possible to adequately remove the contaminants, dump and recharge the system with fresh metalworking fluid. Excessive dirt can "plate out" on parts and lead to rust. Certain settling or filter aid additives may help.
Parts that are still wet with metalworking fluid may be touching other ferrous materials or dissimilar metals.	Avoid metal-to-metal contact in stacking parts after any metal removal operations. Use plastic coated wire baskets rather than metal tote pans. Dry the parts before prolonged storage. Use vapor barrier material between parts during handling and storage. The use of a rust preventive spray or dip may be needed for extended storage.
Hot, humid conditions may accelerate rust problems by slowing the drying action.	Increase the concentration of the mix. Improve plant ventilation. During severe weather conditions, it may be necessary to use a water displacing rust preventive or an additive to the fluid mix.
Fumes from acidic materials may be entering the area.	Improve plant ventilation. Use fans to direct the fumes outside or provide some type of covering for parts and machines.
A high bacterial count may indicate the system has been contaminated from some external source, such as oil leaks from the machines, indiscriminate disposal of cleaners, plating compounds, washer fluids, debris from construction, or the previous fluid.	Locate and repair oil leaks. Determine and eliminate the source of the dumping of any contaminants into the system. It may be advisable to add a biocide to assist in returning the fluid to a normal condition. Aerate the fluid mix to reduce the anaerobic bacteria growth. If the system is extremely dirty from the previous fluid or from construction debris, thoroughly clean according to the recommendations of your supplier, and then recharge with fresh fluid. Contact your supplier for specific recommendations, and perhaps for some laboratory work to determine cause of the problem.

B. Rancidity or Objectionable Odor

Possible Causes	Corresponding Remedies
The concentration of the metalworking fluid mix may be too low.	Make a concentration analysis by one of the described methods and adjust the mix to the recommended concentration. Determine and correct the cause of the low concentration (mixing errors, water leaks, hard water, etc.). For a central system mix, it may also be necessary to check and adjust the pH to the recommended standard.

continued

B. Randicidity or Objectionable Odor (Continued)

Possible Causes	Corresponding Remedies
The recommended concentration range of the fluid may be too low for this application.	Increase the concentration of the mix by 0.5 to 2% increments, depending on the product being used, to find the optimum concentration range.
The fluid tank may be full of chips or grinding swarf, contaminated by tramp oil leakage, or other contaminants, such as food scraps.	Clean the fluid where possible using filtration and/or oil removal equipment. On small tanks, it may be advisable to dump and recharge the fluid. In certain conditions of rancidity, the use of biocide additives to treat for specific bacteria and/or mold problems is the recommended treatment. Use only as recommended by your supplier. It is best to eliminate the source of the contamination.
High dirt content indicates inefficient filtration. This could be due to an incorrect filter setting, a change in a setting, or defective or improper media.	Repair defective media. Restore filter to original settings and adjustments. Experiment to find more effective adjustments and settings. Increase retention time. Contact the filter manufacturer. Investigate the possibility of retaining a higher percentage of the large swarf on the media in order to build a better filter cake. If necessary, thoroughly clean the system according to recommended procedures. Recharge with a fresh mixture. Contact the supplier for recommendations.
Extreme conditions of contamination, excessive dirt load, or both, may require a change in operational procedures.	Aerate the cutting fluid and increase filtration time by running the entire system up to 24 h per day and on weekends, if necessary.
The system may be contaminated from the old cutting fluid or construction debris.	Thoroughly clean the system according to recommended procedures. Recharge with a fresh mixture.
The sulfate content of the water may be too high for this specific product.	Have a water analysis made. If the sulfate content is over 150 ppm, use a higher concentration of metalworking fluid, change to a product that is more compatible with this condition, or use treated water.
Excessive amounts of lubricating oils may be leaking into the system. These oils often contain sulfur or phosphorus, which are ideal foods for bacteria.	Change to a more compatible oil. Prevent leakage into the system. If this is not possible, consider installing oil removal equipment, such as oil skimmers or centrifuges.

C. Excessive Foam

Possible Causes	Corresponding Remedies
The concentration of the cutting fluid mixture may be too high.	Make a concentration analysis and adjust to the recommended concentration. Determine and correct the cause of the mixture being too high. Most frequently, this is due to human error or mechanical problems with metering devices.
The recommended concentration range may be too high for this application.	Decrease the concentration of the mixture by 0.5 to 2.0% increments, depending on the fluid being used, to find the optimum concentration. *Caution*: If the concentration is too low, other problems (rust, rancidity, etc.) may develop.
The level of the cutting fluid in the reservoir may be low, causing air to be drawn into the pump.	Fill the reservoir to the normal operating level with water and concentrate at the recommended concentration.
A crack in the pump housing or intake piping may be allowing air into the system.	Inspect the pump and piping system. Repair or replace defective units.
High outlet pressures, high fluid velocities, sharp corners in the return system, or excessive waterfalls may create high agitation.	Locate any of these foam-producing conditions and reduce or eliminate where possible.

continued

C. Excessive Foam (Continued)

Possible Causes	Corresponding Remedies
The water may be too soft to use with this specific product.	Have a water analysis made. If the total hardness is less than 50 ppm, change to a product that is more compatible with soft water. It is possible to artificially harden the water using a calcium or magnesium salt. Contact your fluid supplier for specific recommendation.
The system may be contaminated from some external source, such as indiscriminate disposal of floor cleaners, washing compounds, etc.	Determine and eliminate the source of indiscriminate dumping of other shop materials into the cutting fluid reservoir. It may be advisable to add antifoaming additives to assist in returning the fluid to a normal condition. Contact the supplier for recommendations. If the system is highly contaminated, thoroughly clean according to the recommended method. Recharge with a fresh mixture.

D. Unsatisfactory Surface Finish or Burn on Parts from Grinding Operation

Possible Causes	Corresponding Remedies
The concentration of the cutting fluid mixture may be too low.	Make a concentration analysis and adjust the mixture to the recommended concentration. Determine and correct the cause of the low concentration (e.g., mixing errors, water leakage, recirculated grit, hard water, etc.).
The recommended concentration may be too low for these specific conditions.	Increase the concentration of the mixture by 0.5 to 2.0% increments, depending on the fluid being used, to find the optimum concentration. At some point, it may be necessary to consider an alternative fluid recommendation.
The flow of the cutting fluid may be inadequate or it may not be reaching the metal removal area.	Increase the volume and readjust the nozzle so that a maximum amount of fluid reaches the metal removal area. Foam or entrained air may be getting into the cut zone in place of fluid. Follow the remedies for reducing foam.
The cutting fluid tank may be full of chips or grinding swarf; contaminated by oil leakage or by other matter.	Drain and thoroughly clean the reservoir and cutting fluid piping system according to recommended procedures. Recharge with a fresh product.
The grinding wheel may be incorrect for this application.	Determine if the wheel is acting too hard or too soft. Change wheel grade accordingly.
The water may be too hard to use with this specific product.	Have a water analysis made. If the total hardness is over 200 ppm, change to a product that is more compatible with hard water, or use treated water. Hard water can cause mix instability problems with emulsion type product.

E. Cutting Tool or Grinding Wheel Life Is not Satisfactory

Possible Causes	Corresponding Remedies
The concentration of the cutting fluid mixture may be too low.	Make a concentration analysis and adjust the mixture to recommended concentration. Determine and correct the cause of the low concentration (e.g., mixing errors, water leakage, recirculated grit, hard water, etc.).

continued

E. Cutting Tool or Grinding Wheel Life Is not Satisfactory (Continued)

Possible Causes	Corresponding Remedies
The recommended concentration range of the fluid may be too low for this application.	Increase the concentration by 0.5 to 2.0% increments, depending on the product used, to find the optimum concentration. At some point it may be necessary to consider an alternative fluid recommendation.
The flow of the cutting fluid may be inadequate or it may not be reaching the metal removal area.	Increase the volume of fluid being used and readjust the nozzle so that the maximum amount of fluid reaches the metal removal area.
The cutter or tool design may be incorrect for this application.	Analyze the tool geometry in relation to the application. Consult the tool engineering specialist. Change the geometry to obtain improved chip formation.
A high bacterial count may indicate the system has been contaminated from some external source, such as oil leaks from the machine tools, indiscriminate disposal of floor cleaners, plating compounds, food remnants, tobacco, etc.; the residual effects of the previous fluid, or construction debris.	Locate and repair all oil leaks. Determine and eliminate the source of indiscriminate dumping of other shop materials into the cutting fluid reservoir. Improve hygienic practices. It may be advisable to add a bactericide to return the fluid to a normal condition. Contact the supplier for recommendations. If the system was extremely dirty from the previous fluid or from construction debris, thoroughly clean according to the recommended method. Recharge with a fresh mixture.
High dirt content indicates inefficient filtration. This could be due to incorrect filter setting, a change in a setting, or improper media.	Repair defective media. Restore filter to original settings and adjustments. Experiment to find more effective adjustments. Increase retention time in settling systems. Contact the filter manufacturer. Investigate the possibility of retaining a higher percentage of the large swart on the media in order to build a better filter cake. If necessary, thoroughly clean the system according to recommended procedures. Recharge with a fresh mixture. It may be advisable to use additives to assist in obtaining a usable mixture without cleaning the system. Contact the supplier for recommendations.

F. Skin Irritation[55]

Possible Causes	Corresponding Remedies
Regardless of the cause, skin irritation is a medical problem and should be treated immediately.	Have the worker report immediately to properly trained medical personnel. Although the ailment may be unrelated to the cutting fluid, it should be investigated and treated by a competent person.
The concentration of the cutting fluid mixture may be too high.	Make a concentration analysis and adjust to the recommended concentration. Determine and correct the cause of the mixture being too high. Most frequently, this is due to human error or mechanical problems with metering devices.
The recommended concentration range may be too high for this application.	Decrease the concentration of the mixture by 0.5 to 2.0% increments, depending on the product used, to find the optimum concentration. *Caution*: If the concentration is too low, other problems (i.e., rust, rancidity, etc.) may develop.
The soap in the washrooms may be too harsh and irritating.	Change to a mild, but equally effective cleaning agent.
The operator's hands may be immersed continually in the metalworking fluid.	Encourage the use of waterproof barrier creams or protective gloves. Use material handling devices where feasible.

continued

F. Skin Irritation (Continued)

Possible Causes	Corresponding Remedies
The operator may be coming into contact with harsh irritating chemicals outside, or even inside the company[56].	Determine if the operator has any activities where he might come in contact with such chemicals (i.e., solvents used in painting, cleaners, and solvents used in automotive repair work, etc.). Substitute the irritating products for ones that will not affect the operator's skin.
The operator may be subject to skin irritation because of poor hygienic conditions.	Encourage washing frequently, wearing freshly laundered work cloths and using protective gloves, aprons, boots, etc., especially if there are excessive splash conditions.
The sump or reservoir may be contaminated from some external source such as indiscriminate disposal of floor cleaners, washing compounds, construction debris, etc.	Determine and eliminate the source of indiscriminate dumping of other shop materials into the cutting fluid reservoir. Improve hygienic practices. If the system is excessively contaminated, thoroughly clean according to the recommended method. Recharge with a fresh product.

G. Eye, Nose, or Throat Irritation

Possible Causes	Corresponding Remedies
Regardless of the cause, these are medical problems and should be treated immediately.	Have the worker report immediately to trained medical personnel. Although the ailment may be unrelated to the cutting fluid, it should be investigated and treated by a competent person.
The concentration of the cutting fluid mixture may be too high.	Make a concentration analysis and adjust the mixture to the recommended concentration. Determine and correct the cause of the high concentration (e.g., mixing errors).
The recommended concentration may be too high for the specific conditions.	Decrease the concentration of the mixture by 0.5 to 2.0% increments, depending on the fluid being used, to find the optimum concentration. *Caution*: if the concentration becomes too low, other problems (rust, rancidity, etc.) may develop.
There may be irritating fumes coming from some other operation in the plant or outside the plant.	Investigate ventilation conditions of heat treating or plating areas, and the plant in general. Improve unsatisfactory conditions with the use of fans until permanent changes can be made. Investigate possible sources outside the plant and take corrective action if required.
There may be excessive splashing or misting of the cutting fluid[57].	Reposition the guards on the machine to contain the splash or mist. Consider enclosing the machines. Encourage the use of safety goggles or glasses.
Microbial contamination of the fluid[58–60].	Consult with the fluid supplier and health professionals to determine the condition of the fluid and the surrounding work environment. New information is continuing to be developed in this area.

H. Objectionable Residue

Possible Causes	Corresponding Remedies
The concentration of the cutting fluid mixture may be too high.	Make a concentration analysis and adjust to the recommended concentration. Determine and correct the cause of the mixture being too high. Most frequently, this is due to human error or mechanical problems with metering devices.

continued

H. Objectionable Residue (Continued)

Possible Causes	Corresponding Remedies
The recommended concentration range may be too high for this application.	Decrease the concentration of the mixture by 0.5 to 2.0% increments, depending on the fluid being used, to find the optimum concentration. *Caution*: if the concentration becomes too low, other problems (rust, rancidity, etc.) may develop.
The cutting fluid reservoir may be full of chips or grinding swarf, contaminated by oil leakage and food remnants, or contaminated by other matter.	Drain and thoroughly clean the reservoir and cutting fluid piping system according to recommended procedures. Recharge with a fresh product.
There may be excessive misting conditions due to inefficient guards.	Design and place the guards, shields, etc., so that misting (especially from grinding operations) is confined to the immediate area of the cut.
The system may be contaminated from some external source such as oil leaks from the machine tools.	Locate and repair all oil leaks. Remove extraneous oil by means of oil skimmers or a centrifuge. If necessary, clean the system thoroughly according to recommended procedures. Recharge the system with a fresh product mixture.
The water may be too hard to use with this specific product.	Have a water analysis made. If the total hardness is over 200 ppm, change to a product that is more compatible with hard water, or use treated water.

VII. CONTRACT FLUID MANAGEMENT[61–63]

Managing metalworking fluids can be a very involved and time consuming process, especially as the products become more complex and the control techniques more sophisticated. As plants recognize the value of properly maintaining their fluids from a productivity and a waste minimization standpoint, they realize that a certain degree of expertise is required to accomplish this. There is also interest in streamlining the purchasing function and minimizing the cost of inventory. For these reasons, contract fluid management or chemical management is in place at many plants.

Under this plan, the metalworking fluid supplier or another vendor has responsibility for the usage and control of the metalworking fluids and other materials, which include lubricants, oils, cleaners, rust preventives, and waste treatment. In some cases, these vendors will put on-site managers and technicians in place to run regular checks on the fluid parameters, calculate fluid concentrate additions, recommend additives and system dumps/recharges, and participate in many other plant activities, such as production meetings, training, and MSDS management. In essence, a partnership is formed between the supplier and the user.

For an agreed upon fee over the course of the contract (typically, 3 to 5 years), the user has the benefit of the expertise of the supplier to manage and control the materials covered by the program. The chemical manager is charged with improving chemical, lubricant, and process performance. This should result in better quality and productivity for the user.

REFERENCES

1. Nachtman, E. S. and Kalpakjian, S., *Lubricants and Lubrication in Metalworking Operations*, Marcel Dekker, New York, pp. 1–61, 1985.
2. Booser, E. R., Ed., *CRC Handbook of Lubrication*, CRC Press, Boca Raton, FL, pp. 335–356, 1984.
3. Springborn, R. K., *Cutting and Grinding Fluids: Selection and Application*, American Society of Tool and Manufacturing Engineers, Dearborn, MI, pp. 5–30, 1967.

4. Zintak, D. C., Ed., *Improving Production with Coolants and Lubricants*, Society of Manufacturing Engineers, Dearborn, MI, pp. 8–49, 1982.
5. Merchant, M. E., Fundamentals of cutting fluid action, *Lubr. Eng.*, 1950, August.
6. Dick, R. M. and Foltz, G. J., How to maintain your coolant system, *Machine Tool Bluebook*, 30, 1988.
7. Ivaska, J., Green management, *Cutting Tool Eng.*, 39, 1991.
8. Ball, A. M., Fluids for metal removal processes, In *Manufacturing Engineering Handbook*, Geng, H., Ed., McGraw-Hill, New York, pp. 33.1–33.8, 2004.
9. Nachtman, E. S. and Kalpakjian, S., *Lubricants and Lubrication in Metalworking Operations*, Marcel Dekker, New York, pp. 63–105, 1985.
10. Independent Lubricant Manufacturers Association. *Waste Minimization and Wastewater Treatment of Metalworking Fluids*, Independent Lubricant Manufacturers Association, Alexandria, VA, pp. 2–4, 1990.
11. Booser, E. R., Ed., *CRC Handbook of Lubrication*, CRC Press, Boca Raton, FL, pp. 361–365, 1984.
12. Hunsicker, D. P. and McCoy, J. S., Compatibility: metalworking fluid, machine tool and lubricant, *Lubr. Eng.*, 366–373, 1996, May.
13. Moon, D. and Canter, N., The seal compatibility problem, *Mfg Eng.*, 2001, June.
14. Barwell, F. T., Lubrication in metalworking, *Tribol. Int.*, June, 171–175, 1982.
15. Nachtman, E. S. and Kalpakjian, S., *Lubricants and Lubrication in Metalworking Operations*, Marcel Dekker, New York, 1985, pp. 55–59.
16. Springborn, R. K., *Cutting and Grinding Fluids: Selection and Application*, American Society of Tool and Manufacturing Engineers, Dearborn, MI, pp. 53–65, 1967.
17. *Machining Data Handbook*, 3rd ed., Vol. 2, Compiled by the Technical Staff of the Machinability Data Center, Metcut Research Associates, Cincinnati, OH, 1980, pp. 16/17–16/96.
18. Yang, C., The effects of water hardness on the lubricity of a semi-synthetic cutting fluid, *Lubr. Eng.*, 133, 1979.
19. Joseph, J., *Coolant Filtration*, Joseph Marketing, East Syracuse, NY, 1987.
20. Opachak, M., Ed., *Industrial Fluids: Controls, Concerns, and Costs*, Society of Manufacturing Engineers, Dearborn, MI, pp. 25–104, 1982.
21. Berger, J. M. and Creps, J. M., *An Overview of Filtration Technology*, *Waste Minimization and Wastewater Treatment of Metalworking Fluids*, ILMA, Alexandria, VA, pp. 63–79, 1999.
22. Abanto, M., Byers, J., and Noble, H., The effect of tramp oil on biocide performance in standard metalworking fluids, *Lubr. Eng.*, 732–737, 1994, September.
23. Zintak, D.C., Ed., *Improving Production with Coolant and Lubricants*, Society of Manufacturing Engineers, Dearborn, MI, pp. 167–212, 1982.
24. *Waste Minimization and Wastewater Treatment of Metalworking Fluids*, Independent Lubricant Manufacturers Association, Alexandria, VA, 1990, pp. 15–159.
25. Childers, J. C., Metalworking fluids — a geographical industry analysis, *Lubr. Eng.*, 542, 1989, September.
26. Deming, W. E., *Quality, Productivity, and Competitive Position*, Massachusetts Institute of Technology, Cambridge, MA, 1982, pp. 267–311.
27. Nachtman, E. S. and Kalpakjian, S., *Lubricants and Lubrication in Metalworking Operations*, Marcel Dekker, New York, pp. 215–222, 1985.
28. *Waste Minimization and Wastewater Treatment of Metalworking Fluids*, Independent Lubricant Manufacturers Association, Alexandria, VA, 1990, pp. 26–30.
29. ASTM E 1302-00, *Standard Guide for Acute Animal Testing of Water-Miscible Metalworking Fluids*, ASTM International, West Conshohocken, PA, 2000.
30. *Metalworking Fluid Standards: Environmental Quality and Safety, Fluid Performance and Condition Monitoring Tests*, ASTM International, West Conshohocken, PA, 2003.
31. Nachtman, E. S. and Kalpakjian, S., *Lubricants and Lubrication in Metalworking Operations*, Marcel Dekker, New York, pp. 107–116, 133–156, 1985.
32. Springborn, R. K., *Cutting and Grinding Fluids: Selection and Application*, American Society of Tool and Manufacturing Engineers, Dearborn, MI, 1967, pp. 83–114.
33. Smith, M. D. and Lieser, J. E., *Laboratory Evaluation and Control of Metalworking Fluids*, SME Technical Paper, Society of Manufacturing Engineers, Dearborn, MI, 1973, MR73–120.
34. Bennett, E. O., The biological testing of cutting fluids, *Lubr. Eng.*, 128, 1974, March.

35. Leep, H. R. and Kelleher, S. J., Effects of cutting conditions on performance of a synthetic cutting fluid, *Lubr. Eng.*, 111, 1990, February.
36. Zimmerman, J. B. et al. Experimental and statistical design considerations for economical evaluation of metalworking fluids using the tapping torque test, *Lubr. Eng.*, 17–24, 2003, March.
37. Clock, J. E., What coolant selection taught us, *Mod. Mach. Shop*, 86, 1986, November.
38. DeChiffre, L. and Belluco, W., Investigations of cutting fluid performance using different machining operations, *Lubr. Eng.*, 22–29, 2002, October.
39. Zintak, D. C., Ed., *Improving Production with Coolants and Lubricants*, Society of Manufacturing Engineers, Dearborn, MI, pp. 167–171, 1982.
40. Opachak, M., Ed., *Industrial Fluids: Controls, Concerns, and Costs*, Society of Manufacturing Engineers, Dearborn, MI, pp. 232–236, 1982.
41. *Metal Removal Fluids: A Guide to Their Management and Control*, ILMA Metalworking Fluid Product Stewardship Group and Organization Resource Counselors, August 1997.
42. *Management of the Metal Removal Environment*, Organization Resource Counselors Website at www.aware-services.com/orc/.
43. *Metalworking Fluids: Safety and Health Best Practices Manual*, U.S. Department of Labor, Occupational Safety & Health Administration, www.osha.gov/SLTC/metalworkingfluids/metalworkingfluids_manual.html.
44. ASTM E1497-00, *Standard Practices for Selection and Safe Use of Water Miscible and Straight Oil Metal Removal Fluids*, ASTM International, West Conshohocken, PA, 2000.
45. Johnston, R. E., Fayer, M., and DeSimone, S., Multicomponent analysis of a metalworking fluid by Fourier transform infrared spectroscopy, *Lubr. Eng.*, 775, 1988, September.
46. Marano, R. S., Cole, G. S., and Carduner, K. R., Particulate in cutting fluids: analysis and implications in machining performance, *Lubr. Eng.*, 376, 1991, May.
47. Opachak, M., Ed., *Industrial Fluids: Controls, Concerns and Costs*, Society of Manufacturing Engineers, Dearborn, MI, pp. 70–84, 1982.
48. Bennett, E. O., The biology of metalworking fluids, *Lubr. Eng.*, 227, 1972, July.
49. Lonan, M. K., Abanto, M., and Findlay, R. H., A pilot study for monitoring changes in the microbiological component of metalworking fluids as a function of time and use in the system, *Am. Ind. Hyg. Assoc. J.*, 30–38, 2002, July/August.
50. Booser, E. R., Ed., *CRC Handbook of Lubrication*, CRC Press, Boca Raton, FL, pp. 371–378, 1984.
51. *Waste Minimization and Wastewater Treatment of Metalworking Fluids*, Independent Lubricant Manufacturers Association, Alexandria, VA, 1990, pp. 31–46.
52. Skells, G., Fluid management skills, *Cutting Tool Eng.*, 52, 1990, October.
53. Opachak, M., Ed., *Industrial Fluids: Controls, Concerns, and Costs*, Society of Manufacturing Engineers, Dearborn, MI, pp. 65–82, 1982.
54. Needelman, W. M., Fiumano, F. A., and Masters, J. A., Controlling grinding coolant contamination in an automotive plant, *Lubr. Eng.*, 479, 1989, August.
55. Bennett, E. O., *Dermatitis in the Metalworking Industry*, ASLE Special Publication SP-11, American Society of Lubrication Engineers, Park Ridge, NJ, 1983.
56. Bennett, E. O., Stop metal dermatitis before it starts, *Mfg Eng.*, 36, 1991, August.
57. Opachak, M., Ed., *Industrial Fluids: Controls, Concern, and Costs*, Society of Manufacturing Engineers, Dearborn, MI, pp. 7–9, 1982.
58. Passman, F. and Rossmoore, H., Reassessing the health risks associated with employee exposure to metalworking fluid microbes, *Lubr. Eng.*, 30–38, 2002, July.
59. *Symposium Proceedings: The Industrial Metalworking Environment: Assessment & Control*, 13–16, November 1995, American Automobile Manufacturers Association, Dearborn, MI, March, 1996.
60. *Symposium Proceedings: The Industrial Metalworking Environment: Assessment & Control of Metal Removal Fluids*, 15–18 September 1997, American Automobile Manufacturers Association, Dearborn, MI, September, 1998.
61. *Tools for Optimizing Chemical Management Manual*, Chemical Strategies Partnership, San Francisco, CA, 1999.
62. *Chemical Management Services: Industry Report 2004*, Chemical Strategies Partnership, San Francisco, CA, 2004.
63. Bierma, T. J. and Waterstratt, F. L., *Chemical Management: Reducing Waste and Cost through Innovative Supply Strategies*, Wiley, New York, 1999.

12 Recycling of Metalworking Fluids

Raymond M. Dick

CONTENTS

I. INTRODUCTION

In 1990, there were approximately 81 million gallons of metalworking fluids manufactured in the U.S.[1] Of all the metal removal fluids consumed in the U.S., 90% was used by the fabricated metal, transportation equipment, and machinery industries.[2] The vast majority of metalworking fluids consumed are found in individual machines rather than central fluid systems. According to the 1989 *American Machinist* Metalworking Survey, published in November 1989, there were 1,870,753 metalworking machines in the U.S. Industry experts at Henry Filter Systems in Bowling Green, OH, estimate that there are approximately 5000 central systems in the U.S. While some machines may use straight oils or no fluids at all, the majority of metalworking machines use water-based fluids.

Based upon the total volume of lubricants used in the U.S., the disposal of these fluids, and oily wastewater in general, has become a much more important issue during the last 20 years.

The metalworking industry has reevaluated its use of lubricants and chemicals as a result of environmental, health and safety, and productivity reasons.

Since the early 1970s, there have been laws enacted to protect surface and groundwater quality. These laws have directly impacted the treatment and disposal methods of oil–water emulsions. These laws include the Federal Water Pollution Control Act (or Clean Water Act), the Clean Water Act Amendments, the Safe Drinking Water Act, and the Resource Conservation and Recovery Act.[3] The end result of this legislation is that stricter methods are required for the proper treatment and disposal of metalworking fluids. Contract hauling costs for oily wastewater have increased by as much as ten to 20 times the cost in the 1970s. Landfilling of oily wastes and sludges containing leachable liquids has been banned. Stricter sewer discharge standards require more effective wastewater treatment methods.

Health and safety issues concerning the use of chemicals in the workplace have also become prominent. Metalworking fluids are carefully studied for operator health and safety characteristics because of the close contact operators have with the fluids. However, these clean fluids become contaminated during use with metal chips, fines, grinding wheel solids, various lubricating oils and greases, cleaners, solvents, microorganisms, and other materials either purposely or accidentally discharged into the sump. These contaminants and lack of proper fluid controls are largely responsible for frequent fluid discharge or dumps.

For the reasons mentioned earlier, the use and control of metalworking fluids have become an important issue for metalworking plants. The purpose of this chapter is to discuss the subject of metalworking fluid recycling as it relates to management and equipment designed to extend fluid life. Some information found elsewhere in this book is repeated here in order to give the reader a comprehensive overview of all aspects of this important subject.

II. BACKGROUND AND HISTORY OF FLUID RECYCLING

In the early 1970s and before, metalworking fluids were considered consumable products, designed for a relatively short life prior to disposal. Many of these fluids were discharged directly to the sewer or contract-hauled to landfills for less than 10 cents/gal.

More recently, the cost of fluid disposal has rapidly increased and, along with liability concerns, there has been a growing importance placed on fluid management. Fluid recycling systems and management techniques have rapidly gained importance as plants seek better fluid life and reduced disposal costs.

Today, greater importance is placed on understanding and managing the metalworking process. The metalworking process consists of the machine, operator, tool or wheel, fluid, and workpiece. Additional process variables may include the water quality, filter systems, and machine variables (such as lubricant systems, sump design, and workpiece handling). In terms of relative cost, the fluid costs have been low, which, in the past, resulted in poor management and frequent disposal.

With today's industry emphasis on productivity, waste minimization, and cost control, many plants have installed fluid recycling equipment. The goal is to optimize fluid performance, reduce oily wastewater volume, and reduce fluid concentrate and disposal costs.

As environmental regulations become stricter, more and more emphasis will be placed on fluid recycling. This chapter identifies the basics of fluid management, as well as fluid recycling technologies and equipment.

III. BASICS OF FLUID MANAGEMENT

For many years, metalworking fluids have been used to increase the productivity of metalworking operations. These fluids provide benefits of lubricity, reduction of friction, and reduction of heat in the metalworking process. In the past, these fluids have been relatively inexpensive to purchase, use, and dispose. More recently, fluid usage costs have increased, primarily due to rising disposal costs. Many metalworking plants are investigating alternative processes to better manage the use and disposal of metalworking fluids.

Since the early 1940s, when the initial research was completed on the metal removal process, water-based fluids have been used to improve metalworking operations.[4] Straight oils have been replaced in many applications due to safety and health reasons. The major problems with straight oils in metalworking plants are fire hazards, slippery floors, and general housekeeping difficulties. There are serious health concerns with breathing oil mist.[5] Compared to straight oils, water-based fluids provide improved machining and grinding characteristics, part finishes, tool life, wheel life, and allow for higher speeds in many operations.[6] In addition, they are cleaner and safer to use.

As metalworking plants seek improvements in productivity and economics, metalworking fluids must be evaluated. Along with improvements in machine tool technology, tooling, materials, and automation, improvements are possible with the use of metalworking fluids.

To improve overall management and control of fluids, each plant must evaluate the following fluid use areas:

Water quality
Fluid selection
Fluid controls
Contaminant removal systems
Management controls

A. Water Quality

The quality of the water mixed with metalworking fluid concentrates is very important to the performance of these fluids. Fluid life, tool life, part finish, foam characteristics, product residue, part or machine corrosion, mix stability, and concentrate usage are all affected by water quality.[7]

During normal fluid use, evaporation and carry-off losses require daily additions of fluid make-up. This process increases the quantity of total dissolved solids (TDS) in the fluid. Figure 12.1 depicts the theoretical increase in total dissolved solids because of evaporation losses and 10% daily make-up. The higher the initial TDS of the water source, the more rapid the TDS increase over

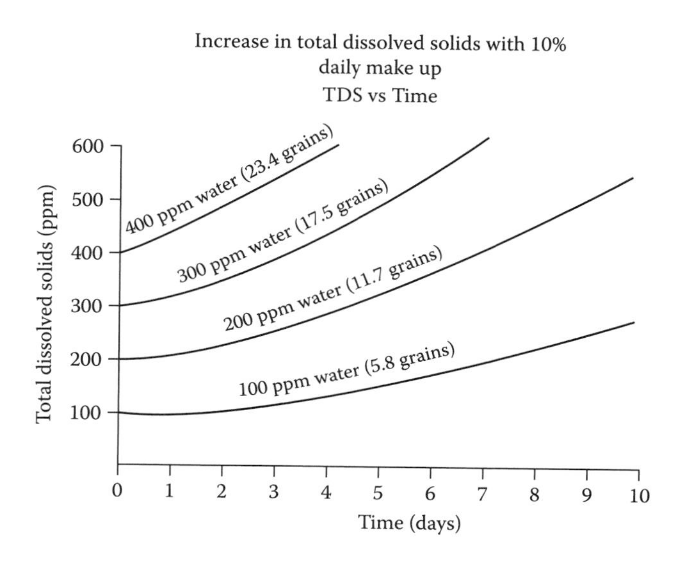

FIGURE 12.1 Increase in total dissolved solids over time with 10% daily make-up additions.

TABLE 12.1
Water-Hardness Comparison

Classification	Parts per Million	Grains per U.S. Gal
Very soft water	Less than 17	Less than 1
Soft water	17–52	1–3
Medium hard water	52–105	3–6
Hard water	105–210	6–12
Very hard water	More than 210	Greater than 12

time. As certain dissolved solids increase in quantity, problems will develop with the fluids. For instance, mineral and hardness salts, particularly chlorides and sulfates, contribute to corrosion at a level of approximately 100 ppm. Sulfates also promote the growth of sulfate-reducing (anaerobic) bacteria in fluids and create a "rotten egg" odor.[7]

It is important to have a water analysis completed on the plant's water. The fluid manufacturer may recommend treated water if the dissolved solids, hardness, minerals, or metals are at a high enough level to cause metalworking fluid application problems. The type of treated water used may be deionized water, distilled water, or reverse osmosis treated water. Water softening will remove the calcium and magnesium hardness ions, however, it can contribute to the corrosiveness of a metalworking fluid since sodium chloride and sodium sulfate are more corrosive than the hardness minerals.[8] A water-hardness comparison is found in Table 12.1.[8]

B. Fluid Selection

Because of the large variety of fluids available, it can be a difficult and time-consuming process to find the best fluid for a given operation. The fluid selected will have the greatest impact on a plant's fluid management program. Recommendations from the supplier will be helpful in narrowing the field of fluids from which to choose.

To understand the basic classes of fluids available, refer to Table 12.2. The four major classes of fluids are straight oil, soluble oil, semisynthetic, and synthetic.[8] The water-based fluid concentrates are typically mixed with water at ratios from 1:10 to 1:50, depending on the specific application and fluid type.

With a knowledge of fluid types and assistance from suppliers, it is important to evaluate the following fluid-related parameters:

Performance: As indicated by laboratory screening tests and field testing.[8]
Health and safety: Per material safety data sheets and other available data from suppliers.[9]

TABLE 12.2
General Classes of Metalworking Fluids

Class	% Petroleum Oil in Concentrate	Appearance of Fresh Fluid
Synthetic	0	Transparent to opaque
Semisynthetic	2–30	Transparent, translucent, or opaque
Soluble oil	60–90	Opaque
Straight oil	100	Transparent

Waste treatment: Using screening tests, field tests, and supplier assistance.
Quality standards of fluid: With ability of supplier to provide consistent quality, using statistical process control methods.
Technical service support: For proper controls and troubleshooting of fluids with supplier's assistance.
Cost: Using overall cost of fluid including purchase price, labor (mixing, transporting, and controlling), machine downtime (for charging and disposing of fluid), waste treatment, and operator safety and health.
Delivery: With supplier providing "just-in-time" delivery.

Each application must be thoroughly evaluated to understand the fluid requirements. For instance, it is important to know the specifics of the operation:

- Type of machining, grinding, etc.
- Central system or individual machine sumps
- Tooling and setup
- Type of metals
- Type of water
- Filtration equipment
- Part requirements (tolerances, finish, and corrosion protection)
- Machine requirements (lubrication, seals, paint, cleanliness, and visibility of work area)
- Operator contact with fluids
- Special requirements, such as high pressure or high volume delivery

While laboratory or screening tests are useful and necessary, the most beneficial information is available from production machines.

In most plants, personnel from manufacturing, maintenance, safety, purchasing, the laboratory, and wastewater treatment areas will have input on the fluid selection process. Ideally, one person or group will have the authority and responsibility to select the best overall fluid. It is best if only one fluid at one concentration can be used in a plant for purposes of management and recycling. If this is not possible, then it is very important to minimize the number of different fluids used in a plant.

C. Fluid Controls

Given the selection of the best fluid for a specific application, it is important to identify the needed controls to maintain optimum fluid performance. Every fluid has a range of parameters within which it is designed to operate. For example, concentration, pH, and contamination level (oil, dirt, and bacteria) are parameters that ideally are controlled for each water-based fluid. The supplier of the fluid must be able to identify the parameters and ranges to be used in controlling the fluid.

The majority of fluid problems arise due to improper concentration, when the fluid mix becomes too rich or too lean. Evaporation and carry-off losses, approximately 2 to 10% per day, alter the fluid's mix ratio.

The concentration will need to be checked on a frequent basis, preferably daily, but at a minimum of once per week. Concentration check methods are discussed in Chapters 7 and 11.

Similarly, it is necessary to check fluid pH on a frequent basis and maintain it in the recommended range. One common occurrence with fluids is that bacteria generate acidic by-products, which lower the fluid's pH. If this situation occurs, it will be necessary to control the bacteria level and readjust the pH of the fluid.

After the initial charge of a fresh mix of fluid into a machine reservoir (or sump), this fluid becomes contaminated with oil, dirt, metals, bacteria, and other materials as a result of its use. For optimum fluid performance and life, fluid contaminants must be controlled. These contaminants can be minimized with good maintenance and housekeeping programs. With many machines, lubricating oils and greases cannot be isolated from the fluid. These contaminants, as well as metal contaminants, are an expected by-product of the machining, grinding, or other metalworking operations. Many of the contaminants that cause fluids to be disposed of frequently are foreign materials, such as floor sweepings, cleaners, solvents, dirt, tobacco, and food. With the goal of improved fluid management, education and revised plant practices will be required to improve housekeeping and sanitation of the fluids.

Contaminant levels of oil, dirt, and bacteria can be monitored to determine how the quantities are changing over time. In many cases, corrective action can be taken to remove the oil, filter solids, or control bacteria to prevent disposal of the fluid. With any product there is a finite life of the fluid. The decision to dispose of the fluid is usually based on an oily and dirty appearance or foul odor. However, by monitoring fluid parameters on a routine basis, the fluid can be better controlled, leading to improved fluid performance and an extended useful life.

D. Contaminant Removal Systems

In many metalworking operations, contaminant removal systems are used to enable the machine to provide a certain finish, tolerance, production rate, etc., on a part. Contaminant removal systems are also used to maintain the fluid in a clean condition to minimize disposal frequency. Two general classes of contaminant removal systems are those for central systems and those for individual machines. Flexible manufacturing systems (or cells) may employ either central system or individual machine reservoirs. Contaminant removal systems are becoming more and more necessary for plants interested in improving fluid management and control.

1. Central Systems

A central system is a large reservoir which supplies fluid to several individual machine tools. The central systems can range in size from a few hundred gallons to over 100,000 gal. Where identical or similar operations are performed on many individual machines, a central system is used to supply one fluid to all the machines. One major advantage of the central system is that it has a contaminant removal system for solids and, in some cases, oil to maintain the fluid in a clean condition. Also, since only one fluid is used, a daily fluid sample will provide a control system for monitoring

TABLE 12.3
Contaminant Removal Equipment for Central Systems

Equipment	Removes		
	Oil	Dirt	Bacteria
Settling/dragout		X	
Multiple weir		X	
Flotation	X		
Positive filters			
Gravity		X	
Pressure		X	
Vacuum		X	
Centrifuge	X	X	
Cyclone		X	
Coalescer	X	X	
Pasteurization			X

concentration, pH, and contamination levels. With proper fluid controls and management techniques, the typical central system fluid will have a life of 1 to 3 years.

Table 12.3 provides a list of the general types of contaminant removal equipment used on central systems.[10] Most of the systems employ some type of filtration to remove solids (metal chips, grinding swarf, and dirt). They can be as simple as settling and dragout systems or more advanced, such as the positive filters. Equipment such as a centrifuge or coalescer may be added to the central system to control tramp oil.

2. Individual Machines

There are a wide variety of contaminant removal systems for individual machines. Table 12.4 provides a list of some of the more commonly used systems.[10] One of the most difficult aspects of controlling fluids in individual machines is that many plants do not monitor these fluids because of the number of samples and tests required. In addition to a large number of individual sumps, there may be different fluids and different concentrations that make the control task more difficult. However, it is recommended that daily checks are made of the concentration, since individual machine fluids can have a rapid change in fluid concentration, even in one day. A simple refractometer test is adequate for the daily checks; however, the chemical titration is a more accurate test and recommended for long-term control.

For many individual machines, contaminant removal systems are provided to handle one particular contaminant. For instance, a grinder may have a combination dragout/paper media filter to keep the fluid clean. A milling machine may have a dragout system/chip conveyor to remove the metal chips. However, few individual machines have contaminant control equipment to control all types of fluid contamination, such as oil, dirt, and bacteria.

TABLE 12.4
Contaminant Removal Equipment for Individual Machine Tools

	Removes		
	Oil	**Dirt**	**Bacteria**
Media-based systems			
Filtration		X	
Pressure		X	
Vacuum		X	
Gravity		X	
Natural force systems			
Settling/gravity		X	
Oil skimmers	X		
Coalescers	X		
Aeration	X		
Mechanical separation systems			
Cyclones		X	
Centrifuges	X	X	
Magnetic separator		X	
Other			
Pasteurization			X

Because of the difficult control situation, many plants are seeking better methods to control fluids in individual machine sumps. Typically, there is economic justification in seeking improved methods since the fluids in individual machines may be disposed of as frequently as once a week.

3. Fluid Recycling

An effective method to extend fluid life for individual machine tools is the use of batch treatment fluid recycling systems capable of removing contaminants such as tramp oil, dirt, and bacteria; and to readjust the fluid concentration before the fluid is returned to the individual machine. The fluid from each machine is treated with the batch treatment equipment on a frequent basis to minimize the contaminants.

Though there are several types of systems on the market, each plant must determine the feasibility of a fluid recycling system for its own purposes. A plant survey is recommended as the first step to identify the number of machines, sump capacities, frequency of disposal, and reason(s) for disposal. Also, data on fluid concentrate cost and gallons purchased, as well as cost of waste treatment (or contract hauling), are important. This data is used to determine the economics of fluid recycling for a particular plant. An example of a fluid survey questionnaire is found in Table 12.5. Based on a thorough evaluation of the current fluid practices and proposed fluid management changes, a study must be completed to select the optimum fluid recycling system. Examples of equipment selection criteria are:

Good economics (capital, operating, maintenance, and energy costs)
Effective removal of contaminants such as oil, dirt, and bacteria
Make-up system to add concentrate or water to recycled fluid

TABLE 12.5
Metalworking Fluid Management

FLUID SURVEY

PLANT:	NUMBER OF MACHINE TOOLS:
LOCATION:	AVERAGE SUMP SIZE:
MANUFACTURER OF:	TOTAL GALLONS OF FLUIDS PURCHASED / YR.:
OPERATIONS:	AVERAGE FLUID CONCENTRATE COST / GAL:
METALS:	COST OF WASTE DISPOSAL / GAL:
FLUID(S) USED:	LABOR COST / HR.:

PLANT SURVEY

DATE: __________

MACHINE	DEPARTMENT	FLUID	FLUID CONCENTRATION	SUMP CAPACITY	DISPOSAL FREQ.?YR.	GALLONS DISPOSED/ YEAR	REASON FOR DISPOSAL (OIL, DIRT, BACT., ETC.)
1.							
2.							
3.							
4.							
5.							
6.							
7.							
8.							
etc.							

Simple operation, low maintenance
Durable, quality equipment
Minimal floor space requirement
Warranty protection
Spare parts and service available from manufacturer

E. Management Controls

In addition to water quality, fluid selection practices, fluid controls, and contaminant removal equipment, management controls are an important part of fluid longevity.

The fluid-use survey previously mentioned is used to identify particular machines or fluids that have high disposal frequencies. In addition, by talking to operators about the specific problems in a department or at a machine, we may learn of particular obstacles to fluid management or recycling. For example, there may not be a "standard practice" for the disposal of cleaners or solvents and these materials may simply be discharged to the metalworking fluid reservoir. These contaminants will directly influence the fluid performance and will make it impossible to recycle the cleaner or the metalworking fluid.

Another obstacle to fluid recycling in many plants is the actual sump design or the machine layout, which prohibits easy access to the fluid. Many sumps are poorly designed, especially in the base of a machine or simply in an inaccessible area, where the fluid can be trapped. Since the cleanout job becomes messy and time consuming, it is seldom completed. Another problem is that machines are placed in close proximity so that the sump cannot be reached with a sump cleaner.

In some cases, it may be necessary to redesign the machine sump or improve the machine layout to minimize the sump cleanout problems. Ideally, the sump is readily accessible to see, smell, sample, and service (add make-up, clean, etc.) the fluid.

Many plants find that equipment changes are necessary to ensure better control of the fluid. This may include the addition of spray hoses to manually flush machines, spray nozzles to minimize stagnant areas, or guarding to prevent carry-off.

The "people management" of fluids is very important as well. It is important that operators and plant personnel keep the fluid clean by obeying good housekeeping and hygiene practices. For example, food, drinks, tobacco, cleaners, solvents, paper, rags, floor-drying compounds, and dirt must be discarded properly and not put into metalworking fluids. Education and training is necessary for all plant personnel if better fluid management is to occur.

It is very helpful to document the fluid condition through simple test methods as discussed earlier (pH and concentration), and observe trends to predict fluid failure. A fluid log at the individual machine is helpful for the operator to complete daily fluid checks and observe fluid changes. If the fluid reaches an "out of control" condition, for example, the concentration is too low, then corrective action can be taken. See Table 12.6 for an example of a fluid log.

For a batch treatment recycling process, it is also very helpful to have a fluid recycling schedule. Every machine can then be cleaned out on a regular basis. Table 12.7 shows an example of a machine cleanout schedule; every machine is cleaned on a 1-, 2-, or 3-week cycle. It is necessary to process used fluids on a frequent basis to avoid contamination problems that overwhelm the fluid and require disposal. While there is no simple rule in terms of a recycling schedule, in many situations once a month is the typical minimum frequency.

Table 12.8 lists the typical factors that determine the required recycling frequency to extend fluid life. The fluid type and water quality have the greatest impact on fluid cleanliness, where some fluids tend to reject oil and dirt better than others. Typically, synthetic and semisynthetic fluids reject tramp oil better than soluble oils and tend to stay in a cleaner condition.

The type of operation and metals used will define the amount of contamination that the fluid receives. For example, grinding cast iron will place a large amount of graphite fines into a fluid, which are difficult to remove, and therefore will reduce the fluid life. Tramp oil leakage from the

TABLE 12.6
Daily Coolant Log

DAILY COOLANT LOG Machine # __________ Coolant in use __________

(Possible Format for Small Plants) Capacity __________ Coolant supplier __________

Date/ Time	Concen- tration	pH	VISUAL CHECKS					Samples Taken for Analysis	ADDITIONS			Remarks
			Rust	Tramp Oil	Machine Build-ups	Rancidity	Color		Coolant Concentrated	Water	Other Additives	

TABLE 12.7
Fluid Recycling Schedule

MACHINE NUMBER	SIZE (GAL)	RECYCLE EVERY WEEKS	WEEK #1					WEEK #2					WEEK #2					WEEK #4					WEEK #5					WEEK #6				
			M	T	W	H	F	M	T	W	H	F	M	T	W	H	F	M	T	W	H	F	M	T	W	H	F	M	T	W	H	F
00081	40	1		X					X					X					X					X					X			
00049	50	1			X					X					X					X					X					X		
00983	40	1		X					X					X					X					X					X			
02564	25	1			X					X					X					X					X					X		
02565	40	1				X					X					X					X					X					X	
00017	15	2									X										X										X	
00018	5	2									X										X										X	
00019	30	2				X										X										X						
00026	80	2					X										X										X					
00043	20	2										X										X										X
00044	60	2										X										X										X
00053	10	3				X															X											
00054	10	3			X															X												
0054	80	3											X															X				
0058	30	3						X															X									
0061	15	3									X															X						
00903	5	3								X															X							
00020	10	3														X															X	
00021	30	3						X															X									
00023	20	3						X															X									
00027	10	3	X															X														
00030	5	3	X															X														
00043	5	3	X															X														
00068	10	3	X															X														
0056	50	3	X															X														
00680	5	3	X															X														

TOTAL GALLONS	WEEK #1	WEEK #2	WEEK #2	WEEK #4	WEEK #5	WEEK #6
MONDAY	85	80	80	85	80	80
TUESDAY	80	80	80	80	80	80
WEDNESDAY	85	80	75	85	80	75
THURSDAY	80	75	80	70	85	70
FRIDAY	80	80	80	80	80	80

TABLE 12.8
Recycling Frequency Determination Factors

Fluid type and water quality
Resistance to oil emulsification
Resistance to bacteria/mold
Contamination
Tramp oil leakage
Solids
Machine usage
Metals/solids loading
Idle fluid
Machine filtration
Sump design/access
Cleanliness of sump and fluid
Control
Monitoring of concentration, pH, volume
Age of fluid

machine is an important source of fluid contamination, which can quickly overwhelm the fluid and require disposal.

The throughput of the machine and its run time will impact the fluid condition. An idle machine may, in fact, cause more problems with fluid rancidity than a continually pumped and aerated fluid.

Many machines are equipped with filters and dragouts to keep the fluid clean. This can greatly extend fluid life. Continuous filtration or routine sump cleanouts will have an important benefit on fluid life.

Obviously, the better the in-plant control of the fluid, the less frequent the need for disposal. Simple tests such as pH and concentration can greatly extend the fluid life. Maintaining the fluid volume at a full level in the sump is important to fluid performance. Finally, age of the fluid will contribute to fluid failure because of oil emulsification, mineral buildup, metals accumulation, and product component depletion.

In addition to the various fluid management techniques we have discussed, one of the most important is simply to have the proper personnel operate and manage the program.

Ideally, one person, or a small number of people, should have the authority to operate the fluid management program. As the responsibility is passed to numerous people, less control of the fluids takes place. For example, a small group responsible for the fluid condition can make sure fluids are tested and properly adjusted. If each operator is responsible for a particular sump, then the tendency is to have fluids frequently discharged rather than worry about fluid testing and filtration.

In many plants, a "coolant committee" is set up to manage fluids. It may include representatives from purchasing, engineering, production, the laboratory, and waste treatment departments. Each group provides different priorities in terms of fluid selection and use, but as a team the best fluid management practices can be employed.

IV. FLUID RECYCLING TECHNOLOGIES

A. Filtration

As previously discussed, the individual sump environment creates many problems for extended fluid life. While many machines are equipped with individual machine filters, routine cleanout and maintenance are critical to long-term cleanliness and performance of the fluid. It is not unusual to find an improperly maintained paper media filter, dragout/conveyor system, or hydrocyclone, which results in fluid problems.

Many individual machines do not have any filtration at all. The sump may include a series of baffles and weirs to trap solids and oils. These machines are more susceptible to recirculating fines, bacteria growth, and fluid problems.

Filtration of metalworking fluid is critical for reasons of part finish, tool life, bacteria control, heat transfer, lubricity properties, etc. Critical operations, such as grinding, may have continuous filtration to minimize contaminant accumulation. Proper metalworking fluid filtration is probably the single largest problem with fluid performance. Typically, the combination of improper filter application (or lack of any filtration) and poor maintenance leads to premature fluid failures.

For these reasons, many companies have opted for a fluid recycling program that can help to improve fluid cleanliness. A set recycling schedule is developed and equipment such as high efficiency sump cleaners are used to routinely clean individual machine sumps. For many plants, this is the best way to guarantee routine cleaning, which is necessary to keep ahead of fluid contamination problems. For more information on fluid filtration see Chapter 10 or Ref. [10].

There is no single absolute level of filtration required for optimum fluid performance. Each process must be evaluated to define the degree of filtration required. For many plants, a level of 200 ppm of total suspended solids or less is commonly found in a used fluid. Though some recirculating fines are always present, the goal should be to minimize these, especially in the critical processes. Central systems offer the best solution for continuous filtration. Where there are numerous machines on similar operations (for example, grinding crankshafts), central systems offer the best method of fluid control. The fluid is maintained on a daily basis to keep it within control limits.

B. Oil Removal

Various types of oils and greases are used for machine tool lubrication, such as hydraulic fluids, gear oils, way oils, etc. Many of these oil-based lubricants either drain or are washed back to the metalworking fluid sump.

Depending on the fluid type (synthetic, semisynthetic, or soluble oil), these lubricating oils and greases can become emulsified with the metalworking fluid or can simply float on the fluid surface, as "free" oil. In either case, "tramp" or "extraneous" oil is a leading cause of fluid failure.

The typical mechanism of failure is due to a free oil layer inhibiting oxygen transfer to the metalworking fluid. This causes anaerobic bacteria to flourish and the well-known side effects (pH is lowered due to acidic by-products of bacteria growth and hydrogen sulfide gas is produced, giving the "rotten egg" odor). Oil components may also act as food sources for the organisms, destabilize emulsion products, and cause increased misting. Therefore, it becomes important to control tramp oil leaks into the fluid. Once the fluid is contaminated, remove as much oil as possible without harming the metalworking fluid emulsion.

Commonly found oil removal devices are skimmers (disk, rope, or belt), coalescers, and centrifuges. Also, there are numerous types of filters and oil sorbent materials used to help remove oil.

The separation of tramp oil, or the rise rate of oil droplets, is dependent on several factors, including droplet size, specific gravity, and temperature.[11] This relationship is expressed by Stokes Law as follows:

$$V_r = \left(\frac{g}{18\mu}\right)(S_w - S_o)D^2$$

where: V_r is the velocity of the rise rate of oil droplets; g is the acceleration due to gravity (981 cm/sec); S_w is the specific gravity of water (metalworking fluid); S_o is the specific gravity of oil; D is the diameter of oil droplet; and μ is the viscosity of water (metalworking fluid) at a specified temperature.

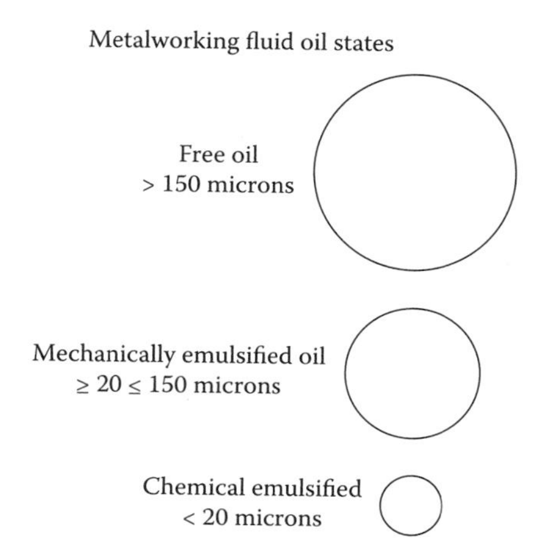

FIGURE 12.2 Metalworking fluid oil states.

In a used metalworking fluid, we define three states of oil. These are chemically emulsified oil, mechanically emulsified oil, and free oil. Figure 12.2 depicts the relative size of these oil states. For the most part, we assume the fluid emulsion is stable and, therefore, we are more concerned with the ever-changing state of the free and mechanically emulsified oil. It is important to understand that the degree of emulsification will impact the separation efficiency of any oil removal equipment. Therefore, the degree of fluid turbulence, type of metalworking fluid, fluid temperature, and mineral content of the fluid will affect the separation of "tramp" or extraneous oil from the product.

In almost all cases, our goal is simply to remove free oil and loosely emulsified oil (such as mechanically emulsified oil) from the used metalworking fluid. We do not want to remove the product oil or other components that would harm the product performance.

As with solids filtration, there is no universally approved standard for the acceptable amount of tramp oil. In many fluids, a small amount of tramp oil may actually have some benefits, such as higher lubricity, decreased foam, and softer residue. However, in most cases the level of free oil and loosely emulsified oil needs to be controlled to a level of 0.5% or less for optimum fluid performance and life. It does not make sense to target chemically emulsified oil, since if this form of oil is removed, we remove valuable product components.

To reduce the tendency for a buildup of chemically emulsified oil, it is necessary to minimize oil contamination, select a product type (such as synthetics and semisynthetics) that is less susceptible to tramp oil emulsification, use high-quality demineralized water, and use oil removal equipment to reduce the level of tramp oil.

As opposed to solids filtration, where central systems offer a distinct advantage to fluid management, the central system typically makes tramp oil removal more difficult. The high pumping recirculation rates with central systems increase the tendency for oil emulsification. However, as mentioned previously, the biggest problem is not the highly emulsified oil, but the free floating and loosely emulsified oil. Routine sump cleanouts and oil filtration are required to keep ahead of oil contamination for individual machine sumps. The two major types of equipment employed for this purpose are the coalescer and centrifuge.

C. Bacteria Control

The most common reason for fluid disposal is rancidity or fluid odor. Typically, the type of chemicals found in metalworking fluids are good food sources for bacteria. The tramp oils and fines also foster bacteria growth due to nutrients and growth sites. Therefore, most metalworking fluid formulations include a bactericide or fungicide to counteract bacteria and/or mold growth in used fluids. Some newer fluid formulations feature the use of chemicals that are "biostatic" or "biostable", which reduce the tendency for bacteria growth and odor development to minimize the need for tankside additives or biocides.

Whatever the fluid formulation, fluid cleanliness is a key to extended fluid life and performance. As discussed previously, the fluid type and water quality have a major impact on the tendency for bacteria to grow. Typically, the higher oil-containing products have more bacteria growth problems, whereas the synthetic-type products have a greater tendency for mold growth. Other common problems leading to increased bacteria growth are improper fluid concentration, lack of pH control, higher mineral content (especially phosphates and sulfates), and reduced fluid volume.

Bacteria use surface and film attachment to reproduce. The ideal sump is easily accessible for frequent cleanouts to prevent the buildup of fines and bacteria colonies. In addition to filtration, pasteurization is employed to control bacteria growth. By raising the fluid temperature to 165°F for at least 15 sec, heat-sensitive bacteria will be killed.[12] Other treatments such as ultrafiltration and ultraviolet radiation have been used to remove and kill bacteria in used metalworking fluids, however, those techniques are not widely used for this purpose.

D. Chemical Additives

Metalworking fluids are consumable products which eventually will require disposal. Up to this point, the primary techniques to extend fluid life mentioned were fluid management and contaminant removal methods. However, if the fluid chemicals are depleted, certain additive "packages" may be used to refortify the product.

As the fluid age increases, there is a greater chance for product losses, which can lead to instability and product failure. Some common chemical additives are pH buffers, biocides, and antifoams. In certain cases, dyes, odorants, lubricants, and surfactants are added to compensate for product changes or losses. These additives are used for central systems or fluid recycling programs to maintain optimal fluid performance and maximize fluid life.

V. FLUID RECYCLING EQUIPMENT

There are numerous approaches to fluid recycling, including the use of existing or "add-on" filtration equipment to either individual machines or central systems. Batch and continuous systems can be used to supplement existing machine filtration. There are many choices of small, portable filters or recycling carts that can be moved from machine to machine for fluid recycling. In addition, many companies now offer a fluid recycling service, where portable recycling equipment is used to process fluids as needed.

The importance of individual machine sump maintenance cannot be overstated. Even with a batch treatment system or a fluid recycling service, many of the fluid spoilage conditions occur as a result of poor sump maintenance.

Sections III and IV reviewed the overall fluid management basics and fluid recycling technologies. This section will discuss available systems to recycle metalworking fluids. For the most part, these systems target the individual machine-type manufacturing facility, which has the greatest need in terms of fluid management. The two primary technologies used for fluid recycling systems are the coalescer and the centrifuge. In each case, these technologies are primarily targeted to separate the tramp oil from the used metalworking fluids. However, to some extent, the coalescer and centrifuge will remove solids but this may become a maintenance problem unless prefiltration is used. As discussed previously, Stokes Law defines the separation of oil droplets from the used metalworking fluid. Figure 12.3 reveals the impact of increasing oil droplet diameter, specific gravity differential, and temperature on the oil droplet rise rate.

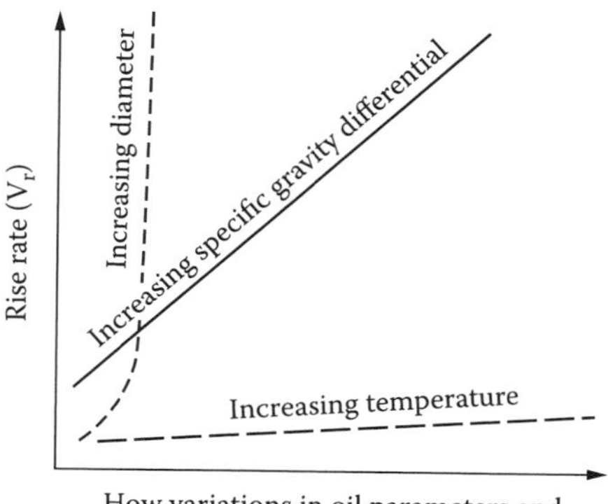

FIGURE 12.3 How variations in oil parameters and application conditions affect an oil droplet's rate of rise (Courtesy of AFL Industries, with permission).

Coalescence uses the property of oil attraction to polypropylene media (or oleophilic, "oil-loving" materials) for removal of tramp oil. Figure 12.4 shows a typical configuration for a coalescer, the vertical tube. These tubes are used in the oil removal tank of a batch treatment recycling system. Other coalescer configurations include inclined plates, loose packed media, or filter cartridges.

The centrifuge uses a series of plates, or a disk stack, spinning at a high rate to physically separate materials of differing specific gravities, that is, oil, water, and solids. The advantages of a centrifuge, typically, are its high throughput rate (approximately 2 gpm) and the centrifugal force, which provides oil separation. One disadvantage can be the amount of maintenance time for certain applications where solids and greases require frequent bowl and disk stack cleaning. Also, the

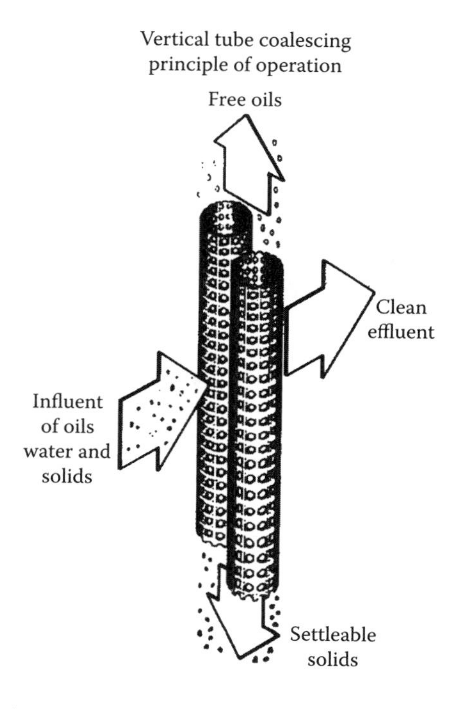

FIGURE 12.4 Vertical tube coalescing: principle of operation (Courtesy of AFL Industries, with permission).

centrifuge has a higher incidence of long-term repairs because of its nature of processing abrasive fluids at high process rates. Figure 12.5 is a schematic of a typical centrifuge.

One newer technique, ultrafiltration (UF), is used on certain fluids to remove contaminants yet reuse the effluent. The UF process seems particularly suited to certain solution or synthetic products where the chemicals pass through the membrane while the contaminants, such as oil, dirt, and certain bacteria are contained.[13,14] This technique provides an excellent quality effluent where practically all the oil and particulates are removed. Because of a very small pore size, the UF process is not suitable to recycle semisynthetics or soluble oil-type emulsions.

Figure 12.6 depicts a typical batch treatment fluid recycling system. The typical procedure is to follow a machine clean-out schedule (for example, cleaning each machine once per month) by removing all the fluids and solids from the sump using a high efficiency sump cleaner. The filtered fluid is then processed through the recycling system, where the clean fluid is tested for concentration and then adjusted. New make-up is added and the reclaimed and refortified fluid is returned to the clean sump. In practice, a split tank sump cleaner is used to clean out the sump using the "dirty fluid" compartment and then recycled fluid from the "clean fluid" tank is immediately returned to the sump. That is, the sump cleaner draws the clean fluid from the recycling unit and has this fluid available for a recharge once the individual machine is cleaned out.

A typical plant using a batch treatment recycling process uses one fluid and one concentration to simplify the clean-out and exchange program. Otherwise, sump cleaners and recycling systems must be cleaned out prior to a fluid change to eliminate cross-contamination.

Figure 12.7 is a more detailed view of a batch treatment unit using coalescer/pasteurization/filtration techniques. A combination of filtration, coalescence, and pasteurization has proven to be an effective and economical method to recycle fluids from individual machines. The equipment may be used in a wide variety of plants and applications to improve fluid management of individual machine fluids. The advantages of such a batch treatment module are:

Excellent removal of contaminants: reduces free oil to 0.5% or less, controls suspended solids levels to 0.1% or less, controls bacteria levels to 100,000 counts/ml.

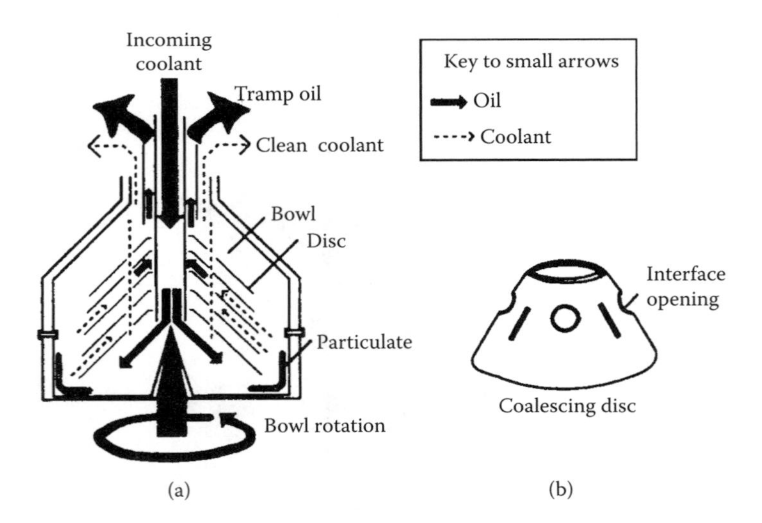

FIGURE 12.5 (a) Schematic of a centrifuge commonly used for tramp oil removal. This centrifuge will remove some fine particulate if it is denser than the fluid. The arrows show the passage of coolant through the unit. (b) Drawing of a typical coalescing disk used to aid in the separation of the tramp oil from the coolant. The opening in the surface of the disk should be placed at the interface of the tramp oil and coolant where separation occurs (Courtesy Henry Filtration Co., Bowling Green, OH. With permission).

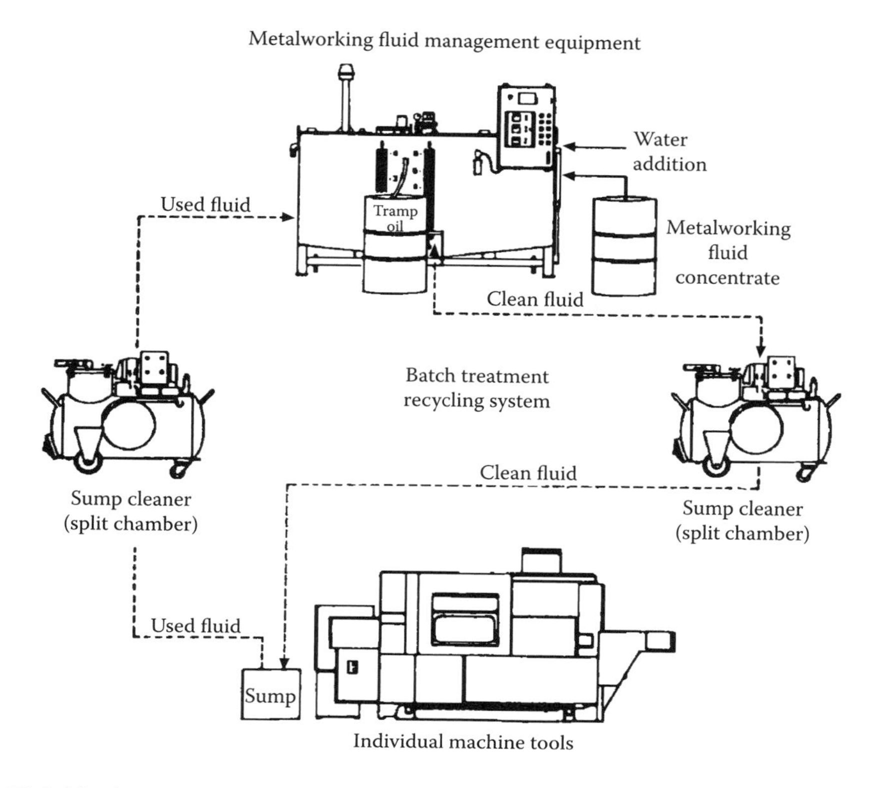

FIGURE 12.6 Metalworking fluid management equipment (Courtesy of Milacron, Cincinnati, OH. With permission).

Low maintenance, 1 h/week to maintain unit
Low operator labor, 1 h/8 h of operation
Simple operation
Minimal floor space
Low costs, total operating and energy costs typically less than 15 cents/gal
Durable pump, valves, heater, coalescer media

The batch treatment module consists of three compartments — the used fluid tank, the oil removal tank (or coalescer tank), and the clean fluid and make-up tank. The used fluid is pumped from the surface of the used fluid tank through the transfer pump, heater, and coalescer. The fluid temperature is raised to 73.8°C (165°F) to lower bacteria counts as a result of pasteurization. Also, by raising the fluid temperature, improved extraneous oil separation occurs in the coalescer tank.

The fluid next passes through the coalescer media, where the media attracts and separates extraneous oil from the metalworking fluid. Coalescence uses the principle of Stokes Law, which states that as the diameter of the oil droplet doubles, the oil droplet rise rate increases by a factor of four. The coalescer media is made of polypropylene, which attracts oil to it in preference to water. The oil separates to the top of the tank and is removed by an oil skimmer. The clean fluid overflows into the clean fluid tank, where the fluid concentration is checked and fluid make-up is added. Typically, 50% by volume of new fluid make-up is added to the clean tank to make up for fluid losses (evaporation, carry-off, and restoring depleted fluid components). The combination of clean recycled fluid and new make-up fluid provides a fluid that can now be returned to the machines. The fluid is returned to the machine with a sump cleaner (clean side) or through a return drop line

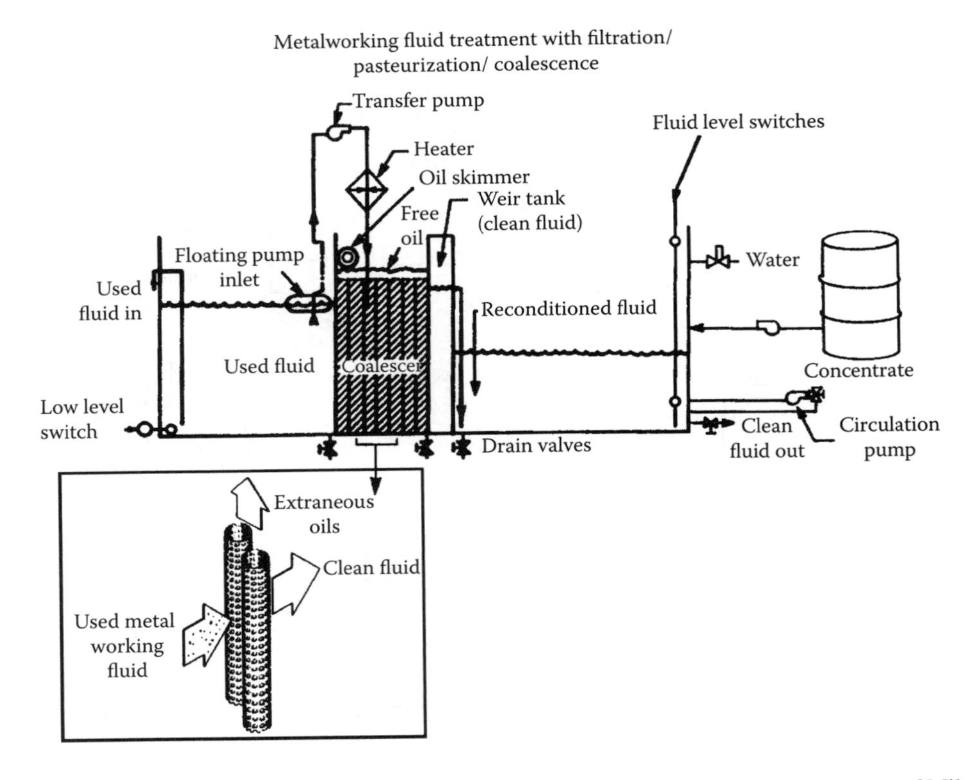

FIGURE 12.7 Metalworking fluid treatment with filtration/pasteurization/coalescence (Courtesy of Milacron, Cincinnati, OH. With permission).

system. The key to success of the fluid recycling program is scheduling the recycling to minimize contaminants and eliminate the frequent disposal of fluids.

VI. PLANT APPLICATION

As an example of a metalworking fluid management application, a company in northeastern U.S. has been successfully recycling fluids for many years. This company manufactures braking systems for transit vehicles. Prior to fluid management, their greatest need was improved manufacturing productivity. A flexible manufacturing cell was installed consisting of 12 machines, including horizontal and vertical machining centers, and lathes. This company manufactures over 50 different parts using the various metals of cast and ductile iron, aluminum, and steel. The plant uses one fluid, a soluble oil, for all the operations.

The fluids are recycled approximately every 1 to 2 weeks, with an average of approximately 500 gal/week recycle. The individual machines are cleaned out using a sump cleaner with a capacity of 90 gal in each compartment. One compartment has filtration for the used fluid, the other compartment is for the clean, recycled fluid.

The fluid management program has provided a typical fluid life of over 1 year, and the disposal costs have been nearly eliminated. The fluid concentrate purchases have been reduced from approximately 24 drums/year to 12 drums/year. The fluid management program has provided benefits of extended fluid life, reduced fluid costs, and improved fluid performance.

Typically, fluids from individual machines will need to be recycled at least once a month. While the simplest schedule for fluid recycling is a set frequency (monthly, weekly, etc.), many plants may want to be more specific as to the exact point at which fluids must be recycled or disposed. In addition, limits based upon laboratory tests can be set up for each fluid with the help of the fluid supplier to indicate at what point the fluid needs recycled or disposed. For instance, if the pH value decreases, extraneous oil value increases, or the bacteria quantity increases, action will be required. Certain fluids and operations can tolerate more contamination, which is the reason specific limits need to be set up by the fluid supplier and the metalworking plant.

There are numerous choices of metalworking fluid batch recycling systems in the market. Typically, the coalesecer and centrifuge form the "heart" of the system. Other "add-ons" may include media filtration, water treatment, concentration control, and equipment to add fresh makeup. Sump cleaners provide an important method to clean out dirty machines. In some plants the recycled fluid is returned to machines by a pipeline system.

Some plants obtain fluid recycling services from an outside company that recycles fluids on a contract basis. For the most part, these companies offer tramp oil removal for central systems. However, some plants may also batch treat tanks of used metalworking fluids to extend fluid life. Fluid recycling services typically are filtration and centrifuge systems mounted on a truck that can be driven plant to plant. These companies will charge by either the gallons processed or based on a service contract.

Metalworking plants with fewer individual machines may favor portable recycling equipment that cleans one machine sump, then is transported to the next machine sump for cleaning. Typically, these units take up minimal floor space, are easily transported, and can be hooked up to a machine sump and "cycled" to allow the coolant to be filtered to remove small fines and tramp oil. These units are not designed to remove large amounts of chips or swarf, which is best accomplished with the machine dragout, filter, or a portable sump cleaner.

Examples of portable fluid recycling equipment include small coalescer/filter tanks or small centrifuge systems. See Figures 12.8 and 12.9 for an example of a portable recycler unit, in this case a centrifuge for removal tramp oil and fines (courtesy of Alfa Laval).

These units rely on setting up schedules to periodically recycle the coolant to keep the metalworking fluid in good condition. Of course concentration control, adding new makeup, and good filtration are keys to maintaining the fluid in good condition. These portable units can offer a less expensive and more flexible method to recycle fluids for the smaller plants.

In summary, there are a variety of effective methods to extend the life of metalworking fluids. Each plant must determine the most effective equipment to suit their purposes for fluid recycling. Some variables to consider are the fluid types to be recycled, water quality, types and quantities of contaminants, machine sump and filter configurations, cleanliness required, operator acceptance, and resources available (capital equipment and personnel). The "bottom line" must be an

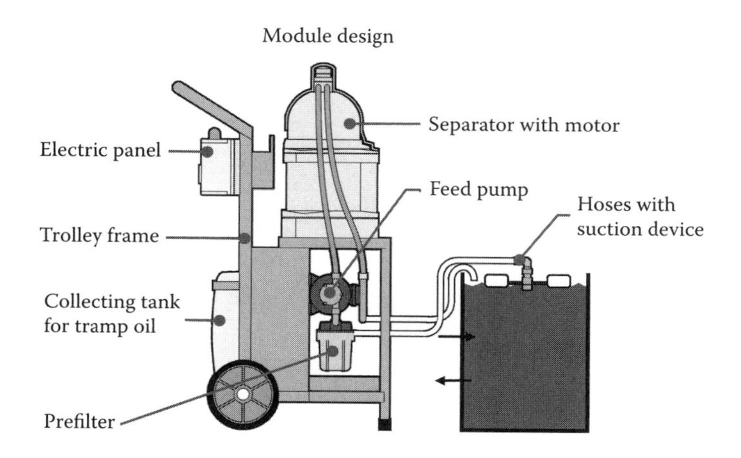

FIGURE 12.8 Portable Centrifuge Module Design (Courtesy of Alfa Laval. With permission).

FIGURE 12.9 Photograph of Portable Centrifuge (Courtesy of Alfa Laval. With permission).

evaluation of the benefits vs. costs of fluid recycling for each plant. Most plants realize the high cost of poor fluid management and are willing to invest in improved fluid management methods and equipment.

VII. WASTEWATER TREATMENT AND DISPOSAL

With an effective fluid management program in place, a metalworking plant can reasonably expect to lower its fluid concentrate purchase costs by 30 to 60%, depending on current fluid management practices at the plant. Reduced disposal volume will vary from 50 to 80% for most plants. However, even with management improvements, daily fluid make-up requirements may represent 20 to 30% of the concentrate purchase costs for most plants. Even the best fluids, used with good fluid management, will need to be disposed eventually. Typical reasons for disposal are:

High bacteria or mold counts, resulting in breakdown of product
Excessive contaminants (oil, dirt, etc.)
Excessive buildup of dissolved minerals and metals
Selective depletion of product components
Mechanical breakdown within either the machine or central system which requires pumping out fluid

The major waste treatment and disposal options available are:

Contract hauling
Chemical treatment
Ultrafiltration
Incineration
Evaporation

The selection of the treatment method for disposal will depend on factors such as volume of wastewater generated, composition of wastewater, classification of hazardous vs. nonhazardous, availability and cost of contract hauling, and whether the plant has access to a sewer system.

Ideally, with a careful fluid management program, the wastewater volume generated will be minimal. The high cost of contract hauling in many areas has resulted in plants opting for in-plant controls and equipment to eliminate the cost of contract hauling. The technology is available to waste-treat fluids, using several stages and types of equipment, to provide water clean enough for reuse for cooling water, parts washer, or metalworking fluid make-up. This technology enables plants to approach the goal of "zero discharge." Even with advanced wastewater treatment equipment, there are waste by-products that must be disposed. Once again, the assistance of the fluid supplier will be helpful in selecting the optimum waste treatment or disposal method.

VIII. SUMMARY

The economics of fluids use is changing rapidly, since improved fluid productivity, environmental safety and health, and proper disposal are major needs. Purchase price, labor, machine downtime, productivity (quality, production rates, scrap, tool, or wheel life), operator safety and health, and disposal costs are all part of the overall fluid use cost.

As metalworking fluid needs change and costs increase, it is increasingly important for plants to implement a fluid management program. The areas of fluid selection, water quality, fluid controls, contaminant removal equipment, wastewater disposal, and overall economics must be evaluated. Through careful fluid management, metalworking plants can generate substantial improvements in fluid performance and economics.

REFERENCES

1. *Report on the Volume of Lubricants Manufactured in the United States*, Independent Lubricant Manufacturers (1990). Presented to the Independent Lubricant Manufacturers Association 1991 Annual Meeting, September 28–October 1, by E. Cleves, Interlube Corporation, p. 2.
2. "Metalworking fluid trends 1991," speech by K. E., Rich, Lubrizol Corporation, November 1, 1991.
3. Leiter, J. L. and Fastenau, R. A., Environmental law, In *Waste Minimization and Wastewater Treatment of Metalworking Fluids*, Kelly, R., Dick, and Dacko, Eds., Independent Lubricant Manufacturers Association, Alexandria, VA, pp. 8–10, 1991.
4. Schaffer, G., The AM award M. Eugene Merchant, *Am. Mach.*, 124(12), 90–97, 1980.
5. Lucke, W. E., Cutting fluid oil mist in the shop, *SME Tech. Paper*, MR78–MR266, 1978.
6. Bennett, K. N., Iron Age's guide to metalcutting fluids, *Iron Age*, 18–26, 1984, November.
7. Springborn, R. K., Ed., *Cutting and Grinding Fluids: Selection and Application*, ASTM, Dearborn, MI, pp. 102–104, 1967.
8. Drozda, T. J. and Wick, C., Eds., *Tool and Manufacturing Engineers Handbook*, Vol. 1, McGraw-Hill, New York, pp. 25–26, see also pp. 29, 361–369, 1983.
9. Pinkelton, B. H., The OSHA hazard communication standard, *ASLE*, 43(4), 236–243, 1987.
10. Joseph, J. J., *Coolant Filtration*, Joseph Marketing, East Syracuse, NY, pp. 27–28, see also pages 37–44, 1985.
11. "A guide to understanding the treatment of oily wastewater," AFL Industries, Form 800138, p. 3, see also page 7.
12. Hill, E. C. and Elsmore, R., Pasteurization of oils and semulsions, *Biodeterioration*, 5, 469, 1983.
13. Sköld, R. O. and Mahdi, S. M., Ultrafiltration for the recycling of a model water based metalworking fluid designed for continuous recycling using ultrafiltration, *J. Soc. Tribol. Lubr. Eng.*, 47(8), 653–659, 1991.
14. Sköld, R. O. and Mahdi, S. M., Ultrafiltration for the recycling of a model water based metalworking fluid: process design considerations, *Lubr. Eng.*, 47(8), 686–690, 1991.

13 Waste Treatment

John M. Burke and William A. Gaines

CONTENTS

I. INTRODUCTION

Increasingly stringent discharge regulations have impacted manufacturing operations globally, making proper disposal of spent metalworking fluids (MWF) imperative. The requirements vary among regions, countries, states and provinces, but generally include both conventional and nonconventional pollutants (Table 13.1). In the U.S., the Environmental Protection Agency (EPA) has passed numerous laws to protect the nation's waterways. The authority for enforcing these laws has been passed down to state and local levels in many cases. These laws are strictly enforced with fines and/or imprisonment, depending on the severity and the intent of the violation. Treatment of the compounds derived from MWFs (the primary source of organic contaminants) poses unique issues when designing a robust treatment system.[1]

TABLE 13.1
Conventional and Nonconventional Pollutants

Abbreviation	Pollutant Name
Conventional Pollutants	
BOD_5	Biochemical oxygen demand — 5 day
COD	Chemical oxygen demand
SS	Suspended solids
NH_3–N	Ammonia as N
TKN	Total Kjeldahl nitrogen
O & GHEM	Oil and grease as hexane extractable materials
Nonconventional Pollutants	
As	Arsenic
Al	Aluminum
Se	Selenium
CN	Cyanide
Hg	Mercury
Pb	Lead
Cu	Copper
Ni	Nickel
Ag	Silver
Zn	Zinc
Cr	Chromium
SO_4	Sulfate
NO_3	Nitrate
NO_2	Nitrite
TTO	Total toxic organics as per U.S. EPA 40CFR 433

MWF wastewater is comprised of many sources, not just "coolants" and washing "soaps." Additional sources include floor cleaners, phosphate wastes, vibratory deburring discharge, impregnation fluids, stamping and drawing compounds, lapping compounds, machine lubricants, test cell blow down, first fill oils, die casting lubricants, and many more. Because of this, the wastewater contains free oils, stable oil–water emulsions, water-soluble organic compounds, dissolved and undissolved metals, inorganic compounds such as nitrates, chlorides, sulfates, and suspended and settleable materials. Table 13.2 identifies some common chemicals found in untreated wastewater from metalworking facilities.

The influent characteristics (concentration of free oil, emulsified oil, and soluble organics) and the volume of wastewater to be treated dictate the most effective wastewater treatment plant (WWTP) design. The plant must provide a robust and flexible treatment system to consistently handle metals, phenols, and oils, as well as to provide treatment for increased loadings of chemical oxygen demand (COD), complex emulsifiers, and organic nitrogens (e.g., alkanolamines) associated with use of MWFs.

Petroleum oil, the most common contaminant, is generally the easiest to remove since it has very limited solubility in water.[2] For reference, "soluble oils" are actually oil in water (O/W) emulsions, and are not truly soluble oil solutions as the name implies.

In summary, there is no perfect method of wastewater treatment for this complex mixture of contaminants. One reason is that treated wastewater, commonly referred to as effluent, can have widely varying discharge limitations depending on the facility's location. Table 13.3 shows several municipal industrial sewer use discharge limits and ranges in the state of Michigan alone.

TABLE 13.2
Typical Contaminants Found in Wastewater from Metal Cutting Manufacturing Facilities

Hydrocarbon compounds that are floatable, suspended, dispersed, emulsifiable, or settleable
Petroleum oils
Vegetable oils
Animal fats
Waxes
Fatty acid soaps such as those of calcium, magnesium, iron, and aluminum
Chlorinated esters and paraffins

Suspended, and settleable solids
Graphite
Microbiological contaminants such as bacteria and fungus
Vibratory deburring particulates
Floor sweepings

Metals (may be dissolved or as micro-particulates less than 100 μm)
Iron
Aluminum
Copper
Chrome
Zinc
Manganese
Molybdenum
Lead
Nickel

Nonmetals (typically dissolved)
Arsenic
Selenium

Dissolved solids
Salts (sodium and potassium salts)

Dissolved organics compounds
Amines
Amides
Esters
Glycols
Surfactants
Detergents
Free fatty acids
Fatty acid soaps such as those of sodium and potassium
Fatty alcohols
Biocides
Phosphate esters
Chelating compounds such as EDTA, citric acid, and NTA

Reviewing these different limits, it is easy to understand why the selection of the treatment method varies depending on the facility's location. An equally important factor is the changing nature of the influent character. Changing chemistries and processes used in a manufacturing facility have a significant impact on the success or failure of a given treatment method. For example, if

TABLE 13.3
BOD_5, COD Discharge Limits by Treatment Plant Size in State of Michigan[a]

Parameter	Class A Plants (29 Plants) 5.8–1200 MGD[b]		Class B Plants (65 Plants) 0.42–8.5 MGD		Class C Plants (14 Plants) 0.28–3.5 MGD	
	Low	High	Low	High	Low	High
BOD_5						
Violation limit[c]	350	10,000	185	2600	250	350
COD						
Violation limit	1670	3000	440	4000	NA	NA

[a] Summarized from a comprehensive table provided by the Michigan Water Environment Federation 2002.
[b] MGD, million gallons per day.
[c] All values in mg/l.

a manufacturing facility changes its floor cleaning formulation to include chelating agents such as EDTA, the effectiveness of an existing chemical treatment system will be negatively impacted. Lastly, each treatment step described in this chapter can have a significant cost and operational impact. Even though a wastewater treatment system can be highly automated, maintenance still requires a human element.

II. A STRATEGIC PLAN FOR WASTE TREATMENT

Because there is no perfect single treatment method, because waste streams vary day to day, and because the discharge limits can be restrictive, careful planning of a successful treatment approach is imperative. Consider the following steps.

A. Waste Minimization

Use all waste minimization strategies where practical in the manufacturing environment. The use of purified water for makeup of MWFs, filtering or centrifuging the fluids, and proper use of microbiocides, coupled with chemical management strategies, can greatly reduce the waste generated from a facility. These and other waste minimization strategies are discussed elsewhere in this book. Waste minimization can be a clear win–win approach. First, the facility saves money by reducing the use of MWFs. Second, cost savings are realized by reducing the wastewater volume requiring treatment. A summary of MWF waste minimization practice guidelines can be found at www.epa.gov or www.osha.gov. Professional societies such as the Society of Tribologists and Lubrication Engineers (www.stle.org) and the Society of Manufacturing Engineers (www.sme.org) offer courses and technical papers on management strategies for proper use and control of MWFs.[3,4]

B. Obtain All Applicable Discharge Regulations

Large municipalities are typically more flexible on industrial discharges than small cities (Table 13.3). It is necessary to have the industrial sewer use ordinance and the industrial sewer use discharge limits before starting a design approach. Some municipalities' sewer use ordinances have such highly restrictive limits that the treatment system can be too expensive to build and operate relative to contract hauling.

C. Outsourcing Processes

If certain metal finishing processes such as electroplating or phosphating can be outsourced, then fewer dissolved metals will be in the discharge. An additional benefit to outsourcing is that it may allow the facility to operate under less restrictive discharge standards. For example, in the U.S., the EPA has established federal limits on manufacturing facilities that make parts using traditional metalworking processes (drilling, boring, reaming, turning, etc.), and also perform metal finishing (such as chemical milling, electroplating, electroless plating, chemical etching, phosphating, and printed circuit board manufacturing). Refer to U.S. federal law, 40CFR 433.[5] Local municipal treatment plants may further establish separate limits on discharges into their systems. Eliminating or outsourcing certain processes may reduce regulatory burden.

D. Contract Hauling

Why treat at all? A tank truck can typically transport 5000 U.S. gals. Paying to haul that volume once a week may be a better economic decision than paying for the time, energy, maintenance, permits, sampling, and floor space for an on-site treatment facility. Fees for hauling include transportation and disposal fees, plus potential demurrage if the tank truck waits more than a given time during loading. Hauling costs increase significantly if certain chemicals are present in high quantities (including lead, hexavalent chromium, organic chlorine, high sulfur) or have a hazardous characteristic (low flash point, corrosivity, reactivity). These chemicals or characteristics can cause the waste stream to be designated as "hazardous" by EPA definitions; and the cost to manifest, store, transport, and dispose of these fluids increases significantly. Some waste oil haulers may not be authorized to transport or dispose of listed hazardous wastes. Hauling costs can vary significantly depending on location. For example, in southeastern Michigan, where rates are highly competitive, hauling costs can be in the range of $0.15 to 0.20 per gallon for nonhazardous wastewater containing 1 to 10% oil by volume. In rural areas of eastern U.S., the same wastewater can cost $1.00 or more per gallon to haul.

E. Commitment

Waste treatment systems can be both labor-intensive and technically demanding. Buying an "off the shelf" treatment system with minimal management support is a recipe for failure. Before proceeding with on-site wastewater treatment, ask the designer or supplier about operator requirements, energy consumption, percentage of waste reduced (concentration factor), necessary spare parts, periodic maintenance requirements, and average cost to treat wastewater based on current cost burdens.

III. Basic Treatment Methods

Depending on the level of treatment required to meet discharge standards, one, two, or three stages of treatment may be necessary. These treatment steps follow a consistent, logical order (Figure 13.1).[6]

A. Primary Treatment
   (1) Flow controls to manage peak hydraulic surges
   (2) Equalization to manage variations in waste concentrations
   (3) Solids/liquids separation to separate free oil and settleable solids
B. Secondary Treatment
   (1) Thermal evaporation
   (2) Distillation with heat recovery; vapor compression distillation (VCD)
   (3) Membrane separation; ultrafiltration (UF) or micro-filtration (MF)
   (4) Chemical treatment, or "emulsion breaking"

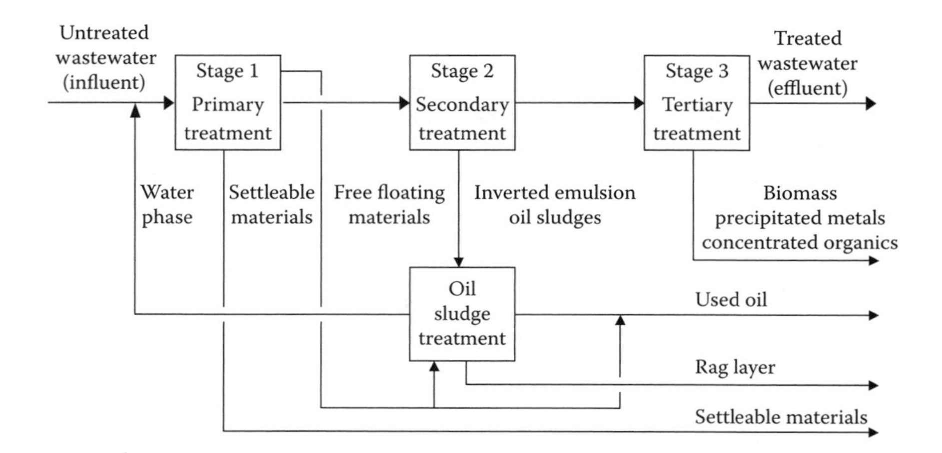

FIGURE 13.1 Basic treatment steps in metalworking fluid wastewater treatment.

C. Tertiary Treatment
   (1) Membrane separation (reverse osmosis (RO) or nanofiltration)
   (2) Carbon adsorption (powdered or granular carbon)
   (3) Biological treatment
   (4) Metals precipitation (chemical)
   (5) Chemical oxidation
   (6) Electrochemical oxidation (ECO)
   (7) Combinations of these steps

A. Primary Treatment

Wastewater rarely occurs at a predicable time or flow rate. Therefore, sufficient storage is necessary to prevent flooding of the WWTP. As a general guideline, the facility should have enough storage capacity to hold the largest waste-generating process tank on the premises.

Wastewater influent concentrations can vary unpredictably. Most wastewater treatment systems do not tolerate significant variations of influent concentrations. Sufficiently large storage facilities can generally buffer these peak variations.

Wastewater containing high total solids may require filtering before routing into storage tanks. Some suspended solids may continue to settle over time in the storage tank, thus requiring periodic tank cleaning. Allowing these solids to settle out will generally improve chemical emulsion breaking, since the treatment objective is to produce a floating precipitate. Ultrafilter treatment processes also benefit because small solids or fines can plug membrane pores. Finally, adequate storage, or quiescent devices such as gravity separators, allow for semi emulsified oils to separate as free-floating oil for easy skimming, thus reducing the loading on the secondary treatment processes.

B. Secondary Treatment Methods

1. Thermal Evaporation

A convenient method to reduce the volume of wastewater from metalworking operations is thermal evaporation. Thermal energy is applied to the waste solution, causing it to boil. The steam/distillate vapors are vented to the atmosphere by way of an exhaust stack. Boiling the wastewater in a

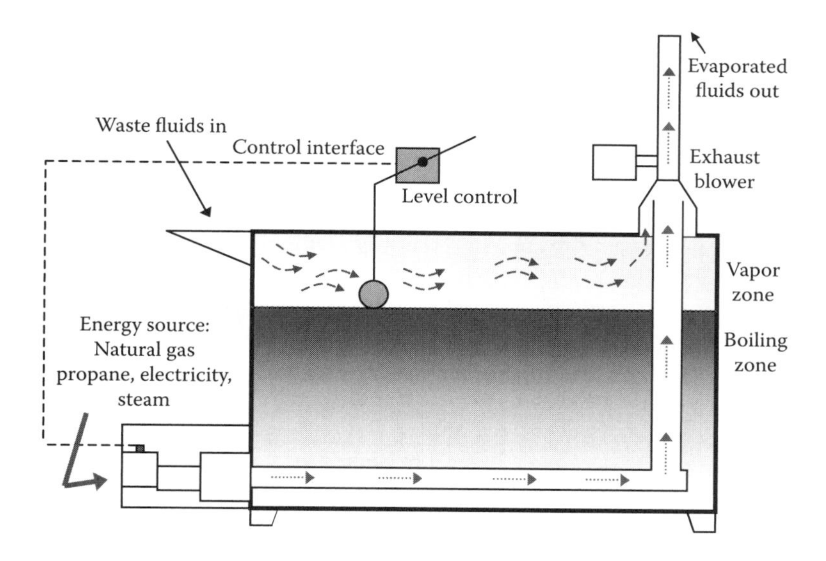

FIGURE 13.2 Schematic of a basic thermal evaporator.

vacuum and then condensing the vapors can gain some efficiency. Figure 13.2 is a schematic of basic thermal evaporator. Figure 13.3 is a photograph of an actual thermal evaporator.

The energy cost to operate an atmospheric thermal evaporator is calculated as follows:

From basic thermodynamics, one British thermal unit (BTU) is the heat required to raise one pound of water 1°F, and 960 BTUs are required to turn one pound (#) of water into steam at 212°F.[7]

FIGURE 13.3 Basic thermal evaporator (courtesy Severn Trent Services, Samsco).

Assume one has 1000 U.S. gallons of wastewater at 80°F containing 1% oil (by volume) to be evaporated, and the goal is to concentrate the solution to 50% oil by volume, or a 50 × reduction. Further assume natural gas with an energy value of 1,000,000 BTUs per thousand cubic feet (MCF) at a cost of $8.00 per MCF is used. Therefore, 1000 gals of wastewater contains 990 gal of water and 10 gal of oil. To achieve final oil concentration of 50% requires 980 gal to be evaporated (leaving 10 gal of water plus 10 gal of oil). To raise the fluid temperature to 212°F requires 1000 gal (212°F − 80°F) = 142,000 BTUs. Plus, to evaporate 980 gal of water the following calculation applies: 980 gal H_2O × 8.34 #/gal × 960 BTUs/# = 7,847,040 BTUs. ∴ Total BTU requirement = 142,000 + 7,847,040 = 7,989,040 BTUs.

However, thermal evaporators do not operate at 100% efficiency. Natural gas is not completely combusted, heat is not completely transferred into the fluid, and heat losses occur throughout the system. Also, the heat transfer rate into the boiling solution decreases significantly as the fluid concentration increases from 1 to 50% oil. For this example, assume the evaporation efficiency from start to end averages 60%: 7,989.040 BTUs ÷ 0.60 = 13,315,067 total BTUs required and (13,315,067 BTUs) ÷ (1,000,000 BTUs/MCF)($8.00/MCF) = $106.52 cost to treat 1000 gal of waste metalworking fluids by conventional thermal evaporation.

Advantages and disadvantages of using thermal evaporation:

Advantages	Disadvantages
Simple overall concept	Creates foam
Concentrates waste	Energy intensive
Eliminates sewer discharge	Possible air pollution source (volatile organics)
Easy operation	Stack corrosion possible
Low capital cost	Odors
Low water in sludge	Economical only for low volumes
Can tolerate solids	Possible fire hazard

2. Vapor Compression Distillation

This process is similar to thermal evaporation except that a significant amount of heat (93%) is recovered and reused within the process, thus decreasing operational costs. This method uses a mechanical blower and several heat exchangers to recover the latent heat of vaporization and the sensible heat of the condensed vapor.[7]

There are many types of VCD systems. The two most common are falling film vapor condensers and forced circulation vapor condensers. They can operate in a pressure or vacuum arrangement. Figure 13.4 shows a schematic of a forced circulation VCD system. Figure 13.5 shows a 10,000-gal per day VCD system. The skid dimensions of the unit pictured are 18 ft long by 8 ft wide by 18 ft tall.

These systems can be used as a secondary or tertiary treatment device. As a secondary device, oil arriving at 0.5% by volume can be concentrated to 60% by volume in one pass using the forced circulation VCD method. System operating costs (assuming electrical cost is $0.05/kW-h) can be less than $0.01/gal. As a tertiary device, the VCD can be used to polish treat wastewater to a very high level. Used as either a secondary or tertiary device, the metals content in the condensed vapor is typically less than 0.01 mg/l per metal.[8,9]

Another novel use of VCD technology is to process oil sludge that arrived at 40 to 50% oil by volume, and concentrate it to 90% by volume in one pass. Energy cost, with electricity at $0.05/kW/h will range from $0.015 to $0.020 per gallon or oily sludge processed.

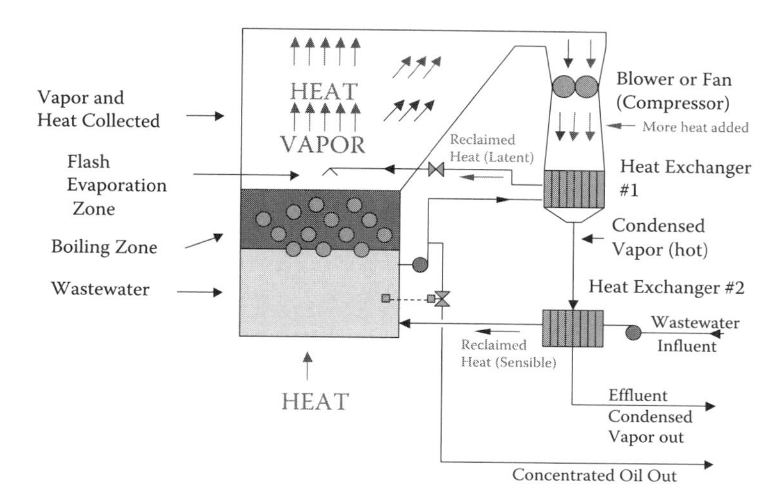

FIGURE 13.4 Schematic of vapor compression distillation forced circulation evaporator.

3. Ultrafiltration

Small UF package systems are increasingly popular for the removal of emulsified oil and other dispersed components found in MWFs from cutting, grinding, and drawing operations. With slight modifications, these systems can also be used to treat the fluids from alkaline and acid parts washing or cleaning baths for recycling. Package treatment systems are also available in which a biological reactor has been coupled to the UF step, providing both secondary and tertiary MWF treatment. With that in mind, UF is considered to be a very versatile process.

UF is a pressure-driven membrane filtration process. It uses molecular size openings or "pores" typically in the range of 0.02 to 0.07 μm to separate emulsions and macromolecules into two

FIGURE 13.5 Force circulation vapor compression distillation system (courtesy VACOM, LLC).

TABLE 13.4
Ultrafiltration Flux Rate Variables[10]

Concentration of oil
Viscosity of the oil
Viscosity of the feed solution
Temperature of the feed solution
Concentration of total suspended solids, inclusive of bacteria and fungus
Feed pressure
Cross flow velocity
Transmembrane pressure drop (pressure on each side of the membrane)
Incompatible materials such as silicone and silicates

phases: a clean permeate phase and a concentrated retentate phase. Microfiltration is essentially identical to ultrafiltration, except that the pore sizes are typically 0.1 μm. UF has been applied in oil separation from wastewaters in many industries including adhesives and sealants, commercial laundries, synthetic rubber manufacturing, timber products processing, and metalworking operations. Unlike RO that provides separation down to the ionic level, UF consists of a more open membrane (larger pore size), and lower pressures are employed.[10] Pressure ranges for ultrafiltration are typically 30 to 70 lb/in.^{2} gauge (PSIG). UF membranes will reject compounds greater than approximately 0.01 μm in effective diameter, since a gel layer forms on the membrane surface, improving separation effectiveness. For reference, the size of a bacterial cell is typically greater than 0.5 μm. UF membranes cannot retain low molecular weight soluble organic and inorganic compounds (see Table 13.2).

Determining the UF membrane area needed for treating oily wastewater is dependent on many factors (refer to Table 13.4). The primary factor is determination of "flux rate" or the rate at which a certain volume of wastewater passes through a membrane area in a given time. It is usually expressed as GSFD, which is U.S. gallons of permeate produced per square foot of (effective) membrane surface area per day.

The next important sizing element is determining the concentration factor. In general design conditions, a UF can concentrate oily wastewater from a range of 0.5 to 1.0% by volume, to a range of 30 to 50% by volume. This range is dependent on the same factors listed in Table 13.4. If an ultrafilter system starts with a concentration of 1% oil and concentrates to 50% oil by volume, it is said to have a concentration factor of 50×. However, the flux rate of 1% oil at the start of a cycle will be dramatically reduced when processing to 50% oil at the end of a cycle. Determination of the average flux over the entire concentration cycle is essential to determine the effective area of membrane surface necessary to treat a given volume of wastewater per day. This determination is best done in a pilot study over several weeks or, in some cases, months. The purpose of extending the length of test time is to evaluate variations of the influent waste stream.

Ultrafilter membrane surfaces remain clean via a concept referred to as "cross flow filtration." Cross flow rate or velocity is the volume or velocity of the process fluid passing across the surface of the membrane per unit of time. The cross flow volume along the membrane surface is many times greater than the permeate flow rate through the membrane. Typical ranges of cross flow are 1 gal/min of cross flow per 1.6 ft^2 of membrane, to 1 gal/min of cross flow per 2.75 ft^2 of membrane. Using other units, the cross flow rate will range from 25 to 75 gal/min for each gallon per minute of permeate produced (assuming an average flux rate of 20 GSFD). Higher cross flow rates are used for fluids that have high fouling potential. Reducing the cross

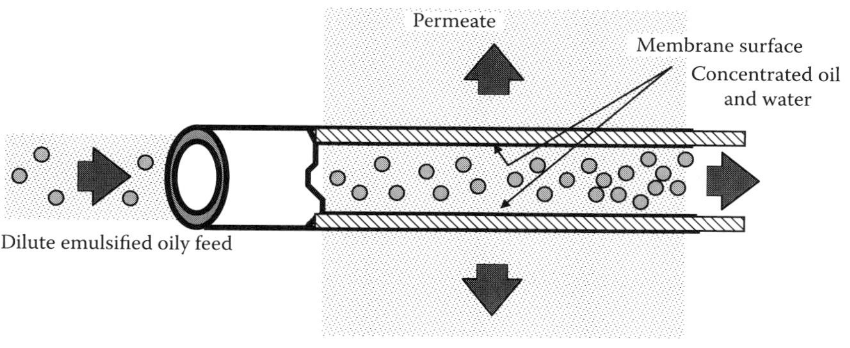

FIGURE 13.6 Ultrafiltration: the cross flow concept.

flow velocity can result in rapid loss of flux rate and premature fouling of the membrane surface[11] (see Figure 13.6).

The term "fouling" is broadly used to describe anything that causes a decrease in flow rate. Fouling can be caused by any combination of solutes, particulates, and precipitates. Fouling can occur above, at, or even inside the pores of the membrane. Usually it is a combination of effects. Examples of foulants include: silt, solids and precipitates, inorganic colloids, scale and metal oxides, free oil (ultrafilter membranes do not tolerate unemulsified oil), biological matter, silicone-based defoamers, silicate-based cleaners, and cationic polymers. Fouling may be either reversible or irreversible. Free oil, for example, causes an immediate and dramatic flux reduction but is completely recoverable. On the other hand, certain cationic polymers cause a sudden and sometimes permanent loss of flow. Cleaning will only recover a flux loss due to reversible fouling. Silicates, for example, cause gradual fouling; however, the fouling is generally considered irreversible.

The rejected materials often form a viscous gelatinous layer on the membrane. This gelatinous layer acts as a secondary membrane, reducing the flux and often reducing the passage of low molecular weight solutes. Surface fouling is a result of the depositing of submicron particles on the surface, as well as the accumulation of smaller materials caused by crystallization and precipitation reactions. Using a membrane with a surface modified by chemical or physical methods can significantly improve its flux characteristics.[12] Specific chemical contaminants can cause rapid fouling of the membrane surface and thus loss of flux. Some of these foulants are:

- Silicone defoamers (all forms of silicone, siloxanes, and silicates should be carefully evaluated before using with UF)
- Paint solvents such a MEK, MIBK, toluene, xylene, and acetone
- Strong acids or solutions with a pH below 3.5
- Temperatures exceeding 120°F (depending upon the specific membrane)[13]

There are two basic membrane systems used on the oily MWF: wide channel and narrow channel. The wide channel is a round tube from 1/4 to 1 in. in diameter. The narrow channel membrane uses either spiral wound flat sheets or very small channels of approximately 0.050 in. in diameter. These narrow channel tubular membranes are also referred to as "hollow fiber" membranes. Membranes can be made of polysulfone (PS), polyacrylonitrile (PAN), polyvinylidene fluoride (PVDF), or ceramic. PAN and PVDF are the most common polymeric membranes used in oily waste treatment.[14]

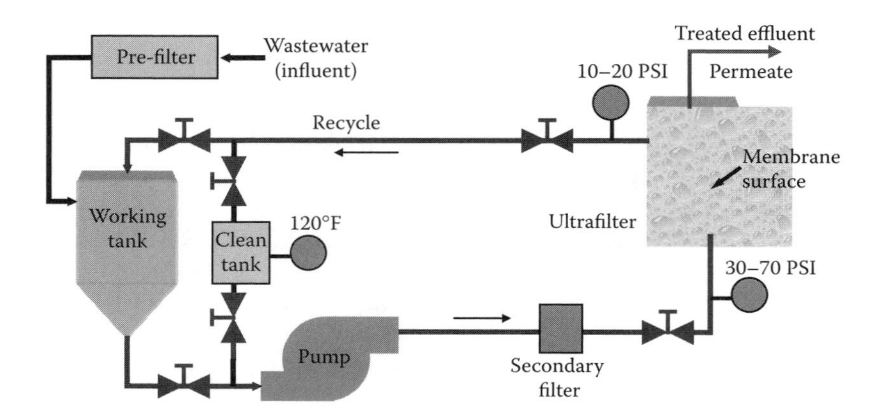

FIGURE 13.7 Basic ultrafilter flow schematic.

Figure 13.7 represents a simple schematic flow diagram of a UF treatment system. Figure 13.8 is a wide channel membrane system.

4. Chemical Emulsion Breaking

a. Modified Bentonite Process

One of the oldest methods of oily waste treatment is clay-based treatment. The most common clay used for this is bentonite. The major constituent of bentonite clay is the mineral montmorillonite. It also contains trace levels of shale, quartz, and sometimes gypsum. The bentonite is a mixture of cations such as sodium, calcium, magnesium, and potassium. Mixtures of bentonite and quaternary amine or other proprietary polymers are the most commonly used for treatment of MWFs. The advantage of the bentonite process is its simplicity in treatment of basic emulsions. A one-step addition of the bentonite mixture and rapid mixing can produce an absorptive floc that generally

FIGURE 13.8 Wide channel ultrafilter (courtesy Koch Membrane Systems, Inc.).

settles, with a clear water phase above the floc. In addition, the oily products in the wastewater bind strongly to the bentonite mixture and can be dewatered in a belt or filter press. Bentonite also can remove a portion of dissolved metals.[15] The disadvantage of the bentonite process is that the oil material is bound in a clay complex and is not recoverable by conventional reclaimers. The clay mixture is generally land filled as a disposal method. For this reason, bentonite has limited acceptance. When oily wastewater is processed by conventional chemical, distillation, or membranes methods, the oil sludges can be further processed to produce salable oil.

b. Inorganic Salt and Organic Polymeric Chemical Treatment

Chemical treatment followed by gravity floatation is employed in a number of industries, including those involving metalworking operations. In order to chemically separate or "break" a stable oil–water emulsion, the surface charges at the oil/water interface must first be destabilized to allow the emulsified droplets to coalesce.[16] The small oil droplets will then combine to form larger droplets. These oil droplets can readily float to the surface if sufficient oil is present. A positively charged cationic emulsion breaker (called a "coagulant") is required since the dielectric, chemically-induced characteristics of water and oil cause emulsified droplets to carry negative surface charges. The coagulant may be an inorganic (e.g., mineral acid or polyvalent metal salt) or an organic (e.g., cationic polymer) chemical. Ideally, two distinct layers would be formed: a free oil supernatant and a clear-water subnatant. In actual practice however, a third layer (called scum or "rag") forms between the oil and water layers as a result of suspended solids having oil occluded to their surfaces.

The typical sequence of chemical additions when breaking emulsions is:

1. Add acid to lower the pH of the wastewater.
2. Add coagulant(s) to destabilize the emulsion.
3. Add base to raise the pH.

This process is classically referred to as the acid–alum method. Occasionally, an anionic polyelectrolyte is added as a flocculent after the addition of a base, to better agglomerate the colloidal solids. The acid (for example, sulfuric) converts the carboxyl ion found in surfactants to carboxylic acids that allow the oil droplets to agglomerate. The salts of aluminum sulfate, calcium chloride, ferric chloride, or ferrous sulfate are the most widely used coagulants to aid in the agglomeration of the oil droplets. The base (typically sodium hydroxide) raises the pH to cause soluble metals to precipitate as their hydroxide.[17] These precipitates become occluded with oil and become part of the intermediate rag layer.

Cationic polymers have largely replaced purely inorganic salt programs for oily waste emulsion breaking. Cationic polymers are organic molecules used as coagulants to destabilize the repulsive forces that maintain a stable emulsion. As such, they are substitute materials for the inorganic cations (for example, Al^{+3}, Ca^{+2}, Fe^{+3}) that have historically been used, and they reduce the dosage of these chemicals. There are two reasons these polymers have received such widespread use in recent years: reduced solids generation and cost. The removal of heavy metals such as copper and zinc by precipitation is desirable since it lowers the effluent concentration of these regulated metals. Unfortunately, inorganic coagulants such as aluminum also precipitate and cause the production of a substantial amount of excess sludge. Excess solids in the rag layer increase disposal costs and slow the rate of separation. In addition, if too much rag layer is formed, this renders separation ineffective. Unlike their inorganic counterparts, polymers are water-soluble materials that do not precipitate under alkaline conditions. The higher charge density of polymers also results in significantly lower dosage rates.

FIGURE 13.9 Gang stirrer (courtesy © 2005 Phipps & Bird, Inc., Richmond, VA).

Cationic polymers cost more per gallon than the inorganics they replace. Even at lower dosage rates a polymer program would cost more than an inorganic program, all other factors being the same. However, inorganic programs are most effective in moderately acidic environments while cationic polymers can function well in neutral to slightly alkaline environments. This is particularly notable since sodium hydroxide is typically the most costly of all the treatment chemicals. Therefore, the most effective method to reduce overall chemical cost would be to minimize acid use with a subsequent reduction in the use of caustic.

It is important to note that different polymer classes and blends have different "windows" of optimum dosage and pH. De-emulsification occurs when positively charged coagulants destabilize the emulsion's negative surface charge. Overdosing coagulants can actually restabilize or resolublize the oil/water emulsion so that it begins to have the appearance of the initial emulsion. Therefore, while a polymer with a very high charge density may have a lower dosage rate than one with less charge density, it may not be the best choice. If a high-charge polymer's functional "window" is too narrow to easily determine during bench testing and/or accurately dose in the treatment tank, then poor treatment separation will result. Duplicating jar testing and then actual dosing in a full size system requires diligent and skilled operators.*

Since wastewater characteristics vary from batch to batch due to changes in manufacturing operations and process chemicals, optimum dosages of these chemicals must be determined experimentally for each batch. A common bench device to assist technicians in proper chemical dosing is a gang stirrer. Here, small dose variations of polymers can be evaluated side by side in the same time frame (see Figure 13.9).

Table 13.5 shows the most common polymer formulations and their relative charge densities. Chemical suppliers also have a number of blends of organic polymers and inorganic salt chemistries. It is less expensive to purchase bulk inorganic salts from commodity vendors rather than from specialty chemical manufacturers, but only if the volumes warrant; otherwise polymer/salt blends are appropriate. Commodity vendors will not offer day-to-day onsite treatment consultation. Specialty polymer suppliers can offer onsite assistance and backup laboratory support if necessary. For that reason, treatment by specialty polymer supplier has gained acceptance.

Oily WWTPs using acid/alum have historically benefited most from the DADMAC, DMDAAC, and EPI formulations (Table 13.5). The high molecular weight polymers usually produce a larger and stronger floc, having improved settling and a more compact sludge at the

* Lower strength emulsions generally require lower dosages and have wider "windows" than higher strength emulsions. This makes the determination of correct chemical dosages crucial to the success of the chemical de-emulsification.

TABLE 13.5
Commonly Used Chemicals for Oily Wastewater Emulsion Breaking

Organic Coagulants	Molecular Weight	Charge Density
Melamine formaldehyde	10,000	Med-high
Polyamine	30,000	Very high
Acid–tannin polymer	500,000	Very high
Poly-DADMAC (dialkyl dimethyl ammonium chloride)	50,000–1,500,000	Low-med
EPI/DMA (epichlorohydrin dimethylamine)	50,000–300,000	High-very high
DMDAAC (dimethyl dialkyl ammonium chloride)	100,000–300,000	Very high
Inorganic Coagulants		**Charge Density**
Aluminum sulfate		Very high
Ferric chloride		High
Ferric sulfate		High
Ferrous sulfate monohydrate		High
Polyaluminum chloride		Very high
Polyaluminum hydroxychloride		Very high
Aluminum chloride		Very high
Aluminum chlorohydrate		Very high
Calcium chloride		Medium
Sodium aluminate		Very high

expense of water clarity. The selection of the "best" chemical program remains both an art and a science, requiring patience to balance cost, settleability, sludge density, water clarity, and the pH and dosage windows. Comprehensive jar testing, followed by a controlled shop trial, is the appropriate technique to determine the most cost-effective program that ensures continuous compliance. Table 13.6 indicates the results obtainable with basic chemical treatment. Once implemented, the optimum program will vary as changes occur in manufacturing fluids, dumping schedules, and oil concentrations. Thus, chemical treatment programs usually require batch-by-batch adjustments. Overall, chemical programs should be revisited on a periodic basis to evaluate cost, performance, and efficiency. A picture of modern emulsified oil before and after treatment with a single dose polymer is in Figure 13.10.

TABLE 13.6
Chemical Treatment Results, Before and After Treatment, Starting Solution 5% v/v

Fluid Type	BOD_5		COD		Oil and Grease[a]	
	Before	After	Before	After	Before	After
Basic emulsified oil[b]	17,500	350	790,000	900	35,000	70
Basic emulsified oil, HWS[c]	18,900	390	810,000	1300	34,000	79
Premium emulsified oil, HWS	24,500	2700	1,100,000	6500	20,500	195
Basic semisynthetic	15,900	2200	35,000	4500	4500	140
Semisynthetic, HWS	18,300	2600	37,000	5200	3900	160
Synthetic, HWS	8200	1600	45,000	24,000	900	35

[a] As hexane extractable material, EPA method 413.1.
[b] All results in mg/l.
[c] HWS, hard water stable.

FIGURE 13.10 Chemical emulsion breaking—polymer method.

Metalworking operations typically generate two types of wastewater: dilute oily wastewater and concentrated spent MWFs. The dilute oily wastewater is produced from parts-washer overflows (detergents and soaps), machine lubricants (hydraulic, gear, and way oils), floor cleaners, and coolant drag-out. Concentrated spent MWFs are generated when a metalworking "coolant" system has reached the end of its useful life and the system is drained, cleaned, and recharged.

There are two design configurations for conveying wastewater from the manufacturing plant to the WWTP, usually referred to as "1-pipe" or "2-pipe" systems. A 1-pipe system carries all wastewater in a single pipe to the WWTP. A 2-pipe system conveys the concentrated spent MWFs separately from dilute wastewaters. The latter configuration allows an operator to blend a desired volume of concentrated MWF with the dilute wastewater at his discretion in order to reduce tank-to-tank variations. Operational skill is very important when blending in concentrated MWFs so as not to create an untreatable batch.

From a facilities perspective, the emulsion-breaking process is conducted in either a continuous treatment system or by a batch process. Many early chemical emulsion-breaking treatment systems were carried out in a continuous mode.[18] A continuous process has certain advantages, most significant being the ability to process large volumes of wastewater in a relatively compact area. The main risk of continuous treatment is that the system can accidentally, and without warning, produce poor water quality. This is a significant drawback to this type of process. Due to increasingly stringent regulations, consistent and reliable high-quality effluent discharge is required. For this reason, many facilities involved in chemical treatment usually conduct their operation in a batch process mode. Whether in a continuous or batch mode, the chemical approach to treatment generally remains the same. A continuous treatment method is shown in Figure 13.11, and a batch treatment schematic is shown in Figure 13.12.

Chemical emulsion breaking requires intimate mixing of the wastewater with the emulsion-breaking chemicals followed by flocculation and flotation. Flotation may be performed simply by quiescent gravity separation in the batch tank or in a separate treatment unit, such as a dissolved air flotation clarifier (DAF). Since a 1-pipe system is straightforward and DAF is discussed elsewhere in this chapter, the following example describes the sequence of events for a 2-pipe system using in-tank separation.

After each batch tank has filled and the tank is isolated from additional influent flow, a laboratory jar test is performed to determine the amount of spent MWFs that can be added. This is carried out to determine the optimum chemical dosages required for treatment. The determined amount of spent MWFs is then blended into the batch tank from the spent MWF storage tank and the tank contents are mixed.

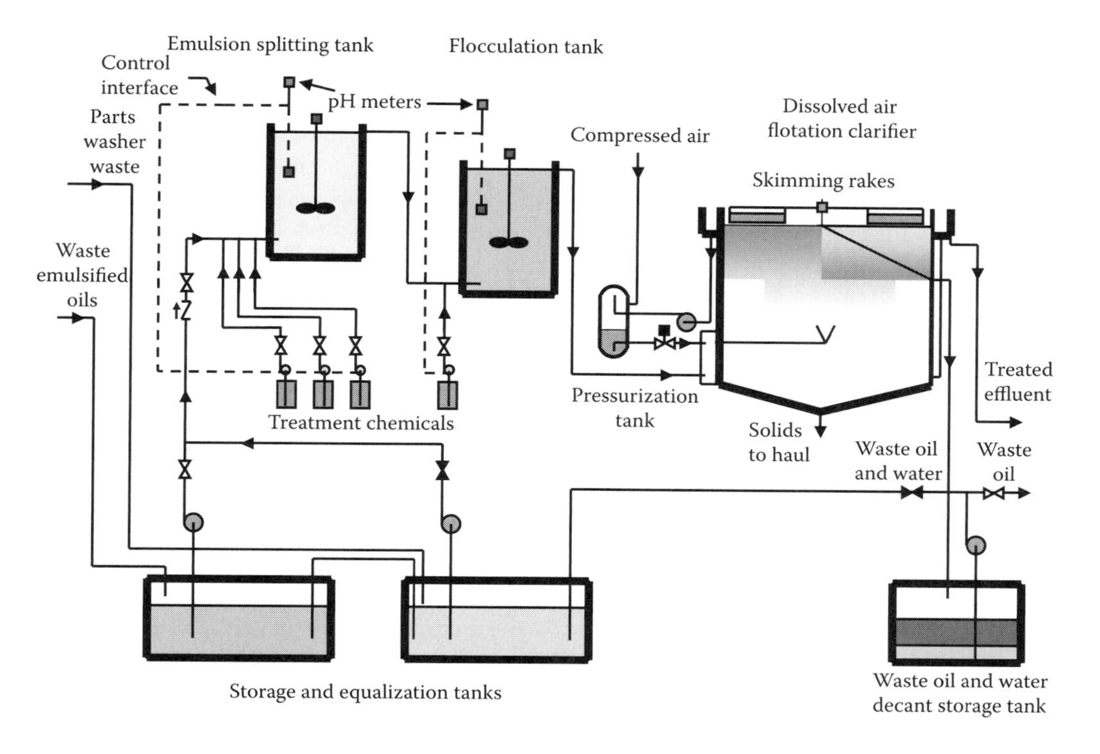

FIGURE 13.11 Basic schematic of chemical treatment — continuous flow method.

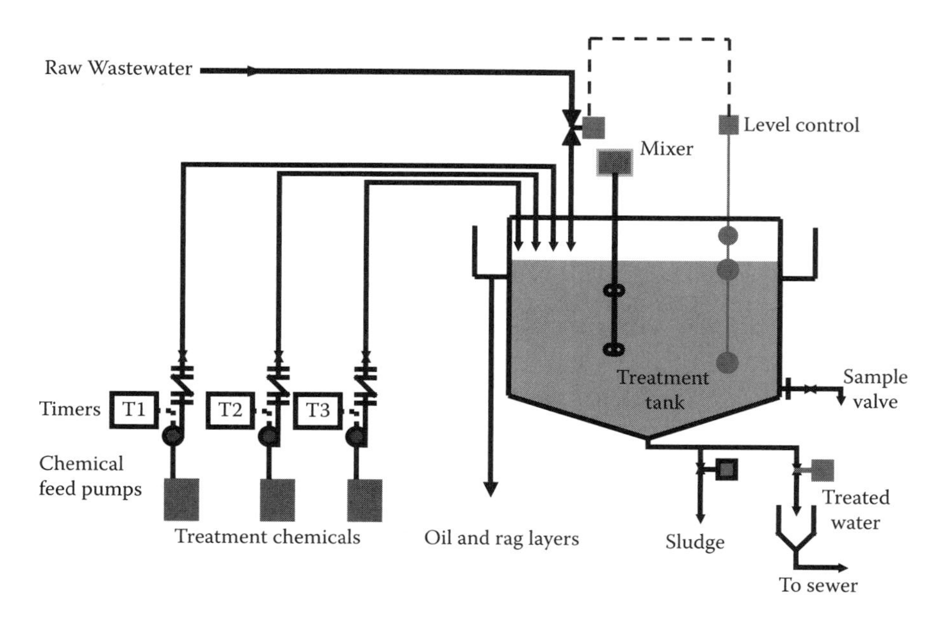

FIGURE 13.12 Basic schematic of chemical treatment — batch treatment method.

Note that the type and degree of mixing can have a significant impact on the quality of separation that occurs. Thorough mixing is required to bring the wastewater and coagulant(s) into intimate contact and to fully destabilize the emulsion. Yet over-mixing occasionally leads to only a small portion of water separating while the majority of the tank remains emulsified, or even re-emulsifies. Because of its appearance, operators term this "making a milkshake." For this reason system designers will often utilize air spargers rather than mechanical mixers to eliminate high shear forces at the prop tip, and operators will often turn the mixers off after sufficient mix time has been achieved for each chemical.

Once the wastewater and spent MWF have mixed, the determined amounts of chemicals are added to destabilize the oil/water emulsion, and sodium hydroxide is added to precipitate heavy metals. The contents are then allowed to sit quiescently for gravity separation of the oil and water phases. The clear, subnatant water phase is sampled and analyzed to ensure that it meets discharge standards before it is drained at a controlled rate to the effluent (most typically the sanitary sewer). The discharge is stopped before the oil cap becomes entrained in the effluent — normally by tank level measurement, since the cap depth is known historically from batch to batch. Turbidity measurement of the effluent is also often employed to automatically halt discharge and alert the operator.

The floating oil is allowed to accumulate over several batches of treatment, and a portion is then skimmed off and sent to a dedicated skim-oil tank. It is important that the volume of oil cap be properly controlled. If all of the oil was skimmed and there was little or no oil in next batch, there would be inadequate buoyancy, and the precipitates might sink or stay suspended in the tank as "stringers" or "floaters." Optimum cap maintenance procedures are site-specific, being dictated by the type and amount of oil and solids. These procedures might read "skim the oil cap to 1 ft depth when it exceeds 2 ft depth."

Skim oil from the batch tank treatment is typically only 10 to 15% oil and contains less than 1% solids. It is cost-effective to ship this material directly to oil reclaimers only if the volume generated is small. Oil reclaimers generally employ a graduated scale where: (1) low oil concentration waste has a high hauled cost per gallon; (2) a breakpoint occurs where there is no cost per hauled gallon; and (3) a credit is provided to the generator for high content oil waste. Oil reclaimers reclaim or rerefine waste oils and either reformulate the oil for resale or sell the recovered oil on the fuels market. Their graduated scales vary widely depending on location and changes in the fuels market.

Disposal economics drive the waste generator to further concentrate skim oil on-site if large volumes of low concentration oil are produced. This is achieved using a combination chemical and thermal process called oil "cooking." The skim oil is sent to a treatment tank (usually 2000 to 10,000 gal capacity) where acid, polymers, and steam concentrate the oil and minimize its volume. Oil concentrations up to 80 to 99% (by volume) can be achieved. The most robust treatment adds 1 to 3% of tank volume of concentrated sulfuric acid, raises the temperature to 160 to 190°F using steam, and holds the temperature for 12 to 24 h.[19,20] Steam may be utilized either by direct sparging using eductors, or by indirect coil heating. Direct sparging provides additional mixing but leads to additional volume due to condensation of steam to liquid water. Following a cooling off period, the highly acidic subnatant water phase is returned to the head of the WWTP for reuse, while the concentrated oil is stored in a tank before being sent to an oil reclaimer. The water from this oil cooking treatment offsets the acid requirement for the next batch in sequence.

The cooking process can produce offensive malodors, so off-gas scrubbing may be required. A recirculating, counter-current, packed bed scrubber with caustic addition is most often used. Occasionally, oxidants such as sodium hypochlorite or hydrogen peroxide are also added along with the caustic as an additional odor reducing method.

Oil concentrating is a process of diminishing returns. Concentrating from 10 to 50% is nearly always economically justified, while from 90 to 100% is rarely justified. There is a "sweet spot" that balances disposal economics with oil cooking treatment costs. Oil cooking does not require individual batch jar testing, but the process economics should be reviewed at least annually to verify the sweet spot. Recent efforts have been directed at concentrating skim oil using polymers/surfactants/heat, and at concentrating at ambient temperatures.

C. Tertiary Treatment: Physical–Chemical Treatment Processes

Physical and physical–chemical treatment processes include RO, granular carbon adsorption (GAC), biological treatment, metals precipitation, chemical oxidation, and ECO.

1. Reverse Osmosis and Nanofiltration

RO membranes provide a barrier to small molecular weight, dissolved organics, and inorganics. Thus they are used to remove such contaminants as water-soluble organics, cations, and anions (for example, chlorides and phosphates). While RO removes much smaller molecules than ultrafiltration, it operates at much higher pressure and has higher initial capital expense and operating costs. RO membranes are easily fouled, so pretreatment is required to ensure that the feed is essentially free of oil and suspended solids. If the influent to the RO process contains calcium or magnesium, a continuous feed of an antiscalent is required. If iron is present in the influent, then antiscalent and chelants will likely be required in combination. Frequent cleaning with sodium hydroxide is also required to maintain a steady flux rate through the membrane. RO membranes are not as durable as ultrafilter membranes with regards to cleaning methods. Over-aggressive cleaning of an RO membrane can lead to complete destruction of the RO membrane integrity in just a few hours.

Nanofiltration (NF) is similar to RO, except that the molecular weight cutoff (MWCO) for NF is typically between 250 and 400 MWCO, and less than 150 MWCO for RO.

2. Carbon Adsorption

GAC is capable of achieving a high degree of posttreatment, provided the dissolved organics following secondary treatment are readily absorbable onto activated carbon.[21] GAC is typically manufactured from specific grades of bituminous coal or coconut shells by high temperature steam activation that provides a highly microporous surface.

This carbonaceous material has a large internal surface area (500 to 1500 m^2/g) that is capable of adsorbing a wide variety of substances to its internal surface if adequate contact time is allowed. It can be impregnated with finely distributed chemicals to improve its ability to adsorb certain target parameters (for example, mercury or cyanide).

When used for liquid phase adsorption, GAC is most commonly used in fixed filter beds. Wastewater is passed downward through cylindrical contactors that have a bottom support bed and drainage system. The contactor may be constructed of plastic, fiberglass, or coated steel depending on the required materials of construction.

Some methods employ the use of powdered activated carbon (PAC) in place of GAC. PAC is injected directly into the feed stream as an absorber and as a settling aid, or filtered out after a period of contact time. In this application, PAC is not reusable.

GAC can be regenerated by simply heating the carbon to 600°F, and held at that temperature for 1 h. This is typically carried out off-site at the suppliers' thermal reactivation kilns. While the reactivation restores activated carbon to near virgin quality, the overall absorbency decreases slightly with each subsequent regeneration. Operational costs to regenerate the carbon may be high depending on the concentration of organics in the secondary feed stream, thus increasing the frequency of regeneration. If the carbon is not regenerated, spent carbon may be land filled. However, this raises additional concerns as the pollutants are just transferred from wastewater to land disposal.

As with RO, pretreatment is critical to ensure that no oil or total suspended solids (TSS) are present in the feed to the activated carbon system. Oil and TSS will quickly bind to the carbon and render further adsorption ineffective. Carbon is particularly effective for removing free chlorine from potable water sources. Carbon is not particularly efficient for removing primary, secondary, and tertiary amines from MWF effluent.

3. Biological Treatment Processes

Biological systems have become an increasingly popular method of treatment of MWF wastewaters. Biological digestion is appropriate treatment for biodegradable organic compounds. Many methods and approaches are available.[22] Biological processes are classified based on whether the bacteria are free-floating (planktonic) or attached to a solid surface (sessile). Those in which the active biomass is suspended as free organisms or microbial aggregates are called "suspended growth" reactors. Those where the growth occurs on a solid medium are called "fixed growth" reactors. Both suspended and fixed growth reactors have been utilized to treat MWF wastewaters following secondary treatment. Suspended growth designs predominate due to their increased ability to handle varying influent loading. Each is designed to treat 5-day biochemical oxygen demand (BOD_5), as well as reduce associated COD. They can also be designed for nitrification and denitrification to remove nitrogenous compounds such as amines. Recent laboratory and pilot studies have also greatly advanced knowledge of the biodegradability of individual MWF components.

Example. The aerobic fluidized bed (AFB) process is used at four locations of a large automotive manufacturer. In the AFB system, wastewater passes upward through a reactor containing a bed of sand or granular activated carbon medium at a velocity sufficient to expand the bed, resulting in a fluidized state. Once fluidized, the medium provides a vast surface area for biological growth, leading to a biomass concentration approximately five to ten times greater than that normally maintained in conventional activated sludge bioreactors. The treatment plant flow sheet consists of conventional primary and secondary treatment followed by tertiary treatment using the AFB process. In a performance evaluation, the two-stage AFB system achieved a median BOD_5 removal of 86%, together with essentially completed ammonium oxidation at a wastewater hydraulic retention time of less than 6 h.[23]

A membrane biological reactor (MBR) system is basically a conventional activated sludge process that utilizes membrane filtration instead of gravity sedimentation for solids separation. MBRs have become widely accepted and utilized on large-scale systems, most recently for the treatment of MWF wastewaters. This is attributable to two main factors. First, MBRs have the ability to sustain high biomass concentrations and solids retention times (SRT). This allows MBRs to treat widely varying influents of high organic loading while still producing a high quality effluent. Second, membrane systems are now less capital intensive and provide for more flexible operation. The MBR process provides quantifiable benefits over conventional activated sludge systems, including: a small footprint, low effluent suspended solids even if the wastewater does not settle well, reduced wasting and sludge production, robust system performance, and improved biological degradation.

The MBR process operates at higher mixed liquor suspended solids levels (12,000 to 30,000 mg/l) than conventional activated sludge plants (2000 to 3000 mg/l). This provides a large biomass with a long sludge age (e.g., frequently over 50 days) and better digestion. The mixed liquor is separated into a concentrate phase and a clean permeate phase by a UF. The concentrated solids are returned to the bioreactor while the clean permeate is discharged.

Several automotive facilities have installed and operated MBR systems without preremoval of oil. For these systems, raw MWF wastewater is fed straight to the aeration tank of the system, where the emulsion is de-emulsified as organic emulsifiers are biodegraded. Sludge in the MBR contains partially degraded oil along with biomass because of nonbiodegradable or difficult-to-degrade compounds. This sludge could interfere with both digestion and membrane separation.[24] This also eliminates the opportunity of recovering waste oil. The oil used in MWFs is typically petroleum

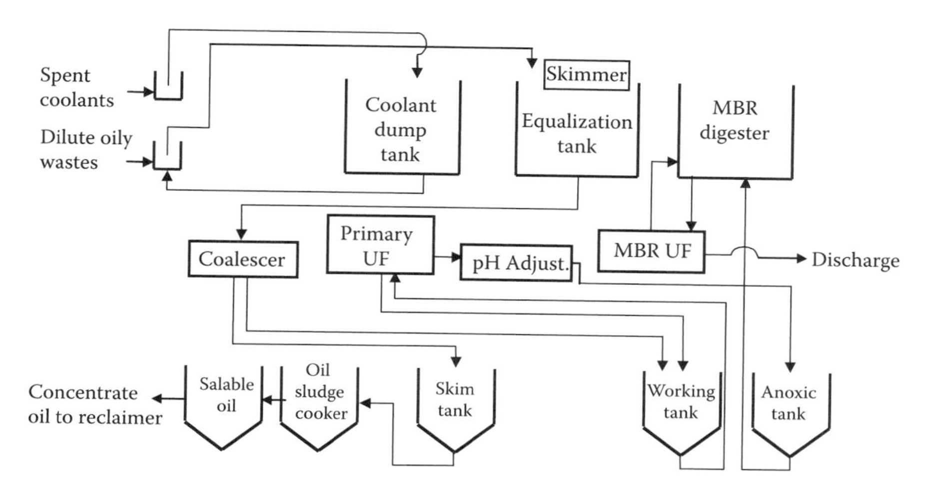

FIGURE 13.13 Basic schematic of membrane bioreactor process flow.

mineral oil, which consists of paraffinic, aromatic, and naphthenic hydrocarbons and which is difficult to biodegrade. Kim reported nonbiodegradable organics after treating a simulated MWF wastewater both anaerobically and aerobically.[25] Taylor et al. also reported that the most difficult compounds to biodegrade in MWFs were polyaromatic compounds having greater than four rings, although these structures are not typically found in automotive MWFs.[26,27]

Another approach utilizes UF to separate oils prior to the MBR. This approach overcomes some of the problems inherent with the previously described configuration. The UF essentially eliminates complicating factors associated with high molecular weight petroleum oils and their competing digestion reactions. Also, adding a separate anoxic tank ensures more consistent nitrogen removal than the previously mentioned design. The flow diagram of the final design consisting of equalization tanks, primary UF, and MBR, is shown in Figure 13.13. Metals are removed by both hydroxide precipitation and separation in the secondary UF, and are retained in the biomass in the MBR.

a. Performance of the MBR System

Approximately 10 to 15% of COD in MWF wastewater was found to be nonbiodegradable in aerobic systems.[28] The amount of nonbiodegradable COD was found to be higher in an anaerobic system. Even with this potentially nonbiodegradable COD, the MBR system showed its capability of achieving approximately 90% COD removal, which is consistent with the earlier findings reported on aerobic systems[29,30] (see Table 13.7).

TABLE 13.7
Membrane Bioreactor (MBR) Performance Data

Parameter	Historic Average	MBR Influent Design Average	MBR Actual Influent Standards	City Effluent Discharge	MBR Effluent
BOD_5	726	1500	1574	<15	<5
COD	1380	3000	3700	NR	500
TKN	73	220	189	<20	19

NR, not regulated.

Manufacturing operations generating large volumes of wastewater often require primary, secondary, and tertiary treatment steps to achieve regulatory compliance. A variety of smaller, pre-engineered and shop-fabricated treatment systems are available, and are often well suited for handling the wastewater generated from small manufacturing operations. These smaller systems all employ combinations of primary, secondary, and/or tertiary treatment operations. Primary free oil removal is accomplished via belt and media skimmers on the surface of sumps. Holding tanks are coupled with primary separator designs to increase oil removal efficiency. These separators are typically designed in stages. The first stage removes large oil droplets and settleable solids, while the second stage removes oil particles down to a few microns in diameter, using a mesh of plastic or metal fibers.

4. Metals Precipitation

If secondary treatment employs an ultrafilter and high soluble metals are present, then a metals removal step may be required. There are many methods of removing soluble metals from wastewater, and a common method is chemical precipitation.

By raising the pH with a hydroxide to a range between 8.5 and 9.5, the soluble metals are converted to insoluble metal hydroxides and co-precipitate with the ferrous salt.[31] If the metals are highly chelated, as a result of a mixture with a detergent, then a sulfide may be added to improve the effectiveness of metals precipitation.

5. Chemical Oxidation

A variety of chemical oxidants may be used separately, or in combination, to lower the BOD_5, COD, or other select organic or nitrogenous compounds of a waste stream after a secondary treatment process. Some chemicals and oxidants used for this purpose are: sodium hypochlorite, hydrogen peroxide, ozone, hydrogen peroxide plus ferrous sulfate (Fenton's Reagent), ultraviolet radiation plus hydrogen peroxide, and ultraviolet plus ozone.[32,33]

Extreme caution must be considered when using chemical oxidation methods. The reasons are: (1) handling of concentrated oxidizers is dangerous; (2) the oxidation process may off-gas certain chemicals which may be malodorous or toxic; and (3) the reaction may evolve heat or produce uncontrolled boiling or effervescing during the reaction process, as in the case with Fenton's Reagent. With that in mind, the use of chemical oxidation has seen limited use in the MWF industry. The BOD_5 and COD reduction using the above methods with sufficient reaction time can vary from 10% reduction to 70% reduction depending on the feed stream characteristics.

6. Electrochemical Oxidation

ECO is similar in concept to chemical oxidation, except that the oxidizing process is created *in situ* rather than by adding oxidizing chemicals into the solution. ECO is simple in concept. The process involves an anode and a cathode in very close proximity to each other (approximately 0.1 in.), and a direct current is applied to the anode/cathode pairs. An electrolyte, such as a conductive salt, is added to improve current transfer from the cathode to the anode. The direct current voltage is controlled from 4.5 to 6.0 V (see Figure 13.14).[34]

The oxidizing reaction occurs on the anode surface. The theoretical oxidizing reactions produced on the anode surface are thought to be primarily oxygen radicals such as the hydroxyl radical (OH^-) or singlet oxygen (O). These reactants are believed to exist only a few seconds, thus a continuous driving voltage and circulation of the wastewater across the anode/cathode pairs is required. Burke et al. reported an 83% reduction of COD over 420 min of a 100:1 dilute solution of three amine mixture. The three amines used in this experiment were monoethanolamine (CAS # 141-43-5), triethanolamine (CAS # 102-71-6), and monoisopropanolamine (CAS # 78-96-6).

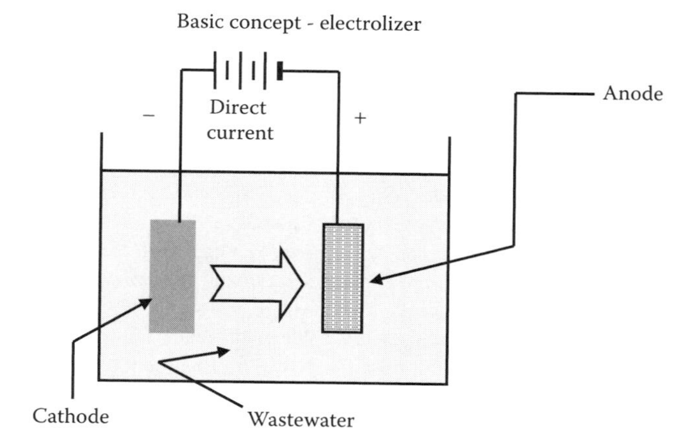

FIGURE 13.14 Basic schematic of electrochemical oxidizing cell.

IV. SUMMARY

There are many options for the waste treatment of MWFs. Free oil and basic oil emulsions are the simplest to treat and provide the best results. Complex emulsions, semisynthetic, and synthetic MWFs, in that order, are the most difficult to treat. Sometimes the receiving environment can restrict any disposal of MWFs, regardless of the treatment approach. Careful planning will ensure the best outcome without costly mistakes, associated penalties, embarrassment, and possible redesigns.

REFERENCES

1. United States Environmental Protection Agency, Codified Federal Regulations 40.
2. Lewis, R. J., *The Condensed Chemical Dictionary*, 13th ed., Wiley, 1997.
3. UAW–GM–Delphi, *Control Technologies Symposium for Metal Removal Fluids*, Summary of Symposium Data, 2001.
4. McClure, T. et al. Institute of Advanced Manufacturing Sciences and Waste Reduction and Technology Transfer Foundation, *Shop Guide to Reduce the Waste of Metalworking Fluids*.
5. United States Environmental Protection Agency, Codified Federal Regulations 40 CFR 433.
6. Burke, J., Waste treatment of metalworking fluids, a comparison of three common methods, *Lubr. Eng.*, 238–246, 1991, April.
7. Van Wylen, G. and Sonntag, R., *Fundamentals of Classical Thermodynamics*, 2nd ed., Wiley, New York, 1973.
8. Burke, J., *Treatment of Metalworking Fluid Wastewater by Mechanical Vapor Recompression*, Society of Manufacturing Engineers, Applications of Metalworking Fluids in Machining Operations, March 2001.
9. Burke, J., *Treating the Wastewater Blues*, Lubes-N-Greases, Falls Church, VA, 2003, October.
10. Sutton, P. and Mishra, P., *Metalworking Fluids*, Marcel Decker, New York, p. 377, 1994.
11. Wagner, J., *Membrane Filtration Handbook*, Osmonics Inc., Minnetonka, MN, 2000.
12. Sutton, P. and Mishra, P., *Metalworking Fluids*, Marcel Decker, New York, p. 377, 1994.
13. *Alkaline Cleaner Recycling Handbook*, Osmonics Inc., Minnetonka, MN, 2000.
14. *Ultrafiltration Handbook*, Romicon Inc., Woburn, MA, 1982.
15. Alther, G. and Ellis, W., Bentonite flocculants helps clean up wastewater, *Clean Tech 2001*, 2001, May.
16. Baum, S. and Prather, B., *Industrial Oily Waste Treatment*, American Petroleum Institute and American Society of Lubrication Engineers, pp. 47–65, 1968.

17. Kemmer, F., *The Nalco Water Handbook, Emulsion Breaking*, McGraw Hill, New York, pp. 11.1–11.18, 1988.
18. Burke, J., *Waste Treatment of Metalworking Fluids*, Society of Tribologists and Lubrication Engineers, Annual Metalworking Fluid Education Course, 2004.
19. Kimball, V., *Waste Oil Recovery and Disposal, Recovery Processes*, Noyes Data Corporation, Park Ridge, NJ, 1975.
20. Baum, S. and Prather, B., *Industrial Oily Waste Treatment*, American Petroleum Institute and American Society of Lubrication Engineers, pp. 97–101, 1968.
21. Kemmer, F., *The Nalco Water Handbook, Adsorption*, McGraw Hill, New York, pp. 17.7–17.12, 1988.
22. Tchobanoglous, G., *Wastewater Engineering; Treatment, Disposal, Reuse*, Metcalf and Eddy/McGraw Hill, Lakefield, MA/New York, 1979.
23. Sutton, P. and Mishra, P., *Metalworking Fluids*, Marcel Decker, New York, 1994.
24. Stroup, D., Sutton, P., Mishra, P., and Janson, A., Novel biotreatment of oily wastewaters: results from full scale operation, paper presented at the *Water Environment Federation Industrial Wastes Technical Conference*, Pittsburgh, PA, 1995.
25. Kim, B., Anderson, S., and Zemla, J., Aerobic treatment of metal-cutting fluid wastewater, *Water Environ. Res.*, 64, 258, 1992.
26. Anderson, J., Kim, B., Mueller, S., and Lofton, T., *Composition and Analysis of Organic Compounds in Metalworking and Hydraulic Fluids*, Ford Research Technical Report SR-2001-0093, 2001.
27. Taylor, G., Knee, N., Keep, M., and Freestone, V., Biodegradation of used machine tool cutting fluids, *Proceedings of Waste Management 99*, February 28–March 4, 1999.
28. Camp R., Garza, G., Salinas, R., Krezowski, L., and Kim, B., *Water Management at G.M. De Mexico Ramos Arizpe Complex*, SAE Technical Paper Series 920190, Soc. Of Automotive Engineers, 1992.
29. Kim, B., Anderson, S., and Zemla, J., Aerobic treatment of metal-cutting fluid wastewater, *Water Environ. Res.*, 64, 258, 1992, *ibid.*
30. Kim, B., Rai, D., Zemla, J., Lipari, F., and Harvath, P., Biological removal of organic nitrogen and fatty acids from metal-cutting-fluid wastewater, *Water Res.*, 28(6), 1453–1461, 1994.
31. Cushnie, G., Jr., *Pollution Prevention and Control Technology for Plating Operations*, National Center for Manufacturing Sciences and National Association for Metal Finishers, pp. 241–253, 1994.
32. Miller, R., and Anderson, J., *Methods of Reducing Chemical Oxygen Demand of Metalworking Fluids After Pre-Treatment by Membrane Separation*, Society of Manufacturing Engineers Technical Paper MR93-156, Metalworking Fluids Clinic, 1993.
33. Hoffmann, M., *Advanced Oxidation Technologies for the Reduction of Triethanolamine-Related Oxygen Demand in Wastewater*, Interim Report, W.M. Keck Laboratories, Caltech, November 1994.
34. Burke, J., Petlyuk, A., Warchol, J., Hardee, K., and Coin, R., *Reduction of Biochemical Oxygen Demand and Chemical Oxygen Demand of Metalworking Fluid Wastewater by Electrochemical Oxidation*, Society of Tribologists and Lubrication Engineers 59th annual meeting, Toronto, Canada, May 2004.

14 Contact Dermatitis and Metalworking Fluids

C.G. Toby Mathias

CONTENTS

I. INTRODUCTION

A. General Considerations

Occupational skin diseases account for a significant proportion of all occupational illnesses reported in the Bureau of Labor Statistics Annual Survey of Occupational Injuries and Illnesses. Machine tool industries are consistently found among Standard Industrial Classification categories with the highest incidence rates and numbers of cases of occupational skin diseases.[1] Contact dermatitis is an inflammatory condition induced by external contact of a substance or material with the skin surface. At least 90% of all occupationally acquired skin disorders are due to contact dermatitis[2]; most of these are irritant rather than allergic reactions. Metalworking fluids (MWF) are important causes of occupational contact dermatitis, especially in machine tool industries, where they may be routinely blamed for any dermatitis that arises in association with exposure, whether or not such blame is ultimately justified. In at least one large epidemiological study of 286 machinists exposed to MWF, the prevalence of contact dermatitis was 27%; only 2.8% of these were allergic reactions to ingredients of MWF, the rest were irritant reactions.[3] A special form of contact dermatitis, contact folliculitis, induces an inflammatory reaction localized to hair follicles. Both contact dermatitis and folliculitis are best understood within the context of the normal structure and function of skin.

B. Structure and Function of Skin

The skin has two principal layers, the epidermis and the dermis. Each layer contains various structural elements that are important not only with regard to specific functions but also with regard to disease processes which may affect them.

The epidermis is a relatively thin layer compared with the dermis, and ranges in total thickness from 100 to 200 μm. The outermost portion of the epidermis is the stratum corneum, a thin membrane 15 to 30 μm thick, and composed of packed, cornified, dead cellular tissue with tight intracellular spaces filled with complex lipids. The stratum corneum constitutes the principal physical barrier against penetration of chemical substances and microorganisms into the living layers of skin and, ultimately, the body in general. The stratum corneum is thickest on the palms and soles, where it may be 75 to 100 μm. The bulk of the epidermal layer is composed of squamous cells (keratinocytes), which actively synthesize keratinous filaments, keratohyaline granules, and membrane-coating granules filled with complex lipids, all of which are ultimately destined to become the principal structural elements of the outermost protective stratum corneum. The innermost portion of the epidermis is a single-cell layer of actively germinating cells, the basal cells. It takes approximately 2 weeks for newly generated squamous cells to mature and transform into the cornified cells of the stratum corneum, and an additional 2 weeks for the newly formed, cornified cells to desquamate into the environment.

Special pigment-producing cells, called melanocytes, are located along the basal cell layer of the epidermis. The pigment (melanin) protects the skin and body against the sun's harmful ultraviolet radiation and is responsible for tanning as well as racial differences in pigmentation. Melanin is packaged into granules within the melanocytes; under suitable ultraviolet light stimulation, these granules are transferred into surrounding keratinocyte cells. Melanocytes may be nonspecifically injured by severe inflammation from burns or chemical irritation, leaving residual postinflammatory changes of increased or decreased pigmentation. Melanocytes may also be selectively inhibited or destroyed by the toxic effects of some phenolic or catecholic chemicals, which resemble the amino acid tyrosine, a precursor of melanin synthesis, leading to depigmentation of the skin.

Another specialized cell, the Langerhans cell, is also found along the basal and suprabasal layer of the epidermis. These cells participate in cutaneous immune surveillance and are responsible for selective antigen uptake in allergic contact dermatitis reactions. Following uptake, the antigen is presented to T-lymphocytes circulating through skin and regional lymph nodes, where further initiation and amplification of the immune response occurs.

The bulk of perceptible skin thickness is a deeper layer of connective tissue, the dermis, composed of fibrous proteins (e.g., collagen, elastin, and reticulin) embedded in an amorphous ground substance. The tensile strength and elasticity of the dermis provide protection from mechanical injury, while the permeability characteristics of the ground substance allow diffusion of nutrients from blood vessels to other cellular elements of the dermis and epidermis. Blood is distributed within dermal tissue through a highly developed and interconnected network of both superficial and deep vessels. A large volume of circulating blood may be brought close to the skin surface when the body temperature is elevated; dissipation of heat from blood vessels at the surface is an important component of the body's thermoregulatory reflexes. Dilatation of these vessels accounts for cutaneous erythema (redness) caused by various physical or chemical irritations to skin. Densely intertwined nerve tissue fibers also transverse the superficial dermis and function as sensory receptors.

Specialized appendage structures originate in the dermis but have ducts which traverse the epidermis to the skin surface. Sweat glands respond to both thermal stimulation and emotional stress. Although the primary function of sweat is evaporative cooling of the body surface when it is stressed by heat, secondary buffering against alkaline substances may be provided by lactic acid, amphoteric amines, and weak bases contained in sweat. Hair follicles are found on all cutaneous surfaces except the palms, soles, and mucous membranes. Hair grows cyclically with alternating periods of growth and quiescence. Follicles grow and rest independently of one another; some shedding occurs daily but is barely noticeable in a normal healthy state. While hair has some protective and insulating function in other mammals, this function is relatively unimportant in man. Instead, hair functions principally as a secondary sensory organ through stimulation of nerve endings in the richly innervated hair bulbs. Sebaceous glands are intimately associated with hair follicles and share the same duct opening to the skin surface. The oily content (sebum) of the sebaceous glands functions as a sexual attractant in some mammals, but its function in man is unclear. The fatty acids derived from sebum have some bacteriostatic and fungistatic properties, and may have a limited function in this regard, once deposited on the skin surface.

II. PATHOGENESIS

A. Irritant Contact Dermatitis

The pathogenesis of contact dermatitis may involve either irritant or allergic mechanisms. Irritant contact dermatitis is an inflammatory reaction provoked by a direct, local toxic effect of MWF constituents or contaminants on epidermal cells; no immunological mechanisms are involved. Damaged epidermal keratinocytes release nonspecific chemical mediators causing dilation of dermal blood vessels (redness), leakage of fluids from blood vessels into the skin (swelling, small blisters, drainage of fluid from skin surface), and a variable cellular response (lymphocytes and polymorphonuclear cells). A strong irritant causes inflammation within minutes to hours following cutaneous exposure, and is sometimes called a chemical burn. MWF, however, are only weak potential irritants and require frequent or prolonged skin contact for days to weeks before visible inflammation occurs. The constituents of MWF responsible for irritant contact dermatitis have not been well defined. The potential to irritate skin is concentration dependent, and diluted MWF are inherently less irritating than concentrates. Contamination of MWF with lubricating oils, dirt, debris, and bacteria may increase irritant potential, presumably through alteration of pH and formation of more irritating breakdown products; some machinists anecdotally report that dirty or rancid MWF are more irritating than clean, fresh MWF. Prolonged entrapment of MWF against skin caused by protective clothing, such as occurs when clothing becomes accidentally saturated or gloves are donned before washing hands contaminated with fluid, will also enhance the irritant potential. Finally, if the outermost protective stratum corneum is injured or damaged by some other mechanism, e.g., microscopic abrasions from swarf (metal chips) or an unrelated skin disorder, the resistance of skin to penetration by MWF is lessened and the potential for irritation increased.

B. Allergic Contact Dermatitis

Allergic contact dermatitis is an inflammatory reaction within skin initiated by antigenic stimulation of a delayed cellular immune hypersensitivity response. Following skin exposure, potentially allergenic chemicals may complex in various ways with tissue protein; such complexes are called hapten–protein conjugates. If receptors on the surfaces of specialized epidermal cells (Langerhans cells) "recognize" and bind these complexes, the allergen (hapten–protein conjugate) is processed and carried to regional lymphatic glands, stimulating the release of sensitized lymphocytes (T-cells) into the circulation. These sensitized lymphocytes recognize the allergen wherever they encounter appropriately formed hapten–protein conjugates in skin and initiate a complex series of biochemical events. The resulting inflammatory reaction is often more intense than, but sometimes indistinguishable from, an inflammatory irritant reaction. Most potential allergens encountered in MWF are relatively weak and, unlike the allergenic resin in poison ivy or oak, usually require months or even years of exposure before sensitization actually develops.

Once sensitization has occurred, only minimal exposure may be necessary to provoke or sustain dermatitis. Biocides added to MWF are the most common causes of allergic contact dermatitis, although sensitization has occasionally been reported from antioxidants, fragrances, rust inhibitors (chromate), or metallic salts (nickel, chromate, cobalt) leached from machined metals.[4] Some biocides containing potential sensitizers are listed in Table 14.1. The risk of sensitization increases when undiluted biocides are handled or added to MWF at higher than recommended concentrations.

TABLE 14.1
Potentially Allergenic Biocides Used or Added to Metalworking Fluids and Their Patch Test Concentrations

Some Common Trade Names[a]	Chemical Names	Patch Test Concentrations[b]
Bioban CS-1135	4,4-Dimethyloxazolidine/ 3,4,4-trimethyloxazoline	1%
Bioban CS-1246	1-Aza-3,7-dioxa-5-ethylbicyclo(3,3,0) octane	1%
Bioban P-1487	4-(2-Nitrobutyl) morpholine/4,4,- (2-ethyl-2-nitrotrimethylene) dimorpholine	1%
Busan 85	Potassium dimethyldithiocarbamate	1%
Captax, Dermacid, Mertax, Thiotax	mercaptobenzothiazole	1%
Dowicide 1	*o*-Phenylphenol	1%
Dowicide A	Sodium *o*-phenylphenate	1%
Dowicil 75 or 200 Quaternium-15	1-(3-Chloroallyl)-3,5,7, triaza-1-azoniaadamantane chloride	2%
Formalin	Formaldehyde	1% aq
Bioban GK, Busan 1060, Grotan BK, Onyxide 200, Triadine 3	Hexahydro-1,3,5-tris-(2-hydroxyethyl)-*s*-triazine	1%
Grotan HD	*N*-Methylol chloroacetamide	0.1%
Hibitane	Chloroacetamide	0.2%
Kathon 886 MW	5-Chloro-2-methyl-4-isothiazolin-3-one/2-methyl-4 isothiazolin-3-one	0.01% aq
Omadine (zinc) Omadine (sodium)	Salts of 2-pyridinethiol-1 oxide	1%
Onyxide 500, Bronopol, Bioban BNPD	2-Bromo-2-nitropropane-1,3-diol	0.25%
Ottafact Preventol CMK	*p*-Chloro-*m*-cresol	1%
Ottasept Extra Nipacide CMX	*p*-Chloro-*m*-xylenol	1%
Proxel CRL	1,2 Benziosothiazolin-3-one	0.1%
Tris Nitro	2-(Hydroxymethyl)-2-nitro-1,3-propanediol	1%
Vancide TH	Hexahydro-1,3,5-triethyl-*s*-triazine	1%
Vancide 51	Sodium dimethyldithiocarbamate/sodium 2-mercaptobenzothiazole	1%

[a] List compiled from a number of worldwide medical and trade journals. May also be available under other trade names.
[b] Allergens prepared in white petrolatum except where otherwise specified; aq = aqueous.

III. CLINICAL FINDINGS

The cutaneous changes of irritant contact dermatitis caused by MWF are often mild, consisting only of dryness and chapping. With more severe irritation, cutaneous changes progress through various stages of clinical redness (erythema), scaling, and crusting; occasional small blisters (vesicles) may be present, particularly on the sides of the fingers. Clinical changes accompanying allergic contact dermatitis are often more severe and associated with more extensive erythema, swelling, and larger vesicles or blisters. Itching is usually the predominant symptom, although varying degrees of stinging or burning may also occur. While allergic contact dermatitis is usually more severe than irritant contact dermatitis, it is impossible to differentiate the two on the basis of appearance alone. A unique clinical form of irritant contact dermatitis, contact folliculitis, is characterized by erythematous papules or pustules surrounding hair follicles. Negative bacterial cultures distinguish this from folliculitis caused by infectious organisms.

Both irritant and allergic contact dermatitis occur on skin surfaces where maximal exposure takes place and, in the case of MWF, almost invariably involve the hands and/or forearms. Dermatitis may be accentuated in the web spaces between the fingers; involvement of the backs of the hands is common. When contact occurs almost exclusively from handling wetted parts, the palmar surfaces may be preferentially affected. Contact dermatitis from MWF does not usually affect skin surfaces covered by clothing unless the clothing becomes noticeably wetted and saturated for extended periods of time.

IV. DIAGNOSTIC TESTING

A. PATCH TESTS

The patch test is the procedure most frequently utilized to establish a diagnosis of contact dermatitis. This test is appropriately performed only when allergic sensitization to a constituent of MWF is suspected. Since the overwhelming majority of cases of contact dermatitis from MWF is due to contact irritation rather than allergy, a significant potential for misuse and misinterpretation of a patch test exists. As it is likely that many workers with contact dermatitis from MWF will undergo patch testing, health and safety personnel in machine tool industries must understand the performance, interpretation, and limitations of the patch test.

The test procedure is basically simple but has been carefully standardized; deviation from recommended guidelines may lead to false positive or false negative test results.[5] Since skin reactivity varies on different body surfaces, the upper and midback regions are recommended as the standard test sites; the upper outer arms are also acceptable. Suspected allergens, prepared at standardized concentrations (usually in white petrolatum or water) specifically for testing on the recommended sites, are placed on small inert disks and taped securely to the skin with adhesive tape. The most widely used test device is the Finn chamber on Scanpor tape, which consists of a series of aluminum disks premounted on special porous, hypoallergenic nonirritating tape. The tape is removed 2 days (48 h) after initial application. If a tested individual is allergic, an inflammatory reaction appears under the disk containing the substance to which the individual is sensitized. The intensity of any observed reaction is generally graded as follows: ?= doubtful reaction; 1+ = weak, nonvesicular, slightly infiltrated reaction; 2+ = strong, edematous, or vesicular reaction; 3+ = extreme, blistered, or ulcerative reaction. Since 30 to 40% of true allergic reactions may not be clearly positive until 3 or 4 days (72 to 96 h) after the initial test application, a second delayed reading, in addition to the initial 48 h reading, is necessary. While some authors maintain that a single 72 h reading is sufficient, this author's experience finds weak reactions extremely difficult to interpret with only a single 72 h reading and prefers two readings at 48 and 96 h. There is general agreement that patch test reaction intensities of 2+ or 3+ at any reading probably indicate true allergic sensitization (assuming that proper test concentrations and procedures have been utilized); a 1+ reaction is a common intensity of false positive irritant reactions, which may occasionally be observed, and does not always indicate true allergic sensitization. Most false positive irritant 1+ reactions are apparent at 48 h but decrease in intensity to a doubtful or negative reaction by 96 h. It is therefore usually safe to regard a 1+ reaction as indicative of allergic sensitization if the 1+ intensity persists for at least 96 h or appears at any delayed reading following an initially negative or doubtful reaction at 48 h. Performing two separate readings at 48 and 96 h after patch test application greatly facilitates the interpretation of these 1+ reactions.

For obvious reasons, a conclusion as to whether allergic sensitization has occurred to any constituent in MWF can be reached only for constituents which were actually patch tested; no conclusions can be reached about constituents which were not tested. Only 20 substances (Table 14.2) are currently recommended and widely available in the U.S. for routine patch testing. This screening series has been designed to detect common contact allergens within the general environment and is not specifically designed to detect allergies to common sensitizers in MWF. However, positive reactions from the routine screening antigens may sometimes indicate the presence of a sensitizer in MWF, either from occasional use of these substances in MWF or cross-reaction with chemically similar substances. Positive reactions to mercaptobenzothiazole and mercapto mix usually indicate hypersensitivity to rubber, since they are most frequently used as accelerators in rubber; but they may also indicate contact allergy to mercaptobenzothiazole used as a biocide and corrosion inhibitor in MWF. Carba mix contains several related carbamates, also used as antoxidants in rubber; positive reactions usually mean hypersensitivity to rubber, but may indicate contact allergy to related carbamates used as biocides in MWF. Positive reactions to formaldehyde may indicate hypersensitivity to any number of formaldehyde-releasing biocides used in MWF; however, the allergen may also be the entire molecular structure of formaldehyde-releasing biocide, and contact allergy will be missed in this latter case if the whole biocide is not tested. Quaternium-15, a formaldehyde-releasing preservative frequently used in cosmetics, is a case in point. Also called Dowicil 200 in industry, it is a frequent sensitizer in its own right and cross-reacts with formaldehyde in only 25% of cases. Ethylenediamine hydrochloride is a sensitizing preservative found in some prescription topical medications, but it is also present in at least one microbiocide (Proxel CRL). Chromate and nickel are common sensitizing metals in the general environment; positive reactions may indicate hypersensitivity to machined metal leached into recirculating MWF, and chromate may be directly added to some MWF as a corrosion inhibitor. Colophony and balsam of Peru are derived from pine tree resins; positive reactions

TABLE 14.2
Routine Screening Patch Test Series

Benzocaine 5%	Formaldehyde 1% (aq)[a]
Mercaptobenzothiazole 1%[a]	Ethylenediamine 1%[a]
Colophony (rosin) 20%[a]	Epoxy resin 1%
p-Phenylenediamine 1%	Quaternium-15 1%[a]
Imidazolidinyl urea 2% (aq)	*p-tert*-Butylphenol formaldehyde 1%
Cinnamic aldehyde 1%[a]	Mercapto mix 1%[a]
Lanolin alcohol 30%	Black rubber mix 0.6%
Carba mix 3%[a]	Potassium dichromate 0.25%[a]
Neomycin sulfate 20%	Balsam of Peru 25%[a]
Thiuram mix 1%	Nickel sulfate 2.5%[a]

Hermal Pharmaceuticals, Oak Hills, NY. All allergens prepared in white petrolatum except where otherwise specified; aq = aqueous.

[a] Positive reactions may indicate hypersensitivity to constituent of metalworking fluids. See Section IV, Diagnostic Testing.

usually indicate hypersensitivity to fragrance but could indicate contact allergy to substances added to MWF as deodorizers (e.g., pine oil). Cinnamic aldehyde, another common sensitizing fragrance, could also indicate hypersensitivity to a fragrance added to MWF.

Wherever possible, patch testing should be supplemented with common sensitizers often encountered in MWF (e.g., Table 14.1), prepared for testing at recommended concentrations. Supplemental patch test allergens specifically prepared to detect contact allergies caused by MWF may be purchased from companies manufacturing patch test antigens abroad, but they are not widely available in the U.S. due to restrictions on importation. Alternatively, raw materials (e.g., biocides) may be supplied or prepared for testing by industries which either manufacture MWF or use them; in such cases, these raw materials must be carefully prepared and thoroughly mixed at recommended patch test concentrations before testing. Where these options are not available, the actual MWF may be tested, but there is no unanimous consensus on appropriate patch test concentrations. Full-strength concentrates of MWF frequently produce false positive irritant reactions; if diluted to working concentrations (e.g., 1:30 to 1:40), the concentration of the allergen may be reduced below the threshold concentration needed to induce a positive reaction on the back, where the test is performed, thereby giving a false negative reaction. Under actual working conditions, however, working dilutions of MWF may still elicit allergic contact dermatitis reactions due to factors which enhance percutaneous absorption at sites of exposure. This author recommends a patch test concentration of 10% (prepared from clean undiluted MWF), with the caveat that even this concentration may occasionally produce a marginal false positive reaction. Used, contaminated MWF should not be patch tested, as they are more likely to cause false positive irritant reactions, and run the risk of secondary skin infection from microbial contamination at the patch test site.

B. Miscellaneous

When a true positive allergic patch test reaction is obtained and the allergen can be traced to a particular MWF, the intellectual craving for proof that dermatitis was caused by MWF is more easily satisfied. However, the patch test alone is not infallible proof. For example, a positive patch test to nickel may be obtained, and trace amounts of nickel may be found in MWF (presumably from leaching into the fluid from machined parts). However, what if the dermatitis had occurred only on the soles of the feet, not on the hands or arms? Would it automatically be logical to assume that contact allergy to nickel in MWF must be causing the dermatitis, considering that contact allergy to nickel is a common occurrence in the general population from sensitization to jewelry or metallic clothing fasteners, regardless of employment? Conversely, completely negative patch test results may be obtained. Is it justifiable to conclude that hand dermatitis was not caused by MWF, considering that most cases are due to contact irritation rather than allergy, and negative patch test results are to be expected? The conundrum can be resolved by defining the role of patch testing in proper perspective, as only one of several criteria which would be considered in establishing causation. The following seven criteria have been proposed[6]:

1. Dermatitis should be clinically eczematous.
2. Noticeable skin exposure to substances capable of causing contact dermatitis should have occurred (e.g., MWF repeatedly contacting skin).
3. Dermatitis should occur on skin surfaces when exposure is principally occurring (e.g., hands and arms in the case of MWF).
4. Temporal relationship between onset of exposure and onset of dermatitis should be consistent with contact dermatitis (i.e., usually within the first three to 6 months of initial exposure or changes in work duties which increase the amount of exposure).
5. Dermatitis improves when exposure ceases (e.g., vacations) and worsens upon reexposure (e.g., return to work).

6. No other diagnoses or causes are likely.
7. Patch tests indicate a specific allergen encountered in the work environment when contact allergy is suspected (e.g., specific biocide in MWF).

All of the proposed criteria have exceptions and only reasonable medical probability, not 100% absolute certainty, can usually be ascertained. It has been suggested that at least four of the seven criteria be satisfied before concluding that a particular worker has contact dermatitis caused by MWF.

V. PROGNOSIS

Several independent follow-up studies of patients affected by occupational contact dermatitis have consistently demonstrated a surprisingly poor prognosis; approximately 25% clear completely, another 50% improve but have periodic exacerbations, while 25% persist unchanged or worse, despite protective measures or job modifications.[7] Contact dermatitis from MWF is no exception. In one large follow-up study of 121 machinists diagnosed with contact dermatitis from MWF, 78% of those who continued to work still had persistent dermatitis after 2 years, while 70% of those who had stopped working (no further exposure to MWF) also had persistent dermatitis.[8] The reasons for this relatively poor prognosis are not well understood. It is often presumed that continued exposure to MWF, type of contact dermatitis (irritant vs. allergic), and underlying endogenous factors have some influence on prognosis. However, no significant differences in prognosis were observed among machinists with contact dermatitis who either continued or stopped exposure to MWF after onset of dermatitis, even after controlling for type of contact dermatitis (irritant vs. allergic) and endogenous factors.

VI. TREATMENT

The primary objective of medical treatment of contact dermatitis is simple and obvious: restore the skin to its previously normal appearance and function. Over-the-counter moisturizers (e.g., hand lotions or creams) may suffice for mild dryness or chapping, but should not be used on visibly inflamed skin; such preparations may actually worsen contact dermatitis from MWF.[9] Once inflammation has occurred, the mainstays of treatment are topical corticosteroid medications which are simply rubbed onto the affected skin at an average frequency of four times per day. An abundant and sometimes bewildering number of preparations are available by prescription and differ principally by their respective potencies. In general, the higher the potency, the greater is the cost. Hydrocortisone, a low-potency topical steroid, is now available over the counter in 1% cream or ointment formulations. It is occasionally effective when contact dermatitis is mild and limited; it may be used by occupational health personnel as initial treatment for affected workers, but is not likely to be effective if contact dermatitis is moderate, severe, or extensive. Triamcinolone acetonide, 0.1% cream or ointment, is an intermediate-strength steroid preparation available by prescription in both generic form and bulk quantity, making it an extremely cost-effective treatment for moderate contact dermatitis when large surface areas (e.g., arms) are involved and frequent applications required. Severe contact dermatitis usually requires high-potency prescription topical steroids; a few generic preparations are available to reduce cost. If palmar surfaces are involved, treatment may occasionally require one of the newer ultrapotent (and ultraexpensive) topical steroids, but they are seldom necessary elsewhere; unfortunately, no generic equivalents are available. Treatment of extensive, severe, or disabling contact dermatitis sometimes requires internal corticosteroid administration by mouth or injection.

Secondary treatment objectives include keeping an affected worker on the job, maintaining safe working conditions, and preventing recurrences. In this author's experience, most cases of contact dermatitis from MWF can be effectively treated without taking the worker off the job. The approach usually requires that the prescribed topical corticosteroid be applied to affected surfaces frequently during the workshift, sometimes as frequently as every 1/2 to 1 h, since working conditions tend to remove applied medications from the skin. Dirty or contaminated skin should be gently cleaned and rinsed before applying the topical steroid. Small containers of water and dry rags may be kept at the workstation for this purpose. If palmar surfaces are affected by vesicular (blistering) dermatitis, this approach is less likely to be effective and the affected worker will need to be removed from duties which necessitate exposure to MWF, at least temporarily until a response to treatment is obtained. If allergic sensitization has actually occurred, it is extremely unlikely that dermatitis will ever be satisfactorily controlled unless all exposure is eliminated; this usually means changing to a new MWF without the offending allergen or removing the affected worker permanently from job duties where exposure will occur.

Treatment must also take into consideration the maintenance of safe working conditions. Although protective gloves are often used in other industries as a means of reducing or controlling skin exposure, safety considerations usually prohibit this approach in most machine tool industries, where a glove accidentally caught by a rapidly turning machine part may pull a finger or hand into the machine, causing severe crush injury or amputation. Ointment (petrolatum)-based topical steroids are generally too greasy for frequent application to the hands and may endanger the worker if a firm grip on a handled part cannot be maintained; cream-based topical preparations are preferable during the workshift.

Workers with contact dermatitis from MWF who remain at their usual jobs are prone to recurrences when treatment is stopped, since exposure to MWF continues. Barrier creams specifically marketed for wet-work have not lived up to their expectations and may actually aggravate contact dermatitis if applied to affected skin before completed healing occurs.[10] White or yellow petrolatum (petroleum jelly, Vaseline) is water impermeable and is theoretically the most effective barrier that can be applied; unfortunately it is greasy and may present a safety concern when applied to the hands, as already mentioned. For these reasons, this author prefers using an equal part mixture of 1% hydrocortisone cream and 1% hydrocortisone ointment, applied to the hands and arms during the workshift as a barrier cream. The consistency of this mixture, halfway between a cream (too water washable) and petrolatum (too greasy) is an excellent compromise; it provides a temporary barrier, does not have to be applied as frequently as straight creams, and contains hydrocortisone which may inhibit any tendency for contact dermatitis to recur.

VII. PREVENTION

Preventive efforts should be directed primarily at those workers exposed to MWF who have never developed contact dermatitis. Workers with contact dermatitis are best managed by medical treatment as outlined above; in some cases, some of the preventive measures discussed below may actually aggravate preexisting contact dermatitis rather than improve it.

Strategies for preventing contact dermatitis from MWF may be based upon an eight-step program.[11] Points of emphasis within any preventive program will vary, depending on specific circumstances and conditions within the workplace. In all cases, prevention begins with recognition and acknowledgment that frequent or prolonged exposure to some MWF may cause irritant contact dermatitis irrespective of manufacturers' claims or reassurances. Some of these reasons have already been discussed. Safety personnel need to identify the ingredients used in their MWF and recognize the additional hazard of allergenicity from some of them, particularly biocides and corrosion inhibitors (see Table 14.1). Allergenicity is a greater concern where concentrated biocides are added to MWF in the workplace to prolong their working lives; such practices may cause exposure to high concentrations of potential allergens, either directly when the concentrated biocides are handled or indirectly when recommended concentrations are accidentally exceeded in the recirculated MWF following measurement error.

After recognizing that some MWF may cause contact dermatitis, additional preventive measures may be implemented. Engineering controls focused on process containment have been very successful, but may be limited by cost and feasibility. Splashguards were one of the first successful preventive measures and are still useful today. The introduction of robotic equipment, which coincidentally decreases worker exposure to MWF while increasing the quality and output of manufactured products, is also likely to decrease the incidence of contact dermatitis. Careful monitoring of recirculating MWF is another important aspect; this should include sump filtering to remove contaminating oils and debris, filter changes at appropriate intervals, maintenance of concentration and pH within manufacturers' guidelines, careful addition of biocides at appropriate intervals to prevent rancidity, regulation of effective biocide concentrations, and a prompt replacement with fresh MWF when irreversible deterioration has occurred. Where feasibility permits, selection of MWF containing biocides without demonstrated allergenicity (Table 14.1) will lessen the risk of allergic contact dermatitis.

Personal protective measures have only a limited role in prevention of contact dermatitis from MWF. Gloves are usually impractical and may become a safety hazard if caught in the moving parts of machinery; when handling concentrated biocides, however, glove wearing should be mandatory. Protective aprons may be helpful around equipment where splashing of MWF onto clothing is likely to occur. Barrier creams have no proven benefit in preventing contact dermatitis from MWF; experimental data suggest that they may actually exacerbate the irritant effects of MWF under some circumstances.[6]

As some degree of skin contact with MWF is inevitable in most machine tool operations, personal and environmental hygiene efforts have added importance. Mild soap and water is sufficient to remove water-soluble MWF from skin and should be used several times during the workshift; most oil-soluble MWF can usually be removed in a similar fashion. Organic solvents should never be used for this purpose. Where oil-soluble MWF (or associated grease stains) cannot be easily removed with mild soap, waterless hand cleaners (which contain an organic solvent dispersed in a cream formulation) are often satisfactory and are substantially less irritating than straight organic solvents. Tenacious stains on the palms may require abrasive soaps (e.g., pumice); these products work by stripping the outermost stained stratum corneum away but may contribute to irritation from MWF if used on the thinner, more sensitive skin of the forearms and dorsal hands. Waterless cleaners and abrasive soaps should never be used to clean skin already affected by contact dermatitis, as they will probably aggravate it. Regardless of the method by which the skin is cleaned, the single most important preventive measure (in this author's experience) is application of a moisturizer immediately after washing. Petroleum jelly is very effective but greasy and may make it difficult to grip objects firmly if not wiped completely off the palms; some workers find both the greasy feeling and slipperiness of palms objectionable and will not use it. Moisturizing lotions are least objectionable but may need to be applied more frequently than is often feasible to keep the skin from drying out under the usual rigors of machine tool operations. For these reasons, this author prefers heavy moisturizing creams, which need to be applied less frequently but vanish quickly enough that they seldom interfere with grip. Hydrophilic ointment is a generic moisturizing cream, available through retail pharmacy outlets, which I have found quite satisfactory for this purpose. In addition to personal hygiene, environmental hygiene provides some additional benefit where significant skin exposure may occur from contaminated work surfaces. Wetted parts should be wiped clean whenever possible prior to handling. Other work areas should be kept clean so that hands and arms will not be resting on surfaces coated with MWF residues.

Education efforts, aimed at promoting safe work practices and awareness of MWF as potential causes of contact dermatitis form another important element of a comprehensive prevention program. Such educational efforts need to be directed at management, safety personnel, and machine tool workers, covering all relevant aspects of contact dermatitis prevention. Simple manuals or booklets (pocket size) should be developed for easy reference. From a practical point of view, most educational training received outside the employees' work areas (e.g., classroom instructional pamphlets) will not be successful unless reinforced by knowledgeable safety personnel who visit employees' workstations. This obviously requires a serious commitment on the part of management.

There are no current federal or state regulatory requirements governing skin exposure to MWF. Any such regulations would have to be based on well-characterized close-response relationships between exposure times and cutaneous irritation, of which there are none. Given the dynamic nature of the machine tool industry and the wide range of variables which may interact to produce contact dermatitis, it is unlikely that any generic guidelines or regulations will be forthcoming. Machine tool manufacturing plants and manufacturers of MWF may voluntarily post signs in work areas where MWF are used, indicating that prolonged or frequent skin exposure to MWF may cause dermatitis; this warning is usually indicated on a material safety data sheet (MSDS). A reminder to follow the company's recommended dermatitis prevention program may accompany such warnings. Manufacturers of MWF may consider voluntarily listing the biocide on the MSDS even if it is present at a level less than 1% (current mandated level for noncarcinogens), since biocides are the most important causes of allergic contact dermatitis from MWF and may induce sensitization at concentrations less than 1%.

Motivation is an often neglected element of prevention programs. Ethical considerations aside, management may be motivated by the simple knowledge that happy and healthy workers are more productive workers, and in the long run this increases profits and reduces workers' compensation costs. Motivating workers to adopt safe work habits is a complex issue at best. Incentive programs which reward departments for no injuries or lost work time over a defined period have been criticized on the grounds that they may intimidate workers from reporting real injuries or illnesses for fear of repercussions. It is doubtful whether caps, T-shirts, or other paraphernalia, often used to promote overall company morale, can be a serious motivational tool for preventing occupational illness such as contact dermatitis. In this author's opinion, any successful motivational approach must be based on a demonstrated sincerity for the workers' well-being and personalized on an individual level. Safety personnel must know their workers as individuals and what motivates other aspects of their lives; employees must feel welcome to express their individual health concerns with safety personnel.

Finally, the role of preemployment screening to prevent contact dermatitis from MWF must be considered. Legal requirements generally preclude discrimination in hiring on the basis of preexisting disease, unless it is likely that a given job will aggravate the preexisting disease. In fact, it is quite likely that any new worker with preexisting hand or arm dermatitis will have this condition aggravated by skin exposure to MWF. Therefore, preemployment physical examinations for positions in machine tool manufacturing plants should, at a minimum, include questions about prior hand or arm dermatitis and a complete skin examination for evidence of active skin disease. In addition, individuals with personal or family histories of atopic allergies (hay fever and related seasonal allergies, asthma, childhood eczema) have increased predispositions to developing irritant contact dermatitis, particularly in wet-work occupations, compared with nonatopic individuals.[12] While the actual probabilities that contact dermatitis will occur are unknown, newly hired workers should be screened for these traits, carefully trained in dermatitis prevention techniques as outlined above, and supervised closely for the development of irritant contact dermatitis within the first several months of employment.

Prevention of contact dermatitis may seem like an idealistic goal in the machine tool industry, where skin exposure to MWF, cleaners, and solvents are still inevitable with current technology, but it is a goal worth pursuing. The elements of a successful program, as outlined above, will require a coordinated effort at all levels of management and employment.

REFERENCES

1. Mathias, C. G. T. and Morrison, J. H., Occupational skin diseases, United States, *Arch. Dermatol.*, 124, 1519, 1988.
2. Occupational Skin Disease in California (with Special Reference to 1977), California Department of Industrial Relations, Division of Labor Statistics, San Francisco, CA, 1982.
3. deBoer, E. M., Van Ketel, W. G., and Bruynzeel, D. P., Dermatoses in metal workers (I). Irritant contact dermatitis, *Contact Dermatitis*, 20, 212, 1989.
4. Fisher, A. A., *Contact Dermatitis*, Lea and Febiger, Philadelphia, PA, p. 531, 1990.
5. Fisher, A. A., *Contact Dermatitis*, Lea and Febiger, Philadelphia, PA, p. 9, 1990.
6. Mathias, C. G. T., Contact dermatitis and workers' compensation: criteria for establishing occupational causation and aggravation, dermatitis, *J. Am. Acad. Dermatol.*, 20, 842, 1989.
7. Nethercott, J. R. and Gallant, C., Disability due to occupational contact dermatitis, *Occup. Med. State Art Rev.*, 1, 200, 1986.
8. Pryce, D. W., Irvine, D., English, J. S. C., and Rycroft, R. J. G., Soluble oil dermatitis: a follow-up study, *Contact Dermatitis*, 21, 28, 1989.
9. Goh, C. L., Cutting oil dermatitis on guinea pig skin (II). Emollient creams and cutting oil dermatitis, *Contact Dermatitis*, 24, 81, 1991.
10. Goh, C. L., Cutting oil dermatitis on guinea pig skin (I). Cutting oil dermatitis and barrier creams, *Contact Dermatitis*, 24, 16, 1991.
11. Mathias, C. G. T., Prevention of occupational contact dermatitis, *J. Am. Acad. Dermatol.*, 23, 742, 1990.
12. Shmunes, E., The role of atopy in occupational skin disease, *Occup. Med. State Art Rev.*, 1, 219, 1986.

15 Health and Safety Aspects in the Use of Metalworking Fluids

John K. Howell, William E. Lucke, and Eugene M. White

CONTENTS

I. INTRODUCTION

Workers in machining environments are exposed to numerous substances and conditions that may affect their health and safety. These exposures are related to chemical, physical, biological, and ergonomic hazards. Occupational safety and health specialists, e.g., industrial hygienists, are trained to prevent, recognize, evaluate, and control inimical exposures in the workplace. Traditionally, a hierarchy of controls is used to protect workers.

A. Engineering Controls

- Process and facility design
- Machine enclosures
- Machine guarding, shields, and barriers

- Ventilation
- Mist collectors
- Material substitution

B. Administrative Controls

- Employee training
  1. Work practices
  2. Hazard recognition
- Task scheduling

C. Personal Protective Equipment (PPE)

- Respirators
- Eye protection
- Hearing protection
- Protective clothing
- Gloves

Regardless of the extent and effectiveness of exposure controls, some workers experience health and safety problems as a result of exposures to both known and unknown agents and conditions in the workplace. Depending on the type and extent of exposures to chemicals, susceptible individuals may experience health problems related to the toxicological, mutagenic, and carcinogenic properties of some of those chemicals or contaminants in those chemicals.

II. CHEMICAL EXPOSURES

As shown in Table 15.1, industrial chemicals can enter the body by various routes, i.e., skin absorption, inhalation, and ingestion.

Exposures to some chemicals in the workplace may be problematic because many molecules exert their damaging effects at the cellular level. Occupational diseases and functional disorders caused by chemicals commonly exhibit classical dose–response or exposure–response relationships. Adverse physiological effects initiated by chemical exposures may be acute (short length of time after exposure) or chronic (prolonged), depending on a variety of factors:

- Route of exposure
- Duration of exposure

TABLE 15.1
Routes of Chemical Exposures

Exposure Route	Facilitators	Potential Health Effects	Comments
Skin absorption	Dermal permeability, damaged skin, prolonged exposure	Contact dermatitis allergic dermatitis	Skin sensitization to certain chemicals occurs in some individuals
Inhalation	Respirable and inhalable aerosols (1–5 μm), vapors, particulates	Upper respiratory tract irritation, pulmonary diseases	Most common exposure route
Ingestion	Accidental swallowing	Mouth, pharyngeal and esophageal damage, digestive tract disorders	Minor exposure

- Type of chemical or contaminant
- Concentration of chemical or contaminant
- Susceptibility of worker to chemical or contaminant
- Interactions of chemicals and contaminants in a mixed exposure

Owing to the multiplicity of mixed chemical exposures encountered in any given workplace environment, positive identification of the etiologic agent of an occupational disease and its source(s) may not be easily discerned. In many cases, after intensive industrial hygiene and epidemiologic studies to find the causes and origins of workplace illnesses, the results are indeterminate. As we will see, overall improvement in local and area ventilation (increased air turnover rates) can often help to reduce or eliminate the effects of problematic exposures.

III. MACHINING ENVIRONMENT AND HEALTH

Throughout this chapter, we refer to the fluids used for cutting and grinding, machining and honing, etc. as metalworking fluids (MWF). In fact, these fluids to which we most often refer are more properly known as metal removal fluids. They are a subset of the broader group of industrial products known as MWF, which include industrial cleaners, metal forming and stamping fluids, corrosion preventives, and heat-treating fluids, as well as metal removal fluids.

Facilities that utilize MWF range from small shops with a few workers to sprawling industrial complexes that cover many acres with hundreds of employees during any given work shift. Regardless of the size of the facility, normal occupational activities produce a variety of exposure hazards, which mainly consist of airborne emissions from industrial processes, aerosols, diesel exhausts, dusts, vapors, and particulates.

MWF fulfill a number of machining functions, but the major benefits are: (1) to provide lubricity at tool-workpiece interfaces, (2) to dissipate heat generated during machining activities, and (3) to flush away metal swarf and fines generated during the fabrication of metal parts. MWF formulations are complex mixtures of chemical constituents blended to suit the requirements of a specific industrial machining process. The National Institute for Occupational Safety and Health (NIOSH) has estimated[1] that over 1.2 million workers were exposed to MWF (water-soluble products) and more than 6 million workers were exposed to mineral oils (nonaqueous industrial lubricants).

As MWF circulate through individual sumps and large distribution systems, they may undergo chemical, physical, and biological changes over time that affect their compositions:

- Constituents and additives deplete
- Metal fines and swarf accumulate
- Dirt and shop debris accumulate
- Tramp oils, hydraulic fluids, way oils, and other exogenous lubricants accumulate
- Mists increase
- Microbial growth occurs
- Organic acids accumulate
- pH fluctuations occur
- Alkalinity changes occur
- Hard water salts increase
- Emulsions break
- Foam increases

Changes in the composition of used MWF can affect the health of workers in various ways, as shown in Table 15.2.

This chapter discusses safety and health aspects of worker exposures to MWF. These complex industrial lubricants have undergone numerous changes and modifications in formulations over a

TABLE 15.2
Potential Health Effects Caused by Physical, Chemical, and/or Biological Changes in the MWF

Physical	Chemical	Biological
Dirt build-up (respiratory irritation ↑)	Depletion of effective biocide concentration (microbial growth ↑)	Uncontrolled bacterial growth (respiratory irritation ↑, aesthetic acceptability ↓)
Increase in tramp oil content (unemulsified) (mists ↑, respiratory exposures ↑, dermatitis ↑)	Excessive biocide addition (sensory and respiratory irritation ↑, skin irritation ↑)	Uncontrolled fungal growth (respiratory irritation ↑, esthetic acceptability ↓)
Metal fines build-up (skin abrasions ↑)	Nitrosamine production (carcinogenesis potential ↑)	Endotoxin build-up (respiratory irritation ↑, pulmonary function ↓)
High alkalinity (skin and respiratory irritation ↑)	Dissolved metal content (skin irritation ↑)	System stagnation/rancidity (respiratory illness ↑)
Unstable emulsions (skin irritation ↑)	Organic acid production (skin irritation ↑)	Presence of human pathogens (rare, poor survival rates, infrequent effects on health)

span of many decades in response to knowledge gained about the potential health effects of various constituents. The development of modern-day MWF involves intensive research and testing. Products must meet or exceed stringent end-user technical specifications for machining performance. Concurrent with R&D activities, fluid formulations must comply with internal and regulatory health and safety requirements that are optimally protective for workers.

As discussed below, poor control of coolant microbiology has emerged as a key consideration in MWF health and safety issues. See Chapter 9 for a detailed discussion of the microbiology of MWF.

Exposure to ingredients found in MWF can be high. In a 1987 toxicology and health data call-in on antimicrobial agents commonly used in these fluids, the Environmental Protection Agency (EPA) found MWF applications to be in the high exposure category. In a typical large automotive metal processing operation, it is estimated that approximately 360,000 gal of straight oil and 621,000 gal of water-miscible MWF concentrate are used per year. Owing to their high volume of use in applications, which can give rise to significant occupational exposures, a thorough understanding of the potential health effects of these materials is critical.

As has been discussed in previous chapters, MWF are complex mixtures. Even those classified as straight oils typically contain additives, such as sulfonated or chlorinated compounds. Water-based (soluble oils, semisynthetic, synthetic) fluids, although diluted in use (typically 1:20 in water, but possibly as low as 1:10 or as high as 1:200 in water), may have a multitude of chemical constituents — mineral oils, sulfonates, amines (typically alkanolamines), fatty acids (which can react with amines or alkali metal hydroxides to form soaps), borates, polyalkylene glycols, esters, dyes, biocides — all of which may contribute to the toxicological profile of the fluid. As described later, fluids in use become contaminated with, for example, hydraulic oils, way oils, metal fines, microbiological growth and microbiological decay products, add to the complex profile.

Information on the potential health effects of chemicals is gleaned from two sources: toxicology or animal test data and epidemiology or studies on human populations. Toxicology studies are typically conducted using a single chemical or mixture with controlled exposure levels, test animals, and exposure environment. Some manufacturers of MWF test their products as a whole, but the utility of these tests is limited because of changes in composition during use. Toxicology data exist on many of the individual components. Results of these animal studies can be extrapolated to effects that might be observed in humans.

Epidemiology studies eliminate the need for species-to-species extrapolation, but they introduce many other variables. Unlike the controlled conditions employed in animal experimentation, real-world exposures to chemicals are often unknown, even in industrial settings. Exposures can occur via multiple routes and be of inconsistent durations and frequencies that may fluctuate over time. A number of epidemiology studies have been conducted on populations working with MWF. These data together provide a comprehensive picture of the potential health hazards of exposure to MWF and serve as a guide for the control of these exposures.

IV. CONCEPTS IN TOXICOLOGY AND TEST METHODS

Descriptive animal toxicity testing under well-defined conditions of exposure forms the foundation of chemical hazard evaluation. These studies are the starting point for evaluating the health and safety ramifications of chemical exposure. Most descriptive toxicity testing follows guidelines prescribed by regulation and is designed to predict possible effects in humans. In general, animal models have been found to be very good predictors of possible human responses.

The single most important factor determining the toxicity of a chemical is dose. As first expressed by Paracelsus in the 1500s, all chemical substances are potentially harmful to living organisms under some conditions. The level of exposure or amount of a chemical to which an individual is exposed will decide the effect.

The total dose received by an individual is a function of the exposure concentration, duration, and, for intermittent exposures, frequency. Animal toxicity testing is classified as acute, subchronic, and chronic based upon duration of exposure. Toxicity is generally a function of the total dose. At lower concentrations of exposure, toxicity is more likely to be manifested by exposures of longer duration or greater frequency.

Acute effects, produced by short-term exposure to high doses of a chemical, often produce different effects to long-term, low-level exposure to the same chemical. Ethanol is a familiar example of a chemical that produces primary acute effects upon the central nervous system or long-term effects in the liver. Toxicity, or the severity of response, tends to decrease as the dose administered is dispersed over time. The effect produced by consumption of six 1-oz servings of alcohol in 6 h is less than that produced by six 1-oz servings in 1 h. If the dose is spread over a great enough length of time, there may be no observable effects (i.e., one tenth of an ounce of alcohol per day for 60 days).

This decrease in severity of effect as the dose is fractionated over time occurs in large part because of detoxification and excretion of the chemical between successive doses. Diminished effects may also be observed when the cumulative effect or injury produced by each administered dose is partially or fully reversed before the subsequent dose is administered.

One of the most frequently monitored toxicological end points is acute lethality. Lethal dose studies act as a starting point for the design of subsequent long-term exposure investigations. The acute or single dose of a substance producing death in 50% of the test animals is termed the "lethal dose 50," denoted LD_{50}. More relevant acute toxicity end points for evaluating potential MWF hazards measure skin, eye and respiratory irritation, and sensitization effects. Results from these studies are expressed with numerical scores indicating the degree of severity of irritation, tissue damage, or response.

Subchronic experiments generally expose animals for 90 days and attempt to mimic potential human exposures. For example, if inhalation is expected to be the primary route of exposure to humans, then the animals will be exposed to the test compound for 6 to 8 h per day, 5 d per week over the 90-d period. Subchronic experiments can be quite informative without also incurring exorbitant costs in conducting the experiments. Included here would be specialty studies to determine target organ toxicity, such as bioassays for evaluating neurotoxicity, reproductive effects, or teratogenicity (effects on the fetus).

Chronic assays expose animals to the material over their lifetime (approximately two years for rats and mice). The end point of primary interest in these studies is carcinogenicity or the potential of exposure to a chemical to cause cancer or tumor formation. Chronic studies are very unusual for MWF as a whole, and even fairly unusual for the components, except for some of the biocides, ethanolamines, or oils.

Lastly, as a substitute for chronic studies, short-term bioassays are used as indicators of mutagenicity and potential carcinogenicity. These assays use bacterial or mammalian cell culture systems and can be useful when conducted in a battery of tests providing complementary information. They can be run quickly with minimal cost. ASTM E1302, *Guide for Acute Animal Toxicity Testing of Water-Miscible Metalworking Fluids*, reviews acute animal toxicity tests applicable to water-miscible MWF, as manufactured.[2]

V. HEALTH EFFECTS ASSOCIATED WITH EXPOSURE TO METALWORKING FLUIDS

A. Acute Effects

Acute effects associated with exposure to in-use MWF include both contact and allergic dermatitis and acute respiratory effects.

Skin contact is prevalent in MWF operations and often difficult to control. Primary irritant contact dermatitis and allergic contact dermatitis are more common today than oil-induced dermatitis, known as folliculitis. Causative factors are influenced by the ingredients and nature of the MWF, concentration and pH, duration of exposure, and other factors such as age, contamination, skin type, previous exposure, presence of other skin disease, and personal hygiene. Most often, more than one causative factor is in play with cases of contact dermatitis. Additives such as amines, petroleum sulfonate, and some of the biocides have also been associated with contact dermatitis. Skin sensitization, which is an allergic response to a chemical or a component in the material, has also been reported. Some biocides and corrosion inhibitors, including isothiazalones, formaldehyde, and mercaptobenzothiazoles, have been reported to have sensitization potential. Metal allergy dermatitis may also occur due to the dissolution of small amounts of metallic ions from some alloys being worked in the system. Nickel and chromium, found in stainless steels, and cobalt, used as a binder in tungsten carbide tooling, are three of the most common metal skin sensitizers.[3] The previous chapter addresses in detail the effects of MWF on the skin.

Many components in MWF have been tested in animals for acute, oral, and dermal lethality (LD_{50}), as well as for skin and eye irritation. For example, mineral oils are classified as relatively nontoxic, the LD_{50} being greater than 10 g/kg on oral exposure in rats and greater than 3 g/kg following dermal application in rabbits. They are also classified as mild to moderate skin and eye irritants. Most other components in MWF are classified as moderately toxic to nontoxic when evaluated in short-term, acute, LD_{50} studies.

Besides potential skin and eye contact, inhalation exposure is also a common route of occupational exposure. Workplace aerosols — mist — in an industrial metalworking environment are produced as machining fluids continuously flooding the cutting and grinding tools and the part being produced. Mist consists of suspended liquid droplets formed by breaking up a liquid into a dispersed state, such as by splashing or foaming, or by condensation from the gaseous state.

In the metalworking environment, it is generally recognized that there are three mechanisms of mist formation that may operate simultaneously:

- Evaporation at elevated temperatures and condensation
- Mechanical motion
- Bubbling of the MWF

The component distribution in the shop air is dependent on which mist formation mechanisms come into play during the machining or grinding process.

Heat is generated during the cutting or grinding process. As fluid enters the hot cutting zone, volatile components vaporize and then recondense to form small, less than 1 μm, particles. The chemical composition of recondensed vapors is a reflection of the more volatile components of the formulation. Importantly, these droplets are of such small diameter that they tend to migrate through a large area of the shop.

Mechanical motion, that occurs when the fluid strikes the rotating part or tool, generates larger particles, generally in the 1 to 10 μm range. These droplets will have the same composition as the bulk fluid. The larger particles will not migrate far from the point of generation.

Aeration of the MWF, such as when a fluid foams or is excessively agitated, creates other mist particles. In this case, when the fluid comes to rest, entrained air can escape causing mist to form as bubbles break. White and Lucke[4] have shown that some components, such as volatile short-chain organic acids, can selectively concentrate in this fraction.

Turchin and Byers[5] have shown that tramp oil, a frequent contaminant of an in-use MWF, increases the amount of mist.

Aside from mist created in the machining or grinding operation itself, mist can be generated in the shop from fluid circulating in open troughs in central system returns, from compressed air blow-off of parts before inspection or packaging, or from parts washers. Compositions of products used for parts washing may be similar to MWF used in the shop or they may be more closely related to more traditional alkaline cleaners, and comprised of nonionic and anionic surfactants and alkaline detergent builders, such as sodium or potassium polyphosphates and carbonates, or sodium or potassium hydroxide.

Finally, MWF are usually not the only products in the shop atmosphere. Depending on the facility, the shop atmosphere may also contain welding fumes, tow motor exhaust, and other metallic and nonmetallic particles.

Beginning in the 1980s, a large body of research into acute respiratory effects resulting from exposure to MWF in both laboratory animals and in human cohorts has emerged.

In 1989, Kennedy et al.[6] reported changes in pulmonary functionality in 89 machine operators at two automotive manufacturing facilities, comparing the results with 42 unexposed assembly workers in the same factories. Pulmonary function was measured on Mondays and Fridays, pre- and postshift. A 5% cross-shift decrease in forced expiratory volume (FEV_1), a pulmonary function measure indicative of air flow obstruction, occurred in 23.6% of machine operators and in only 9.5% of assembly workers. Exposures ranged from 0.07 to 0.44 mg/m^3 for assembly workers and from 0.16 to 2.03 mg/m^3 for machinists. Exposure levels were similar across the various different machining fluid types used. The authors categorized exposures as low ($<$0.20 mg/m^3), medium (0.20 to 0.55 mg/m^3) or high ($>$0.55 mg/m^3). After adjusting for a history of childhood asthma, for smoking prior to lung function testing, and for race, odds ratios for an FEV_1 response were found to be 4.4 among workers exposed to aerosols of straight oils, 5.8 for oil emulsions, and 6.9 for synthetic fluids. While no decrement was reported between Monday and Friday, the authors concluded that the medium exposure group had significantly increased rates for reversible, cross-shift FEV_1 responses relative to the low exposure group and that the concentration of inhalable (less than 9.8 μm) particles at which no FEV_1 response would be seen was less than 0.20 mg/m^3 of aerosol. The authors speculated that the decrease in FEV_1 reported may be due to multiple causes, including biological contamination, because of the diversity and chemical complexity of the fluids studied.

Acute respiratory effects from exposure to MWF have been manifested as irritation or alteration of pulmonary function in laboratory animals. For example, in 1991, Schaper and Detweiler[7] evaluated the sensory and pulmonary irritation potential of ten aerosolized MWF in mice. The animals were exposed by inhalation for 3 h to 20 to 2000 mg/m^3 of fluid. Six of the fluids represented new and used pairs. All of the fluids at some dose were capable of producing sensory and pulmonary irritation with little or no change in pulmonary histopathology following examination immediately after exposure. At 24 h postexposure, evidence of mild to moderate interstitial pneumonitis and bronchopneumonia was observed. For the fluids tested, the irritancy potential was as follows: synthetics > solubles > straight oils. There were no significant differences in irritation potential of the three used fluid samples vs. the corresponding unused fluids. Although some formulation information was available, it was not possible, the authors concluded, to determine the components(s) in each fluid responsible for the respiratory effects induced in mice.

Gordon[8] noted in 1992 that the agents responsible for potential adverse effects of inhalation of in-use MWF mist comprised not only constituents, such as oils and other additives including biocides, but their overall alkalinity (pH approximately 9), frequent hypoosmolarity (differing osmolarity of mist aerosol and conducting airways), and contamination with a wide variety of bacterial and fungal agents. Gordon studied the respiratory impact of used machining fluids and new machining fluids contaminated with endotoxin on guinea pigs and found that specific airway conductance (sGaw) decreased in a dose-dependent manner with increasing exposure to used fluids. sGaw declined from preexposure baseline values by 0 ± 2%, 7 ± 5%, and 40 ± 3% after a 3-h exposure to 1, 10, and 100 mg/m^3 of soluble oil machining fluid. These exposures were associated with airborne endotoxin concentrations of 0.3, 1.9 and 5.3 μg/m^3. Exposures, under similar conditions, of nebulized water, or 10 and 100 mg/m^3 of unused machining fluid produced decreases in sGaw of 3 ± 2%, 3 ± 2%, and 19 ± 4%, not significantly different from water exposures. Exposures to 10 and 100 mg/m^3 of aerosols of unused machining fluid spiked with endotoxin (0.9 and 8.2 μg/m^3 endotoxin) produced decreases in sGaw of 14 ± 8% and 38 ± 10%, statistically significant for the high dose exposure. Gordon found that the dose-dependent effects on airway conductance were accompanied by more sensitive cellular and biochemical changes in bronchial lavage fluid as well. Gordon noted that the pH of the new and used machining fluids was 10.2 and 9.1, respectively, and that the osmolarity of water and unused and used machining fluids were 19, 18, and 86 mOsm, respectively. He concluded that the airway obstruction and acute lung injury observed in guinea pigs resulting from exposure to used machining fluid aerosols meant that microbial contamination during use produced a machining fluid product which was significantly more toxic than the unused, clean machining fluid.

Ameille and coworkers,[9] in 1995, found in a study of 308 male workers exposed to either straight or soluble MWF (or both) at a large French car-making plant with mist levels of 2.6 ± 1.8 mg/m^3 straight oil (soluble oil exposure was not measured, and compared with 78 unexposed workers used as a control group) that while the prevalence of respiratory symptoms did not differ significantly among the four groups, prevalence of cough and/or phlegm increased with increasing duration of exposure to straight oils, after adjustment for smoking. While no increase in respiratory symptoms was observed with exposure to soluble oil, an inverse synergistic effect was observed in workers who smoked exposed to straight oils.

That same year, Gordon and Harkema[10] found that exposure of rats to 10 mg/m^3 used soluble oil, machining fluid (consisting of approximately 70% hydrotreated light naphthenic distillates, 20% sodium fatty acid and petroleum sulfonates and other additives, and containing 0.8 μg/m^3 endotoxin) for 3 h per day for 3 d produced a significant increase in intraepithelial mucosubstances (Vs) in the epithelial lining of both the nasal septum and intrapulmonary airways. These changes were also accompanied by a significant increase in total cells and neutrophils in the lavage fluid. A significant increase in Vs also occurred in the nasal septum only of animals similarly exposed to unused machining fluid aerosol with no measurable endotoxin. The results suggested that in addition to endotoxin, nonendotoxin components of machining fluid aerosols might contribute to an increase in phlegm and chronic bronchitis sometimes reported for workers exposed to machining fluid aerosols.

Krystofiak and Shaper,[11] in 1996, further examined a semisynthetic fluid, one of the MWF examined in an earlier study, and its components for sensory and pulmonary irritation in a mouse model. Mice were exposed for 3 h to the fluid and separately to its components, which included two alkanolamides, a boric acid amide, petroleum oil, a sodium petroleum sulfonate, hexahydro-1,3,5-tris(2-hydroxyethyl)-s-triazine, and a potassium soap. Sensory irritation was evoked by all of the components of the fluid, most almost immediately. Pulmonary irritation was observed with all components except the boric acid amide. Except for sodium petroleum sulfonate, which produced pulmonary irritation within 30 min of exposure, pulmonary irritation evoked by the other components generally occurred between 2 and 3 h of exposure. Immediately following exposure, a moderate recovery of respiratory frequency occurred with petroleum oil and the potassium soap but little recovery was seen immediately following exposure to the alkanolamides, sodium sulfonate, and triazine. The authors concluded that the alkanolamides, potassium soap, sodium petroleum sulfonate, and triazine components were similar in irritancy potential both to one another and to the semisynthetic fluid itself and that, based on potency and fractional composition, those five components largely contributed to the irritancy of the fluid.

Detweiler-Okabayashi and Schaper[12] continued this line of work by evaluating a previously studied synthetic MWF. In this case, the authors concluded that the fatty acid alkanolamide condensates, tolutriazole, and the triazine biocide components were similar in irritancy to one another and to the fluid itself. From the potency and the fractional composition, the authors concluded that the alkanolamides and triazine biocide largely contributed to the irritancy of the product.

Ball and Lucke,[13] using the same procedure, evaluated 17 commercially available MWF for sensory irritation potential. When their data were combined with the data generated by Schaper and Detweiler, no significant differences were found between the irritancy potentials of soluble oils, semisynthetics, or synthetics. There appeared to be as much variation in irritancy potential within fluid classes as there was between classes.

The irritancy potentials of the fluids measured by Ball and Lucke were used to develop regression models using the compositions of the fluids as input variables. An initial model developed predicted an inherent irritation potential that could not be assigned to the presence of one or more components, but which could be reduced by incorporating both amines and acids. The authors noted that the failure to find an assignable cause of irritation was a major flaw for this model and that the model's failure to predict the known irritancy of the short-chain organic fatty acids made the model further suspect.

An alternative model was developed using the raw respiratory rate reduction data. That model predicted that irritancy potential would depend on the concentrations of sodium petroleum sulfonates, ethanolamine, alkanolamides, and tall oil fatty acids. This model made sense from a chemical standpoint and identified the alkanolamides as the most potent class of irritants used in formulating MWF.

Thorne and DeKoster,[14,15] using established guinea pig and mouse inhalation models, showed that in-use MWF were more toxic than the corresponding neat MWF. Adjusting the pH of the neat fluid or spiking the fluid with active cultures of *Pseudomonas pseudoalcaligenes* (which was the predominant Gram-negative bacterium in many of the in-use MWF samples) did not increase the potency of the fluid to the level of the in-use fluid with the guinea pigs evaluated. Additional studies with guinea pigs revealed that there was significant inflammation resulting from a 3-h exposure to MWF aerosol at 50 to 77 mg/m^3. The inflammatory response was marked by significant change in the bronchoalveolar lavage (BAL) fluid from 3% neutrophils to 60 to 79% in the MWF-exposed guinea pigs. Total cells in the lung lavage increased from 0.4×10^6 cells/ml in the controls to between 7.7×10^6 and 11.3×10^6 cells/ml in the exposed groups. The in-use MWF used for the studies ranged in endotoxin concentration from 280 to 170,000 endotoxin units (EU)/ml.

Additional studies carried out by Thorne and DeKoster with normal mice and endotoxin sensitive mice revealed a dose-dependent 10,000-fold increase in BAL neutrophils in the lavage, a 100-fold increase in concentration of interleukin-6 and tumor necrosis factor-α in sensitive mice following exposure to in-use MWF. Such an inflammatory response was not observed with exposure to neat MWF or with sham exposure. In addition, there was no increase in total cells, neutrophils, interleukin-6 or tumor necrosis factor-α to inhaled, in-use MWF, by resistant mice. Removal of microorganisms by filtration of the MWF did not change the responses observed in either strain. The authors concluded that lung inflammation may be an important outcome from exposure to in-use MWF and that endotoxin was a toxicant of importance.

Milton et al.[16] exposed hamsters by intratracheal instillation to two soluble oils (both used and neat, and all adjusted to 3.7% petroleum oil content). At the highest doses instilled, the used soluble oil A contained 149,000 EU/ml, while the used soluble oil B contained 190,000 EU/ml. After 24 h exposure, the animals were sacrificed. While all the MWF samples produced a toxic response, exposure to the used MWF resulted in greater toxicity as compared with the unused fluids, as measured by a number of parameters. In additional experiments in which endotoxin was added to unused soluble A, the toxic response was similar to that of the used soluble oil A MWF. In another experiment, during which lipopolysaccharide (LPS) neutralizing protein was added to reduce the endotoxin concentration, toxicity was reduced, although not all parameters were lowered.

In another study, Woskie et al.[17] found that workers exposed in an automotive plant to MWF aerosol had higher exposures to inhalable aerosol (0.181 vs. 0.046 mg/m^3) than unexposed workers. Exposed workers also had higher mean endotoxin exposures than unexposed workers (7.1 vs. 1.9 EU/m^3).

Thorne et al.[18] studied exposures at an engine plant with two engine lines, which used three MWF in eight central systems. They found that exposures averaged 1.24 mg/m^3 on the older line and 0.74 mg/m^3 on the new line. Endotoxin units ranged from 39 to 166,000 EU/ml in the bulk MWF and from below detection (<4 EU/m^3) to 790 EU/m^3 in air. Airborne endotoxin concentrations were significantly correlated with bulk, viable mesophillic bacteria, and Gram-negative bacteria. In this environment, airborne endotoxin also correlated with gravimetric aerosol concentration. The levels of endotoxin in the bulk MWF samples as analyzed by Thorne et al. in the engine plant were the same order of magnitude as those found by Milton and by Thorne and DeKoster to cause toxic responses in guinea pigs.

Sprince et al., in 1997,[19] assessed differences in prevalence of respiratory symptoms between machine operators and assemblers in an automobile transmission plant and exposure–response relationships with MWF type, total aerosol, endotoxin, culturable bacteria, and fungi. They found that machine operators had significantly more usual cough, usual phlegm, work-related chest tightness and postshift symptoms of chest tightness, throat irritation, and cough compared with assemblers. They found exposure–response relationships between respiratory symptoms and total aerosol, as well as culturable fungi and bacteria, but associations with endotoxin were not strong or consistent, possibly due to low levels found.

In the study, Sprince et al. found total aerosol exposures among machine operators to average 0.33 vs. 0.08 mg/m^3 for assemblers. Exposure to total culturable bacteria was 17,200 (range of 740 to 148,500) vs. 318 CFU/m^3(44 to 1830) for assemblers. Exposure to total culturable fungi was 1060 CFU/m^3 (228 to 9420) for operators vs. 191 CFU/m^3 (70 to 837) for assemblers. Endotoxin exposure averaged 31 EU/m^3 (31 to 984) for machine operators vs. 3.1 (2.5 to 3.5) for assemblers. Total culturable bacteria exposures were significantly higher with one soluble oil studied.

Robbins et al.[20] studied personal exposures in an automotive transmission plant to thoracic particulate, viable plus nonviable thoracic bacteria, and vapor phase nicotine among 83 machinists exposed to soluble oils vs. 46 dry assemblers, looking for acute respiratory effects. Mean concentration of thoracic particulate was 0.13 mg/m^3 among assemblers, 0.32 mg/m^3 in the valve body machining area, and 0.56 mg/m^3 in the case machining area. Average personal exposures to thoracic bacterial were 0.38 bacteria/cc in assembly, 0.87 bacteria/cc in valve body, and 2.66 bacteria/cc in case. Average personal endotoxin levels, collected in only one exposure monitoring round, were 16.4 EU/m^3 in assembly, 34.7 EU/m^3 in valve body, and 234 EU/m^3 in case. Machinists and assemblers were grouped into low (<0.157 mg/m^3), medium 0.157 to 0.468 mg/m^3), and high (>0.468 mg/m^3) exposure categories. The percentages of workers with at least a 5% decrease, as well as a 10% decrease, in FEV_1 and for forced vital capacity (FVC) were calculated so as to be able to compare with Kennedy et al.'s data.

Pulmonary function testing showed that the percentage of machinists experiencing Monday cross-shift decrements of 5% or greater vs. assemblers in at least one round of testing was 42.9 vs. 26.8%; for 10% cross-shift decrements, the percentage was 16.9 vs. 7.3%. With respect to thoracic particulate, for both FEV_1 and for FVC, there was a trend of increasing risk of 10% or greater cross-shift decrements with increasing exposure among obstructed subjects. Current smokers were substantially more likely to be obstructed than other subjects (36.6 vs. 9.4%). As compared with the risk for low-exposure, nonobstructed, subjects, the risk at higher exposures among obstructed subjects was found to be substantial and highly significant.

In the earlier Kennedy study, median exposures to soluble oils were about 0.55 mg/m^3 of thoracic particulate, similar to the Robins study. In the Kennedy study, machinists were two and a half times more likely to have a 5% or greater decrease in FEV_1 on Monday, as compared with nonexposed assembly workers, as compared with about 1.6 times (42.9 vs. 26.8%) in the Robins study. The Robins study differed from the Kennedy study in two ways. First, the exposure assessment strategy for Robins was more comprehensive. Robins hypothesized that the thoracic bacteria or associated products, such as endotoxin would be probable causative agents for acute respiratory effects. Second, Kennedy et al. did not consider baseline obstruction as a covariate. Among obstructed subjects, but not unobstructed subjects, Robins et al. found a consistent association between increasing thoracic particulate exposure and greater cross-shift decrements in both FEV_1 and FVC in all models studied. Findings with thoracic bacteria used as the exposure measure were quite similar. The results were consistent with assignment of responsibility for the observed effect to thoracic bacterial concentrations themselves or some other parameter correlated with bacterial concentration (e.g., the correlation coefficient with endotoxin measures in the Robins study was 0.9). The authors noted, however, that the findings were equally consistent with the conclusion that other constituents of MWF aerosols, which may also be correlated with bacterial concentrations, are responsible. The authors could not determine from the data whether the association was with total bacteria or with viable bacteria, nor with any particular bacterial family.

Robins et al. noted that it was biologically plausible and consistent with the data produced, that the primary reason for the observed association was that the pulmonary effects associated with the experience of repeated cross-day or cross-week decrements over a period of years causally leads to more permanent, irreversible adverse pulmonary changes that are reflected in the observed lower baseline ratios. An alternative explanation was that individuals with lower pulmonary function, or who had been more susceptible to the effects of other pulmonary irritants, such as cigarette smoke, are more sensitive to the irritating or bronchoconstrictive effects of MWF aerosols. A finding that 6 of the 83 subjects working in machining areas, but none of the 46 workers in assembly had greater than 19% cross-shift decrements in FEV_1 or FVC strongly suggested that exposures in machining using soluble oils are associated with clinically significant adverse acute pulmonary effects in some individuals.

Robins et al. concluded that the study offered substantial evidence that MWF aerosol levels found in present-day U.S. machining operations were associated with clinically significant acute decrements in pulmonary function in some individuals. The authors recommended that exposure levels in machining areas be reduced to the lowest levels feasible. However, the data suggested that it would be impractical to consider an exposure limit below 0.2 mg/m^3 of thoracic particulate (equivalent to about 0.28 mg/m^3 total particulate) in that, among workers in assembly, in which there were no discernable point sources for particulate exposure, the median thoracic exposure was 0.14 mg/m^3, the 75th percentile was 0.15 mg/m^3, and the maximum was 0.31 mg/m^3. The authors further recommended a medical surveillance program and a smoking cessation program be instituted for workers exposed to MWF aerosols.

Greaves et al.[21] reviewed respiratory symptoms in a population of 1811 autoworkers at three General Motors facilities, which included 1042 then currently working as machinists; 769 assembly workers without direct exposure to MWFs served as a reference group. Workers were grouped into several exposure categories: for current exposure to straight oils, machinist exposures were categorized at 0.16, 0.30, or 0.74 mg/m^3 (a value of zero was assigned to each current assembly worker); for exposure to soluble oils, the three possible exposure values were 0.34, 0.59, and 0.72 mg/m^3; and, for synthetics (which also included semisynthetics), exposures were 0.31, 0.42, and 0.48 mg/m^3. Machinists exposed to any type of fluid had significantly elevated odds ratios for cough, phlegm, wheeze, and chronic bronchitis, relative to assembly workers. By specific type of fluid, cough, phlegm, and wheeze were significantly elevated for straight and synthetic exposure and chest tightness was also significantly associated with exposure to synthetic fluids. Interestingly, none of the slight elevations in risk observed among those exposed to soluble oils reached statistical significance. In all the models, current smoking was the strongest and most significant predictor of respiratory problems. Workers currently exposed to synthetic fluids had the highest rates of improvement away from work for all symptoms: 63 to 74% of workers exposed to synthetic fluids stated their symptoms improved away from work, while 37 to 55% of those exposed to straight oils, and 42 to 54% exposed to soluble oils, responded similarly.

Eisen et al.[22] reported, in a reanalysis of the above data, that exposure to some types of MWF (synthetics, which included semisynthetics, and possibly straight oils), may cause occupational asthma. Their results provided no evidence that exposure to soluble oils were related to asthma. Six of 29 posthire asthmatics had worked in jobs with exposures to synthetic (two) and semisynthetic (four) fluids sometime in the two years prior to diagnosis. For those six individuals, the average aerosol concentration was 0.60 mg/m^3.

Kriebel et al.[23] examined 216 machinists at an automotive parts manufacturing plant exposed to soluble or straight MWF and 170 nonmachinists, performing spirometry at the beginning and end of their shift. In this case, most machines had individual circulation systems. The authors observed an approximately threefold increase in the incidence of 5% decrease in FEV_1 among those with exposures above about 0.15 mg/m^3, as compared to those with exposures below about 0.08 mg/m^3. Endotoxin exposures were low in the plant (mean, EU/ml^3, in machining: 12.6 for straight oil; 16.0 for soluble) as were culturable bacteria (mean, CFU/m^3, in machining: 123 for straight oil; 753 for soluble oil). Only a weak association between personal airborne endotoxin exposure and cross-shift change in FEV_1 was observed, and, interestingly, there was a reduced incidence of 5% FEV_1 decrement in machinists in the highest third of the population with respect to exposure to airborne concentrations of culturable bacteria. The authors, comparing their data to that of Kennedy et al., noted that machinists in the Kennedy study experienced average aerosol exposure levels about four times higher than in the plant Kriebel et al. studied. (In Kennedy et al.'s study a 5% cross-shift decrease in FEV_1 occurred in 23.6% of machine operators vs. 9.5% of assembly workers.)

Sama et al.,[24] looking at the same data, found an association of exposure to sulfur with decrease of FEV_1. The authors found a more than threefold relative risk of cross-shift decrement in FEV_1 in those with sulfur exposure above the median (3.4 $\mu g/m^3$) compared with exposures below the median. The authors, in examining material safety data sheets for the products used in the plant, determined that soluble oil coolants contained 10 to 30% sodium petroleum sulfonate while straight oils contained 0 to 1.5% elemental sulfur or $<10\%$ sulfurized mineral oil. Since the chemical analysis determined elemental sulfur, it was not possible to determine which sulfur compounds workers were actually exposed. If it was sodium petroleum sulfonate, the authors estimated that exposures to that compound would range from 34 $\mu g/m^3$, for the low-exposure group, to 101 $\mu g/m^3$

for the high-exposure group. More specifically, the authors found about a 50% increase in the incidence of 5% FEV_1 decrement in those with medium sulfur exposure (range 2.5 to 4.4 $\mu g/m^3$) and an approximately fourfold increase in risk associated in those with high sulfur exposure (>4.4 $\mu g/m^3$). Comparing their data with Kriebel et al.'s, Sama et al. found, among machinists, that airborne sulfur concentration appeared to have a clearer dose–response relationship and a stronger association with cross-shift decline in FEV_1 than inhalable mass concentration. While the authors cited Schaper and Detweiler-Okabayashi's findings that sodium petroleum sulfonate was the most irritating component of a soluble oil tested, because other irritants were known to occur in MWF (such as endotoxins), it was suspected that sulfur represented a marker of MWF exposure conditions at this plant that were particularly irritating.

In 1998, Zacharisen et al.[25] reviewed respiratory illnesses presented by 30 workers over a 15-month period (March 1996 to May 1997) from a southeast Wisconsin automotive engine manufacturing plant that employed nearly 1600 persons. Of the 30, hypersensitivity pneumonitis affected seven workers, while occupational asthma and industrial bronchitis affected 12 and 6 workers, respectively. Airborne bacterial concentrations ranged from 525 to 4200 CFU/m^3. At the time of the outbreak, six synthetic and one soluble oil MWF were used in the plant, along with several biocides and other additives. All MWF were changed to a soluble oil product in March 1996. Additionally, existing air filtration system filters were equipped with sensors to indicate the need for filter replacement, new mist collectors with high-efficiency particulate air filters were installed, and biocide usage was adjusted so as to maintain less than 10^4 CFU/ml bacterial count in the MWF. Steam cleaning of machines to eliminate fungi and potentially infectious residues was performed as needed. Finally, a new engine assembly line was enclosed to decrease mist dispersion. That renovation exceeded $5,000,000 in costs.

No cases of work-related respiratory symptoms had been reported to the plant's occupational medicine department prior to the outbreak and no new cases were reported after August 1996. The authors commented that any or all of the changes made in the plant environment might have been responsible for or associated with the end of the outbreak, or, it could have ended spontaneously for unknown reasons. The authors noted that alterations in plant maintenance, cautious use of biocides, splash guards on machines, increased automation, and exhaust ventilation enhancing modifications may decrease and potentially prevent respiratory disorders related to MWF.

NIOSH, in 1998, published *Criteria for a Recommended Standard: Occupational Exposure to Metalworking Fluids*.[26] Based on review of all the literature available at that time, NIOSH said that workers currently exposed to MWF aerosols had an increased risk of nonmalignant respiratory disease and skin disease. To eliminate, or greatly reduce the risk of adverse health effects resulting from MWF exposure, NIOSH recommended an exposure limit of 0.4 mg/m^3 (thoracic) or 0.5 mg/m^3 (total particulate). The recommended exposure limit (REL) was based on an evaluation of health effects data, sampling, and analytical feasibility and technological feasibility.

Becklake,[27] in the same year, discussed the question of whether acute airway responses to repetitive occupational exposures lead to chronic airway dysfunction. In her discussion, Becklake noted selection bias, in which susceptible individuals leave the workplace early on, causing a healthier survivor population to become the subject of study, which may play a role in this question. She also noted information bias (difficulty in selecting the most appropriate exposure metric) as another factor.

Milton[28] explained why endotoxin is an important factor to consider in water-based MWF exposures. Endotoxin, the term applied to the LPS portion of the outer membrane of Gram-negative bacteria, is well-recognized as having the ability to affect multicellular animals adversely, such as fever produced in rabbits after injection. Studied in other areas, such as cotton dust exposure, it was now well recognized, Milton said, that exposure to inhaled endotoxin can potentially have adverse effects, depending on individual tolerance and frequency and amount of exposure. Milton showed evidence that endotoxin exposures as low as 45 EU/m^3 can have an acute impact on lung function in humans in the fiberglass production industry, and cited data suggesting that exposures to as low as

150 EU/m^3 in the machining industry also had an adverse impact. Milton noted that evidence for chronic airflow obstruction as a result of prolonged endotoxin exposure in other industries was mixed. Further, there was some evidence that endotoxin exposure might be protective against lung cancer. Finally, Milton noted that all chemical forms of LPS are not equally toxic (and thus may have different biological effects) as well as the issue of consistency of endotoxin measurement, now mostly overcome through introduction of two ASTM standardized measurement methods.

In 1999, a committee working under the auspices of Organization Resources Counselors, a human relations consulting firm whose Washington, DC office specializes in Occupational Safety and Health Administration (OSHA) affairs, published an internet-based guide called *Management of the Metal Removal Fluid Environment*.[29] Written by industry experts, the guide included information on MWF themselves, facilities and equipment, ventilation and controls, metal removal fluid systems, filtration systems, an MWF systems management program, including how to manage the fluid itself, such as testing and maintenance, environmental issues, health issues, and information for employers and employees. The guide was unanimously recommended by the OSHA's Metalworking Fluid Standards Advisory Committee, discussed below, as the definitive resource for systems management.

Kennedy et al.[30] investigated early pulmonary responses to MWF exposure by enrolling first-year machinist apprentices and apprentices in three other trades into a 2-yr longitudinal study. Full-shift, personal exposure samples for total particulate for the machinists varied from 0.3 to 3.65 mg/m^3 (mean, 0.46 mg/m^3). At follow-up, average change in bronchial responsiveness was double for machinists compared with control subjects, and machinists were more likely to have developed new bronchial hyperresponsiveness (BHR) with asthma-like symptoms. Analysis showed that increased BHR was associated with duration of exposure to both synthetic and soluble MWF. The authors concluded that exposure to water-based MWF, especially synthetics, was associated with increasing BHR during the first 2 yr of exposure.

Gordon[31] and Gordon and Galdanes[32] observed, in guinea pigs, changes in pulmonary function and in cellular and biochemical indices in BAL lavage fluid after a 3-h exposure at approximately 50 mg/m^3 to both fresh and used semisynthetic, soluble, and synthetic fluids (changes: semisynthetic $>$ soluble $>>$ synthetic). Greater toxicity was observed with used machining fluid aerosols and, within the used machining fluid types, significantly greater adverse effects were observed with the more poorly maintained fluids (i.e., those with heavy microbial contamination). Changes in biochemical and cellular parameters in BAL fluid occurred after a single exposure to 5 mg/m^3 of the poorly maintained used machining fluid aerosols. Further, changes in inflammation, but not in lavage fluid lactate dehydrogenase (LDH) or protein, were observed in animals repeatedly exposed to semisynthetic machining fluid aerosols. A statistically significant increase in lavage fluid neutrophils was observed in guinea pigs exposed to 5 mg/m^3 used semisynthetic machining fluid aerosols for four weeks. In separate experiments, adjustment of the unused semisynthetic machining fluid to isotonicity and to pH 7 significantly reduced adverse affects.

In this study, Gordon and Galdanes found that the increases in adverse effects produced by the used, dirty machining fluid aerosols were associated with the number of viable bacteria in the fluids, as well as the presence of endotoxin in the aerosols. The authors noted that an increase in polymorphonuclear leukocytes (PMNs) in lavage fluid, a sensitive marker of adverse effects after exposure to the low concentration of the used, dirty machining fluid aerosols, was a well-known result of endotoxin inhalation. The increase in PMNs at 24 h after the final exposure was the only significant change in animals repeatedly exposed to the used, dirty machining fluid aerosols over 30 d. Thus, despite an apparent development of tolerance to other acute changes in lavage fluid protein and in LDH, inflammatory changes persisted. The increase in PMNs was absent in animals examined at 72 h after the final exposure, suggesting, the authors said, recovery from the 30-d exposure regimen.

The alkaline pH and the hypotonicity of fluids may play a role in the adverse effects associated with inhalation of machining fluid aerosols. When the authors adjusted the pH and osmolarity of a sample of new semisynthetic fluid to 7.0 and 300 mOsm, respectively, the effects of exposure to 50 mg/m^3 of the adjusted fluid were markedly inhibited. However, when a sample of the used, dirty, semisynthetic fluid, which had produced significant adverse effects, was adjusted similarly, no reduction of adverse effects occurred.

OSHA impanelled a Standards Advisory Committee (SAC) in 1997 to make recommendations to the Assistant Secretary of Labor for OSHA regarding MWF. As background for this action, in December 1993, the International Union, United Automobile Aerospace and Agricultural Implement Workers of America (UAW) had petitioned OSHA to take emergency regulatory action to protect workers from the risks of occupational cancers and respiratory illnesses due to exposure to MWF. OSHA sent an interim response to the UAW stating that the decision to proceed with rulemaking would depend on the results of the OSHA Priority Planning Process. Following the Priority Planning Process report, which identified MWF as an issue worthy of agency action, the assistant secretary asked the National Advisory Committee on Occupational Safety and Health (NACOSH) for a recommendation about how to proceed with MWF. NACOSH unanimously recommended that OSHA form a SAC to address the health risks caused by occupational exposure to MWF. The Assistant Secretary accepted the recommendation of NACOSH. OSHA then established a 15-member SAC that made recommendations regarding a standard, a guideline, or other appropriate response to the dangers of occupational exposures to MWF. The membership of the committee was balanced and included individuals appointed to represent the following affected interests: industry; labor; federal and state safety and health organizations; professional organizations; and national standards-setting groups.

The SAC published a final report in 1999.[33] In that report, the committee recommended that OSHA act to mitigate the adverse health effects associated with exposure to MWF. This decision was based on the demonstrated health effects: asthma, hypersensitivity pneumonitis, other respiratory disorders, and dermatitis. In the opinion of the committee, each of these health effects was a material impairment of health, presented a significant risk, and occurred throughout the industry. The committee also recognized that there were other health conditions, including cancer, related to MWF exposure for which the evidence was still evolving.

The committee supported the use of a defined occupational exposure limit and unanimously supported the use of systems management to control exposure. The committee unanimously recommended that the scope of any OSHA action include fluids used in the machining environment including the operations of cutting, machining, grinding, and honing. While the committee recommended that the scope be limited to these operations and their fluids, the exclusion of other MWF or related processes or environments did not imply the lack of a potential problem in those related fluids, processes, or environments.

A majority of the committee recommended that OSHA promulgate a comprehensive 6(b) standard to protect employees from the adverse effects of MWF and material impairment of health. A dissenting minority recommended several nonregulatory alternatives.

The majority of the committee recommended that a standard for MWF should include a permissible exposure limit (PEL), systems management, medical surveillance, and training. This approach for a standard should control exposure and achieve a meaningful reduction of disease, the report said.

A majority of the committee recommended a new 8-h time-weighted average PEL of 0.4 mg/m^3 thoracic particulate (0.5 mg/m^3 total particulate). The scientific rationale for the recommended PEL was based on studies of asthma and diminished lung function. This research was provided in the NIOSH Criteria Document, and in the record and report of the committee.

In addition to a PEL, the committee recommended systems management of the MWF environment to protect employee health further. As noted in the report, systems management included a comprehensive programmatic approach with enclosure, ventilation, fluid management, and other actions to control exposure and minimize contact with the fluid. The committee also recommended an active medical surveillance program as an essential component of the proposed

6(b) standards, since there was evidence of adverse health effects below the PEL. Lastly, the committee noted that training and outreach activities were essential components of any action involving MWF.

The Committee believed that the recommended PEL, a systems management approach, active medical surveillance, and training were technologically and economically feasible for employers affected by this recommendation and recognized that the recommendations were substantial and would require a phase-in period.

The committee's report, while accepted by the assistant secretary of labor for OSHA, did not result in promulgation of an OSHA 6(b) standard. Instead, OSHA published a *Metalworking Fluids Best Practices Manual* in 2001.[34] The manual provided general information about MWF and recommended a systems management approach to control exposure and minimize contact with the fluid. That strategy included engineering and work practice controls such as machine enclosure, ventilation, and the use of PPE.

The guide also recommended that employers establish a fluid management program that included designating responsibility of the system to one or more persons knowledgeable in the chemistry involved in metalworking processes. The program should also include standard operating procedures for testing fluids, a data collection and tracking system, employee participation in setting up and operating the overall system, and a continuing training program.

Also included in the guide were recommendations for instituting an exposure monitoring program (air sampling) on, at minimum, an annual basis. The manual recommended that employee exposures be reevaluated whenever there were significant changes in production, equipment, or processes that might cause new or additional exposures to MWF. Finally, the manual recommended a proactive medical monitoring program for exposed employees that would help identify early evidence of respiratory impairment or skin disease. That early identification would prompt corrective action, which would help reduce the incidence and severity of MWF-associated diseases, according to the manual.

OSHA was careful to point out that the *Metalworking Fluids Best Practices Manual* was not a new standard or regulation, and it created no new legal obligations. It was to be advisory in nature, informational in content, and was intended for use by employers in providing a safe and healthful workplace for workers exposed to MWF.

Abrams et al.[35] in 2000, in a companion study to Robins et al. (1997), characterized personal exposures in an automotive transmission plant over a 15-month period. Mean personal air concentrations were 0.13 mg/m^3 in final assembly, 0.32 mg/m^3 in valve body, and 0.56 mg/m^3 in case. Average personal exposures to thoracic fraction bacteria were 0.38 bacteria/cc in final assembly, 0.87 bacteria/cc in valve body, and 2.66 bacteria/cc in case. Personal endotoxin exposures, collected in round 3, were 16.4 EU/m^3 in final assembly, 34.7 EU/m^3 in valve body, and 234 EU/m^3 in case. Sump bacteria levels were on the order of 10^8 CFU/ml, with the case line consistently having the highest levels ($5-8 \times 10^8$ CFU/ml). Addition of biocide to the valve grinding central system, which contained one soluble oil at the time of the first round of sampling, had an immediate effect on bacterial concentrations, but levels returned to normal within weeks of treatment. (By the time of the second and third rounds of sampling, the valve grinding central system had been changed to the same soluble oil product that had already been used in valve machining and case.) With that product, no biocide additions were used to control bacteria; since 1989, bacterial growth had been controlled by aeration and pH adjustment (to pH of approximately 9). The authors noted that the acridine orange staining method used to count both viable and nonviable bacteria indicated substantially higher bacteria counts than the more traditional dipstick method. Culturable bacteria were overwhelmingly Gram-negative species, including *Pseudomonas*, *Brucella*, *Aeromonas*, *Methanobacter*, *Comamonas*, and *Deleya*. In machining areas, personal endotoxin and bacterial concentrations were highly correlated, and significantly correlated for thoracic particulate, thoracic bacteria, and total endotoxin.

The exposures measured were in the same range as observed by Kennedy et al. Owing to the significant correlation, reducing airborne particulate levels should also reduce ambient bacteria and endotoxin, both suspect agents of respiratory impairment, according to the authors. Finally, the authors noted that addition of biocide to the first product used in valve grinding had no apparent effect on sump endotoxin concentrations and only a transient effect on sump bacterial concentrations. The authors speculated that addition of biocide on an intermittent basis may not prevent the occurrence of adverse health effects if those effects were due to endotoxin.

Brown et al.,[36] in 2000, compared three extraction methods (pyrogen-free water (PFW); PFW with a polyoxyethylene sorbitan monolaurate; and a sonicated PFW containing that same surfactant) for their effectiveness in the quantitative removal of endotoxin from unused and used bulk water-soluble MWF samples. Results showed that vigorous recovery methods yielded higher amounts of recovered endotoxin.

Virji et al.[37] identified factors leading to increased exposure to microorganisms at an automotive plant. Their final full multivariate model predicted a significant reduction in bulk microbial levels by increasing pH of the fluid and reducing the amount of tramp oil leaking into the fluid. For airborne exposure, the final multivariate model predicted a significant control of airborne microorganisms by increasing worker distance from the machine, reducing the number of machines within 10 ft of the worker, decreasing the bulk microbial levels, and adding machine enclosures.

Eisen et al.,[38] in 2001, measured lung spirometry and estimated current and past exposures for a cohort of 1811 male automobile workers. In their study, current exposure was not found to be a factor with either FEV_1 or FVC, nor was past exposure to either water-based MWF (soluble or synthetic). The authors did find that past exposure to straight oils was significantly associated with FVC. The association, they said, was more obvious among older workers and among workers who never transferred from MWF-exposed jobs to assembly. The magnitude of the association between FVC and lifetime exposure to straight oil MWF was slightly larger than the estimated cigarette effect, suggesting that the impact of an additional year of exposure to 1 mg/m^3 of straight oil particulate in the thoracic particle size range, had the same effect on FVC as smoking one pack a day for one more year.

Piacitelli et al.[39] in a study of 79 small machine shops, found that 62% of 942 personal samples collected were less than the NIOSH REL of 0.5 mg/m^3. However, the authors said, at least one sample collected exceeded the NIOSH REL in 61 of the 79 facilities studied and that 100% of the samples collected exceeded the REL in ten shops. While no measurements of acute health effects were made during these studies, the authors suggested that workers in these small shops may have risks of adverse health effects similar to those observed in the automotive industry.

Bracker et al.,[40] in 2003, described a longitudinal assessment of intervention effectiveness at a metalworking plant after an outbreak of hypersensitivity pneumonitis. In this case, 29% of the plant's 120 production workers were given a clinical diagnosis of HP during the 2 yr of the investigation. Recommended interventions included: improving MWF management practices, enclosing selected MWF machining operations; eliminating mist cooling; exhausting two additional water-based industrial processes (a wet dust collector and an abrasive blasting process); increasing general ventilation; and, training. Samples of bulk fluid from the machining sumps showed bacteria in the range of 10^5 to 10^8 CFU/ml, endotoxin in the range of 7200 to 200,000 EU/ml and oil mist in the range of 0.05 to 0.18 mg/m^3 (mean, 0.09 mg/m^3 TWA), well below NIOSH's REL of 0.5 mg/m^3. After implementation of the fluid management program, the log of the mean concentration of sump bacteria remained at least two orders of magnitude below the log of the initial mean sump bacteria concentration. *Mycobacterium* sp. was found in only one machine tool, but also found in samples from the water-based dust collector and from the vapor blast slurry. The investigator's experience with managing the response to this outbreak underscored the importance of a systems approach, including a fluid management program, to exposure assessment and control.

Linnainmaa et al.[41] studied the effects of triazine use and machine enclosure on worker's exposure to bacteria, endotoxin, and formaldehyde in the use of MWF. Two machine tools were cleaned using a disinfectant and new emulsified fluids containing triazine as a biocide were introduced. In the experiment, one machine tool was replenished with water and only some MWF containing triazine, while the other was also replenished with triazine to keep the triazine level above 500 ppm, the minimum recommended. Fluids were monitored every two weeks for eight months. Endotoxin levels increased rapidly, to over 18,000 ng/ml, when triazine levels in the one machine fell below 500 ppm. Bacterial count increased as well, reaching 1.55×10^7 CFU/ml. In the other machine, in which triazine was maintained between 500 and 2500 ppm, endotoxin levels were found to be between 200 and 400 ng/m^3. Mean formaldehyde exposure in the breathing zone of workers was found to be 0.052 mg/m^3. The authors found that enclosure of the machines substantially reduced workers' exposure to endotoxin.

Zeka et al.[42] reanalyzed a previously published study, which had found evidence for associations between exposures to MWF and pulmonary function impairments restricted to workers who exhibited obstruction at baseline. In this reanalysis, the authors found increased evidence of cross-shift decrement in FEV_1 as MWF exposure increased in the full cohort, not just with those obstructed.

Bukowski[43] reviewed much of the epidemiological literature involving exposure to MWF. In his view, while machinists may have experienced slightly higher prevalence of common respiratory symptoms and mild and reversible cross-shift changes in some measures of pulmonary function, the inconsistency and potential for both random and systemic error in the body of literature argued against drawing definitive conclusions. There was no substantive evidence, he said, that any of these effects led to permanent disease or impairment. Bukowski noted that the most likely causal agents for respiratory effects in these workers were microbial contaminants in the water-based MWF.

Oudyk et al.[44] investigated the association between semisynthetic and soluble oil MWF exposure and respiratory symptoms in a cross-sectional survey in a large automotive machining facility in Canada, using a self-administered respiratory symptom-screening questionnaire, worker assignment, and exposure mapping measurements, both average and peak. With a response rate of 81%, the authors found high symptom prevalence: 29% of workers reported weekly or daily phlegm; 23%, dry cough; 42%, runny or plugged nose. Average aerosol concentration in departments with exposure ranged from 0.02 to 0.84 mg/m^3 and peak levels from 0.2 to 2.84 mg/m^3. Average exposures ranging from 0.25 to 0.84 mg/m^3 were statistically significantly associated with wheezing, chest tightness, sore throat, and hoarse throat. Wheezing was also associated with average exposures of 0.10 to 0.16 mg/m^3. When peak exposures were included in the regression, statistically significant effects of exposure in the range of 0.20 to 0.47 mg/m^3 were associated with dry cough, phlegm, wheezing, chest tightness, and sore and hoarse throat. Current smokers had higher odds of having certain symptoms, but no smoking-exposure term was significant in their models. Workers in departments having no exposure to MWF had average department aerosol concentrations of 0.02 to 0.06 mg/m^3. The authors noted that the exposure measurements were not personal exposure measurements and may have underestimated actual worker exposure. Nonetheless, they said, they had observed an association of increasing upper and lower respiratory symptoms with estimated MWF exposure well below the NIOSH REL of 0.5 mg/m^3.

Ross et al.[45] in 2004, in a study of personal exposure in small machine shops in Vancouver, Canada, found average aerosol exposure to be 0.32 mg/m^3 (range, 0.06 to 2.19 mg/m^3). Factors associated with increasing exposure included time spent grinding, operating an enclosed computer-controlled machine, the presence of welding in the shop, and the number of machines. Factors associated with reduced exposure included machining aluminum, milling, the height and shape of the shop roof, and the presence of shop ventilation.

Veilette et al.[46] followed microbial growth in an MWF system that had been associated with prior cases of HP after system dump, cleaning, and recharge (DCR). The fluid system was maintained by the supplier, and fluid maintenance procedures were typical of the industry, the authors said. The authors found rapid progression in total bacteria counts, as determined by fluorescence microscopy, from no measurable bacteria in the neat fluid, to 6.9×10^6 cells/ml after 12 h and to 2.2×10^6, 3.6×10^8, and 6.1×10^8 cells/ml after one, three, and six months. There were 5.7×10^7 cells/ml in the preDCR used fluid. Plate count analyses dramatically underestimated the total bacterial population as provided by the direct count method. Polymerase chain reaction (PCR) analyses showed the presence of *M. chelonae* and other mycobacteria strains in the used fluid at three and six months. Other strains of *Mycobacterium* were detected at 12 h and one month as well. The authors concluded that dumping, cleaning, and recharging a problematic MWF system without removing all sources of subsequent contamination led to rapid bacterial regrowth, including Gram-positive strains.

Gordon[47] reviewed the toxicity of MWF and found that the majority of evidence strongly suggested that microbial changes that occur during use in the workplace are responsible for the pulmonary effects reported for workers exposed to MWF aerosols. Gordon noted that bacterial endotoxin appeared to be the most important contributor to the effects of in-use MWF aerosols, but observed that it was likely that other microbial agents also had potent effects.

Lim et al.[48] in 2005, in a study of the inflammatory and immunological responses in rats to subchronic exposure to endotoxin-contaminated metalworking aerosols, found that lung inflammatory responses were induced in rats without changing lung pulmonary function after repeated exposures to MWF contaminated with endotoxin.

In the Lim study, F344 rats were exposed for 6 h/d to a 10 mg/m^3 concentration of a nonadditized soluble oil for 5 d/week for 8 weeks with endotoxin at concentrations of 1813 and 20,250 EU/m^3 or with no endotoxin. There were no changes in body weight or in respiratory function (tidal volume, respiratory frequency) during the eight weeks exposure to the endotoxin-containing MWF, but lung weight increased significantly in the rats exposed with and without endotoxin. Increases in lung weight, number of PMN cells, and levels of extracellular cytokines and NOx were all more significant in rats exposed to endotoxin-contaminated MWF as compared with rats exposed only to MWF. The levels of several endotoxin-specific antibodies (IgG2a and IgE) increased significantly in the rats exposed to endotoxins.

The authors noted that in some workplaces in Korea, concentrations of endotoxins in bulk fluids of 35,000 EU/ml have been found, and, in a situation in which MWF were present at a geometric mean concentration of 0.94 mg/m^3, endotoxin concentration in the air was measured at 11,911 EU/m^3. As a result, the authors said, the concentrations of endotoxin used in their study effectively represented actual workplace exposure to water-soluble MWF. The authors further noted that, in contrast to studies such as that of Thorne and DeKoster in which the lung functionality parameters of experimental animals had changed, in their study, animals were checked for lung functionality after weekly exposure during rest, in contrast with Thorne and DeKoster where the lung functionality of the animals was measured during exposure. The authors suggested that repeated exposure to MWF contaminated with endotoxin induces lung inflammation and downregulates production of extracellular cytokines and the migration of PMN cells to inflammatory sites.

Results of the acute studies listed above show that exposure to used MWF, contaminated with endotoxin and other microbiological contaminants, produces acute adverse respiratory effects in laboratory animals to a greater degree than exposure to neat MWF. Epidemiological studies show that exposure to in-use MWF also result in acute adverse respiratory effects, in some cases at exposures lower than the NIOSH REL of 0.5 mg/m^3. While endotoxin exposure was not always measured, those studies resulting in observation of adverse, acute respiratory effects, were often accompanied by exposure to endotoxin. Thus, many of the effects may have been due to microbial contaminants rather than the MWF. Further, field studies show that microbial contamination of in-use MWF is difficult to control, even after dumping, cleaning, and recharging. Microbiology of MWF is reviewed in detail in Chapter 9.

B. Chronic Effects

Extensive data exist on the subchronic and chronic effects of exposure to MWF and their components. These studies typically focus on respiratory effects or cancer potential.

1. Respiratory Effects

a. Asthma and Other Respiratory Effects

NIOSH, in their criteria document,[26] reviewed the literature published up to that time, regarding asthma associated with exposure to MWF.

NIOSH noted that symptoms of airways irritation (e.g., cough) occur with sufficient exposure to airborne irritants. Further, they said, the acute airways' response to an inhaled irritant often involves short-term, apparently reversible decrements in measured pulmonary function. Repeated exposure to an irritant can evolve into chronic bronchitis, a condition characterized by chronic production of phlegm. NIOSH went on to say that the inflammation associated with chronic airways irritation may also cause accelerated decline in lung function, which can result in symptomatic functional impairment and pulmonary disability.

Asthma, NIOSH stated, is an airways disease with a marked variability in airflow limitation. It can be induced by exposure to an immune sensitizer or an irritant. Whether initially induced by a sensitizer or irritant, symptomatic episodes of immunologic or irritant asthma can be triggered by subsequent exposure to the specific causative agent or any irritant, even at concentrations substantially lower than those tolerated by nonasthmatic persons. NIOSH noted that there was increasing evidence to suggest that a worker's occupational asthma is more likely to become chronic (with irreversible airflow limitation and continuing airways hyperresponsiveness even after removal from exposure) the longer the worker continues to be exposed after onset of asthma.

Savonius et al.[49] in 1994, reported that two metal workers who, after several years of exposure, developed asthma attributed to the triethanolamine (TEA) in the MWF they had used. Exposure conditions were not reported.

Eisen,[50] in 1995, described an inverse exposure–response relationship between aerosol concentration of synthetic MWF used and the prevalence of self-reported, physician-diagnosed asthma. She demonstrated that the incident asthma cases were more than twice as likely to be exposed to synthetic MWF as the nonasthmatic machinists in the year of asthma onset. She also observed some evidence of transfer bias away from jobs with exposure to synthetic MWF.

Greaves et al.[21] whose work was reported earlier, looking at the same cohort in 1997, suggested there was no clear relationship between self-reported, physician-diagnosed asthma and current aerosol exposure to straight oil, soluble, or synthetic MWF, but did suggest that there was a relationship between cumulative exposure to soluble fluids related to asthma among these workers. As discussed above, Eisen, using a cohort approach and proportional hazards model, concluded exposure to synthetics during the 2 yr prior to diagnosis with asthma was statistically significantly associated with diagnosis of asthma. Again, aerosol measurements for the six asthmatics who worked with synthetics during the 2 yr prior to diagnosis averaged 0.60 mg/m^3, with a range of 0.36 to 0.91 mg/m^3.

Kriebel et al. whose work was also reviewed earlier,[23] also found evidence for an association between self-reported, physician-diagnosed asthma and work as a machinist. In their work, machinists exposed to soluble oil reported asthma twice as often as nonmachinists (OR $= 2.1$; 95% CI $= 0.9$–6; $p < 0.10$).

Rosenman et al.[51] reported the results of a mandatory surveillance system for occupational illness in Michigan and noted 86 occupational disease reports of work-related asthma (WRA) secondary to exposure to MWF. The authors reported that, during this time, 1988 to 1994, MWF were the second most common cause of WRA in the state.

NIOSH, in its document, said that, considered in the aggregate, the studies summarized provided evidence indicative of an elevated risk of asthma among workers exposed to MWF aerosol exposure concentrations currently found in large automotive shops.

NIOSH, in its document, also reviewed prevalence of other symptoms of airways disorders, some of which is reviewed in detail under acute effects, above. NIOSH's overall conclusions for epidemiological studies of respiratory symptoms were that the studies presented were generally consistent, and in the case of more recent studies, presented compelling evidence that occupational exposure to MWF aerosols caused symptoms consistent with airways irritation, chronic bronchitis, and asthma. NIOSH also noted that there was no compelling evidence that identified any particular component or components of MWF aerosol as the predominant cause of the symptoms.

In 2002, Petsonk[52] reviewed the topic of WRA and the implications for the general public. Petsonk noted that asthma had been increasing over the past two decades in the U.S., that the onset of asthma had been reported as a result of occupational exposure to over 350 agents, and that WRA had become the most frequently diagnosed occupational respiratory illness. Petsonk reported WRA incidence rates of 29 to 710 cases per million workers per year and that 10 to 25% of adult asthma is work related. Thus, asthma related to exposure to MWF aerosols is part of a larger problem observed in workplaces.

b. *Hypersensitivity Pneumonitis*

Hypersensitivity pneumonitis (HP), also referred to as extrinsic allergic alveolitis, is an immunologically mediated inflammatory lung disease. The inflammatory process occurs in susceptible individuals as a result of repeated exposures to inciting substances (antigens) such as organic dusts, animal proteins, microorganisms, and low molecular weight chemicals. Upon examination of affected individuals, HP may present a variety of medical symptoms ranging from flu-like illness (acute phase), recurring pneumonia (subacute phase), and unusual weight loss, dyspnea, and coughing (chronic phase). Early medical intervention with good prognosis includes moving patients away from problematic environments. Affected individuals who continue to be exposed to problematic aerosols may experience gradual decrements in pulmonary function. It has been determined that HP occurs as a result of antigen-antibody mechanisms that lead to an accumulation of PMNs in alveoli and airways, and the subsequent influx of mononuclear cells into these sites results in nonnecrotizing granulomas. The chronic stage of HP is characterized by irreversible pulmonary fibrosis, i.e., the occlusion of respirable areas of lung tissue. Depending on the origin of exposure, HP is known by various names, two of which are bird breeder's lung and farmer's lung.

HP, as a result of exposure to contaminated, in-use MWF, is a relatively new phenomenon, and usually appears in clusters, that is, small groups of affected machinists in a larger occupational setting. HP was first reported in 1993 by Muilenberg et al.[53] in which ten machinists were diagnosed with HP at a machining plant where the in-use MWF revealed levels of 10^6 to 10^7 mycobacteria per milliliter of bulk MWF. The species was not identified.

Bernstein et al.[54] in 1995, reported on a situation in which six auto parts manufacturing workers were diagnosed with HP, with four of the six workers presenting pulmonary interstitial infiltrates in their lungs, causing decreased diffusing capacity. After removal from the metalworking area, all six workers recovered. Serum precipitins to one or more microbiological isolates were identified in all six workers, but not in eight of nine nonexposed control subjects. The most frequent precipitin response was against antigens of *P. fluorescens*, cultured from the in-use MWF. The authors termed the HP disorder associated with exposure to MWF aerosols machine operator's lung.

In 1997, a workshop, jointly sponsored by NIOSH and the UAW–Chrysler National Joint Committee on Health and Safety, was held during which eight clusters of HP in the automotive machining industry were reviewed. Kriess and Cox-Ganser, in their review of the meeting,[55] concluded that a risk existed for HP where water-based MWF were used and unusual microbiological contaminants predominate. The authors noted that strong candidates for microbial etiology were nontuberculosis mycobacteria and fungi, and that these cases occurred among other cases with other work-related respiratory symptoms and chest diseases. Further, they noted, reversibility of disease has occurred in many cases with exposure cessation, allowing return to work to jobs without MWF exposures, or, in some cases, to jobs without the same MWF exposures. All told, a total of 98 cases of physician-diagnosed HP had occurred in these eight plants; the first case had been reported in 1991. In three plants, cases were reported at the time of the workshop. *M. chelonae* was found in four of the plants. Total particulate exposures were often 0.5 mg/m^3 or less.

Shelton et al.[56] in 1999, reviewed three case reports of HP outbreaks in machining plants. In the first case, one worker out of a group of six was diagnosed with HP. In the second case, two machinists (out of some 700 machine workers) were diagnosed by a treating pulmonologist with HP. In the third case, five machinists with physician-diagnosed HP were reported. In all three cases, mycobacteria in the *M. chelonae* complex were identified in the bulk fluids, in colony counts up to over 10^6 CFU/ml sample.

Moore et al.[57] in 2000, investigated contamination of air and MWF systems in a manufacturing plant at which recent cases of HP had been diagnosed, and found the presence of rapidly growing mycobacterium species in 100% of the MWF central systems and 75% of the individual machines, with counts in the range of 10^2 to 10^7 CFU/ml. With one exception, all contamination was limited to a then-recently introduced semisynthetic fluid at the plant. The investigators found that samples obtained from diluted (5%) but unused MWF, a replenishment line with 2% unused MWF, an MWF pasteurizer, city water, and deionized water all were culture-negative for this species of mycobacterium.

In another report,[58] three cases of HP were diagnosed at an automobile brake manufacturing facility which employed 150 machinists and 250 others in nonmachining areas. Centers for Disease Control investigators found that 107 workers had been placed on work restriction during the year in which the HP cases were diagnosed. Medical records reviewed for 32 workers who remained on medical leave during the year showed 14 meeting the definition for occupational asthma and 12 meeting a definition for HP. The other six workers had illnesses consistent with work-related bronchitis or rhinosinitus or dyspnea. The plant used a semisynthetic MWF in four central systems, which included a formaldehyde-releasing biocide. An isothiazolinone-based biocide was also used to control microbiological growth. *M. immunogenum* was detected in bulk fluid samples, at levels of up to 10^6 CFU/ml. Personal samples from the machining area were in the range of 0.1 to 0.9 mg/m^3 (median: 0.6 mg/m^3) with two samples above the NIOSH REL of 0.5 mg/m^3.

Wallace et al.[59] in 2002, reported results in a study in which they collected 107 MWF isolates of mycobacteria from ten machining plants in the U.S. and Canada; 102 of the 107 isolates were identified as *M. immunogenum.*

Falkinham,[60] in 2003, reviewed the extent of respiratory disease in a variety of settings associated with exposure to environmentally opportunistic mycobacteria. Falkinham noted that exposure to aerosols was a common feature of the disease outbreaks and suggested that, as mycobacteria are very resistant to the disinfectants used in water treatment, their growth in MWF systems may also be related to resistance to typical MWF biocides used.

Primm et al.[61] in 2004, reviewed the health impacts of environmental mycobacteria and predicted increasing incidence of interactions between humans and mycobacteria in coming years. The authors identified three major factors driving this increase: disinfection of drinking water with chlorine selecting mycobacteria by reducing competition, disinfection attempts in medical and industrial settings which may likewise select for mycobacteria, and an increasing percentage of human population with predisposing conditions.

Selvaraju et al.[62] in 2005, reviewed the microbicidal activity of four biocides in a synthetic MWF against *M. immunogenum* and *P. fluorescens*, a predominant Gram-negative bacterial contaminant often found in in-use MWF. The authors found in these experiments that an isothiazolone biocide was more effective than a phenolic biocide, and both of these were more effective than two formaldehyde-release biocides against *M. immunogenum*.

Lastly, Beckett et al.[63] reviewed the case of a previously healthy man working as a machinist in an automotive factory who developed HP in 1995, and whose medical condition was followed for 9 yr. He was removed from the metalworking environment but symptoms reoccurred when he was later reexposed to MWF, with further decrement in lung functionality occurring. MWF aerosol in the area in which he was working was 0.42 mg/m^3. While standard bacterial and fungal counts were below 10 CFU/ml, investigators found 1.6×10^5 *M. immunogenum*/ml in the MWF. The patient, when completely removed from exposure to MWF, stabilized.

As noted in Chapter 9, there is still much discussion and speculation among occupational health specialists about the etiology of cases of HP that have occurred among workers employed at machining facilities. Some researchers assert that MWF aerosols that contain species of acid-fast, nontuberculosis bacteria (NTB) of the genus *Mycobacterium* are causative agents of HP. As noted above, three species, i.e., *M. abcessus*, *M. chelonae*, and *M. immunogenum*, have been prominently mentioned. These species are relatively fast-growing bacteria (compared with slower growing NTB) that may proliferate in MWF distribution systems along with Gram-negative and Gram-positive bacteria, and fungi. However, incidences of HP are rare among machinists, and numerous epidemiological investigations have not demonstrated a strong and consistent correlation between mycobacteria exposures and the onset of HP symptoms. In many MWF distribution systems that have had high concentrations of mycobacteria in fluids, there have been no reported cases of HP among hundreds of workers exposed to mycobacteria-laden aerosols. Also, there are numerous confounding factors, which may have etiological significance for HP in manufacturing environments, including antigenic substances present in fugitive dusts, environmental endotoxin, ventilation systems, chemical emissions, cooling towers, extramural (make-up) air, etc.

MWF end users have enlisted a variety of procedures to minimize exposures to antigens that may cause HP, including the provision of improved area ventilation (increased air changes, better air filtration), effective facility housekeeping, the use of bioresistant fluids, and close monitoring of coolant microbiology. These methods are simple yet effective ways of protecting workers against a variety of inimical workplace exposures. Research continues to ascertain the cause(s) of HP.

2. Carcinogenicity

Certainly the health effect that has caused controversy in using MWF is the possibility of cancer caused by exposure to fluids. One of the initial concerns in this area arose with the report[64] that diethanolnitrosamine, which has been shown to be a liver carcinogen in laboratory animals, could form in MWF that contained both nitrites and diethanolamine (DEA). Additionally, there has been concern over the potential carcinogenicity of some petroleum-based mineral oils. The nitrosamine problem was addressed relatively easily in the industry by simply avoiding the combination of secondary amines and nitrites. The oil concern is more complex and involves a critical component of all but synthetic MWF.

One of the major difficulties in an evaluation of the literature on mineral oils is the poor definition of the material under study. In the past, the term mineral oil has been used to describe oils derived from coal, shale, petroleum crude oil, and even animal and vegetable sources. There has been little recognition of the vast differences in the production, uses, chemical, and physical and toxicological characteristics of mineral oils.

This was the case when, in 1973, the International Agency for Research on Cancer (IARC) cited various reports in the literature and stated that there was sufficient evidence of carcinogenicity of some mineral oils in experimental animals and humans.[65] IARC did, however, acknowledge that mineral oils vary in their composition, which may also affect their carcinogenicity.

In order to clarify the mineral oil issue, IARC convened a group of scientists to evaluate the carcinogenicity of mineral oils derived from petroleum crude oils, which are then further refined and used as base oils in fuels and lubricants.[66]

a. *Refining History and Carcinogenic Potential of Mineral Oils*

The important factors in the production of lubricating oil products are the petroleum crude oil type, the manufacturing or refining process, and the formulation of the final product. Petroleum crude oils are classed as paraffinic or naphthenic. Lubricant refining and, correspondingly, product formulations have changed considerably over the years. Until about 1940, processing consisted of acid refining with clay finishing and subsequent dewaxing by chilling. Solvent refining (and solvent dewaxing) was first introduced in the U.S. and Europe in the 1930s. This is an extraction process which, following solubilization of the polycyclic aromatic hydrocarbons (PAHs), selectively removes olefins, naphthenes, and then paraffins, depending on the severity of the process. Hydrotreating, a newer, more severe process than hydrofinishing, was introduced in the 1960s. Through a catalytic hydrogenation process, the lubricant base oil is made more paraffinic by the saturation of olefins. The severity of the hydrogenation dictates the degree of conversion of aromatics to naphthenes.

Particularly since 1985, the trend has been towards more highly refined oils with removal of unwanted impurities including PAHs, constituents believed to be major factors in imparting carcinogenic activity to these products. Consequently, animal studies have been conducted on refined mineral oils derived from these newer processing techniques in order to evaluate carcinogenic potential; these are primarily mouse skin-painting studies. These studies are fairly common in toxicology and considered to be relatively accurate in predicting skin carcinogenic potential in man. Data from these animal carcinogenicity studies are the primary basis for the following conclusions drawn by IARC:

- There is sufficient evidence of carcinogenicity for:
  - Untreated vacuum distillates
  - Acid-treated oils (which includes caustic neutralization, dewaxing, or clay treating)
  - Aromatic oils
  - Mildly solvent-refined oils
  - Mildly hydrotreated oils
- There is no evidence of carcinogenicity for:
  - Severely solvent-refined oils
  - White oils (when administered by routes other than intraperitoneal injection)
- There is inadequate evidence (one study) to evaluate severely hydrotreated oils or oils that have been mildly solvent refined with subsequent mild hydrotreatment; however, to date, the data have indicated these are not carcinogenic.

Therefore, if the oil in a MWF has been severely refined, carcinogenic risk from this component would be unlikely. From this information, a question that naturally arises is: How is mild or severe refining defined? Unfortunately, there is considerable controversy as well as data gaps in defining these terms, and the Chemical Abstract Services number (CAS number) indicates only the final refining process — not the severity or earlier processing. Owing to this, OSHA published in the Federal Register on December 20, 1985, *Hazard Communication; Interpretation Regarding Lubricating Oils*. One of the purposes of this document was to define mild hydrotreatment. The critical factors for defining this process are pressure and temperature. OSHA has decided that an oil has been mildly hydrotreated if it has been processed at a pressure of 800 lb/in.^{2} or less, at temperatures of up to 800°F. Unfortunately, OSHA did not define any of the other refining parameters.

Since the OSHA publication, other research or reviews of health affects of oil mist have appeared.

Mackerer,[67] in 1989, reviewed the health effects of mineral oils and concluded that dermatitis remained the most prevalent health effect of exposure to cutting fluids. He noted that in most cutting fluid applications, polycyclic aromatic (PCA) levels do not increase substantially; severely treated virgin oils, for which there is no evidence of carcinogenicity, should be used in cutting oil formulations; rerefined oils (see discussion below), if used, should be tested and certified, on a batch by batch basis if necessary, to assure low levels of PCA; good personal hygiene was important in minimizing potential adverse health effects; and proprietary additives should be tested for toxicity by the supplier and results evaluated by the user's health professionals.

McKee et al.[68] reviewed a severe hydrotreatment refining process that was optimized for PAHs. As compared with mild and unrefined hydrocarbon distillates, the severely hydrotreated distillates did not produce skin tumors in mice during a 2-yr study. Further, McKee and Scala,[69] in 1990, tested several solvent-extracted base oils and found that neither the solvent-extracted base oils themselves, or new or used cutting fluids and industrial oils formulated from them, were dermal carcinogens during a 2-yr test with mice.

Dalbey et al.[70] in 1991, exposed rats to aerosols of one of three base stocks used to formulate lubricating oils: solvent-refined oil, a hydrotreated oil, and a hydrocracked oil. Exposures were for 6 h/d, 5 d/week for 4 weeks at levels of 0, 50, 210, and 1000 mg/m^3. The only treatment-related changes were in the lung and associated lymph nodes, in which both the wet weight of the lung and the dry:wet ratio increased in a concentration-related manner. Associated with the increased weights were accumulations of foamy alveolar macrophages, particularly in alveoli close to alveolar ducts. Mild infiltration by neutrophils was observed with the hydrotreated oil and with the solvent-refined oil, and thickened alveolar walls were noted with the highest concentration of hydrocracked oil. The mild responses at very high concentrations, the authors concluded, indicated a low degree of toxicity for these aerosols.

In 1993, Chasey and McKee[71] evaluated 94 samples of oil in a mouse epidermal cancer bioassay, along with several biological and chemical test methods, including a dimethyl sulfoxide (DMSO) extraction test, a UV absorbance test of the DMSO extract, and the modified Ames test (see below), and compared these results to refining history. The authors concluded that, while all the alternative test methods showed similar agreement with the dermal bioassay data, the short-term tests should not be used in preference to dermal carcinogenicity data for hazard identification or labeling purposes.

In 2003, Mackerer et al.[72] reviewed petroleum mineral oil refining and evaluation of cancer hazard. The authors noted that processes that can reduce PAC levels in refined petroleum base oils are known, but that the operating conditions for the processing units (e.g., temperature, pressure, catalyst type, and residence time) to achieve adequate PAC reduction are refinery specific. Mackerer reviewed the short-term test data available, and compared them with mouse bioassay data available for the tested oils, concluding that currently produced base oils were noncarcinogenic.

Also in 2003, Dalbey and Biles[73] reviewed subchronic and chronic inhalation studies with mineral oils with emphasis on studies published since 1990, but also including inhalation studies of mineral base oils prior to 1985. The authors concluded, from all the work reviewed, that reactions of laboratory animals to exposure were limited to the lung and appeared to be primarily nonspecific responses to an aerosol with low toxicity.

Since 1964, the American Conference of Governmental Industrial Hygienists (ACGIH) have published a threshold limit value (TLV) for mineral oil, oil mist particulate, oil mist, or mineral oil mist of 5 mg/m^3, as a time-weighted average. Beginning in 1992, ACGIH has placed mineral oil mist, and more recently, mineral oil on its notice of intended changes list. ACGIH, for example, in their 2003 publication *TLVs® and BEIs®*,[74] proposed lowering the TLV for mineral oil from 5 to 0.2 mg/m^3, a 25-fold reduction. ACGIH also proposed a designation of "A2 — suspected human carcinogen" for poorly and mildly refined mineral oil, and "A4 — not classifiable as a human carcinogen" for highly and severely refined mineral oil.

ACGIH has adopted categories for carcinogenicity as follows:

- A1 — *Confirmed human carcinogen*. The agent is carcinogenic to humans based on the weight of evidence from epidemiologic studies.
- A2 — *Suspected human carcinogen*. There is limited evidence of carcinogenicity in humans and sufficient evidence of carcinogenicity in experimental animals with relevance to humans.
- A3 — *Confirmed animal carcinogen with unknown relevance to humans*. The agent is carcinogenic in experimental animals at a relatively high dose, by route(s) of administration, at site(s), of histologic type(s), or by mechanism(s) that may not be relevant to human exposure.
- A4 — *Not classifiable as a human carcinogen*. Agents which cause concern that they could be carcinogenic for humans but which cannot be assessed conclusively because of the lack of data.
- A5 — *Not suspected as a human carcinogen*. The agent is not suspected to be a human carcinogen on the basis of properly conducted epidemiologic studies in humans.

ACGIH assigns no carcinogenicity designation to substances for which no human or experimental animal carcinogenic data are reported.

The proposed significantly lower TLV for *both* poorly/mildly refined mineral oils and for highly/severely refined mineral oils has caused concern, particularly since previously discussed data presented by Mackerer, and by Dalbey and Biles, do not support such a significant reduction of TLV for highly/severely refined mineral oils. Debate on this issue is likely to continue.

b. The Modified Ames Assay: Predicting Mineral Oil Carcinogenicity

Another technique for evaluating potential carcinogenicity of oils is the modified Ames assay based on the standard Ames bacterial mutagenicity assay. Blackburn et al.[75] reported that the modified Ames assay has a correlation coefficient of 0.92 for oils with median boiling points between 5000 and 107°F when compared with the results of the long-term mouse skin-painting studies. Roy et al.[76] reported a significant correlation between the 3- to 7-ring PAH compounds and both mutagenic and carcinogenic potency. The two major advantages of this type of assay vs. conducting mouse skin-painting tests are time and money. The modified Ames can be run and evaluated in a few days, whereas mouse skin painting takes approximately 2 yr of testing and another year or so to analyze the results; obviously at a much greater cost. In addition, the modified Ames can be used quickly to screen oils of unknown refining history in order to predict potential carcinogenicity. The assay, described in detail in ASTM E1687,[77] is now widely used in the oil industry, often for quality control.

c. Human Studies on Metalworking Fluids

Numerous epidemiology studies have evaluated the carcinogenic potential of occupational exposure to oil mists and MWF. The majority of cancer types reported have been skin (scrotal), respiratory, or gastrointestinal. Cases of scrotal cancer have been extensively reported in the literature, primarily associated with use of poorly refined oil in conjunction with poor personal hygiene.[78] Possible causal relationships between exposure to oil mist and cancer of the respiratory or gastrointestinal tracts is difficult to establish as tumors in these organs are common and many other factors, such as smoking and diet, could be responsible. Additionally, few of these studies provide analytical data on either the bulk fluid, including the critical question of refining history of

TABLE 15.3
Epidemiology Studies of Metalworking Fluids

Period of Exposure	Fluid Class	Location		Ref.
1938–1968	Straight oil, soluble oil, synthetic	North central U.S.		(83)
1950–1967	Cutting oil	Iowa		(84)
1920–1984	Straight oil, soluble oil	Michigan	Plant 1	(80), et seq.
1939–1984	Straight oil, soluble oil synthetic[a]	Michigan	Plant 2	
1920–1984	Straight oil	Michigan	Plant 3	
1951–1984	Unknown	Ohio	Male	(82)
1951–1984	Unknown	Ohio	Female	
1917–1984 (deaths between 1940 and 1984)	Straight oil, soluble oil, synthetic[b]	Michigan		(38)
1917–1984 (deaths between 1985 and 1994)	Straight oil, soluble oil, synthetic[b]	Michigan		
1935–1973	Cutting oil	Connecticut		(85)
1950–1966	Synthetic or semi-synthetic	Sweden		(86)
1950–1966	Straight oils, soluble oils	Sweden		(87)
1950–1978	Cutting oils	British Columbia		(88)
1938–1979	Unknown	New York		(89)
1911–1982	Mostly soluble oils	Connecticut		(90)
1920–1982	Various	Connecticut		(91)
1945–1960	Various	Finland		(92)
1950–1966	Straight oils, soluble oils	Sweden		(93)
	Cutting oils	Illinois		(94)
1935–1991	Unknown	Connecticut		(95)
1966–1988	Unknown	Ohio		(96)
1954–1959	Straight oils	Sweden		(97)
1950–1987	Cutting oils	Iowa		(98)

[a] Semi-synthetic fluids classed as synthetics.
[b] Semi-synthetic fluids classed as soluble oils.

the oil, or air contaminant levels associated with the effects reported, making difficult extrapolation of the results to other settings.

After a systematic and comprehensive review of epidemiologic studies conducted prior to 1997, Calvert[79] reported substantial evidence of an association between exposure to MWF and increased risk of cancer of the larynx, rectum, pancreas, skin, scrotum (straight oils), and bladder (synthetic fluids) for "at least some MWF used prior to the mid-1970s." A detailed summary of the more important studies is given in Table 15.3. Note that none of these studies reflects exposures after 1988. Only two of the studies were concerned only with exposures subsequent to 1950, the date when semisynthetic and synthetic fluids and the use of nitrites in MWF were introduced.

A large body of data has been developed under a study of three plants supported by General Motors and the UAW, covering exposures from 1917 to 1984[80] and deaths from 1940 to 1994,[81] expressed as standard mortality ratios (SMR). A comparison of the results of this study with those of Rotimi,[82] DeCouflé,[83] and Aquavella[84] can be made for esophageal cancer (Table 15.4), stomach cancer (Table 15.5), rectal cancer (Table 15.6), liver cancer (Table 15.7), pancreatic

TABLE 15.4
SMR Values for Esophageal Cancer

Deaths Observed	SMR	LCL	UCL	Comments	Ref.
31	1.25	0.85	1.77	Plant 1 white males	(80)
10	0.68	0.33	1.25	Plant 1 black males	(80)
11	0.84	0.42	1.51	Plant 2 white males	(80)
6	1.38	0.5	3.01	Plant 3 white males	(80)
104	1.09	0.89	1.31	Pre-1994	(81)
45	1.23	0.95	1.37	1985–1994	(81)
1	0.26	0.07	1.46		(84)
4	1.14	0.31	3.34		(83)

TABLE 15.5
SMR Values for Stomach Cancer

Deaths Observed	SMR	LCL	UCL	Comments	Ref.
72	1.08	0.84	1.36	Plant 1 white males	(80)
17	0.96	0.56	1.54	Plant 1 black males	(80)
33	1.26	0.87	1.77	Plant 2 white males	(80)
4	0.59	0.16	1.50	Plant 3 white males	(80)
151	1.04	0.89	1.23	Pre-1994	(81)
50	1.26	0.93	1.66	1985–1994	(81)
0	1.35	0.90	11.90	Females	(82)
28	1.35	0.90	1.96	Males	(82)
6	0.88	0.32	1.91		(84)
17	1.25	0.73	2.00		(83)

TABLE 15.6
SMR Values for Rectal Cancer

Deaths Observed	SMR	LCL	UCL	Comments	Ref.
37	1.01	0.71	1.39	Plant 1 white males	(80)
3	0.64	0.13	1.86	Plant 1 black males	(80)
17	1.09	0.63	1.74	Plant 2 white males	(80)
7	1.70	0.68	3.50	Plant 3 white males	(80)
87	1.07	.85	1.33	Pre-1994	(81)
23	1.21	0.77	1.82	1985–1994	(81)
Not reported				Females	(82)
Not reported				Males	(82)
Not reported					(84)
8	1.25	0.54	2.46		(83)

TABLE 15.7
SMR Values for Liver Cancer

Deaths Observed	SMR	LCL	UCL	Comments	Ref.
27	1.06	0.7	1.54	Plant 1 white males	(80)
11	1.60	0.80	2.87	Plant 1 black males	(80)
11	0.99	0.49	1.77	Plant 2 white males	(80)
9	2.77	1.26	5.25	Plant 3 white males	(80)
94	1.40	1.14	1.72	Pre-1994	(81)
36	1.65	1.15	2.29	1985–1994	(81)
Not reported				Females	(82)
Not reported				Males	(82)
0	0.00	0.00	1.14		(84)
Not reported					(83)

TABLE 15.8
SMR Values for Pancreatic Cancer

Deaths Observed	SMR	LCL	UCL	Comments	Ref.
44	0.76	0.55	1.02	Plant 1 white males	(80)
21	1.70	1.05	2.61	Plant 1 black males	(80)
25	0.85	0.55	1.26	Plant 2 white males	(80)
8	0.87	0.37	1.71	Plant 3 white males	(80)
179	1.06	0.92	1.23	Pre-1994	(81)
80	1.41	1.12	1.76	1985–1994	(81)
Not reported				Females	(82)
30	1.11	0.75	1.58	Males	(82)
13	1.36	0.73	2.33		(84)
8	1.05	0.45	2.07		(83)

TABLE 15.9
SMR Values for Laryngeal Cancer

Deaths Observed	SMR	LCL	UCL	Comments	Ref.
16	1.02	0.58	1.66	Plant 1 white males	(80)
7	1.63	0.65	3.36	Plant 1 black males	(80)
15	1.85	1.03	3.05	Plant 2 white males	(80)
2	0.77	0.09	2.79	Plant 3 white males	(80)
55	1.18	0.89	1.55	Pre-1994	(81)
15	1.00	.56	1.65	1985–1994	(81)
Not reported				Females	(82)
5	0.62	0.20	1.45	Males	(82)
0	0.00	0.00	1.82		(84)
Not reported					(83)

TABLE 15.10
SMR Values for Colon Cancer

Deaths Observed	SMR	LCL	UCL	Comments	Ref.
95	0.92	0.74	1.13	Plant 1 white males	(80)
9	0.59	0.27	1.12	Plant 1 black males	(80)
34	0.69	0.48	0.97	Plant 2 white males	(80)
22	1.47	0.92	2.22	Plant 3 white males	(80)
280	0.94	0.83	1.06	Pre-1994	(81)
116	1.05	0.87	1.27	1985–1994	(81)
Not reported				Females	(82)
43	0.93	0.67	1.25	Males	(82)
Not reported					(84)
17	1.31	0.76	2.09		(83)

TABLE 15.11
SMR Values for Lung Cancer

Deaths Observed	SMR	LCL	UCL	Comments	Ref.
320	1.02	0.91	1.14	Plant 1 white males	(80)
77	1.05	0.83	1.31	Plant 1 black males	(80)
213	1.16	1.01	1.32	Plant 2 white males	(80)
60	0.91	0.70	1.17	Plant 3 white males	(80)
1155	1.06	1.00	1.12	Pre-1994	(81)
475	1.11	1.01	1.22	1985–1994	(81)
3	1.32	0.27	3.85	Females	(82)
224	1.13	0.99	1.29	Males	(82)
70	1.31	1.02	1.65		(84)
Not reported					(83)

TABLE 15.12
SMR Values for Skin Cancer

Deaths Observed	SMR	LCL	UCL	Comments	Ref.
10	0.61	0.29	1.13	Plant 1 white males	(80)
1	0.78	0.01	4.34	Plant 1 black males	(80)
11	1.06	0.53	1.89	Plant 2 white males	(80)
7	1.27	0.51	2.62	Plant 3 white males	(80)
38	0.66	0.46	0.90	Pre-1994	(81)
9	0.38	0.18	0.72	1985–1994	(81)
Not reported				Females	(82)
Not reported				Males	(82)
Not reported					(84)
Not reported					(83)

TABLE 15.13
SMR Values for Leukemia

Deaths Observed	SMR	LCL	UCL	Comments	Ref.
65	1.57	1.21	2.00	Plant 1 white males	(80)
4	0.71	0.19	1.81	Plant 1 black males	(80)
20	0.94	0.57	1.45	Plant 2 white males	(80)
9	1.07	0.49	2.02	Plant 3 white males	(80)
162	1.33	1.14	1.56	Pre-1994	(81)
62	1.48	1.14	1.92	1985–1994	(81)
Not reported				Females	(82)
17	0.89	0.52	1.42	Males	(82)
3	0.36	0.07	1.04		(84)
3	0.56	0.11	1.62		(83)

cancer (Table 15.8), laryngeal cancer (Table 15.9), colon cancer (Table 15.10), lung cancer (Table 15.11), skin cancer (Table 15.12), leukemia (Table 15.13), and brain cancer (Table 15.14).

In making these comparisons, it should be noted that the MWF being used were not characterized and the inconsistencies found may reflect differing carcinogenic potentials for different fluids. Further, the number of expected deaths can be based on national, state, county, or ethnic/gender subgroups. Having said that, the results reported do show inconsistencies; but the reported SMR values tend to be consistently near the null value of 1.00 (no effect).

The data from the GM–UAW cohort have also been analyzed by means other than the SMR. Tolbert[99] estimated SMR values according to the type of fluid exposure, as well as Poisson regression analysis, and reported associations of exposure to straight oils with rectal, laryngeal, and prostate cancer, and a weak association between straight oil and esophageal cancer. Poisson regression analysis found limited evidence for an association between pancreatic cancer and exposure to synthetic fluids that was not reflected in the SMR analysis.

Case–control studies were also reported for laryngeal cancer,[100] lung cancer,[101] pancreatic cancer,[102] esophageal cancer,[103] stomach cancer,[104] and breast cancer.[105] The results of the case–control studies were presented as showing an association between straight oil and laryngeal cancer, a negative association between exposure to synthetic fluids and lung cancer, a strong association

TABLE 15.14
SMR Values for Brain Cancer

Deaths Observed	SMR	LCL	UCL	Comments	Ref.
33	1.34	0.92	1.88	Plant 1 white males	(80)
2	0.72	0.08	2.59	Plant 1 black males	(80)
15	0.90	0.51	1.49	Plant 2 white males	(80)
7	0.85	0.34	1.75	Plant 3 white males	(80)
88	1.11	0.90	1.38	Pre-1994	(81)
30	1.16	0.78	1.66	1985–1994	(81)
Not reported				Females	(82)
Not reported				Males	(82)
Not reported					(84)
2	0.30	0.04	1.08		(83)

between exposure to synthetic fluids and pancreatic cancer, and associations between exposures to synthetic fluids and soluble oils in combination with grinding and esophageal cancer.

All the GM–UAW studies group fluids into three classes: straight, soluble, and synthetic. The fourth class of fluids, the semisynthetics, is sometimes classified as soluble oils, sometimes as synthetics, or their classification is not mentioned. For example, in the pancreatic and esophageal cancer studies, it is revealed that the semisynthetic fluids were classed as soluble oils. In the laryngeal and lung cancer studies, the classification of semisynthetic fluids is not mentioned. In another analysis for respiratory effects, semisynthetics were grouped with synthetics. If semisynthetics were properly treated as a separate class throughout the analysis of the studies, the conclusions reached by these studies could have been changed significantly. In any event, the misclassification would have inflated the number of deaths reported for either soluble oils or synthetic fluids, and seriously limits the ability of the studies to produce valid conclusions about fluid classes.

A more sophisticated analysis using a Cox proportional hazards model and penalized splines found an association between larynx cancer incidence and cumulative straight MWF exposure. The results for esophageal cancer were less consistent. For stomach cancer there was no evidence of excess risk.[106]

The relevance of these results to modern MWF is questionable. In the largest study,[6] 64% of the deaths came from a plant that opened in 1917. The average length of exposure at this plant is listed as 29 years, so that half of all first exposures would have been prior to 1954. For the only plant where semisynthetic or synthetic fluids were used, operations began in 1938. The average length of exposure was 19 yr; half of all exposures would have started in the period between 1938 and 1964. For the studies from these plants to be applicable to MWF used at present, the findings would need to be extrapolated from the 1940s.

The composition of MWF has not been static during this time. Changes in the refining processes for mineral oils have been noted above. Formulation changes in straight oils included discontinuing the use of kerosene (cancer and fire safety), sperm oil (conservation concerns), and polychlorinated biphenyls (environmental concerns). The practice of using chlorinated solvents (cancer concerns) as tapping fluids was eliminated.

Changes in water-based fluids between 1970 and 1984 included the elimination of alkali metal nitrites (and the potential for contamination by nitrosamines), chromates (cancer concerns), and para-*tert*-butylbenzoic acid (testicular atrophy) in semisynthetic and synthetic fluids, and the replacement of naphthenic acids (cancer concerns) by petroleum sulfonates in soluble oils. There was also a reduced use of phenolic biocides (environmental concerns) in fluid formulations and as tankside additives, although this trend has been reversed recently as a result of concern over fluid contamination by mycobacteria.

With the promulgation of the Hazard Communication Standard by OSHA in 1985, the use of severely solvent-refined or severely hydrotreated base stock oils became general practice. As discussed later, the use of short-chain chlorinated paraffins (cancer concerns) was also sharply reduced.

Changes in MWF formulations have not been driven entirely by health or environmental concerns. The metalworking manufacturing industry has changed drastically over the last 50 yr. Improvements in process efficiencies, conversions from iron and steel to nonferrous metals, ceramics and plastics, and reductions in the size of central systems have completely changed the definition of an acceptable MWF. A fluid from the 1950s would be as relevant in today's market as the original Model T Ford would be on an interstate highway.

Based on the complexity and diversity of the composition of MWF, the changes in formulations to remove suspect chemicals, and the changes in the technical specifications for machining over decades, it is impossible to make generalized conclusions based on the old data, which studied only a small fraction of the fluids that were in use in past decades.

These epidemiology and toxicology studies are not adequate to make general conclusions about all MWF. Workplaces were randomly chosen based on availability of funding or a perceived need to study a particular worksite. The few toxicity studies used fluids that were available at the times those studies were conducted. The fluids employed in most toxicology and epidemiology studies are insufficiently characterized by composition and type to make any valid conclusions that can be extended to MWF in general. Moreover, most epidemiology studies involved exposures to fluids significantly different in composition than those used today. This does not even consider the question of identifying and controlling confounding factors.[107] Updates on these epidemiology studies will be similarly limited.

Given the marked variation in the composition of MWF across worksites and over time, and the limited ability of available epidemiologic studies to isolate effects of critical components, the present studies can only attempt to describe workplace factors that were relevant 20 years ago and would not reflect the significant changes in MWF formulations since 1986. Avoiding hazards of future exposures must be based on careful and continuing review of the toxicity of the fluids, their components, and their possible contaminants.

d. Carcinogenicity of Metalworking Fluid Components

Limited data are available on the carcinogenicity of individual components used in MWF.

Knaak et al.[108] reviewed the toxicology of mono-, di-, and triethanolamine, including studies conducted by National Toxicology Program (NTP) on DEA and TEA. DEA was seen as the most toxic of the three alkanolamines. For example, a subchronic study of DEA applied to mice in varying dosages over a 13-week period, yielded a calculated dosage of 32 mg/kg/d as a no observed adverse effect level in male rats. An increased incidence of nephropathy, tubule mineralization, and skin histopathological changes that accompanied organ weight changes in low-dose females, suggested a more significant effect of dermally administered DEA in female rats and a lack of a clear NOAEL.

NTP[109] completed a 2-yr dermal assay of DEA. Under the conditions of these 2-yr dermal studies, there was no evidence of carcinogenic activity of DEA in male F344/N rats administered 16, 32, or 64 mg/kg DEA or in female F344/N rats administered 8, 16, or 32 mg/kg. There was clear evidence of carcinogenic activity of DEA in male and female B6C3F1 mice based on increased incidences of liver neoplasms in males and females and increased incidences of renal tubule neoplasms in males. Nonetheless, DEA has not been included in the NTP listing of carcinogens, nor has it been declared a carcinogen by IARC.

Dermal administration of DEA to rats was associated with increased incidences of acanthosis (males only), hyperkeratosis, and exudate of the skin and increased incidences and severities of nephropathy in females. Dermal administration of DEA to mice was associated with increased incidences of cytoplasmic alteration (males only) and syncytial alteration of the liver, renal tubule hyperplasia (males only), thyroid gland follicular cell hyperplasia, and hyperkeratosis of the skin.

Based on the subchronic results, most formulators of MWF moved away from including DEA or DEA alkanolamides in their MWF formulations during the 1990s.

Knaak et al. reviewed NTP's subchronic results for TEA for rats and mice, which were administered doses of between 0 and 2000 or 4000 mg/kg/d TEA for 13 weeks. Significant decreases in body weight gains occurred in high-dose group male and female rats and there were statistically significant increases in the kidney weights of high dosage males and females. While renal tube mineralization in female rats occurred at >500 mg/kg/d, the effect was not dose related. Based on the data, a NOAEL for subchronic dermal administration of TEA in rats was 250 mg/kg/d.

NTP also performed a companion subchronic study on mice. From the study results, a NOAEL for dermally applied TEA in mice appeared to be between 1000 and 2000 mg/kg/d.

Knaak et al. also reviewed the NTP chronic toxicity studies on TEA in rats and mice. In the dermal rat study, TEA was applied in acetone solution at dosages of 0, 32, 64, or 125 mg/kg/d (male) and 0, 63, 125, or 250 mg/kg/d (female) for 5 d/week for 2 yr. Control groups were administered acetone only; no untreated controls were included. Systemic effects were primarily limited to kidney tissues. Step-sectioning of kidneys failed to detect a greater number of tumors in treated rats than standard methods and, combining all proliferative lesions, resulted in an almost identical incidence between treated and control groups of rats, suggesting a lack of tumorigenic response in the kidneys of male rats. Overall, according to Knaak et al., the study failed to generate clear evidence of a carcinogenic response in rats and that the male kidney tumor data were equivocal. In the companion mouse study, similar issues between treated animals and acetone-only treated controls were observed.

Despite the relatively high doses of TEA used in the mouse assay, Knaak et al. observed that there were no significant treatment-related effects in mortality and body weight, and relatively few systemic effects. The most significant nonneoplastic changes found in the male mice were consistent with a chronic bacterial hepatitis in these animals. Selective staining of hepatic tissues revealed the presence of a *Helicobacter* species, presumably *H. hepaticus*. Infection of mice with these bacteria had earlier been shown to be associated with higher incidence of hepatocellular neoplasms in mice. While the incidence of hepatic adenomas in male and female mice were higher than controls at the 2000 and 1000 mg/kg/d levels, the incidence of carcinomas was elevated only in female mice and not in a dose-related manner. As 94% of the mice with evidence of infection also had tumors, the authors of the report concluded that the data did not present clear evidence of carcinogenicity and that the male mouse response was equivocal evidence.

TEA has not been determined to be a carcinogen either by NTP or by IARC.

Since secondary alkanolamines can react with nitrites in MWF to form nitrosamines, which are liver carcinogens in laboratory animals, combinations of such alkanolamines and nitrites in MWF have been effectively banned by the publication of U.S. Federal Significant New Use Rule in 1993.[110]

A formerly common family of compounds in MWF that are now considered both a carcinogenic concern and an environmental concern are chlorinated paraffins. These compounds were often included in fluids as extreme-pressure additives and antiwear agents. In 1984, NTP[111,112] completed carcinogenicity testing of two of the chlorinated paraffins, a C12, 58–60% chlorine product and a C23/24, 40–43% chlorine (Cl) material. Rats and mice were fed the chemical over their lifetimes. The C12, 58–60% Cl was positive for carcinogenicity in both rats and mice; however, equivocal results were obtained following exposure to the C23/24, 40–43% Cl compound. Therefore, the C12, 60% Cl, chlorinated paraffins were classified as suspect carcinogens. These compounds are not, however, absorbed through the skin,[113] eliminating this potential route of exposure.

e. *Reclaimed, Rerefined, and Used Mineral Oils*

Reclaimed, rerefined, and used oils represent a growing class of mineral oils that are increasingly used in manufacturing in response to pollution prevention and waste minimization initiatives. The potential hazards of these materials result from contaminants associated with their initial use. In large part, the toxicity of these fluids has not been studied. The IARC monograph did not address reclaimed oils and has only a brief section on used oils, which is not defined. McKee et al.[113] assessed the dermal carcinogenic potential and PAH content of refined oils in new and used MWF. None of the solvent-extracted base oils induced skin tumors, and similarly, the cutting fluids prepared from these oils were not carcinogenic in mouse skin-painting bioassays. Additionally, there was no evidence that use increased PAH content or carcinogenic potential.

VI. CONCLUSIONS

The complexity of MWF and their patterns of use result in a complex health and safety profile for these materials. Key considerations include both the toxicity of the starting components, as well as the impact of contaminants, particularly microbiological contaminants and decay products, which enter the fluid in use. Routes of exposure include inhalation and dermal contact.

Epidemiology studies have provided useful insights on the potential health effects of exposure to MWF as a whole, but evaluated effects that may have been due to exposures from fluids used 20 to 30 years earlier. In these studies, determination of specific causative agents is difficult because of a lack of chemical data both on the bulk fluids as well as on employee exposures that might have occurred historically. It is certainly clear that the chemical nature of MWF has changed dramatically over time, especially with respect to the oils often contained in these products, since changes have occurred in refining practices, aimed at removal of impurities such as PAH. This makes extrapolation of the results from many epidemiology studies to current MWF applications difficult.

Toxicology testing on MWF components provides both qualitative and quantitative data on the potential health effects from exposure, and can assist in interpreting results of epidemiology studies. We now know, for example, that only severely treated mineral oils can be used in MWF.

In all applications where MWF are used, good industrial hygiene practice must be maintained. Consideration must be given to reducing exposure through use of impervious gloves and protective clothing, barrier creams, and local and general exhaust ventilation, as well as other industrial hygiene control measures. Absent an OSHA Permissible Exposure Limit (PEL), NIOSH's REL of 0.5 mg/m^3 should be the minimum exposure target level. A mist control program, based on a total systems approach, including control of tramp oil contamination, is very important. Practical steps users can take to control mist include, for example:

- Optimize machine tool feeds and speeds. Excessively high feeds and speeds increase the amount of aerosol generated.
- Minimize coolant flow rates, consistent with desired part finish and dimensions, and with movement of generated chips or swarf.
- Use flooding instead of spray application, wherever possible.
- Control tramp oil contamination. Research shows excess tramp oil increases mist.
- Installation and maintenance of local and general exhaust ventilation to reduce exposure to equal or less than the NIOSH REL is essential.

Recent research has shown that acute respiratory effects, and some chronic effects such as HP, are the result of microbial growth in water-miscible MWF. As a result, besides reducing exposure to 0.5 mg/m^3 or less, both users and the chemical managers who manage MWF must assure a MWF management program is in place, which includes good microbiological control of fluids. If this is accomplished, along with control of skin irritation and dermatitis, risk of any potential chronic effects can be minimized.

REFERENCES

1. NIOSH, *National Occupational Hazard Survey*, Vol. 3, National Institute of Occupational Safety and Health, Cincinnati, Publication no. 78-114, pp. 216–229, 1977.
2. ASTM E1302. *Guide for Acute Animal Toxicity Testing of Water-Miscible Metalworking Fluids*, available from ASTM International, West Conshohocken, PA 19428, 1942, http://www.astm.org.
3. Zugerman, C., Cutting fluids: their use and effects on the skin, *Occup. Med.: State of the Art Rev. I*, 245, 1986.
4. White, E. and Lucke, W., Effects of fluid composition on mist composition, *Appl. Occup. Environ. Hyg.*, 18, 838–841, 2003.
5. Turchin, H. and Byers, J., Effect of oil contamination on metalworking fluid mist, *Lubr. Eng.*, 56(7), 21–25, 2000.
6. Kennedy, S., Greaves, I., Kriebel, D., Eisen, E., Smith, T., and Woskie, S., Acute pulmonary responses among automobile workers exposed to aerosols of machining fluids, *Am. J. lnd. Med.*, 15, 627–641, 1989.
7. Schaper, M. and Detweiler, K., Evaluation of the acute respiratory effects of aerosolized machining fluids, *Fundam. Appl. Toxicol.*, 16, 309–319, 1991.

8. Gordon, T., Acute respiratory effects of endotoxin-contaminated machining fluid aerosols in guinea pigs, *Fundam. Appl. Toxicol.*, 19, 117–123, 1992.
9. Ameille, J., Wild, P., Choudat, D., Ohl, G., Vancouler, J., Chanut, J., and Brochard, P., Respiratory symptoms, ventilatory impairment, and bronchial reactivity in oil mist-exposed automobile workers, *Am. J. Ind. Med.*, 27, 247–256, 1995.
10. Gordon, T. and Harkema, J., Mucous cell metaplasia in the airways of rats exposed to machining fluids, *Fundam. Appl. Toxicol.*, 28, 274–282, 1995.
11. Krystofiak, S. and Schaper, M., Prediction of an occupational exposure limit for a mixture on the basis of its components: Application to metalworking fluids, *Am. Ind. Hyg. Assoc. J.*, 57, 239–244, 1996.
12. Detweiler-Okabayashi, K. A. and Schaper, M. M., Respiratory effects of a synthetic metalworking fluids and its components, *Arch. Toxicol.*, 70, 195–201, 1996.
13. Ball, A. M. and Lucke, W. E., *Evaluation of Sensory Irritation Potential for Commercially Available Metalworking Fluids, Metalworking Fluids Symposium I, The Industrial Metalworking Environment: Assessment and Control,* Detroit, MI, November 1995, *Symposium Proceedings*, The American Automobile Manufacturers Association, pp. 95–97, 1996.
14. Thorne, P. and DeKoster, J., *Respiratory Health Effects of Machining Fluids in Laboratory Animals, Metalworking Fluids Symposium I, The Industrial Metalworking Environment: Assessment and Control,* Detroit, MI, November 1995, *Symposium Proceedings*, The American Automobile Manufacturers Association, pp. 98–105, 1996.
15. Thorne, P. and DeKoster, J., Pulmonary effects of machining fluids in guinea pigs and mice, *Am. Ind. Hyg. Assoc. J.*, 57, 1168–1172, 1996.
16. Milton, D., Brain, J., and Rees, D., *Acute Effects of Metal Working Fluids in a Respiratory Inflammation Model, Metalworking Fluids Symposium I, The Industrial Metalworking Environment: Assessment and Control,* Detroit, MI, November 1995, *Symposium Proceedings*, The American Automobile Manufacturers Association, pp. 106–107, 1996.
17. Woskie, S., Virji, M., Kriebel, D., Sama, S., Eberiel, D., Milton, D., Hammond, S., and Moure-Eraso, R., Exposure Assessment for a Field Investigation of the Acute Respiratory Effects of Machining Fluids. I. Summary of Findings, *Am. Ind. Hyg. Assoc. J.*, 57, 1154–1162, 1996.
18. Thorne, P., DeKoster, J., and Subramanian, P., Environmental assessment of aerosols, bioaerosols, and airborne endotoxins in a machining plant, *Am. Ind. Hyg. Assoc. J.*, 57, 1163–1167, 1996.
19. Sprince, N., Thorne, P., Poppendorf, W., Zwerling, C., Miller, E., and DeKoster, J., Respiratory symptoms and lung function abnormalities among machine operators in automobile production, *Am. J. Ind. Med.*, 31, 403–413, 1997.
20. Robbins, T., Seixas, N., Franzblau, A., Abrams, L., Minick, S., Burge, H., and Schork, M., Acute respiratory effects on workers exposed to metalworking fluid aerosols in an automotive transmission plant, *Am. J. Ind. Med.*, 31, 510–524, 1997.
21. Greaves, I., Eisen, E., Smith, T., Pothier, L., Kriebel, D., Woskie, S., Kennedy, S., Shalat, S., and Monson, R., Respiratory health of automobile workers exposed to metal-working fluid aerosols: Respiratory symptoms, *Am. J. Ind. Med.*, 32, 450–459, 1997.
22. Eisen, E., Holcroft, C., Greaves, I., Wegman, D., Woskie, S., and Monson, R., A strategy to reduce healthy worker effect in a cross-sectional study of asthma and metalworking fluids, *Am. J. Ind. Med.*, 31, 671–677, 1997.
23. Kriebel, D., Sama, S., Woskie, S., Christiani, D., Eisen, E., Hammond, S., Milton, D., Smith, M., and Virji, M., A field investigation of the acute respiratory effects of metal working fluids, I: Effects of aerosol exposures, *Am. J. Ind. Med.*, 31, 756–766, 1997.
24. Sama, S., Kriebel, D., Woskie, S., Eisen, E., Wegman, D., and Virji, M., A field investigation of the acute respiratory effects of metal working fluids, II: Effects of airborne sulfur exposures, *Am. J. Ind. Med.*, 32, 767–776, 1997.
25. Zacharisen, M., Kadambi, A., Schleuter, D., Kurup, V., Shack, J., Fox, J., Anderson, H., and Fink, J., The spectrum of respiratory disease associated with exposure to metal working fluids, *J. Occup. Environ. Med.*, 40, 640–647, 1998.
26. *Criteria for a Recommended Standard: Occupational Exposure to Metalworking Fluids*, National Institute for Occupational Safety and Health, U.S. Department of Health and Human Services, DHHS Publication 98–102, Cincinnati, OH, 1998. Available at: http://www.cdc.gov/niosh/pdfs/98-102.pdf

27. Becklake, M., *The Relationship Between Acute and Chronic Airway Responses to Occupational Exposures, Metalworking Fluids Symposium II, The Industrial Metalworking Environment Assessment and Control of Metalworking Fluids,* Detroit, MI, September, 1997, *Symposium Proceedings*, The American Automobile Manufacturers Association, pp. 49–54, 1998.
28. Milton, D., *Biological Responses to Endotoxin in the Airways and Alveoli, Metalworking Fluids Symposium II, The Industrial Metalworking Environment: Assessment and Control of Metalworking Fluids,* Detroit, MI, September, 1997, *Symposium Proceedings*, The American Automobile Manufacturers Association, pp. 59–65, 1998.
29. *Management of the Metal Removal Fluid Environment*, ORC Worldwide, Washington DC, 1999. Available at: http://www.aware-services.com/orc/
30. Kennedy, S., Chan-Yeung, M., Teschke, K., and Karlen, B., Change in airway responsiveness among apprentices exposed to metalworking fluids, *Am. J. Respir. Crit. Care Med.*, 159, 87–93, 1999.
31. Gordon, T., *Exposure to MRF Aerosols: Acute and Subchronic Effects in Animals, Metalworking Fluids Symposium II, The Industrial Metalworking Environment Assessment and Control of Metalworking Fluids,* Detroit, MI, September, 1997, *Symposium Proceedings*, The American Automobile Manufacturers Association, pp. 91–94, 1998.
32. Gordon, T. and Galdanes, K., Factors Contributing to the acute and subchronic adverse respiratory effects of machining fluid aerosols in guinea pigs, *Toxic. Sci.*, 49, 86–92, 1999.
33. *Final Report of the OSHA Metalworking Fluids Standards Advisory Committee*, Occupational Safety and Health Administration, Washington, DC,1999. Available at: http://www.osha.gov/SLTC/metalworkingfluids/mwf_finalreport.html
34. *Metalworking Fluids: Safety and Health Best Practices Manual*, Occupational Safety and Health Administration, Washington, DC. Available at: http://www.osha.gov/SLTC/metalworkingfluids/metalworkingfluids_manual.html
35. Abrams, L., Seixas, N., Robbins, T., Burge, H., Muilenberg, M., and Franzblau, A., Characterization of metalworking fluid exposure indices for a study of acute respiratory effects, *Appl. Occup. Environ. Hyg.*, 15, 492–502, 2000.
36. Brown, M., White, E., and Feng, A., Effects of various treatments on the quantitative recovery of endotoxin from water-soluble metalworking fluids, *Am. Ind. Hyg. Assoc. J.*, 61, 517–520, 2000.
37. Virji, M., Woskie, S., Sama, S., Kriebel, D., and Eberial, D., Identifying the determinants of viable microorganisms in the air and bulk metalworking fluids, *Am. Ind. Hyg. Assoc. J.*, 61, 788–797, 2000.
38. Eisen, E., Smith, T., Kriebel, D., Woskie, S., Myers, D., Kennedy, S., Shalat, S., and Monson, R., Respiratory health of automobile workers and exposures to metal-working fluid aerosols: lung spirometry, *Am. J. Ind. Med.*, 39, 443–453, 2001.
39. Piacitelli, G., Sieber, W., O'Brien, D., Hughes, R., Glaser, R., and Catalano, J., Metalworking fluid exposures in small machine shops, *Am. Ind. Hyg. Assoc. J.*, 62, 356–370, 2001.
40. Bracker, A., Storey, E., Yang, C., and Hodgson, M., An outbreak of hypersensitivity pneumonitis at a metalworking plant: A longitudinal assessment of intervention effectiveness, *Appl. Occup. Environ. Hyg.*, 18, 96–108, 2003.
41. Linnainmaa, M., Kiviranta, H., Laitinen, J., and Laitinen, S., Control of workers' exposure to airborne endotoxins and formaldehyde during the use of metalworking fluids, *Am. Ind. Hyg. Assoc. J.*, 64, 496–500, 2003.
42. Zeka, A., Kriebel, D., Kennedy, S., and Wegman, D., Role of underlying pulmonary obstruction in short-term airway response to metal working fluid exposure: A reanalysis, *Am. J. Ind. Med.*, 43, 286–290, 2003.
43. Bukowski, J., Review of respiratory morbidity from occupational exposure to oil mists, *Appl. Occup. Environ. Hyg.*, 18, 828–837, 2003.
44. Oudyk, J., Haines, A., and D'Arcy, J., Investigating respiratory responses to metalworking fluid exposure, *Appl. Occup. Environ. Hyg.*, 18, 939–946, 2003.
45. Ross, A., Teschke, K., Brauer, M., and Kennedy, S., Determinants of exposure to metalworking fluid aerosol in small machine shops, *Ann. Occup. Hyg.*, 48, 383–391, 2004.
46. Veilette, M., Thorne, P., Gordon, T., and DuChaine, C., Six month tracking of microbial growth in a metalworking fluid after system cleaning and recharging, *Ann. Occup. Hyg.*, 48, 541–546, 2004.

47. Gordon, T., Metalworking fluid — the toxicity of a complex mixture, *J. Toxicol. Environ. Health, Part A*, 67, 209–219, 2004.
48. Lim, C-H., Yu, I., Kim, H-Y., Lee, S-B., Kang, M-G., Marshak, D., and Moon, C-K., *Inflammatory and Immunological Responses to Subchronic Exposure to Endotoxin-contaminated Metalworking Fluid Aerosols in F344 Rats*, Wiley Interscience, 2005, www.interscience.wiley.com, DOI 10.1002/tox.200097.
49. Savonius, B., Keskinen, H., Tupperainen, M., and Kanerva, L., Occupational asthma caused by ethanolamines, *Allergy*, 49, 877–881, 1994.
50. Eisen, E., Healthy worker effect in morbidity studies, *Med. Lav.*, 86, 125–128, 1995.
51. Rosenman, K., Riley, M., and Kalinowski, D., Work-related asthma and respiratory symptoms among workers exposed to metal-working fluids, *Am. J. Ind. Med.*, 32, 325–331, 1997.
52. Petsonk, E., Work-Related Asthma and Implications for the General Public, *Environ. Health Persp.*, 110, 569–572, 2002.
53. Muilenberg, M., Burge, H., and Sweet, T., Hypersensitivity pneumonitis and exposure to acid-fast bacilli in coolant aerosols, *J. Allergy Clin. Immunol.*, 91, 311, 1993.
54. Berstein, D., Lummus, Z., Santilli, G., Stiskosky, J., and Bernstein, L., Machine operator's lung: A hypersensitivity pneumonitis disorder associated with exposure to metalworking fluid aerosols, *Chest*, 108, 636–641, 1995.
55. Kriess, K. and Cox-Ganser, J., Metalworking fluid — associated hypersensitivity pneumonitis: A workshop summary, *Am. J. Ind. Med.*, 32, 423–432, 1997.
56. Shelton, B., Flamders, D., and Morris, G., *Mycobacterium* sp. as a possible cause of hypersensitivity pneumonitis in machine workers, *Emerg. Inf. Dis.*, 5, 270–273, 1999.
57. Moore, J., Christensen, M., Wilson, R., Wallace, R., Zhang, Y., Nash, D., and Shelton, B., Mycobacterial contamination of metalworking fluids: involvement of a possible new taxon of rapidly growing mycobacteria, *Am. Ind. Hyg. Assoc. J.*, 61, 205–213, 2000.
58. Centers for Disease Control and Prevention: Morbidity and Mortality Weekly Report, Respiratory illness in workers exposed to metalworking fluid contaminated with nontuberculosis mycobacteria, *J. Am. Med. Assoc.*, 287, 3073–3074, 2002.
59. Wallace, R., Zheng, Y., Wilson, R., Mann, L., and Rossmore, H., Presence of a single genotype of the newly described species mycobacterium immunogenum in industrial metalworking fluids associated with hypersensitivity pneumonitis, *Appl. Environ. Micro.*, 68, 5580–5584, 2002.
60. Falkinham, J. III, Mycobacterial aerosols and respiratory disease, *Emerg. Inf. Dis.*, 9, 763–767, 2003.
61. Primm, T., Lucero, C., and Falkinham, J. III, Health impacts of environmental mycobacteria, *Clin. Micro. Rev.*, 17, 98–106, 2004.
62. Selvaraju, S., Khan, I., and Yadav, J., Biocidal activity of formaldehyde and nonformaldehyde biocides toward *Mycobacterium immunogenum* and *Pseudomonas fluorescens* in pure and mixed suspensions in synthetic metalworking fluid and saline, *Appl. Environ. Micro.*, 71, 542–546, 2005.
63. Beckett, W., Kallay, M., Sood, A., Zuo, Z., and Milton, D., Hypersensitivity pneumonitis associated with environmental mycobacteria, *Environ. Health Persp.*, 113, 767–770, 2005.
64. DHEW (NIOSH), Current Intelligence Bulletin 15. No. 78-127: *Nitrosamines in Cutting Fluids*. October 6, 1976. Rockville, MD.
65. *IARC Monographs on the Evaluation of the Carcinogenic Risk of the Chemical to Man*, Vol. 3. International Agency for Research on Cancer, WHO, Lyon. France, 1973.
66. *IARC Monographs on the Evaluation of Carcinogenic Risk of Chemicals to Humans*, Vol. 33, International Agency for Research on Cancer, WHO, Lyon, France, pp. 87–168, 1984.
67. Mackerer, C., Health effects of oil mists: A brief review, *Toxic. Ind. Health*, 5, 429–440, 1989.
68. McKee, R., Daughtrey, W., Freeman, J., Frederici, T., Phillips, R., and Plutnik, R., The dermal carcinogenic potential of unrefined and hydrotreated lubricating oils, *J. Appl. Toxicol.*, 9, 265–270, 1989.
69. McKee, R. and Scala, R., An evaluation of the epidermal carcinogenic potential of cutting fluids, *J. Appl. Toxicol.*, 10, 251–256, 1990.
70. Dalbey, W., Osimitz, T., Kommineni, C., Roy, T., Feuston, M., and Yang, J., Four-week inhalation exposures of rats to aerosols of three lubricant base oils, *J. Appl. Toxicol.*, 11, 297–302, 1991.

71. Chasey, K. and McKee, R., Evaluation of the dermal carcinogenicity of lubricant base oils by the mouse skin painting bioassay and other proposed methods, *J. Appl. Toxicol.*, 13, 57–65, 1993.
72. Mackerer, C., Griffis, L., Grabowski, J. Jr., and Reitman, F., Petroleum oil refining and evaluation of cancer hazard, *Appl. Occup. Environ. Hyg.*, 18, 890–901, 2003.
73. Dalbey, W. and Biles, R., Respiratory toxicology of mineral oils in laboratory animals, *Appl. Occup. Environ. Hyg.*, 18, 921–929, 2003.
74. *TLVs and BEIs, and Documentation of the Threshold Limit Values and Biological Exposure Indices*, available from American Conference of Governmental Industrial Hygienists (ACGIH), Cincinnati, OH, 45340, 2003.
75. Blackburn, G., Deitch, R., Schreiner, C., and Mackerer, C., Predicting carcinogenicity of petroleum distillation fractions using a modified Salmonella mutagenicity assay, *Cell Biol. Toxicol.*, 2, 63–84, 1986.
76. Roy, T., Johnson, S., Blackburn, G., and Mackerer, C., Correlation of mutagenic and dermal carinogenic activities of mineral oils with polycyclic aromatic compound content, *Fundam. Appl. Toxicol.*, 10, 466–476, 1988.
77. ASTM E1687, *Method for Determining of Carcinogenic Potential of Virgin Base Oils in Metalworking Fluids*, available from ASTM International, West Conshohocken, PA 19428, or http://www.astm.org
78. Bingham, E., Trosset, R., and Warshawsky, D., Carcinogenic potential of petroleum hydrocarbons: a critical review of the literature, *J. Environ. Pathol. Toxicol.*, 3, 483–563, 1979.
79. Calvert, G., Ward, E., Schnorr, T., and Fine, L., Cancer risks among workers exposed to metalworking fluids: a systematic review, *Am. J. Ind. Med.*, 33, 282–292, 1998.
80. Eisen, E., Tolbert, P., Monson, R., and Smith, T., Mortality studies of machining fluid exposure in the automobile industry I: A standardized mortality ratio analysis, *Am. J. Ind. Med.*, 22, 809–824, 1992.
81. Eisen, E., Bardin, J., Gore, R., Woskie, S., Hallock, M., and Monson, R., Exposure–response models based on extended follow-up of a cohort mortality study in the automobile industry, *Scand. J. Work. Environ. Health*, 27, 240–249, 2001.
82. Rotimi, C., Austin, H., Deszell, E., and Day, C., Retrospective follow-up study of foundry and engine plant workers, *Am. J. Ind. Med.*, 24, 485–498, 1993.
83. DeCouflé, P., Further analysis of cancer mortality patterns among workers exposed to cutting oil mists, *J. Natl. Cancer Inst.*, 61, 1025–1030, 1978.
84. Aquavella, J. and Leet, T., A cohort study among workers at a metal components manufacturing facility, *J. Occup. Med.*, 33, 896–900, 1991.
85. Roush, G. C., Kelly J., Meigs, W., and Flannery, J. T., J Scrotal carcinoma in Connecticut metal workers: sequel to a study of sinonasal cancer, *Amer. J. Epidemiol.*, 116, 76–85, 1982.
86. Jarvholm, B., Lavenius, B., and Sällsten, G., Cancer morbidity in workers exposed to cutting fluids containing nitrites and amines, *Brit. J. Ind. Med.*, 43, 563–565, 1986.
87. Jarvholm, B., Lillienberg, L., Sällsten, G., Thringer, G., and Axelson, O., Cancer morbidity among men exposed to oil mist in the metal industry, *J. Occup. Med.*, 23, 333–337, 1981.
88. Gallagher, R. P. and Threlfall, W. J., Cancer mortality in metal workers, *Can Med Assoc J.*, 129, 1191–1194, 1983.
89. Vena, J. E., Sulz, H. A., Fiedler, R. C., and Barnes, R. E., Mortality of workers in an automobile engine and parts manufacturing complex, *Brit. J. Ind. Med.*, 42, 85–93, 1985.
90. Park, R. M., Wegman, D. H., Silverstein, M. A., Maizlich, N. A., and Mirer, F. E., Causes of Death Among Workers in a Bearing Manufacturing Plant, *Am. J. Ind. Med.*, 13, 569–580, 1988.
91. Silverstein, M., Park, R., Marmor, M., Maizlish, N., and Mirer, F., Mortality among bearing plant workers exposed to metalworking fluids and abrasives, *J. Occup. Med.*, 30, 706–714, 1988.
92. Tola, S., Kalliomaki, P. L., Pukkala, E., Asp, S., and Korkala, M. L., Incidence of cancer among welders, platers, machinists and pipe fitters in shipyards and machine shops, *Br J Ind Med.*, 45, 209–218, 1988.
93. Jarvholm, B. and Lavenius, B., Mortality and cancer morbidity in workers exposed to cutting fluids, *Arch. Environ. Health.*, 42, 361–366, 1987.
94. Mallin, K., Berkeley, L., and Young, Q., A proportional mortality ratio study of workers in a construction equipment and diesel engine manufacturing plant, *Amer. J. Ind. Med.*, 10, 127–141, 1986.

95. Russi, M., Dubrow, R., Flannery, J. T., Cullen, M. R., and Mayne, S. T., Occupational exposure to machining fluids and laryneal cancer risk: Contrasting results using two separate control groups, *Am. J. Ind. Med.*, 31, 166–171, 1997.
96. Park, R. A., Krebs, J., and Mirer, F. E., Mortality at an automotive stamping and assembly complex, *Am. J. Ind. Med.*, 26, 449–463, 1994.
97. Jarvholm, B., Fast, K., Lavenius, B., and Tomsic, P., Exposure to cutting oils and its relation to skin tumors and premalignant skin lesions on the hands and forearms, *Scand J. Work Environ. Health.*, 11, 363–369, 1985.
98. Acquavella, J., Leet, T., and Johnson, G., Occupational experience and mortality among a cohort of metal components manufacturing workers, *Epidemiology*, 4, 428–434, 1993.
99. Tolbert, P., Eisen, E., Pothier, L., and Monson, R., Mortality Studies of Machining fluid Exposure in the Automobile Industry II. Risks Associated with Specific Fluid Types, *Scand. J. Work Environ. Health*, 18, 351–360, 1992.
100. Eisen, E., Tolbert, P., Hallock, M., and Monson, R., Mortality studies of machining fluid exposure in the automobile industry III: A case-control study of larynx cancer, *Am. J. Ind. Med.*, 26, 185–202, 1994.
101. Schroeder, J., Tolbert, P., Eisen, E., and Monson, R., Mortality studies of machining fluid exposure in the automobile industry IV: A case-control study of lung cancer, *Am. J. Ind. Med.*, 31, 525–533, 1997.
102. Bardin, J., Eisen, E., Tolbert, P., Hallock, M., Hammond, S., Woskie, S., Smith, T., and Monson, R., Mortality studies of machining fluid exposure in the automobile industry. V: A case-control study of pancreatic cancer, *Am. J. Ind. Med.*, 32, 240–247, 1997.
103. Sullivan, P., Eisen, E., Woskie, S., Kriebel, D., Wegman, D., Hallock, M., Hammond, S., Tolbert, P., Smith, T., and Monson, R., Mortality studies of metalworking fluid exposure in the automobile industry: VI. A case-control study of esophageal cancer, *Am. J. Ind. Med.*, 34, 36–48, 1998.
104. Sullivan, P., Eisen, E., Kreibel, D., Woskie, S., and Wegman, D., A nested case-control study of stomach cancer mortality among automobile machinists exposed to metalworking fluid, *Ann. Epidemiol.*, 10, 480–481, 2000.
105. Thompson, D., Kriebel, D., Quinn, M., Wegman, D., and Eisen, E., Occupational exposure to metalworking fluids and risk of breast cancer among female autoworkers, *Am. J. Ind. Med.*, 47, 153–160, 2005.
106. Zeka, A., Eisen, E., Kriebel, D., Gore, R., and Wegman, D., Risk of upper aerodigestive tract cancers in a case-cohort study of autoworkers exposed to metalworking fluids, *Occup. Environ. Med.*, 61, 426–431, 2004.
107. Knaak, J., Leung, H-W., Stott, W., Busch, J., and Bilsky, J., *Toxicology of mono-, di-, and triethanolamine*, *Reviews of Environmental Contamination and Toxicology*, Vol. 149, Ware, G., Ed., Springer, New York, pp. 1–86, 1997.
108. *Toxicology and Carcinogenesis Studies of Diethanolamine (CAS 111-42-2) in F344/N Rats and B6C3F1 Mice (Dermal Studies)*, Tech. Report TR-478, National Toxicology Program, Research Triangle Park, NC, July, 1999.
109. *Title 40, Code of Federal Regulations, Part 721, Significant New Uses of Chemical Substances, Part 721.4740, Alkali Metal Nitrites*, Government Printing Office, Washington, DC, May 12, 1993, as amended June 23, 1993.
110. *Toxicology and Carcinogenesis Studies of Chlorinated Paraffins (C12, 60% Chlorine) (CAS No. 63449-39-8) in F344/N Rats and B6C3FI Mice*, Tech. Report TR-308, National Toxicology Program, Research Triangle Park, NC, May 1986.
111. *Toxicology and Carcinogenesis Studies of Chlorinated Paraffins (C23, 43% Chlorine) (CAS No. 63449-39-8) in F344/N Rats and B6C3FI Mice*, Tech. Report TR-305, National Toxicology Program, Research Triangle Park, NC, May 1986.
112. Yang, J., Roy, T., Neil, W., Krueger, A., and Mackerer, C., Percutaneous and oral absorption of chlorinated paraffins in the rat, *Toxicol. Ind. Health*, 3, 405–412, 1987.
113. McKee, R., Scala, R., and Chauzy, C., An evaluation of the epidermal carcinogenic potential of cutting fluids, *J. Appl. Toxicol.*, 4, 251–256, 1990.

16 Generation and Control of Mist from Metal Removal Fluids

Jean M. Dasch, Carolina C. Ang, and James B. D'Arcy

CONTENTS

I. INTRODUCTION

In most machining operations, metal removal fluids (MRF) are sprayed on the part to cool and lubricate the process. MRF also protect parts against corrosion and help remove chips. A side effect of using machining fluids is that a portion of the spray from MRF becomes airborne as mist or aerosols. The mist that is not contained within the machine enclosure settles to surfaces making them slippery, reduces visibility, and may present a health concern when absorbed into the body through the skin or respiratory system. Owing to these concerns, mist levels in the plant are regulated by the Occupational Safety and Health Administration (OSHA) and mist exhausted from the plant is regulated by the Environmental Protection Agency (EPA) and the states through environmental permits.

Mist reduction and control represents a major challenge for the machining industry. This chapter will describe the nature of mist, including the mechanisms that form mist, and the regulations that relate to mist, the processes that generate and influence mist levels in plants, and methods to either prevent the generation of mist or control mist after generation. To a great extent, information will be based on experiments performed at the GM R&D Center as well as experiences at automotive plants where engines and transmissions are machined.

II. NATURE OF MIST

Mist, also referred to as aerosols, is the liquid droplets or solid particles that become airborne as they separate from the bulk MRF. Mist may be either a liquid or solid, but not a gas. Mist consists of a complicated mixture of chemicals, which probably is not identical to the chemistry of the MRF. In addition to mist, MRF can also release gaseous material into the air as the more volatile components evaporate. The gases consist of many individual molecules, such as decane. Since the gases are primarily organic, they may be referred to collectively as volatile organic compounds (VOC).

A. Mist-Forming Mechanisms

In most cases, the MRF is applied in flood or through-tool application. Thornburg and Leith[1] identified three mechanisms that lead to mist formation: impaction, centrifugal force, and evaporation/condensation. Impaction refers to the fluid spray striking a surface and then bouncing off as droplets. Centrifugal force refers to mist generated as a fluid-covered surface rotates releasing droplets. In the final mechanism, the fluid evaporates under the heat of machining and then recondenses as the airstream cools. In terms of the size of the mist droplets generated, they found the largest drops were from the centrifugal process and the smallest from evaporation/condensation. Bell et al.[2] arrived at similar mechanisms, although named differently as splash, spin-off and evaporation. Yue et al.[3] modeled two formation mechanisms, atomization and vaporization/condensation, and predicted ambient concentrations after considering various removal mechanisms. Their model was validated by Sun et al.[4] Childers[5] also described aerosol generation from the bursting of bubbles encountered with foaming of the bulk fluid.

The different mechanisms of mist generation associated with fluid application are illustrated in Figure 16.1 as measured in a machining center. When the fluid was turned on, a small increase in mist levels occurred over background levels due to impaction of the fluid against the tool. As the tool started spinning, the mist significantly increased due to centrifugal effects. Then as actual

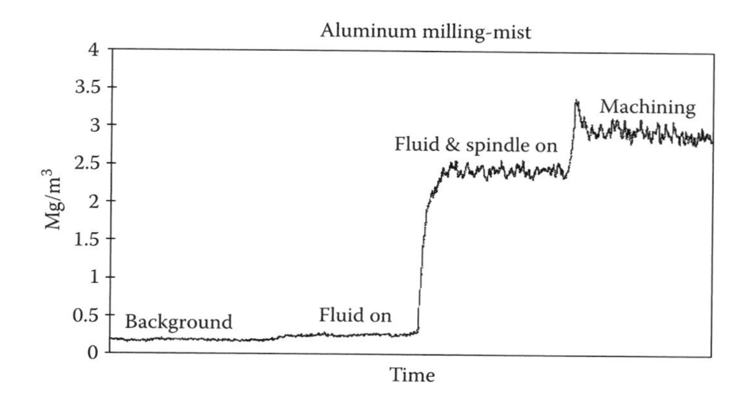

FIGURE 16.1 Mist generated during different machining stages illustrating mist formation mechanisms (*Source*: From Dasch, J. M., Ang, C. C., Mood, M., and Knowles, D., *Lubr. Eng.*, 58, 10, 2002. With permission).

machining started, mist increased again, perhaps due to mist formed from evaporation/condensation owing to the heat of machining. As the aerosols move into the airstream, the more volatile components quickly evaporate. Water with a vapor pressure of 20-mm Hg evaporates more quickly than oil with a vapor pressure less than 0.01-mm Hg.[6] As water evaporates from water-based fluids, the original droplet volume will decrease by 90 to 95% forming particles with diameters 50 to 60% of the original diameter. In addition, some of the organic material will also evaporate. Ilgner et al.[7] have shown that organic components will distribute between the liquid (mist) and vapor phase depending on their volatility. Owing to the stripping out of volatiles, the chemical composition of the mist is not the same as the composition of the bulk fluid. Mist droplets become enriched in heavier, less volatile materials, compared with the bulk fluid. Similarly, the MRF bath gradually changes composition as the more volatile constituents strip out of the bulk fluid during recirculation.[8]

B. Particle Size

The size of the aerosols generated impacts their lifetime in the air, their ease of removal by collection devices, and their fate when inhaled. Large particles (several micrometers in size) settle out quickly whereas submicron particles remain airborne for long periods. Removal processes include sedimentation, impaction, and diffusion. Large particles are more easily removed by sedimentation and impaction and very small particles (<0.1 μm) are removed by diffusion. The most difficult particle size to remove is around 0.3 μm. The D_{50} point (particle diameter at which 50% of the particles will enter) is shown below for particles that enter the body:

Inhalable particles	$D_{50} = 100$ μm	Particles that enter the nose
Thoracic particles	$D_{50} = 10$ μm	Particles that penetrate into lungs
Respirable particles	$D_{50} = 4$ μm	Particles that penetrate to deep lungs

MRF aerosol size distributions have been measured both in plants[9–12] and directly in ductwork from machining centers.[13–15] Size distributions from air sampling in two machining plants are shown in Figure 16.2 and Figure 16.3 below: the first using a soluble oil for machining and the second using a straight oil for grinding. The straight oil produced a somewhat larger aerosol with a mass median diameter (MMD, the diameter at which half the mass is on larger droplets and half on smaller) of 4.8 μm compared to 3.9 μm from the water-based fluid. Larger drops are to be expected from straight oil since there is no water in the droplet to evaporate.

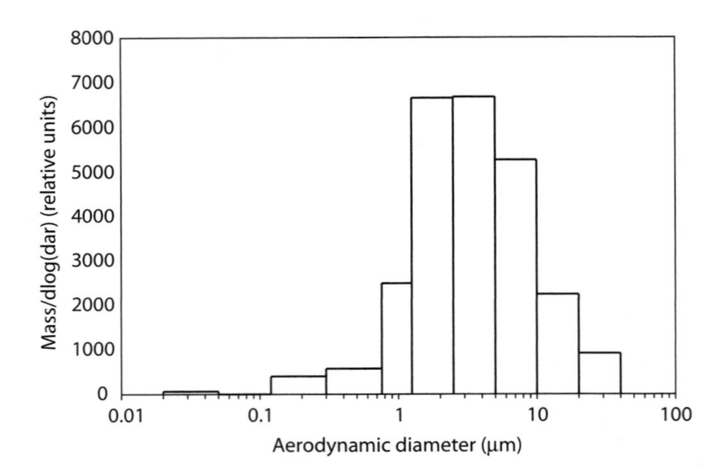

FIGURE 16.2 Particle size distribution of mist generated from a soluble oil MRF.

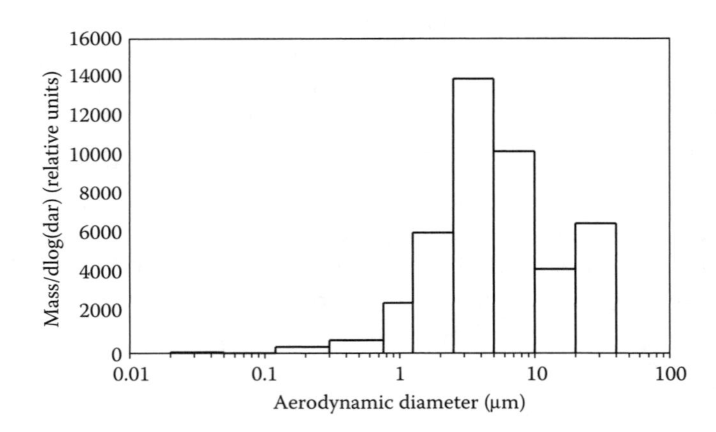

FIGURE 16.3 Particle size distribution of mist generated from a straight oil MRF during grinding.

MRF, but particularly straight oils, sometimes show another particle size mode, believed to result from the evaporation of fluid and then recondensation into very small droplets. Figure 16.4 shows a particle size distribution from high-speed grinding with a straight oil. In addition to the mode between 1 and 10 μm as seen above, another size mode appears between 0.1 and 1 μm, which can be attributed to the evaporation of fluid resulting from the high grinding temperatures followed by recondensation as fine particles less than 1 μm in diameter.

III. REGULATIONS RELATING TO MIST AND VOLATILE ORGANIC COMPOUNDS

A. Indoor Air Regulations

Mist regulations pertaining to indoor air have been under review for over a decade. Since regulations are continuously being modified, the information in this chapter represents the situation as of 2005.

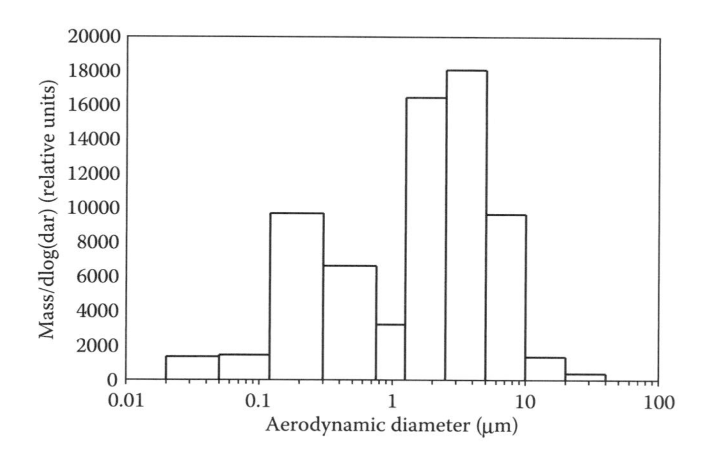

FIGURE 16.4 Particle size distribution of mist generated by a straight oil during high speed grinding showing the fine aerosol produced by evaporation/condensation.

OSHA has a permissible exposure limit (PEL) of 5 mg/m^3 for mineral oil mist in the air, averaged over an 8-h period. They also have a nuisance dust PEL of 15 and 5 mg/m^3 for total and respirable particulate matter, respectively.

The United Auto Workers (UAW) petitioned OSHA in 1993 to establish a new mist standard of 0.5 mg/m^3 for metalworking fluid, regardless of the mineral oil content. In response to the UAW petition to OSHA, the National Institute for Occupational Safety and Health (NIOSH) issued a *Criteria for a Recommended Standard: Occupational Exposure to Metalworking Fluids* in 1998, where they included a recommended exposure level (REL) of 0.5 mg/m^3 of total metalworking fluid aerosol or 0.4 mg/m^3 of metalworking fluid aerosol meeting the thoracic particulate size definition. OSHA formed a Standards Advisory Committee in 1997 that also recommended in their final report a lower exposure standard that would apply to all components of the metalworking fluid aerosol.[16] However, to date OSHA has concentrated on increasing awareness of good metalworking fluid management rather than lowering the mist standard. In December of 2004, OSHA issued a letter to the UAW stating that there was insufficient scientific information to justify a separate metalworking fluid standard at this time.

Meanwhile, the American Council of Governmental Industrial Hygienists (ACGIH) filed a "Notice of Intended Change" in 2001 and proposed lowering the threshold limit value-time-weighted average (TLV©-TWA) for mineral oil mists from 5 to 0.2 mg/m^3, and assigning mists the designation of Suspected Human Carcinogen, A2. This designation could trigger hazard notification, extra training, etc. in various states or nations. ACGIH currently has the general category of metalworking fluids on their list of materials under study and may issue a TLV in the future.

Vegetable oil mists are regulated at higher levels of 15 mg/m^3 as a PEL by OSHA and at 10 mg/m^3 as a TLV©-TWA by ACGIH.

Some states also have regulations pertinent to mist levels. Michigan OSHA (MIOSHA) has a regulation that cleaned process air from local exhaust ventilation systems can only be recirculated back into the plant if the concentrations of contaminants are less than 10% of the regulated values and the plant is supplied with a minimum of 1 cfm/ft^2 of outside air. Air from mist collectors is recirculated back into plant air in many plants. If air is recirculated, the mineral oil content of the mist should be less than 0.5 mg/m^3 (10% of 5 mg/m^3), the total aerosol should be less than 1.5 mg/m^3, and the respirable aerosol level should be less than 0.5 mg/m^3.

Irrespective of levels set by OSHA or ACGIH, many companies have negotiated with labor unions to reduce worker exposures to mist. In the auto industry, many UAW contracts establish an exposure guideline of 1 mg/m^3 for metalworking fluid with existing operations and a commitment to purchase new equipment that would meet a 0.5-mg/m^3-purchase specification. Worker exposure concentrations are based on measurements made with filter cassettes worn by plant personnel over the course of an 8-h shift. Mass is determined gravimetrically and may include oil mist as well as other aerosols, such as metallic dust from machining. In some instances, extraction methods are employed to improve the specificity of the measurements.

During MRF use, solid or liquid particles are emitted into the air (mist), but gases are also emitted, which are referred to as VOC. No OSHA regulations relate to total VOC. However, one specific organic compound, formaldehyde, may be an issue in plants using MRF. Formaldehyde is a potential/suspected human carcinogen that may be released from some biocides, such as triazines and oxazolidines.[16] Although it is unlikely that the PEL of 0.75 ppm for formaldehyde would be exceeded due to MRF use, it is possible that the training level of 0.1 ppm would be exceeded. If exceeded, the OSHA formaldehyde standard training procedures would need to be implemented for employees (29 CFR 1910.1048).

B. Outdoor Air Regulations

VOC emissions need to be calculated as part of the environmental permitting process whether they are exhausted back into the plant or externally. In contrast, particles (mist) are included in the permitting process when mist collectors are exhausted externally, but not when exhaust is recirculated in the plant.

VOC are typically low from MRF due to the low vapor pressure of oils.[6] Prior to 1987, VOC were only considered from fluids with vapor pressures >0.1 mm Hg, which excluded MRF. In 1987, the regulations were changed to include any organic compound that could participate in photochemical reactions regardless of volatility. Based on this definition, MRF became potential sources of VOC. When mist collectors are exhausted outside the plant, particles also must be evaluated as emission sources for environmental permitting.

IV. MIST LEVELS IN PLANT AIR

The levels of mist in plant air are affected by the largest scale systems in the plant to the smallest event at an individual machine. For the purpose of this paper, we will work from the large scale to the small scale: plant, machine, fluid, and process.

A. Plant Practices

Ventilation levels in the plant affect the mist concentration. If the same amount of generated mist is diluted into a larger volume of air, mist concentrations will be lower. MIOSHA rule 3101 states that general ventilation should be >1 cfm/ft^2 in areas where contaminants are generated. Ventilation is frequently higher in the summer as doors and windows are left open.

The fluid handling system for the plant can also be a source of mist. In general, fluid movement will lead to mist generation and the higher the energy associated with that movement, the more mist will be generated. Fluid movement can be generated from skimmers in the tank area, from flumes that carry mist, or from compressed air used to clean parts. Open flumes carrying fluid at high speeds will generate mist. It has been estimated that fluids moving through flumes can generate one third of the mist in a plant. Closed flumes or piping help to contain this source of mist. Compressed air is often used to blow off fluid or chips, but this will generate high levels of mist for short periods of time.

Many plants recirculate air from mist collectors internally. Although a mist collector should capture $>90\%$ of the mist in most situations, residual mist and VOC will be returned to the plant air.

B. Machine Issues

In older plants machining is done on open transfer lines where the part is transferred from one machining operation to the next. Typically, the line is not enclosed as constructed, but hoods are placed above machining areas. Mist levels are difficult to control since nothing limits the movement of mist and overhead hoods are not very efficient. Various engineering controls can be added to transfer lines, such as the addition of Plexiglas panels around machining operations or fluid nozzles that turn off the fluid flow during nonmachining periods. In some cases, enclosures are retrofitted around transfer lines. Unfortunately, Hands et al.[17] showed that retrofitted enclosures are not as effective as OEM enclosures purchased as part of the machine tool.

Newer facilities typically have agile machining centers that are fully enclosed and ventilated to mist collectors. Enclosures alone will not prevent escape of mist, but must be ventilated and the mist routed to a collector. This will be covered in greater detail later in the chapter.

C. Fluid Issues

The characteristics of the fluid, contaminants in the fluid, and fluid application also impact mist generation. The effect of fluid concentration, type, volatility, temperature, age, tramp oil levels, velocity of application, and method of application on mist generation were tested. The results are documented fully elsewhere and summarized below.[18,19]

1. Fluid Concentration

A soluble oil was tested at four concentrations: 0 (tap water), 2.5, 5, and 10% during aluminum milling. The mist concentration and droplet MMD are shown in Table 16.1. Total mist and droplet size increased with fluid concentration. The effect of increasing concentration was not linear in that a doubling of concentration resulted in less than a doubling of mist. One explanation could be that as the particle size shifts to larger particles, more droplets may settle out before reaching the sampler.

2. Fluid Type

Several studies examined the effect of the type of fluid on mist generation. MRF can be divided into straight oils and water-based fluids. Water-based fluids can then be subdivided into soluble oils, semisynthetics, and synthetics. Cutcher and Goon[20] found that the type of water-based fluid affected mist generation in the order: solubles > semisynthetics > synthetics. Turchin and Byers[21] saw the same trend with fluid type with solubles highest and synthetics lowest in mist generation.

TABLE 16.1
Effect of Fluid Concentration on Mist Generation during Aluminum Milling

Fluid Concentration (%)	Total Mist (mg/m^3)	MMD (μm)
0	0.72	2.3
2.5	4.4	5.4
5.0	6.0	7.5
10	8.4	10.4

Source: Adapted from Dasch, J. M., Ang, C. C., Mood, M., and Knowles, D., *Lubr. Eng.*, 58, 10, 2002. With permission.

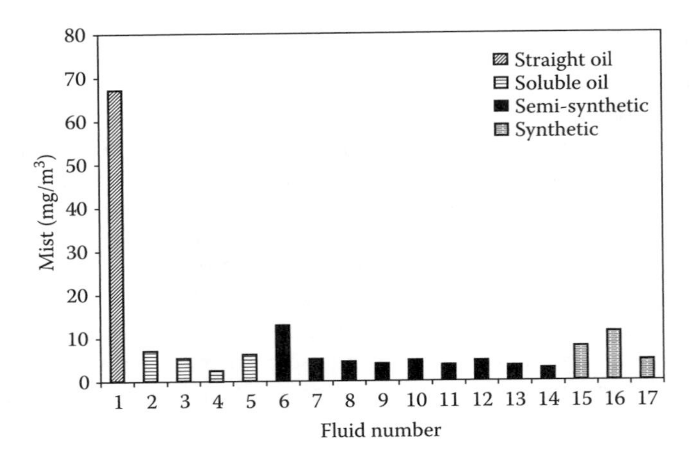

FIGURE 16.5 Comparison of mist generated from a variety of fluids during milling of aluminum, including one straight oil, four soluble oils, nine semisynthetics, and three synthetics. (*Source*: From Dasch, J. M., Ang, C. C., Mood, M., and Knowles, D., *Lubr. Eng.*, 58, 10, 2002. With permission.)

Total mist was measured from 17 fluids including one straight oil, four soluble oils, nine semisynthetics and three synthetics. The fluids were tested at the manufacturer's recommended concentration: 5 to 6% for most fluids, but 10% for fluids 6, 16, and 17, and 100% for the straight oil as shown in Figure 16.5. During aluminum milling, far higher mist levels were generated from the straight oil, presumably owing to the fact that it consisted of 100% oil as opposed to 5 to 10% from the water-based fluids.

In the case of the water-based fluids, the total mist generated ranged from 2.5 to 6.9 mg/m^3 for the solubles, 2.7 to 12.7 mg/m^3 for the semisynthetics, and 4.6 to 11.1 mg/m^3 for the synthetics. These tests did not show the expected trend of solubles > semisynthetics > synthetics found by Cutcher and Goon[20] and Turchin and Byers,[21] but rather a wide range of mist generation from different fluid types.

Rather, our observation was that mist levels were related to the mass of nonaqueous material present in the fluid. The aqueous portion of the fluid, whether present in the original concentrate or added as dilution water, quickly evaporates from the mist droplets. The remaining nonaqueous material forms the mist in the air. The nonaqueous material was measured by evaporating 5 ml of fluid to dryness and weighing the residue. As shown in Figure 16.6, the mist level was proportional to the amount of nonaqueous residue. In general, the water content of undiluted fluids is in the order of synthetics > semisynthetics > solubles,[5] which could explain the observation made by others of higher mist levels from solubles than synthetics.

3. Fluid Volatility

The volatility of the fluid would be expected to affect the misting in situations where the evaporation/condensation mechanism is important. This was demonstrated to be the case when milling aluminum with three straight oils of differing volatilities, but similar viscosities, shown in Figure 16.7.[14] The three oils were a naphthenic, paraffinic group II, and paraffinic group III, where the naphthenic was the most volatile and the paraffinic III the least. As the volatility of the oil increased, the mechanically generated aerosol (>10 μm) stayed constant, but the heat-generated mode (<1 μm) increased. The effect would be expected to be less with water-based fluids where all fluids are similarly volatile since they are >90% water.

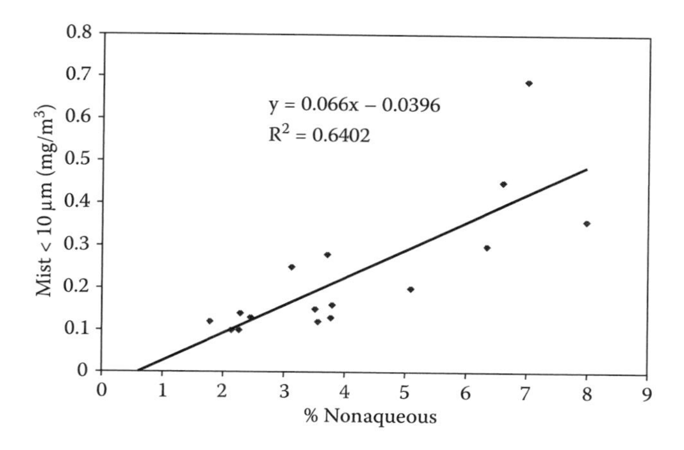

FIGURE 16.6 Relationship of mist to the nonaqueous proportion of the fluid during aluminum drilling. (*Source*: From Dasch, J. M., Ang, C. C., Mood, M., and Knowles, D., *Lubr. Eng.*, 58, 10, 2002. With permission.)

4. Fluid Temperature

The effect of the fluid temperature was investigated using a small 20-gal fluid tank and cooling the fluid using ice packs. As steel was milled, the temperature gradually rose to 25°C. As shown in Figure 16.8, the VOC released increased with temperature as expected since vapor pressure increases with fluid temperature.[6] However, the mist levels remained basically the same over the 9°C (16°F) temperature range. Although the temperature range might have been expected to increase evaporation, the fluid temperature may have had a minor effect compared with the far higher temperatures at the work–tool interface.

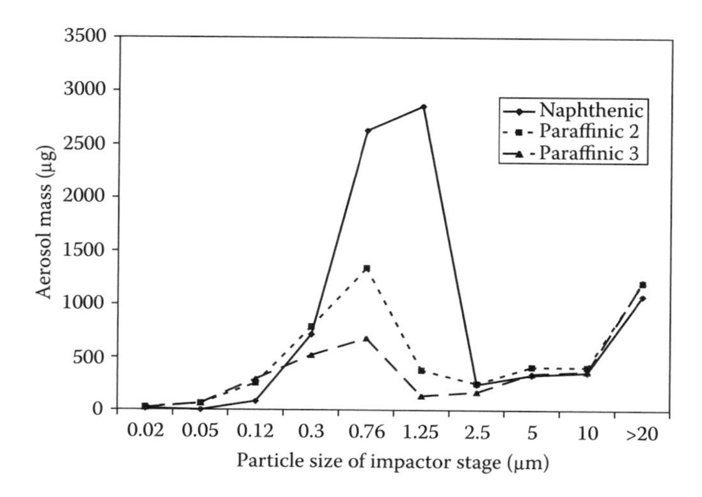

FIGURE 16.7 Size distribution of mist generated from three straight oils of different volatilities. The small-particle mist followed the oil volatility with naphthenic oil being the most volatile and paraffinic III being the least volatile. (*Source*: From Dasch, J. M., Ang, C. C., Mood, M., and Knowles, D., *Lubr. Eng.*, 58, 10, 2002. With permission.)

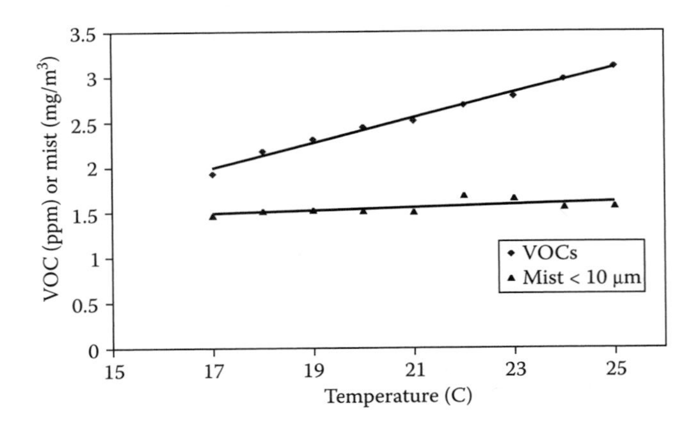

FIGURE 16.8 Effect of fluid temperature on the VOC and mist concentration. (*Source*: From Dasch, J. M., Ang, C. C., Mood, M., and Knowles, D., *Lubr. Eng.*, 58, 10, 2002. With permission.)

5. Fluid Age

The effect of fluid aging on mist generation was investigated using a fresh 8% semisynthetic fluid compared with a used fluid from a manufacturing facility. The plant had analyzed the fluid as having no tramp oil and 12 mg/l of suspended solids. The results for mist and VOC are shown in Table 16.2. The VOC level was obviously lower in the used fluid indicating that the volatile organics had been stripped out with time. However, the mist generation was very similar from fresh and used fluid indicating that there was nothing inherently different in the mist generation from a used fluid.

6. Tramp Oil

Although a used fluid generated similar mist to a fresh fluid in the example above, the used fluid was not contaminated by tramp oil. Tramp oil is the unavoidable contamination of the MRF resulting from leakage of hydraulic oils, way oils, etc. Turchin and Byers[21] used a benchtop grinder to show that mist generation increased approximately linearly with tramp oil. However, the effect of the tramp oil was different in different fluids, varying from 10 to 200% higher mist at a 3% tramp oil level compared with the same fluids without tramp oil. A similar test was conducted in this study with tramp oil consisting primarily of hydraulic fluid with some way oil. Using a 5% semisynthetic fluid, tramp oil was added at levels of 0, 2, 5, and 10%. As shown in Figure 16.9, during aluminum drilling, the mist <10 μm increased with the tramp oil level, although not linearly above 5% tramp

TABLE 16.2
Effect of Fluid Age on Mist and VOC Generation during Milling

Workpiece	Fluid Condition	Mist <10 μm (mg/m^3)	Total Mist (mg/m^3)	VOCs (ppm C3)
Steel	Fresh	0.08	0.60	0.15
	Used	0.11	0.50	0.03
Aluminum	Fresh	0.88	4.6	0.26
	Used	0.92	4.2	0.05

Source: Dasch, J. M., Ang, C. C., Mood, M., and Knowles, D., *Lubr. Eng.*, 58, 10, 2002. With permission.

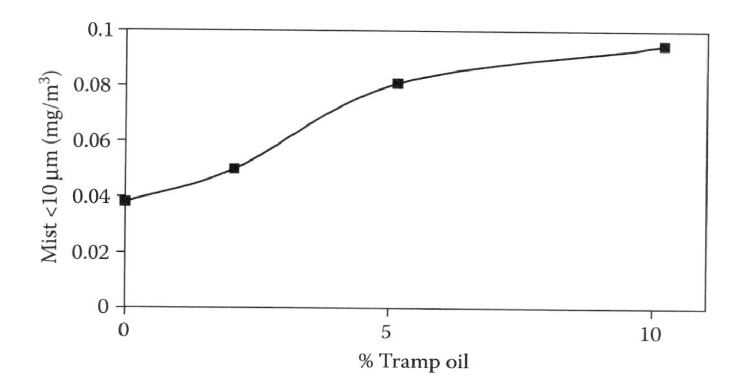

FIGURE 16.9 Effect of tramp oil on mist generation during drilling of aluminum. (*Source*: From Dasch, J. M., Ang, C. C., Mood, M., and Knowles, D., *Lubr. Eng.*, 58, 10, 2002. With permission.)

oil. The reason for the nonlinearity was that only some of the tramp oil was incorporated into the fluid with the remainder floating on the top.

7. Fluid Velocity during Flood Application

During aluminum milling, mist levels were measured as the flow velocity was varied over a factor of 10. Adjusting the MRF flow rate or the nozzle size varied the flow velocity. The results are shown in Figure 16.10. As the fluid velocity increased, the mist droplets < 10 μm increased. However, the effect is not large with the mist increasing by only a factor of 2.5 as the velocity increased by a factor of 10.

8. Through-Tool Application

In some machine tools, the MRF can be routed through the spindle and then through small channels in the tool. Through-tool application is very useful for operations such as deep-hole drilling in which it is difficult for fluid to reach the cutting point during flood application. The fluid may be applied under very high pressures, as high as 1000 lb/in.^{2} In this test, through fluid was applied at

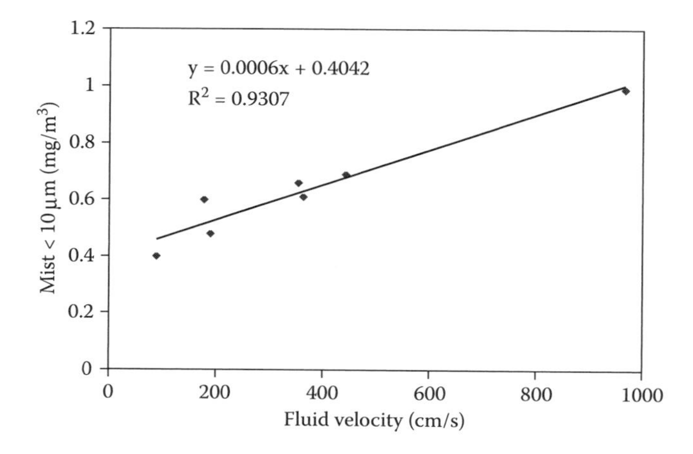

FIGURE 16.10 Effect of fluid velocity on mist generation. (*Source*: From Dasch, J. M., Ang, C. C., Mood, M., and Knowles, D., *Lubr. Eng.*, 58, 10, 2002. With permission.)

TABLE 16.3
Effect of Through-Tool Fluid Application on Mist Generation during Drilling

Material	Application	Pressure (lb/in.^{2})	Flow (gal/min)	Velocity (cm/sec)	Mist <10 μm (mg/m^3)	Total Mist (mg/m^3)	MMD (μm)
Alum	Flood	60	9	398	0.22	0.7	5.9
Alum	Through-tool	200	1.3	2320	0.36	3.7	9.7
Steel	Flood	60	9	398	0.17	0.8	7.7
Steel	Through-tool	200	1.3	2320	0.45	4.7	10.2

Source: Dasch, J. M., Ang, C. C., Mood, M., and Knowles, D., *Lubr. Eng.*, 58, 10, 2002. With permission.

lower pressure and also compared with flood coolant operating at a higher flow rate, but a lower velocity. The results are shown in Table 16.3 for aluminum and steel drilling experiments. In both cases the mist was higher with the through-tool application than with flood coolant even through the total fluid flow rate was much lower. Presumably this effect would be much greater at higher pressures.

D. Processes

In addition to the fluid characteristics and the method of fluid application, the machining process also affects mist generation. During a machining operation such as milling, three parameters determine the metal removal rate, the speed that the tool is rotating, the feed rate, and the depth of the cut. All three were evaluated as to their effect on mist generation. In addition, the effects of the tool diameter and tool wear were evaluated.

1. Tool Speed and Diameter

The effect of tool speed on mist generation was tested using a 5-cm diameter face-mill over the range of 0 to 7850 sfm (0 to 15,000 rev/min). Owing to the large range of conditions, the mill was rotated 2.5 cm above the workpiece, but without actually machining. The results for aerosols <10 μm (Figure 16.11) show a dramatic increase in mist as the tool speed increases. The best fit to

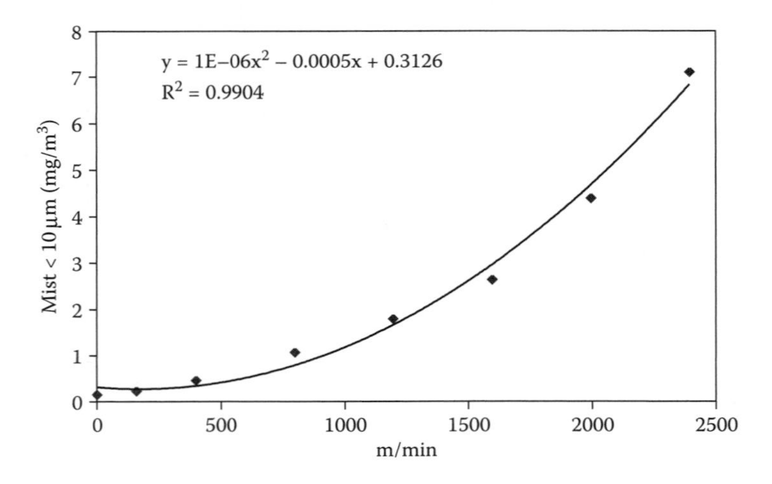

FIGURE 16.11 Influence of tool speed on mist generation. Solid line indicates best polynomial fit to the experimental data points. (*Source*: From Dasch, J. M., Ang, C. C., Mood, M., and Knowles, D., *Lubr. Eng.*, 58, 10, 2002. With permission.)

the data points is a polynomial with mist related to the square of the tool speed. Similarly, the mist generation was also shown to increase as the tool diameter was increased at a fixed spindle speed, owing to the increase in surface speed of the tool. For that reason, a face mill will typically generate more mist than a small-diameter drill. Heitbrink et al.[13] showed that the tool speed affected the mist levels with two to three times higher mist levels at a mill speed of 3820 rev/min than at 1910 rev/min.

2. Feed and Depth-of-Cut

Five machining conditions were used to test the effect of depth-of-cut and feed rate on the generated mist. The depth-of-cut was varied over an order of magnitude from 0.01 to 0.1 in. and the feed rate was increased by 2.5 times from 0.01 to 0.025 in./min. In both cases, mist levels were unaffected.

3. Tool Wear

Cutcher and Goon[20] found large increases in mist during milling as tools became worn, although the amount of tool wear was not quantified. Drilling and milling tests were conducted to try to reproduce this effect. In each case, the test was carried out to the end of tool life or 0.3 mm of wear, as specified under ISO 3685.[22] The effect of worn tools using three different fluids is shown in Table 16.4.

Overall, there was no increase in misting during the drilling tests to the end of drill life. However, the three milling tests all showed increases in mist levels, from 30 to 100% higher with worn tools than with new tools. This is far below the tenfold increase in misting reported by Cutcher and Goon,[20] but they may have gone far beyond end-of-life. A possible mechanism for the increase in misting is that the high heat generated during milling with a dull tool may lead to evaporation and condensation of MRF components. In contrast, little if any fluid would be at the point of tool contact during drilling.

E. Summary of Mist Generation

As shown above, mist generation is affected by many factors, as summarized in Figure 16.12. Factors that affected mist generation were the concentration, fluid type (if straight oil), volatility of straight oils, tramp oil levels, fluid velocity, through-tool application, tool surface speed, and tool wear. The greatest effect was from the tool surface speed. Conversely, the age or temperature of the fluid and the feed and depth-of-cut had no effect on mist generation.

TABLE 16.4
Effect of Tool Wear on Mist Generation during Steel Drilling and Milling

Operation	Fluid	Mist <10 μm (mg/m³)	
		New Tools	End of Life
Drilling	Soluble	0.16	0.17
Drilling	Semisynthetic 1	0.05	0.05
Drilling	Semisynthetic 2	0.04	0.01
Milling	Soluble	0.18	0.34
Milling	Semisynthetic 1	0.10	0.13
Milling	Semisynthetic 2	0.09	0.18

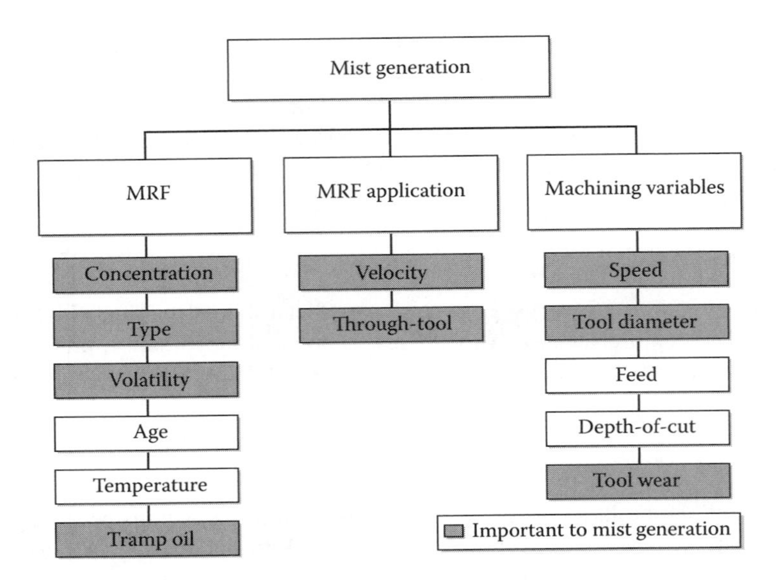

FIGURE 16.12 Factors that influenced mist generation (shaded boxes) and those that had no effect (unshaded boxes).

V. MIST PREVENTION AND CONTROL

Mist reduction can be achieved through two routes: reducing the generation of mist (mist prevention) or eliminating the mist once generated (mist control). Methods in each category are summarized in Figure 16.13 and described below.

A. Mist Prevention

1. Process Changes

Figure 16.12 separated the factors that lead to mist generation from those that do not. Changing the fluid, the fluid application, or the machining parameters can reduce mist. Fluid changes would include avoiding straight oils and highly volatile oils and reducing tramp oil levels. Fluid application changes would include reducing the fluid velocity, the use of through-tool application, or the fluid pressure. Machining changes would include reducing surface speed and avoiding worn tools. Surface speed had the largest effect on mist levels so substituting higher feed rates for high surface speeds could reduce mist generation.

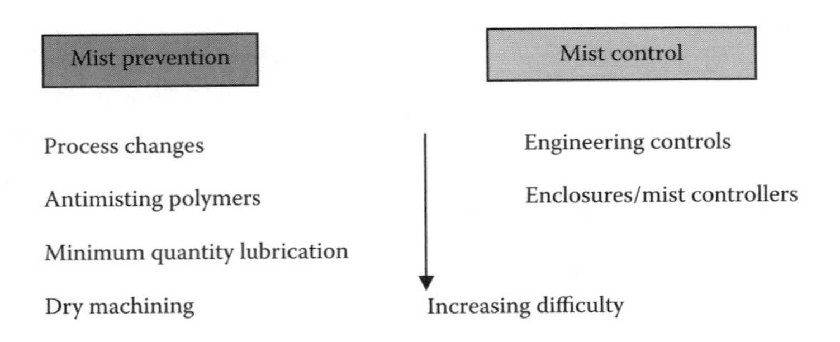

FIGURE 16.13 Hierarchy of methods to prevent and control mist.

2. Antimisting Polymers

Another resource in dealing with high mist levels is the addition of antimisting polymers to the fluid system. Antimisting polymers are organic molecules with a molecular weight of 500,000 to 2,000,000 that add elongational viscosity to the fluid so that it stretches rather than breaking apart into droplets.[23,24] Polymers can decrease misting by 30 to 70%. The downside is that the polymers shear down during fluid use so they must be replenished on a daily to weekly basis.

Three polymers were studied in pilot-scale and plant studies: polyisobutylene (PIB) for straight-oil systems, and polyethylene oxide (PEO) and associative polymers for water-based systems.[25] Results from using 50 ppm of PIB in six different straight oils are shown in Figure 16.14. Mist was reduced by 27 to 54% among the different fluids.

All polymers shear to smaller, ineffective molecules as they are pumped through the system. Of the two water-based polymers, PEO is more likely to shear and needs replacement every day or more often. Associative polymers also shear, but then reassociate into the original polymer, so they are longer lived.[26–28] Figure 16.15 shows a comparison of the two polymers' performance in machining tests emphasizing the effect of the shearing.

Costs were compared for polymer systems in actual plant situations.[25] Although costs are dependent on the system, such as fluid volume, turnover rate, pump pressure, etc., the costs in these particular plants were $18/week/1000 gal for PIB, $183/week/1000 gal for PEO, and $11/week/1000 gal for associative polymers. Antimisting polymers result in increased costs and maintenance requirements, but are a possible solution for difficult mist situations.

3. Minimum Quantity Lubrication

A recent innovative approach to fluid usage is minimum quantity lubrication or MQL, in which very small volumes of MRF are applied as a mist. Typically, a compressed air stream is mixed with the fluid to create a mist with a fluid usage rate of only 10 to 100 cm^3/h. Since one of the functions of the MRF is to remove chips, other arrangements are required for chip removal. In the most sophisticated systems, special machines are made expressly for MQL. Fluid is aerosolized with a compressed air stream in the spindle and delivered to the part as a mist. Chips and mist are automatically collected with vacuum systems. Complete units are expensive and large, but they appear to perform very well, at times even superior to flood coolant.[29]

Retrofitted MQL systems are also available and the lubricant can be applied through either the spindle (if the machine has through-spindle coolant capability) or from nozzles. Generally performance equal to or better than flood coolant application has been noted.[30,31] Advantages of MQL systems include low fluid usage, dryer chips, and no need for tanks, filtration, or fluid

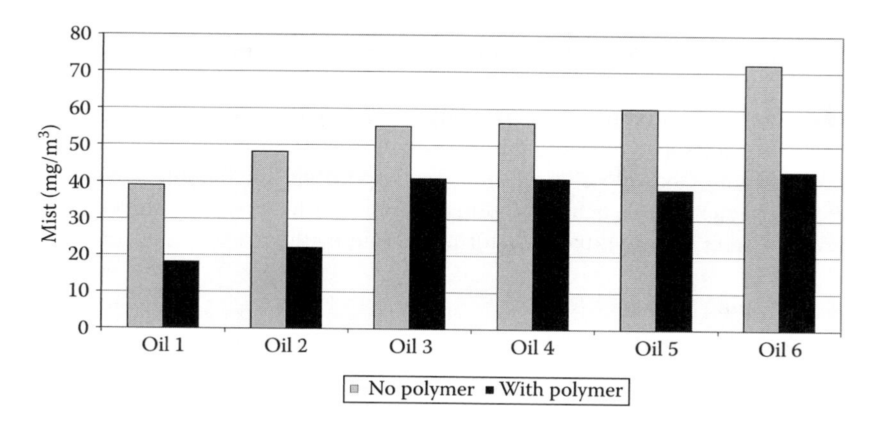

FIGURE 16.14 Mist levels from six different oils before and after addition of antimisting polymer, PIB.

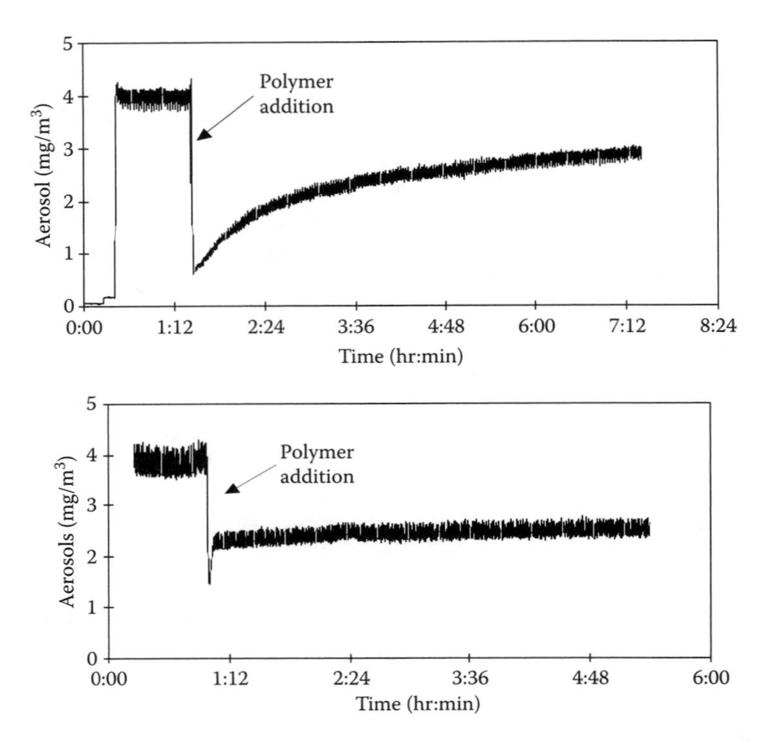

FIGURE 16.15 Comparison of the performance of polyethylene oxide (top) and associative polymers (bottom) showing that the polyethylene oxide initially reduces mist, but then becomes less effective as the polymer shears. In contrast, the associative polymer maintained effectiveness throughout the test. (*Source*: From Dasch, J. M., D'Arcy, J. B., Smolenski, D., *Lubr. Eng.*, 60, 38, 2004. With permission.)

disposal. However, applying the fluid as a mist could conceivably lead to high airborne mist levels. Certainly, MQL machines need fully ventilated enclosures. The effect of MQL on airborne mist levels has not been resolved, with various studies reporting both higher mist levels[30,32] and lower mist levels[33] than with flood coolant.

4. Dry Machining

The ultimate solution to mist and fluid issues is dry machining, but the challenge is to be able to machine efficiently without the use of MRF to cool and lubricate the process. Some materials are dry machined more easily than others. Cast iron contains graphite that acts as an internal lubricant allowing dry machining. Aluminum, on the other hand, becomes gummy when hot, leading to a built-up edge on the tool. Open machining processes such as face milling or turning are more easily adapted to dry machining than closed operations such as drilling and tapping due to issues with heat buildup and chip evacuation. To achieve dry machining, it is necessary to reduce heat buildup and the associated tool wear and workpiece distortion. Some methods used to accomplish this are:

- Low machining speeds or feeds
- Lubricious tool coatings
- Novel tool geometries
- Chip breakers
- Free machining alloys
- Alternate cooling with air or liquid nitrogen

A consortium of industries studied dry drilling of aluminum through the National Center for Manufacturing Sciences.[34] They investigated the effect of workpiece material, drill geometry, operating parameters, tool coatings, and enhancements such as cold air. They found poor performance without any modifications achieving only 25 holes per drill dry. Considerable improvement was achieved with cold air and tool coatings, although they were not as effective as using MQL.

In addition to the technical difficulties with attaining efficient dry machining, there is some argument as to the benefit of dry machining to cleaner air. The mist would drop to zero, but airborne metal particles were shown to increase in one study without the MRF present to wash them down.[35] Another study showed lower aerosol levels near dry operations than wet operations, and the aerosols were larger so would settle out faster.[12]

B. Mist Control

1. Engineering Controls

Engineering controls are defined here as methods to control mist by methods other than mist collectors and chemical additions. A good starting point is to conduct periodic mapping of mist levels throughout the machining area,[36] which provides a "snapshot" of mist concentrations at a point in time. Mapping usually entails measuring the mist concentrations with a handheld monitor at designated locations. The data are recorded, transferred to a computer and two- or three-dimensional maps are generated showing the areas with high mist concentrations. High spots might be remapped using a more intensive sampling grid. Alternating mapping during warm and cold seasons can also give information on seasonal effects.

The mist map will help identify problem operations, after which, a variety of engineering controls can be applied to reduce or control mist levels. Some examples of controls are:

- Decrease the density of machining operations
- Increase building ventilation
- Add air ducts or hoods over problem areas
- Add panels around machines to reduce splash
- Wall off problem areas with dedicated mist collectors
- Turn fluid off when not needed
- Direct fluid nozzles to minimize mist
- Cover trenches and tanks
- Reduce foaming
- Reduce compressed air usage

2. Enclosures and Mist Collectors

The next line of defense is to purchase or retrofit machines with enclosures and add mist collectors. Mist collectors are generally designed with two to four stages, with the first stage removing the largest particles and later stages removing smaller particles. The principle of operation could be impaction (metal screens, filters, packed beds), centrifugal removal, or electrostatics. Early stages typically allow for fluid to drain so they do not require frequent replacement. This is aided if the collector can be turned off periodically allowing for better drainage. The drained material may be reusable in the case of straight oils, but should be disposed of in the case of a water-based fluid. Most collectors have a final high-efficiency filter that is either a HEPA (high-efficiency particulate air) filter that is 99.7% efficient or a 95% DOP (dioctyl phthalate) HEPA-like filter. Both high-efficiency filter types are almost 100% efficient for all particle sizes, such that the air out of a mist collector should have very low particle levels. As they are expensive to replace it is important to have most of the mist removed in the early stages to prolong the life of HEPA filters.

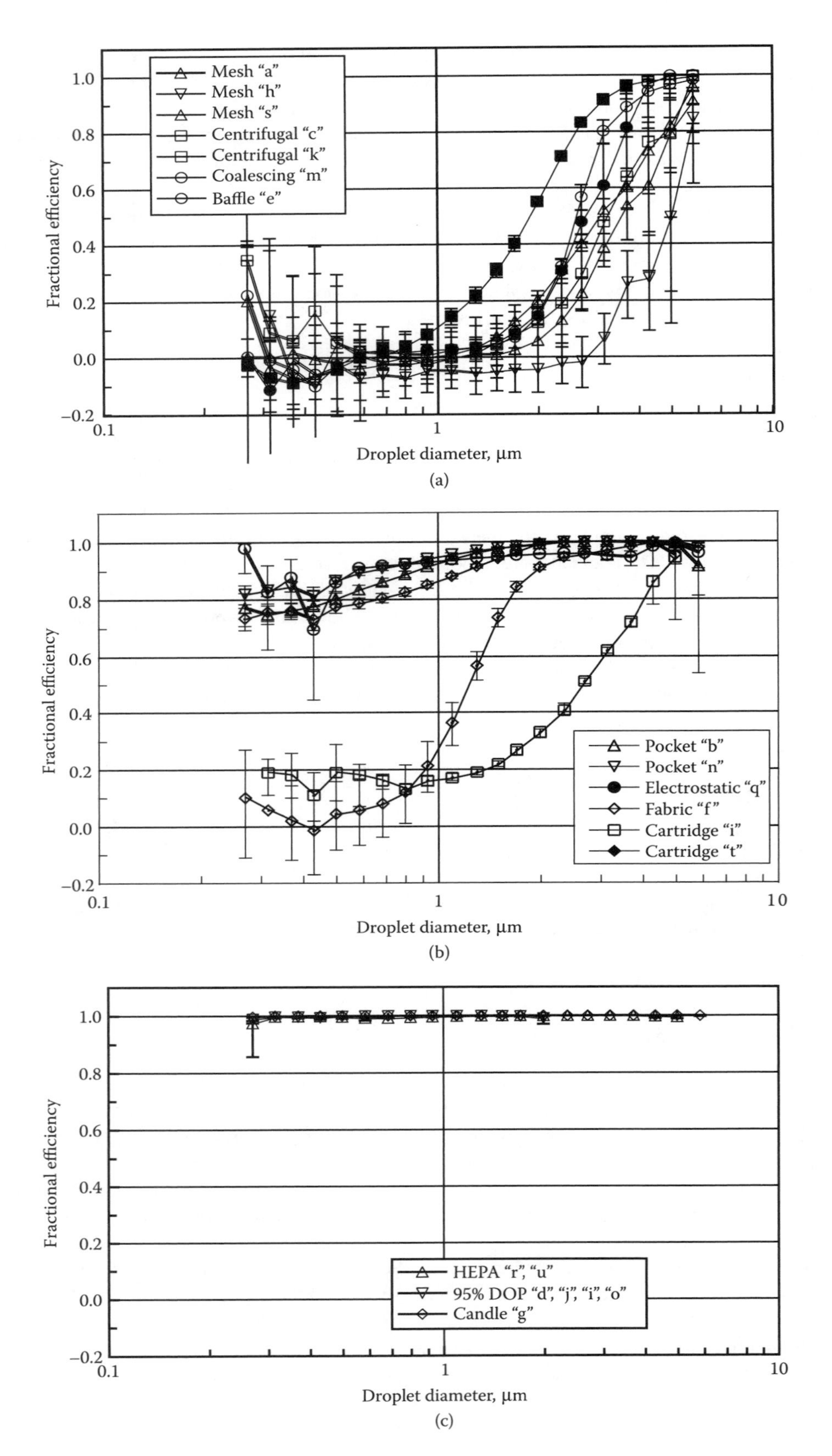

Mesh "a"
Mesh "h"
Mesh "s"
Centrifugal "c"
Centrifugal "k"
Coalescing "m"
Baffle "e"
Fractional efficiency
Droplet diameter, μm
(a)
Pocket "b"
Pocket "n"
Electrostatic "q"
Fabric "f"
Cartridge "i"
Cartridge "t"
(b)
HEPA "r", "u"
95% DOP "d", "j", "i", "o"
Candle "g"
(c)

Leith et al. have studied mist collectors extensively, including capture efficiency of each stage for different size particles and pressure drops across filters.[37] Their results of efficiency of particle removal are shown for the first stage of seven collectors (Figure 16.16a), the second stage (Figure 16.16b), and the final stage (Figure 16.16c). The first stage efficiently removes the large particles, but not the small ones. The second stage of some collectors had higher efficiencies for small particles than the first stage. Final stages were typically very efficient across all particle sizes, but also very expensive.

In choosing and installing mist collectors, a number of issues need to be considered.

a. *Number of Units*

Mist collectors can either be small units dedicated to a single machine or larger units collecting air from many machines. The dedicated unit can often be purchased with the machining center allowing for validation of the entire unit prior to purchase. Centralized units have the advantage of reducing the maintenance requirements to one unit rather than many, but they require ductwork for routing the air from the machine.

b. *Ductwork*

Ductwork, which bridges the distance between the machine enclosure and the mist collector, is expensive and longer runs increase both the ductwork cost and the fan power requirements. A 1996 report estimated that a reasonable cost for ductwork was $3.50/cfm.[38] However, ductwork also serves as a "stage 0" mist collector, with mist settling out over the run, which will increase the time between filter changes on the collector.[39] Optimally, ductwork should be designed with low places where fluid will collect and drainage ports to remove the fluid. Mist collectors and ductwork for straight oil systems may require fire retardant systems.

c. *Maintenance/Energy Costs*

Internal parts and filters need to be cleaned or replaced when plugged. The collector should be equipped with a pressure gauge to monitor the pressure drop across the filters, which indicates the need for filter changes. Alternatively, if the pressure drop is lower than when the filter was new, this could signify a leak in the system. Another useful method to troubleshoot collectors is to hold a continuous mist monitor at the collector exhaust. Readings should be very low (<0.2 mg/m^3). High values could indicate that the filters need changing or that mist-laden air is bypassing the filter.

The pressure drop across the collector is also an indicator of the energy and cost required to move air through the system. In comparison of nine collectors by Leith et al.,[37] they found that the pressure drop ranged over a factor of six between collectors.

d. *Vapor Phase*

During operation, the filters in a mist collector become saturated with fluid. As air continues to flow through the filters, some of the more volatile components in the fluid may go into the vapor state. In one study, 10% of the mineral oil in a collector became vapor downstream.[40] The vapor could then recondense into mist downstream of the mist collector. This would be expected to be an issue only with highly volatile oils.

FIGURE 16.16 Efficiency curves for a variety of mist collectors: (a) first stage of collectors; (b) second stage of collector; (c) final stage of a collector. (*Source*: From Leith, D., Leith, F. A., Boundy, M. G., *Am. Ind. Hyg. Assoc. J.*, 57, 1142, 1996. With permission.)

TABLE 16.5
Mist Reduction Options

Mist Prevention	Mist Control
Process Changes	*Engineering Controls*
Reduce straight oil usage	Increase building ventilation
Decrease oil volatility	Add ventilation in problem areas
Reduce tramp oil levels	Cover flumes, tanks
Reduce fluid velocity	Increase space between machines
Reduce through-tool application	Wall off problem areas
Reduce tool surface speed	Retrofit enclosures/barriers
Avoid worn tools	Turn fluid off when not needed
Antimisting polymers	*Optimize nozzle orientation*
PIB for straight oils	Reduce foaming
Associative poly. for waterbased	Reduce compressed air usage
MQL	*Enclosures/mist collectors*
OEM MQL machines	Test product before purchase
Add-on systems	Match collector to situation
Dry machining	Minimize pressure drops, new filters
Easier: cast iron, open operations	Monitor pressure drop on filters
Harder: aluminum, closed operations	Exhaust out of plant

e. Small-Particle Mist

A special situation that can arise is the generation of fine-particle mist by the evaporation/condensation mechanism. This situation is more likely in high temperature situations, such as using straight oils for high-speed grinding. As shown in Figure 16.16, some mist collectors are not efficient for submicron particles, thus allowing the mist to pass directly to the final HEPA filter for collection. For this special application, the mist collector must have early stages that are efficient in removing small particles.

f. Exhaust

Exhaust from mist collectors can be recirculated to the plant air or exhausted outside the plant. Exhausting air to the outside removes residual mist and VOC in the exhaust stream from the workplace. The downside is that when air is removed from the building, additional energy may be required to heat or cool the replacement air. However, for tempered air facilities the latent heat of recirculated air is much greater than the latent heat of outdoor air and there can be a major energy saving during the cooling season by exhausting the warm, moist process air. Environmental permitting may also be required when mist collectors are exhausted out of the building.

Two strategies were presented to reduce airborne mist, to prevent the generation of mist, or to control the mist once generated. The options in these two categories are summarized in Table 16.5.

ACKNOWLEDGMENTS

The authors would like to thank Doug Knowles from GMs Prototype Shop for his enormous help with machining.

REFERENCES

1. Thornburg, J. and Leith, D., Size distribution of mist generated during metal machining, *Appl. Occup. Environ. Hyg.*, 15, 618, 2000.
2. Bell, D. D., Chou, J., Nowag, L., and Liang, S. Y., Modeling of the environmental effect of cutting fluids, *Tribol. Trans.*, 42, 168, 1999.
3. Yue, Y., Sun, J., Gunter, K. L., Michalek, D. J., and Sutherland, J. W., Character and behavior of mist generated by application of cutting fluid to a rotating cylindrical workpiece, part 1: model development, *J. Manuf. Sci. Eng.*, 126, 417, 2004.
4. Sun, J., Ju, C., Yue, Y., Gunter, K. L., Michalek, D. J., and Sutherland, J. W., Character and behavior of mist generated by application of cutting fluid to a rotating cylindrical workpiece, Part 2: experimental validation, *J. Manuf. Sci. Eng.*, 126, 426, 2004.
5. Childers, J. C., The chemistry of metalworking fluids, In *Metalworking Fluids*, Byers, J. P., Ed., Marcel Dekker, New York, p. 165, 1994.
6. Dasch, J. M., Ang, C. C., and Williams, D., Screening industrial fluids for VOC content: a new vapor pressure method, In *Measurement of Toxic and Related Air Pollutants Proceedings*, Air and Waste Management Association, 1997.
7. Ilgner, R. H., Palausky, A., Jenkins, R. M., Ball, A. M., and Lucke, W. E., Distribution of fatty acids and triethanolamine in synthetic metalworking fluid aerosols generated in the laboratory and field, In *Proceedings of the American Automobile Manufacturer's Association Symposium II: The Industrial Metalworking Environment: Assessment and Control*, AAMA, New York p. 173, 1998.
8. Dasch, J. M., Ang, C. C., D'Arcy, J. B., Powell, C. A., Rogers, J. D., Smolenski, D. J., Tung, S. C., Gagnon, E. G., Lee, J. M., Knowles, D., and Beattie, P., General Motors clean machining program, In *Proceedings of the American Automobile Manufacturer's Association Symposium II: The Industrial Metalworking Environment: Assessment and Control*, AAMA, New York, p. 336, 1998.
9. O'Brien, D. M., Piacitelli, G. M., Sieber, W. K., Hughes, R. T., and Catalano, J. D., An evaluation of short-term exposures to metalworking fluids in small machine shops, *Am. Ind. Hyg. Assoc. J.*, 62, 342, 2001.
10. D'Arcy, J. B., Wooley, R. G., and Chan, T. L., Size distributions and concentrations of aerosols in industrial machining environments, In *Proceedings of the American Automobile Manufacturer's Association Symposium: The Industrial Metalworking Environment: Assessment and Control*, AAMA, New York, p. 186, 1996.
11. Chan, T. L., D'Arcy, J. B., and Siak, J., Size characteristics of machining fluid aerosols in an industrial metalworking environment, *Appl. Occ. Environ. Hyg.*, 141, 1136, 1990.
12. Dasch, J. M., D'Arcy, J. B., Gundrum, A., Sutherland, J., Johnson, J., and Carlson, D., Characterization of fine particles from machining in automotive plants, *J. Occup. Environ. Health*, 2, 609, 2005.
13. Heitbrink, W. A., D'Arcy, J. B., and Yacher, J. M., Mist generation at a machining center, *Am. Ind. Hyg. Assoc. J.*, 61, 22, 2000.
14. Dasch, J. M., Ang, C. C., Wei, L., and Rossrucker, T., The influence of the base oil on misting in metal removal fluids, *Lubr. Eng.*, 57, 14, 2001.
15. Dasch, J. M., D'Arcy, J. B., Kinare, S. S., Yin, Y., Kopple, R. G., and Salmon, S. C., Mist generation from high-speed grinding with straight oils, presented at *International Manufacturing Technology Show*, Chicago, IL, September 10, 2004.
16. NIOSH, Criteria for a recommended standard, *Occupational Exposure to Metalworking Fluids*, NIOSH Publication No. 98-102, 1998.
17. Hands, D., Sheehan, M. J., Wong, B., and Lick, H., Comparison of metalworking fluid mist exposures from machining with different levels of machine enclosure, *Am. Ind. Hyg. Assoc. J.*, 57, 1173, 1998.
18. Dasch, J. M., Ang, C. C., Mood, M., Cheek, C., and Knowles, D. P., Protocol for metal removal fluid evaluation, presented at *Metalworking Fluids Conference*, Society of Manufacturing Engineers, Farmington Hills, MI, November 2000.
19. Dasch, J. M., Ang, C. C., Mood, M., and Knowles, D. P., Variables affecting mist generation from metal removal fluids, *Lubr. Eng.*, 58, 10, 2002.
20. Cutcher, J. A. and Goon, D., Factors affecting mist generation in milling steel, In *Proceedings of the American Automobile Manufacturer's Association Symposium II: The Industrial Metalworking Environment: Assessment and Control*, AAMA, New York, p. 327, 1998.

21. Turchin, H. and Byers, J. P., Effect of oil contamination on metalworking fluid mist, *Lubr. Eng.*, 56, 21, 2000.
22. ISO 3685, *Tool-Life Testing with Single-Point Turning Tools*, International Organization for Standardization, Geneva, Switzerland, 1993.
23. Gulari, E., Manke, C. W., Smolinski, J., Marano, R. S., and Toth, L., Polymer additives as mist suppressants in metalworking fluids: laboratory and plant studies, In *Proceedings of the American Automobile Manufacturer's Association Symposium: The Industrial Metalworking Environment: Assessment and Control*, AAMA, New York, p. 294, 1996.
24. Gulari, E., Manke, C. W., Yurgelevic, S., and Smolinski, J., Suppression and management of mist with polymeric additives, In *Proceedings of the American Automobile Manufacturer's Association Symposium II: The Industrial Metalworking Environment: Assessment and Control*, AAMA, New York, p. 291, 1998.
25. Dasch, J. M., D'Arcy, J. B., and Smolenski, D., Effectiveness of antimisting polymers in metal removal fluids: laboratory and plant studies, *Tribol. Lubr. Technol.*, 60, 38, 2004.
26. Kalhan, S., Twining, S., Denis, R., Marano, R., and Messick, R., Development of shear-stable mist suppressants for aqueous metalworking fluids, In *Proceedings of the American Automobile Manufacturer's Association Symposium II: The Industrial Metalworking Environment: Assessment and Control*, AAMA, New York, p. 185, 1998.
27. Kalhan, S., *An End to Mist in our Time?*, Lubes N Greases, Falls Church, VA, 1998.
28. Kalhan, S., Twining, S., Denis, R., Marano, R., and Messick, R., Polymer additives as mist suppressants in metalworking fluids, Part IIa preliminary laboratory and plant studies — water-soluble fluids, *Soc. Auto. Eng.*, 1998, Paper 980097.
29. McCabe, J. and Ostraff, M. A., Performance experience with near-dry machining of aluminum, *Lubr. Eng.*, 57, 22, 2001.
30. Machado, A. R. and Wallbank, J., The effect of extremely low lubricant volumes in machining, *Wear*, 210, 76, 1997.
31. Braga, D. U., Diniz, A. E., Miranda, G. W. A., and Coppini, N. L., Using a minimum quantity of lubricant (MQL) and a diamond coated tool in the drilling of aluminum–silicon alloys, *J. Mater. Process. Technol.*, 122, 127, 2002.
32. McClure, T. F., Adams, R., Gugger, M. D., and Gressel, G., Comparison of flood vs microlubrication on performance, presented at the *56th Annual Meeting Society of Tribologists Lubricant Engineering*, Orlando, FL, 2001.
33. Anderson, J. E. et al., Particle and vapor emissions from dry and minimum quantity lubrication (MQL) machining of aluminum, In *Proceedings, Air and Water Management Association Annual Meeting*, Indianapolis, IN, 2004.
34. McCabe, J., Dry Holes, *Cutting Tool Eng.*, 54, 2002.
35. Woskie, S. R., Smith, T. J., Hammond, S. K., and Hallock, M. H., Factors affecting worker exposures to metal-working fluids during automotive component manufacturing, *Appl. Occup. Environ. Hyg.*, 9, 612, 1994.
36. D'Arcy, J. B., Air contaminant visualization for cost-effective engineering controls, In *Proceedings of the American Automobile Manufacturer's Association Symposium: The Industrial Metalworking Environment: Assessment and Control*, AAMA, New York, p. 284, 1996.
37. Leith, D., Leith, F. A., and Boundy, M. G., Laboratory measurement of oil mist concentrations using filters and an electrostatic precipitator, *Am. Ind. Hyg. Assoc. J.*, 57, 1142, 1996.
38. Burns, R. E., Harshfield, V. W., Johnston, W. J., and Kramer, R. C., Costs of mist control installations, In *Proceedings of the American Automobile Manufacturer's Association Symposium: The Industrial Metalworking Environment: Assessment and Control*, AAMA, New York, p. 321, 1996.
39. Peters, T. M. and Leith, D., Measurement of particle deposition in industrial ducts, *Aerosol Sci.*, 35, 529, 2004.
40. Cooper, S. J., Raynor, P. C., and Leith, D., Evaporation of mineral oil in a mist collector, *Appl. Occup. Environ. Hyg.*, 11, 1204, 1996.

17 Regulatory Aspects of Metalworking Fluids

Eugene M. White

CONTENTS

I. INTRODUCTION

Metalworking fluids (MWFs) are complex and diverse chemical mixtures used in various types of machining processes to fabricate metal parts. An estimate by the National Institute for Occupational Safety and Health (NIOSH) states that some 1.2 million workers are exposed to MWF during machining operations. These exposures are related to the inhalation of MWF aerosols and skin contact with metal parts, tools, and work surfaces that are coated with used fluids.

Due to the large variety of production-line fluids and special order MWF formulations produced by numerous manufacturers in the U.S., there are literally thousands of MWF products on the open market. Therefore, there is no "typical" MWF *per se* in any given MWF category, i.e., straight oils, soluble oils, semisynthetic fluids, and synthetic fluids. (Unless otherwise stated, the term "metalworking fluids" in this chapter refers to water-soluble fluids.) The Chemical Abstract Service (CAS) — a division of the American Chemical Society — has over 25 million records for chemical substances but none specifically for MWFs. However, there are approximately 700 unique CAS number assignments for chemicals found in various fluid formulations. MWFs are difficult to classify and regulate, due to their chemical diversity and proprietary compositions. Chemical regulations are usually applicable to pure chemical substances or defined compounds that can be sampled and analyzed by scientifically validated methods. Another barrier to the regulation of MWFs is that, during normal usage, they undergo physical, chemical, and biological changes. Thus, it is not always clear whether regulations address issues caused by the inherent properties of fluids (unused), changes that occur in used fluids, or extrinsic conditions in the fluid/machining environment.

Industries and businesses that utilize or come into contact with MWFs are impacted by various regulations. A few of these regulations are listed below:

- MWF Raw Material Producers/Suppliers
  - OSHA Hazard Communication Standard (HCS)
  - DOT Hazardous Materials Transportation Act (HMTA)
  - EPA Toxic Substances Control Act (TSCA)
  - EPA Federal Insecticide, Fungicide and Rodenticide Act (FIFRA)
  - EPA Emergency Planning and Community Right-to-Know Act (EPCRA)
  - EPA Clean Air Act (CAA)
  - EPA Clean Water Act (CWA)
  - EPA Resource Conservation and Recovery Act (RCRA)
  - EPA Toxic Release Inventory (TRI)
- MWF Formulators
  - Hazard Communication Standard
  - Toxic Substances Control Act
  - Clean Air Act
  - Clean Water Act
  - Resource Conservation and Recovery Act
  - Emergency Planning and Community Right-to-Know Act
  - Federal Insecticide, Fungicide and Rodenticide Act
  - Toxic Release Inventory
- MWF Transporters
  - Hazard Communication Standard
  - Hazardous Materials Transportation Act
- MWF Product Distributors
  - Hazard Communication Standard
  - Resource Conservation and Recovery Act
  - Hazardous Materials Transportation Act
  - Emergency Planning and Community Right-to-Know Act

- MWF End-users
  - Emergency Planning and Community Right-to-Know Act
  - Hazard Communication Standard
  - Resource Conservation and Recovery Act
  - Toxic Substances Control Act
- MWF Waste Haulers/Recyclers/Disposers
  - Comprehensive Environmental Response, Compensation, and Liability Act (CERCLA/Superfund)
  - Emergency Planning and Community Right-to-Know Act
  - Hazard Communication Standard
  - Hazardous Materials Transportation Act
  - Resource Conservation and Recovery Act

II. ASPECTS OF CHEMICAL REGULATIONS

Chemical raw materials and products manufactured and utilized in U.S. industries are subject to federal, state, and local regulatory oversight. In general, chemical regulations are largely based on (1) human health and safety concerns, and (2) impacts that chemicals may have on the ambient environment:

- *Human health and safety concerns.* Workers in chemical or related industries, and the general public
- *Environmental impacts.* Pollution control (surface and ground waters, air emissions, waste effluents, solid wastes) and conservation management (including the protection of endangered animal species)

The Environmental Protection Agency (EPA) monitors over 75,000 industrial chemicals (domestically produced or imported) by authority of the Toxic Substances Control Act (TSCA) (40 CFR Parts 700–789). These chemicals are subject to mandatory testing and reporting requirements, especially when scientific data demonstrates that they are hazardous to the environment or to the health of humans. The Occupational Safety and Health Administration (OSHA) administers regulations that protect workers from hazardous chemicals in the workplace. The U.S. Department of Transportation (DOT) regulates the transport of hazardous chemicals through the Hazardous Materials Transportation Act (HMTA) (49 CFR Parts 100–185).

Industrial chemicals are highly regulated in practically every aspect of their manufacture, transportation, use, and ultimate disposal:

- Manufacture of chemicals
  - Employee health and safety (exposure avoidance and reduction)
  - Hazard communication
  - Environmental impacts
  - Air emissions
  - Zoning permits and regulations
  - Chemical storage
  - Spill control measures
  - Effluent discharges
    - On-site storage and treatment
    - Publicly owned treatment works (POTW)
  - Prevention of fires and explosions
  - Waste disposal and recycling
  - Right-to-know (RTK) requirements

- Transportation of chemicals
  - Packaging
  - Hazard communication
    - MSDS
    - Labels
    - Training
  - Truck placarding
  - Spill control measures
- Use of chemicals
  - Employee health and safety (exposure avoidance and reduction)
  - Hazard communications
    - MSDS
    - Labels
    - Training
  - Environmental impacts
  - Air emissions
  - Zoning permits and regulations
  - Chemical storage
  - Spill control measures
- Recycling and disposal of chemicals
  - Waste hauler licensing
  - Waste transport
  - Waste disposal procedures
  - Waste site management

III. PROMULGATION OF REGULATIONS

U.S. federal regulations are published in the Federal Register — a publication of the Office of the Federal Register, National Archives and Records Administration (NARA). The Federal Register consists of interim and final rules, advance notices of proposed rulemaking, executive orders, meeting notices, etc. The general public has an opportunity to comment on proposed regulations by submitting written comments and/or voicing opinions during scheduled public hearings.

The process for creating federal laws follows established procedures:

1. A bill (draft of a proposed law) is sponsored by a member of Congress.
2. The bill is assigned to the appropriate congressional committee (and subcommittee) for consideration (hearings, debates, fact-finding, etc.).
3. The committee will recommend or reject the bill.
4. A committee-approved bill is debated (or considered in both Houses of Congress).
5. Upon approval of the bill by Congress, it is sent to the President, who either approves or vetoes it.
6. An approved bill is called an act (the text is a public statute).
7. The act undergoes standardization by the House of Representatives and is then published in the United States Code (USC). The USC is the official record of all federal, general, and permanent U.S. laws. Responsibility for the publication of the USC is delegated to the Government Printing Office (GPO).
8. The appropriate federal agency creates regulations (specific rules) from the code.
9. The proposed regulations are published in the Federal Register for consideration and comments by the general public.
10. Upon modification of the proposed regulation, the responsible agency presents the regulation as a final rule that is published in the Federal Register.

11. The completed regulation is then codified (detailed) and published in the Code of Federal Regulations (CFR) by the GPO.

The CFR is divided into 50 parts called Titles (numbered 1 to 50) that represent different areas of federal regulations.

Subsequent to its approval, a law undergoes a process of codification in which its broad objectives and principles are transformed into the minute details that are relevant to its implementation. A codified document is assigned to the appropriate title of the Code of Federal Regulations (CFR). Each CFR title (volume) is broad and comprehensive in scope.

Various portions of the CFR may directly or indirectly affect the business practices of MWF suppliers, manufacturers, and end-users. Three specific CFR titles, especially, have much significance for MWF products and, thus, are discussed at length in this chapter:

- 29 CFR (Labor)
- 40 CFR (Protection of Environment)
- 49 CFR (Transportation)

IV. 29 CFR (LABOR)

On December 29, 1970, the Occupational Safety & Health Act (OSH Act) (Public Law 91-59) was signed into law. In part, the Act states:

> To assure safe and healthful working conditions for working men and women; by authorizing enforcements of the standards developed under the Act; by assisting and encouraging the States in their efforts to assure safe and healthful working conditions; by providing for research, information, education, and training in the field of occupational safety and health; and for other purposes.

Under authority granted under the Act, the Secretary of the Department of Labor (DOL) delegates responsibilities to the Assistant Secretary of Labor (OSHA administrator) to protect the health and safety of working men and women by means of regulatory and enforcement powers codified in 29 CFR (Labor). Prior to 1971, the assessment and enforcement of workplace hazards was largely arbitrary and deemed by many as ineffective. The involvement of federal agencies to inspect unhealthy and unsafe workplaces was minimal, and enforcement activities were generally tentative, due to personnel and resource limitations. In many cases, individual states were left to regulate their own workplaces with little federal oversight or accountability. However, on April 28, 1971, as a result of the Act, OSHA was activated.

Since its inception, OSHA has provided the impetus in the U.S. for the protection of a steadily expanding workforce that has doubled in just three decades:

(1971) 56 million workers at 3.5 million worksites

↓

(2005) 115 million workers at 7.2 million worksites

A. The General Duty Clause

The cornerstone of OSHA regulations is broadly stated in Section 5(a)(1) of the OSH Act, i.e., the General Duty Clause (GDC):

(a) "Each employer —

(1) shall furnish to each of his [SIC] employees employment and a place of employment which are free from recognized hazards that are causing or are likely to cause death or serious physical harm to his employees;

(2) shall comply with occupational safety and health standards promulgated under this Act."

(b) "Each employee shall comply with occupational safety and health standards and all rules, regulations, and orders issued pursuant to this Act which are applicable to his own actions and conduct."

The enforcement powers of OSHA under the GDC are profound and sweeping, due to the fact that GDC grants authority to OSHA inspectors to levy citations in situations where there are no specific or applicable occupational health and safety standards. Violations of the GDC occur when a "recognized hazard" is not abated. In situations where applicable standards exist but are inadequate to protect workers, enforcement is sanctioned under the GDC.

The OSHA administrator can use the GDC, federal regulations, and various health and safety standards such as the OSHA Permissible Exposure Limit (PEL) for a substance(s) as criteria to determine the necessity for and extent of enforcement action.

B. OSHA Hazard Communication Standard (29 CFR 1910.1200)

The OSHA Hazard Communication Standard (HCS or HazCom Standard) mandates the dissemination of information to workers and employers about the hazards of chemicals in the workplace and protective measures that can be taken to eliminate or minimize health-threatening or unsafe occupational exposures. A hazardous chemical is defined as a substance that presents a physical or health hazard.

A physical hazard is a chemical with the following characteristics (as determined by validated scientific studies):

- Combustible liquid
- Compressed gas
- Explosive
- Flammable
- Organic peroxide
- Pyrophoric
- Unstable (reactive) or water-reactive

The terms "flammable liquids" and "combustible liquids" have different and distinct meanings under OSHA 29 CFR 1910.106 (Flammable and Combustible Liquids):

1910.106 (a) (18). A combustible liquid is defined as "...any liquid having a flashpoint at or above 100°F (37.8°C)..." (See complete standard for classes of combustible liquids.)

1910.106 (a) (19). A flammable liquid is defined as "...any liquid having a flashpoint below 100°F (37.8°C), except any mixture having components with flashpoints of 100°F (37.8°C) or higher, the total of which make up 99% or more of the total volume of the mixture... Flammable liquids shall be known as Class I liquids..." (See complete standard for classes of flammable liquids.)

A health hazard is a chemical that has been determined to cause adverse effects (acute or chronic) on the health of exposed workers:

- Causes damage to eyes, lungs, mucous membranes, skin
- Carcinogenic — a material that is known to cause cancer
- Corrosive — a material that is highly reactive and causes damage to living tissue
- Hematopoietic system affector — a chemical that adversely affects the blood-forming mechanism of the human body

- Hepatotoxin — a toxin that affects the liver
- Irritant — a chemical that is not corrosive but causes a reversible inflammatory effect on living tissue
- Nephrotoxin — a toxic agent that inhibits, damages or destroys cells and/or tissues of the kidney
- Neurotoxin — a toxic agent that inhibits damages or destroys the tissues of the nervous system
- Reproductive toxin — a substance that causes adverse effects on the reproductive system of men and women
- Toxic — a term used to describe a chemical agent that has a harmful effect on a biological/physiological system (see 29 CFR 1910.1200 App. A)

Relevant portions of the HCS are as follows:

- Hazard Determination [29 CFR 1910.1200 (d)]
  - Requires employers to identify and keep records of hazardous chemicals.
- Implementation Program [29 CFR 1910.1200 (e)]
  - Employers must write a Hazard Communication Plan.
- Chemical Labeling [29 CFR 1910.1200 (f)]
  - All workplace chemicals must be in labeled containers, and pipes that carry hazardous chemicals must be labeled.
- Material Safety Data Sheets [29 CFR 1910.1200 (g)]
  - An MSDS for every chemical in a facility must be readily available (no location or administrative barriers) to employees with information about the hazard(s) of each chemical.
- Employee Training [(29 CFR 1910.1200 (h)]
  - Employees must be trained about the HCS, the hazards of materials, and how to interpret information in MSDSs and labels.
- Trade Secrets [(29 CFR 1910.1200 (i)]
  - Under certain conditions, the manufacturer of a chemical may withhold proprietary information about its formulation, but there are stated exceptions when information must be provided to health care providers in order to facilitate medical treatment of workers.

C. The Material Safety Data Sheets (MSDS)

Probably no other aspect of the HCS has had more impact on U.S. chemical industries than the requirement for the creation of the Material Safety Data Sheets (MSDS). Since the federally mandated inception of the MSDS on May 26, 1986, manufacturers and vendors of chemicals have devoted many resources (financial, personnel, MSDS generation systems, etc.) to ensure that the information contained in a MSDS is current, accurate, and compliant with HCS requirements. An MSDS must contain information about the hazardous chemicals (defined in 29 CFR 1910, Subpart Z) in a product. Chemical manufacturers or businesses that sell products that contain chemicals are under no obligation to provide information in a MSDS about nonhazardous constituents, although they may voluntarily opt to do so.

OSHA has no required format for the MSDS. However, a suggested MSDS format is found in OSHA Form 174 (OMB#1218-0072). The sections of this form are listed below:

- Identity (as used on label and list)
- Section I. Manufacturer's Identification and Contact Information
- Section II. Hazardous Ingredients, Identity Information

- Section III. Physical and Chemical Characteristics
- Section IV. Fire and Explosion Hazard Data
- Section V. Reactivity Data
- Section VI. Health Hazard Data
- Section VII. Precautions for Safe Handling and Use
- Section VIII. Control Measures

Many companies exceed the suggestions of OSHA Form 174 and commonly incorporate 16 sections in a given MSDS:

1. Identity (as used on label and list).
2. Manufacturer's Identification and Contact Information.
3. Hazardous Ingredients, Identity Information.
4. Physical and Chemical Characteristics.
5. Fire and Explosion Hazard Data.
6. Reactivity Data.
7. Health Hazard Data.
8. Precautions for Safe Handling and Use.
9. Control Measures.
10. First Aid Measures.
11. Accidental Release Measures.
12. Product Stability and Reactivity.
13. Transportation Information.
14. Regulatory Information.
15. Waste Disposal.
16. Product Handling and Storage. Other MSDS sections may include toxicological and ecological information.

Currently, due to the rapid expansion of global commerce, there is a concerted movement in the U.S. and abroad for MSDS "harmonization." This initiative advocates the creation of a standardized MSDS format that would be adopted by numerous countries, as a means to facilitate international hazard communication. In this regard, the American National Standards Institute (ANSI) is a major participant through its activities involving the ANSI Z400.1-2003 standard.

D. Right-to-Know Labels

The first observable source of information about a chemical product is contained in its label. Under the OSHA HCS, chemical manufacturers have a legal duty to apply labels onto containers that provide information (in English) about a given product; other languages may be used in addition to English at the discretion of the manufacturer. Identification of the hazardous chemical(s) in the product should contain the following label information:

- CAS number(s)
- Target organ warnings
- Name of the manufacturer
- Address of the manufacturer
- Emergency phone number(s)
- Hazard warnings
- First aid instructions
- Signal words, e.g., "Poison," "Danger"

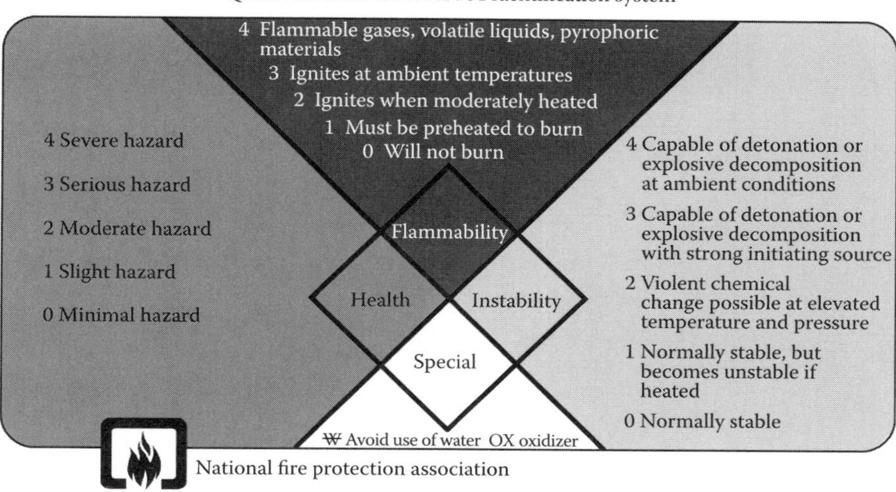

FIGURE 17.1 Example of NFPA hazardous material information label. Reprinted with permission from NFPA 704-2001, *System for the Identification of the Hazards of Materials for Emergency Response*, Copyright © 2001, National Fire Protection Association, Quincy, MA 02269. This reprinted material is not the complete and official position of the NFPA on the referenced subject, which is represented only by the standard in its entirety. This symbol is intended for use in emergency response only. The NFPA cannot be responsible for the use or classification of any chemical, whether such classification was carried out by the NFPA or the user of the symbol.

The information on container labels is not only necessary and helpful for the users of a given product, but it is essential in cases of accidents, spills, and fires when emergency responders need to know what potential hazards confront them.

OSHA has no required label format for the communication of hazard information. However, labels commonly utilize one of the following two systems.

1. National Fire Protection Association (NFPA®) System

The NFPA system, shown in Figure 17.1, uses a combination of numbers and a diamond-shaped, color-coded symbol to provide information about a chemical based on the NFPA 704 Hazard Rating System. NFPA, a nonprofit organization, has a database of hazard ratings for over 3000 chemicals. This organization's diamond-shaped hazard symbol is frequently found on chemical labels. The system is mainly geared to providing information to emergency responders on fire protection issues. The diamond has four sections: blue for health hazard rating, red for fire hazard rating, yellow for reactivity rating, and a white section at the bottom for information about any specific hazard (i.e., oxidizer, corrosive, acid, alkali, etc.).

2. Hazardous Materials Identification System (HMIS®)

The HMIS system was created by the National Paint and Coatings Association (NPCA). This system uses a three-color, horizontal-bar-shaped symbol to indicate health, fire, and physical

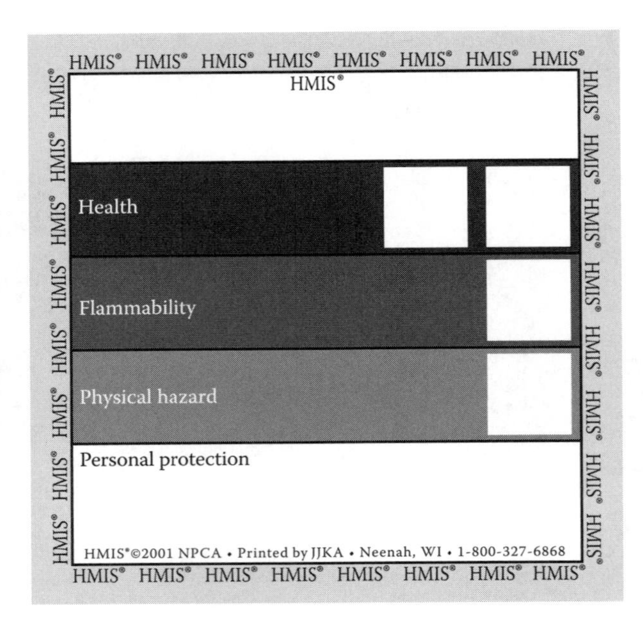

FIGURE 17.2 Example of HMIS III label, courtesy of National Paint and Coatings Association. HMIS is a registered trademark of the National Paint and Coatings Association and is used with permission. HMIS products may be obtained exclusively from J.J. Keller and Associates Inc. at www.jjkeller.com/hmis or 877-564-2333.

hazard information about a chemical. This system is mainly geared to providing information to workers. The latest HMIS III label is shown in Figure 17.2. The top blue bar has a box for a health rating number plus a box to indicate by asterisk if the material is a chronic health hazard; the second red bar is for a flammability rating, and the bottom orange bar is for physical hazard rating.

Although the above two systems are equivalent, there are important differences between them. The NFPA system is more applicable to the informational needs of emergency response personnel and firefighters, while HMIS is specifically designed to relate hazard information to workers. Both systems use a rating scale ranging from zero to four as materials increase in hazard potential. Both use blue and red regions for health and flammability, respectively. However, the third color zone is different between the two systems: NFPA uses yellow for reactivity, while the latest HMIS III label uses orange for physical hazard ratings. The white area of the NFPA diamond provides information about "special hazards" of a chemical, but the white area for HMIS indicates personal protective equipment (PPE) that should be worn under certain conditions. The NFPA and HMIS systems carry no regulatory endorsement or enforcement authority. However, both systems facilitate communication of HCS required information.

E. 29 CFR Relevance for Metalworking Fluids

- MWF raw materials suppliers, manufacturers, and end-users are subject to all applicable OSHA/HCS requirements.
- OSHA has no nominal permissible exposure limit (PEL) for water-soluble MWFs. However, mineral oil — a constituent of many MWFs — has a PEL of 5.0 mg/m^3:
  29 CFR 1910.1000 (Table Z-1, Limits for Air Contaminants)
  - Oil mist, mineral, 5 mg/m^3

- Particulates Not Otherwise Classified (PNOC), 15 mg/m^3 (total) (PNOC refers to particulates with no known toxic effects, but which may be toxic at certain airborne concentrations)
- PNOC, 5 mg/m^3 (respirable)

29 CFR 1915.1000 (Table Z, Air Contaminants, Shipyards)
- Oil mist, mineral, 5 mg/m^3
- PNOC, 15 mg/m^3 (total)

29 CFR 1926.55 App A (Gases, Vapors, Fumes, Dusts, and Mists)
- Oil mist, mineral, 5 mg/m^3
- PNOC, 15 mg/m^3

V. 40 CFR (PROTECTION OF ENVIRONMENT)

The EPA was signed into law on December 2, 1970, and was charged with the mission "...to protect human health and the environment." Regulations of EPA are contained in 40 CFR. Selected sections of 40 CFR are discussed below.

A. 40 CFR 700–799 (Toxic Substances Control Act) (TSCA)

EPA/TSCA tracks over 75,000 industrial chemicals produced or imported into the U.S. The EPA regularly screens these chemicals for human health and environmental hazards. If a chemical is deemed to be hazardous, its manufacturer can be required to submit information and/or test results. TSCA also monitors the properties of new chemicals that enter the industrial stream. A new chemical that is not on the TSCA list is subject to premanufacture notification (PMN); the manufacturer must notify EPA 90 days prior to manufacturing the chemical. Nuclear waste is not regulated by TSCA, unless it is mixed with nonnuclear waste.

The facility reporting aspects of TSCA are as follows:

- Reporting threshold is 25,000 lbs
- Data reported on inorganic chemicals
- Information about use(s) of chemicals
- Information about number of exposed workers
- Concentration of chemicals leaving a facility
- Physical form(s) of the chemical leaving a facility
- Production volumes of chemicals

TSCA also interacts with federal regulations dealing with EPCRA and the Clean Air Act (CAA).

B. 40 CFR 152–186 (Federal Insecticide, Fungicide, Rodenticide Act)

EPA/FIFRA regulates commercial pesticides, including their registration and label language. A pesticide (biocide) cannot be manufactured or distributed unless the manufacturer has registered the product and label with the EPA. The application for registration process for new biocides is rigorous and expensive, and is analogous to the Federal Drug Administration (FDA) approval process for prescription pharmaceuticals:

- Comprehensive application process
- Submission of proposed label

- Statement of claims about the product
- Formulation statement
- Submission of test data

Biocide registration requirements involve extensive toxicological testing of the pesticide for a single use. Other considerations are environmental, social, economic, environmental costs, and the efficacy of the product. Once a biocide registration is granted, it is effective for 5 years. Pending the acquisition of new data that warrants a reexamination of a biocide, the EPA may review a product's registration prior to its expiration date.

Biocides are used in water-soluble MWFs to control the growth of bacteria and fungi. Only biocides that are EPA-registered and approved for use in MWFs can be used in MWF distribution systems. Currently, there are ten categories of biocides that are approved by EPA for use in MWFs:

- Bromonitriles
- Dioxanes
- Formaldehyde condensates (largest category)
- Glutaraldehydes
- Isothiazolones
- Mixtures
- Phenols
- Polymeric quats
- Pyridinethione
- Thiocyanobenzothiazole

Due to the expense and time involved in applying for a biocide registration, and undergoing costly toxicological and environmental impact studies, no new registered biocides have appeared on the market for many years.

C. 40 CFR 260–279 (Resource Conservation Recovery Act)

EPA/RCRA regulates hazardous waste from "cradle-to-grave":

- Waste generation
- Waste transportation
- Waste treatment
- Waste storage
- Waste disposal

The functions of RCRA deal solely with active and future waste facilities and not abandoned waste sites (a function of the Comprehensive Environmental Response, Compensation, and Liability Act (CERCLA or "Superfund").

D. High Production Volume (HPV) Chemical Testing Program

Under the auspices of the TSCA of 1976, the EPA has the authority to collect information from chemical manufacturers on new and existing chemicals. Information regarding a given chemical consists of various data sets relating to physicochemical properties, toxicity, environmental fate, environmental releases, production volumes, and usage applications. Subsequently, the EPA may use this information to limit or prohibit the production of certain chemicals deemed to be inimical to the environment and/or human health.

In 1998, former vice president Gore fostered a partnership of the EPA, the Chemical Manufacturers Association (now the American Chemical Council), and the Environmental Defense Fund (now the Environmental Defense) to form the High Production Volume (HPV) Chemical Testing Program. This is a Chemical Right-to-Know Initiative that involves the following toxicological testing criteria:

- Acute toxicity
- Chronic toxicity
- Developmental/reproductive toxicity
- Mutagenicity
- Ecotoxicity
- Environmental fate

A goal of the HPV Program was to compile a health and environmental database regarding high production industrial chemicals that would be made available to the public by 2005. Over 300 chemical manufacturers voluntarily provide hazard data on nonpolymeric, organic chemicals to the program. The program has established a standardized public database of the toxicity profiles of approximately 2800 chemicals that are produced or imported into the U.S. in amounts equal to or greater than one million pounds per year. Chemicals that are on the HPV list represent over 90% of the chemicals produced and used in the U.S. and data about them is collected under the TSCA/Inventory Update Rule. Companies that manufacture or import organic chemicals in amounts equal to or exceeding 10,000 pounds per year are subject to TSCA/IUR reporting requirements every 4 years. Specific details about the HPV chemical list can be found at http://www.epa.gov/chemrtk/hpvchmlt.htm.

Some of the hundreds of chemicals used in various metal working fluid formulations, including additives, may be subject to HPV testing and reporting requirements. The major U.S. MWF trade association, the Independent Lubricant Manufacturers Association (ILMA), representing over 80% of the MWF manufacturers in North America, stated its position on HPV in a 2005 report: "ILMA supports the public's right-to-know about chemicals found in their environment, homes, workplaces, and the products that they buy; however, the Association is not convinced that EPA's testing program is the most efficient way to obtain the necessary information."[1]

Some critics of the HPV Program feel that it is unnecessary and duplicates other information sources, due to the abundance of existing data on the health and environmental impacts of industrial chemicals. Other detractors cite that HPV animal test data do not always correlate with adverse affects on human health, and that the effort and time required for HPV testing may actually forestall regulatory processes.

E. 40 CFR Relevance for Metalworking Fluids

- Chemicals on the TSCA Inventory that are utilized in MWF formulations are subject to TSCA reporting requirements. [Certain substances on the TSCA Inventory are exempt from reporting requirements (40 CFR 710.30) such as polymers, inorganics, naturally occurring chemical substances, and microorganisms.]
- Biocides that are used in MWF must be registered under FIFRA specifically for that application. Biocides registered for other applications only may not be used for MWF.
- Spent MWFs may be classified as hazardous waste and are subject to applicable treatment/disposal methods as required by federal, state, and local regulations in accordance with RCRA.

VI. 49 CFR (TRANSPORTATION)

The U.S. DOT was established on October 15, 1966. The DOT mission is to:

> Serve the United States by ensuring a fast, safe, efficient, accessible and convenient transportation system that meets our vital national interests and enhances the quality of life of the American people, today and into the future.

The DOT Hazardous Materials Transportation Act (HMTA) of 1975 regulates numerous functions involving the transit of hazardous materials by air, ground, and rail:

- Labeling
- Packaging
- Placarding
- Manifesting (listing of shipment contents)
- Spill reporting

Under HMTA, workers involved in the handling, transporting, shipping, and storing of hazardous waste must undergo HAZMAT training. The packaging of hazardous materials involves performance-oriented packaging standards: shipping name, material identification number, hazard class, package requirements, and vehicle placarding.

HMTA recognizes nine classes of hazardous materials:

Class 1. Explosives.
Class 2. Gases.
Class 3. Flammable liquids.
- ☐ Division 3.1 Flashpoint below −18°C (0°F).
- ☐ Division 3.2 Flashpoint −18°C and above, but less than 23°C (73°F).
- ☐ Division 3.3 Flashpoint 23°C and up to 61°C (141°F).

Class 4. Flammable solids, spontaneously combustible materials, and materials that are dangerous when wet.
Class 5. Oxidizers and organic peroxides.
Class 6. Poisons and infectious materials.
Class 7. Radioactive materials.
Class 8. Corrosives. A material, liquid, or solid that causes visible destruction or irreversible alteration to human skin or a liquid that has a severe corrosion rate (usually tested on steel and aluminum). Some MWFs may fall into this category.
Class 9. Miscellaneous hazardous materials. A material which presents a hazard during transport, but which is not included in any other hazard class (such as a hazardous substance or a hazardous waste).

The Hazardous Materials Table (49 CFR Part 172.101) lists materials that are hazardous for transportation and requirements pertaining to labeling, packaging, and quantity limits.

49 CFR Relevance for MWFs:

- The alkaline properties ($pH > 8.5$) of many MWFs may cause these products to be classified as corrosive, however, they may be nonhazardous depending upon evaluation of analytical testing data and other criteria.
- MWF may be included into HMTA Hazard Class 8.
- The shipping name for some MWFs is "Corrosive Liquid, n.o.s." The n.o.s. designation ("not otherwise specified") is a catch-all term that is used when substances are not specifically named in a regulation.

VII. STATE AND LOCAL REGULATIONS

A. State

Individual state regulations for chemicals closely resemble their federal counterparts. A state law may be more stringent (but not less) than a corresponding federal law. Some states are delegated authority by federal agencies to run programs dealing with air emissions (CAA), water pollution (CWA) and hazardous wastes (RCRA). All states have environmental protection agencies with functions similar to the U.S. EPA that deal with a variety of environmental issues. Contiguous states that share waterways, e.g., rivers and lakes, and are affected by common air pollution problems, may enjoin in interstate cooperative agreements to protect their respective environments. States also have occupational safety and health agencies similar to the federal OSHA that protect workers throughout a given state.

Each state maintains a State Emergency Response Commission (SERC) subject to EPA/EPCRA requirements. SERCs facilitate the gathering of information about facilities that store, produce, use, and release hazardous chemicals as a means to aid emergency responders. Companies that manufacture MWFs are required to file annual EPCRA and SERC reports. States also engage in activities related to Superfund (CERCLA) and cooperate with federal agencies in site remediation operations.

B. Local

Regional municipalities are granted authority from state legislatures and agencies to manage regional air emissions, water quality, wastewater management, sewer systems, and land use (zoning) issues. Industries that operate within a municipality must adhere to local laws and ordinances, industry reporting requirements, and inspections of industrial facilities.

MWF manufacturers file annual EPCRA reports with local fire fighting headquarters, and emergency response offices. Responsibility for monitoring the effluents from various industrial sites is usually placed on local sewer districts. Companies must report the contents, condition, and amounts of their waste discharges, and any unusual conditions that might affect waste treatment plants downstream. Companies within a given locality may be required to secure permits that stipulate limits on air emissions, wastewater effluents, and the generation of solid wastes.

Regional occupational safety and health functions are managed by local health departments and/or state OSHA offices. Local Emergency Planning Committees (LEPCs) are appointed by SERCs. LEPCs are responsible for contingency planning for responses to emergency situations involving hazardous materials. Metalworking manufacturers, and other businesses that use chemicals, must file annual reports according to LEPC requirements.

VIII. CONSENSUS (NONREGULATORY) STANDARDS

Nonregulatory standards or consensus standards are promulgated by private organizations that have no legislative mandate or enforcement authority. These standards are "guidelines" proposed by a consensus of experts in a given discipline. Consensus standards guidelines are commonly written in the following forms:

- Occupational Exposure Limits (OELs)
- Standard Methods
  - Sampling
  - Analytical

Consensus standards organizations are commonly comprised of dedicated committees populated by uncompensated volunteers. Draft standards may undergo review by members of the organization, and they are eventually voted upon and then published. Some examples of nonfederal consensus standards organizations involved in the writing of standards for MWFs are as follows:

- American Society for Testing and Materials (ASTM)
  - ASTM (founded in 1898) is a nonprofit organization and recognized as the largest voluntary consensus standards organization in the world. A wide range of ASTM standards have been written specifically for MWFs.
- American National Standards Institute (ANSI)
  - ANSI (founded in 1918) is a nonprofit organization that administers and coordinates the U.S. voluntary standardization and conformity assessment system.
- American Conference of Governmental Industrial Hygienists (ACGIH)
  - ACGIH (originally founded in 1938 as "National Conference of Governmental Industrial Hygienists," then changed its name to ACGIH in 1946) is a major industrial hygiene and occupational safety and health organization that publishes the well-known Threshold Limit Values (TLV) booklet. In spite of "Governmental" in its name, ACGIH is not a federal- or state-related organization.

IX. NIOSH MWF RECOMMENDED EXPOSURE LIMIT

A federal agency that publishes nonregulatory OELs for chemicals is the National Institute for Occupational Safety and Health (NIOSH). NIOSH is a federal occupational safety and health research institute that was created by the same OSH Act of 1970 that formed OSHA. NIOSH is an institute within the Centers for Disease Control and Prevention (CDC). CDC is under the auspices of the U.S. Department of Health and Human Services (DHHS). NIOSH states its objectives as follows:

- Conduct research to reduce work-related illnesses and injuries
- Promote safe and healthy workplaces through interventions, recommendations, and capacity building
- Enhance global workplace safety and health through international collaborations

NIOSH maintains a list of OELs called Recommended Exposure Limits (RELs). Even though RELs carry no regulatory authority, they are highly influential in federal and nonfederal occupational safety and health organizations. In this regard, it is not unusual for federal regulatory agencies to use REL values as guidelines for their activities.

The NIOSH REL for MWFs is stated as follows:

REL for MWF:
0.4 mg/m^3 (thoracic particulate mass)
0.5 mg/m^3 total particulate mass

Duration. Time-weighted average (TWA) for up to 10 h/day, 40-h workweek. A TWA is the concentration of a substance (usually airborne) that a person is exposed to during a typical workday (8 to 12 h).

Measurement. NIOSH Method 0500 (for mineral oil) — a gravimetric (filter weight) method that measures total aerosol mass.

The 0.5 mg/m^3 MWF REL is stated in the 1998 NIOSH publication, Criteria for a Recommended Standard: Occupational Exposure to Metalworking Fluids. This comprehensive document has been widely circulated and cited. The NIOSH MWF REL is regularly cited in MWF MSDSs.

NIOSH publishes the NIOSH Manual of Analytical Methods (NMAM) — a comprehensive collection of sampling and analytical methods developed by NIOSH. The NMAM 500 method is used for the sampling of MWFs. NIOSH is currently conducting research on methods that are applicable for MWF analysis.

X. REGULATORY REPORTING

Businesses that manufacture, distribute, and use chemical products (and MWFs) are required to file various reports during any given year. Reporting requirements and forms are frequently changed and revised. A brief description of some reports that are applicable to MWFs is given below.

A. SERC (Tier II) Report

This is an annual report chemical businesses must submit that is mandated by the 1986 U.S. EPCRA. EPCRA is also known as Superfund Amendments and Reauthorization Act (SARA), Title III. EPCRA requirements cover over 600 chemicals. Under EPCRA (Section 313), each state establishes a SERC that administers EPCRA in designated districts. Each district has an LEPC. Planning committees have a broad membership consisting of fire and health departments, community organizations, community leaders, government officials, etc. In most cases, states require the following information on Tier II Forms:

- Facility identification
- Contact information
- Information about chemicals*
  - The chemical name or the common name as indicated on the MSDS
  - An estimate of the maximum amount of the chemical present at any time during the calendar year and the average daily amount
  - The location of the chemical at the facility
  - An indication of whether the owner of the facility elects to withhold location information

The annual SERC report is due by March 1 of each calendar year. Some states have initiated electronic filing.

B. Toxics Release Inventory Report

TRI is also known as Title III of SARA. Under EPCRA, the EPA maintains a database of almost 700 chemicals that are released by facilities that manufacture and use them. TRI came into existence in 1987, after the catastrophic incident in Bhopal, India, in 1984 when thousands of people were killed by the release of methyl isocyanate from a manufacturing facility into the air of a contiguous community. Subsequently, U.S. citizens began to demand information about chemicals in their communities. Under TRI, the following information is collected:

- Identification of chemicals in inventories of companies in localities
- Information on locations and quantities of chemicals at facilities
- Information on air emissions and effluents from facilities
- Quantity of chemicals taken away from sites for disposal and recycling
- Information about on-site waste treatment

* http://Yosemite.epa.gov/oswer/CeppoWeb.nsf/tier2.htm

The TRI report is submitted on or before July 1 of each year on Form R or Form A (shortened version for companies exempt from Form R). The EPA has facilitated this filing process by providing computer software to enable electronic filing.

C. Pesticide Report for Pesticide-Producing Establishments

This annual report is required for companies that manufacture pesticides. Once a biocide has been registered with the EPA, FIFRA stipulates that only a registered pesticide-producing establishment can produce the biocide (or pesticide device). Each FIFRA approved production site is issued a pesticide-producing establishment number. An annual report (the EPA Form 3540-16) is submitted to EPA by March 1 of each year. The basic requirements of Form 3540-16 report are as follows:

- Identification information of establishment
- Product name and classification
- EPA product registration number
- Amount produced or repackaged (previous year)
- Amount sold or distributed (previous year), domestic and foreign
- Amount to be produced, repacked, relabeled (present year)

D. State Pesticide Registrations

Each state has registration requirements for EPA-registered pesticides. Regarding pesticides (biocides) used in MWFs, states require that a manufacturer of a biocide(s) must file an annual report with status information about the registered biocide(s). Applications must be filed for new biocides that need state registrations. A registration fee is usually required for each registered biocide. Many state pesticide registration applications for the coming year are due by December 31. If production of a biocide product is discontinued, most states require a phasing-out period to allow time for the product to leave the channels of trade (usually 2 to 3 years). During this period, registration applications must be filed along with applicable fees. Some states have no discontinuation policy.

XI. CONCLUSION

Regulations that govern the manufacture, distribution, transportation, use, and disposal of chemical products are comprehensive and voluminous. MWF production comprises a small subset of the total volume of chemical products manufactured in the U.S. For instance, the number of MWF manufacturers in the U.S. is estimated to be more than 350, although less than ten companies account for nearly one half of an approximately $520 million dollar market. However, because of their chemical compositions, MWFs are subject to many of the same regulatory requirements as higher volume industrial chemicals.

There has been a gradual evolution over many decades in the compositions of MWFs in response to the technological needs of machining operations and health concerns. (The history of MWF usage is discussed in an earlier chapter.) Unlike commodity chemical products, MWFs are technologically advanced industrial lubricants that are subject to continuous research and development. The demands of modern machining technologies and advances in materials science, e.g., new metal alloys, require that fluids meet increasingly stringent manufacturing specifications. Industries engaged in metal fabrication operations require MWF formulations that facilitate the production of high quality metal parts and, concurrently, foster acceptable machine tool life. Though these concerns are effectively outside of the realm of regulatory agencies, they can eventually have profound influences on the promulgation of regulations. Traditionally, the regulatory process follows technological advances.

There is currently much discussion in regulatory circles regarding what constitutes a truly protective OEL for MWFs. However, matters related to reliable sampling and analytical methods, and the selection of "representative" MWFs for testing, are under consideration by federal agencies. Also, future regulations affecting MWFs are dependent on such issues as (1) the determination of the health effects of mixed exposures, (2) the role of microorganisms in MWFs on the respiratory health of workers, and (3) the effects that machining environments have on the health of workers. These are challenging subjects that will require concerted scientific studies.

APPENDIX A. SELECTED REGULATIONS

Comprehensive Environmental Response, Compensation, and Liability Act (Superfund) (CERCLA): 42 U.S.C. s/s 9601 *et seq.* (1980)

The Emergency Planning and Community Right-to-Know Act (EPCRA): 42 U.S.C. 11011 *et seq.* (1986)

Federal Insecticide, Fungicide and Rodenticide Act (FIFRA): 7 U.S.C. s/s 135 *et seq.* (1972)

The Occupational Safety and Health Act (OSHA): 29 U.S.C. 651 *et seq.* (1970)

The Resource Conservation and Recovery Act (RCRA): 42 U.S.C. s/s 321 *et seq.* (1976)

The Superfund Amendments and Reauthorization Act (SARA): 42 U.S.C. 9601 *et seq.* (1986)

The Toxic Substances Control Act (TSCA): 15 U.S.C. s/s *et seq.* (1976)

APPENDIX B. SELECTED STANDARDS AND REPORTS

American Conference of Governmental Industrial Hygienists (ACGIH), *Threshold Limit Values for Chemical Substances and Biological Exposure Indices*, ACGIH, Cincinnati, OH, 2005.

American National Standard Institute (ANSI), *Material Safety Data Sheets — Preparation*, ANSI Z400.1-2003, ANSI, New York, 2003.

American National Standards Institute (ANSI), *Mist Control Considerations for the Design, Installation and Use of Machine Tools Using Metalworking Fluids*, ANSI B11 TR 2-1997, ANSI, New York, 1997.

American Society for Testing and Materials (ASTM), *Practice for Evaluating Water-Miscible Metalworking Fluid Bioresistance and Antimicrobial Pesticide Performance*, E 2275-03, ASTM, Conshohocken, PA, 2003.

American Society for Testing and Materials (ASTM), *Standard Method for Determination of Endotoxin Concentration in Water Miscible Metal Working Fluids*, E 2250-02, ASTM, Conshohocken, PA, 2002.

American Society for Testing and Materials (ASTM), *Standard Practice for Selecting Antimicrobial Pesticides for Use in Water-Miscible Metalworking Fluids*, E 2169-01, Conshohocken, PA, 2001.

American Society for Testing and Materials (ASTM), *Standard Practice for Personal Sampling and Analysis of Endotoxin in Metalworking Fluid Aerosols in Workplace Atmospheres*, E 2144-01, ASTM, Conshohocken, PA, 2001.

American Society for Testing and Materials (ASTM), *Standard Practice for Selecting Antimicrobial Pesticides for Use in Water-Miscible Metalworking Fluids*, E 2169-01, ASTM, Conshohocken, PA, 2001.

American Society for Testing and Materials (ASTM), *Standard Practice for Safe Use of Water-Miscible Metal Removal Fluids*, E 1497-00, ASTM, Conshohocken, PA, 2000.

American Society for Testing and Materials (ASTM), *Provisional Practice for Personal Sampling and Analysis of Endotoxin in Metal Removal Fluid Aerosols in Workplace Atmospheres*, PS 94-98, ASTM, Conshohocken, PA, 1998.

American Society for Testing and Materials (ASTM), *Standard Practice for Safe Use of Water-Miscible Metalworking Fluids*, E 1497-94, ASTM, Conshohocken, PA, 1997.
American Society for Testing and Materials (ASTM), *Determining Carcinogenic Potential of Virgin Base Oils in Metalworking Fluids*, E 1687-95, ASTM, West Conshohocken, PA, 1997.
American Society for Testing and Materials (ASTM), *Determining Carcinogenic Potential of Virgin Base Oils in Metalworking Fluids*, E 1687-95, ASTM, Conshohocken, PA, 1995.
American Society for Testing and Materials (ASTM), *Safe Use of Water-Miscible Metalworking Fluids*, E1497-94, ASTM, Conshohocken, PA, 1994.
National Institute for Occupational Safety and Health (NIOSH), *Criteria for a Recommended Standard: Occupational Exposure to Metalworking Fluids*, DHHS (NIOSH) Publication No. 98-102, NIOSH, Cincinnati, OH, 1998.

APPENDIX C. BEST PRACTICES, GUIDELINES, REPORTS

American Society of Heating, Refrigerating and Air-Conditioning Engineers, Inc. (ASHRAE), *ASHRAE Guidelines: Minimizing the Risk of Legionellosis Associated with Building Water Systems*, ASHRAE Guideline 12-2000, ASHRAE, Atlanta, GA, 2000.
Chemical Manufacturers Association (CMA), *Comments of the Chemical Manufacturers Association, Alkanolamines Panel on Criteria for a Recommended Standard: Occupational Exposures to Metalworking Fluids*, June 6, 1996, CMA, Arlington, VA, 1996.
DHHS (NIOSH), *What You Need to Know about Occupational Exposure to Metalworking Fluids*, Publication No. 98-116, NIOSH, Cincinnati, OH, 1998.
Health and Safety Executive (HSE), *Management of Metalworking Fluids. A Guide to Good Practice for Minimising Risks to Health*, HSE, U.K., 1994
Independent Lubricant Manufacturers Association (ILMA), *White Paper: Hypersensitivity Pneumonitis: Is there an association with Triazine Biocides and Mycobacteria in Metalworking Fluids?*, ILMA, Alexandria, VA, 2003.
National Center for Manufacturing Sciences (NCMS), *Metalworking Fluids Evaluation Guide*, Report 0274RE95, NCMS, Ann Arbor, MI, 1996.
OSHA, *OSHA/ILMA Alliance*, http://www.osha.gov/dcsp/alliances/ilma/ilma.html, 2004.
OSHA MWF Standards Advisory Committee, *Final Report of the OSHA Metalworking Fluids Standards Advisory Committee* (Sheehan, M.J., chairperson and editor), West Chester University, PA, 1999.
OSHA, *Metalworking Fluids: Safety and Health Best Practices Manual*, http://www.osha.gov/SLTC/metalworkingfluids.
Organization Resources Counselors, Inc. (ORC), *Management of the Metal Removal Fluid Environment: A Guide to the Safe and Efficient Use of Metal Removal Fluids*, ORC, Washington, DC, 1999.

APPENDIX D. EPA REGIONAL OFFICES

Region 1 (CT, MA, ME, NH, RI, VT) Environmental Protection Agency 1 Congress St. Suite 1100 Boston, MA 02114-2023 http://www.epa.gov/region01 Tel.: (617) 918-1111; fax: (617) 565-3660 toll free within Region 1: (888) 372-7341	*Region 2* (NJ, NY, PR, VI) Environmental Protection Agency 290 Broadway New York, NY 10007-1866 http://www.epa.gov/region02 Tel: (212) 637-3000 fax: (212) 637-3526

continued

APPENDIX D. (CONTINUED)

Region 3 (DC, DE, MD, PA, VA, WV)
Environmental Protection Agency
1650 Arch Street
Philadelphia, PA 19103-2029
http://www.epa.gov/region03
Tel.: (215) 814-5000; fax: (215) 814-5103
toll free: (800) 438-2474
e-mail: r3public@epa.gov

Region 4 (AL, FL, GA, KY, MS, NC, SC, TN)
Environmental Protection Agency
Atlanta Federal Center 61 Forsyth Street
SW Atlanta, GA 30303-3104
http://www.epa.gov/region04
Tel.: (404) 562-9900; fax: (404) 562-8174
toll free: (800) 241-1754

Region 5 (IL, IN, MI, MN, OH, WI)
Environmental Protection Agency
77 West Jackson Boulevard
Chicago, IL 60604-3507
http://www.epa.gov/region5
Tel.: (312) 353-2000; fax: (312) 353-4135
toll free within Region 5: (800) 621-8431

Region 6 (AR, LA, NM, OK, TX)
Environmental Protection Agency
Fountain Place 12th Floor, Suite 1200
1445 Ross Avenue
Dallas, TX 75202-2733
http://www.epa.gov/region06
Tel.: (214) 665-2200; fax: (214) 665-7113
toll free within Region 6: (800) 887-6063

Region 7 (IA, KS, MO, NE)
Environmental Protection Agency
901 North 5th Street
Kansas City, KS 66101
http://www.epa.gov/region07
Tel.: (913) 551-7003; toll free: (800) 223-0425

Region 8 (CO, MT, ND, SD, UT, WY)
Environmental Protection Agency
999 18th Street Suite 500
Denver, CO 80202-2466
http://www.epa.gov/region08
Tel.: (303) 312-6312; fax: (303) 312-6339
toll free: (800) 227-8917
e-mail: r8eisc@epa.gov

Region 9 (AZ, CA, HI, NV)
Environmental Protection Agency
75 Hawthorne Street
San Francisco, CA 94105
http://www.epa.gov/region09
Tel.: (415) 947-8000 (866)
EPA-WEST (toll free in Region 9)
fax: (415) 947-3553
e-mail: r9.info@epa.gov

Region 10 (AK, ID, OR, WA)
Environmental Protection Agency
1200 Sixth Avenue
Seattle, WA 98101
http://www.epa.gov/region10
Tel.: (206) 553-1200; fax: (206) 553-0149
toll free: (800) 424-4372

Source: http://www.epa.gov/epahome/postal.htm

APPENDIX E. SELECTED INTERNET RESOURCES

American Chemistry Council (ACC), http://www.americanchemistry.com/cmawebsite.nsf
American Conference of Governmental Industrial Hygienists (ACGIH), http://acgih.org
American National Standards Institute (ANSI), http://www.ansi.org
American Society for Testing and Materials (ASTM), http://www.astm.org
Final Report of the OSHA Metalworking Fluids Standards Advisory Committee, http://www.osha.gov/SLTC/metalworkingfluids/mwf_finalreport.html
Hazard Material Identification System (HMIS®), http://paint.org/hmis/index.cfm
Independent Lubricant Manufacturers Association (ILMA), http://www.ilma.org
LEPC/SERC NET, http://www.rtknet.org/lepc
National Fire Protection Association (NFPA®), http://www.nfpa.org
National Institute for Occupational Safety&Health (NIOSH), http://www.cdc.gov/niosh
Organization Resources Counselors, Inc. (ORC), http://www.orc-dc.com
Occupational Safety and Health Administration (OSHA), http://www.osha.gov

Standards Advisory Committee on Metalworking Fluids. OSHA 1997, http://www.osha.gov/pls/oshaweb

Summary: Final Report of the OSHA Metalworking Fluids Standards Advisory Committee, http://www.osha.gov/dhs/reports/metalworking/MWFSAC-FinalReportSummary.html#START

U.S. Department of Transportation (DOT), http://www.dot.gov

U.S. Environmental Protection Agency (EPA), http://www.epa.gov

REFERENCE

1. Leiter, J. L., Cramer, A. B., and Rider, A. R., *Objectives on Government Affairs Issues: Background and Current Status*, Prepared for Distribution at the Independent Lubricant Manufactures Association, Management Forum, Lake Las Vegas, NV, March 31–April 2, 2005.

18 Costs Associated with the Use of Metalworking Fluids

Lloyd J. Lazarus

CONTENTS

I. INTRODUCTION

There are many misconceptions about the costs of using metalworking fluids (MWF) in machining operations. Unlike a perishable cutting tool used to make a particular part feature, the MWF can be used for multiple parts with different features that are machined on that particular machine tool. The costs of the MWF are shared by all the parts machined by that particular charge of MWF. The costs should be spread over the number of parts or on an annual basis. Rarely does a customer fill the sump of a machine tool with an MWF and discard it after running a few parts, so the fluid may continue to be used for many more months. In many cases MWF costs are small compared with the benefits — increases in efficiencies, mitigating hazards, and aiding in chip removal from the work zone.

In order to understand the costs of using MWF, one must understand the complete machining system and how these factors interact. Figure 18.1 is a fishbone diagram of the machining system. The main branches represent how the machine tool, perishable cutting tool, workpiece material, process parameters, MWF and their Environmental, Safety, and Health (ES&H) impact overall cutting performance, the quality of the part generated, and the perishable cutting tool life. Each main branch has smaller branches that represent the individual factors and how they can interact with each other and affect the machining operation. This diagram can be tailored to individual

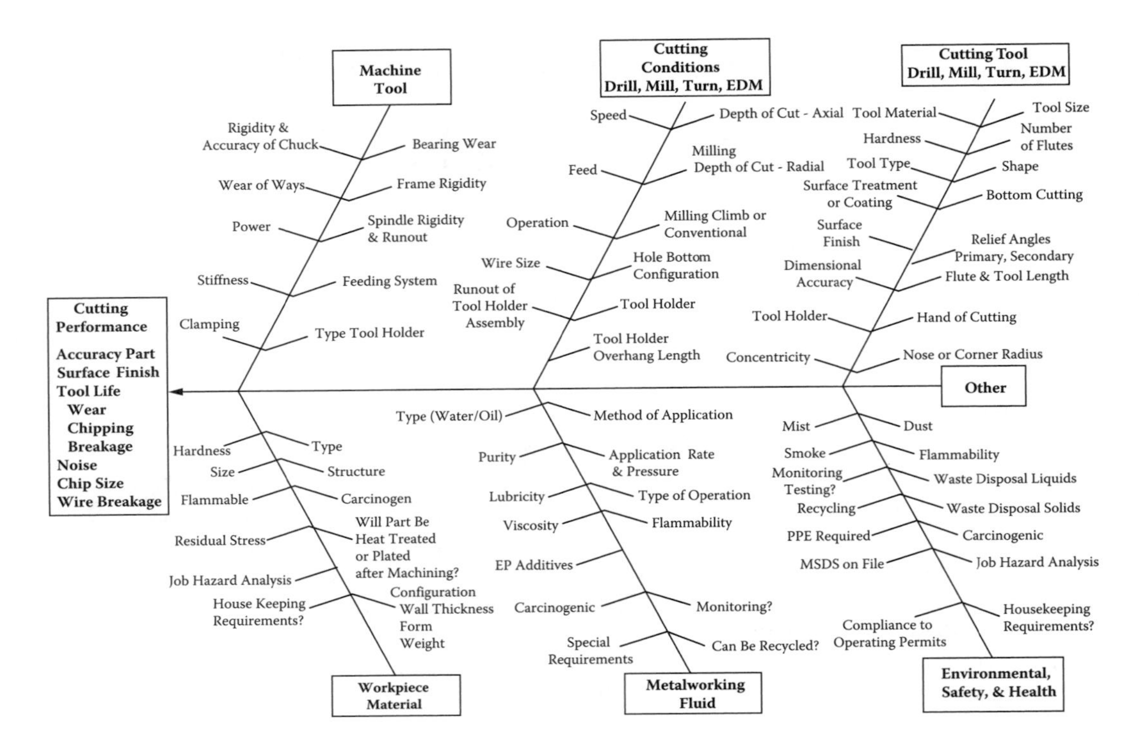

FIGURE 18.1 The machining system.

operations and workpiece materials. It is useful for troubleshooting problems in the machining process, and has been adapted for MWF.

In some cases an item is listed on multiple branches; for example, flammability. Titanium is a flammable metal. The insurance institute requires that either a water-based MWF or an inert atmosphere be used to mitigate the fire hazard; therefore, flammability should be listed on the workpiece branch and the ES&H branch. In the case of plunging a shape with an electro discharge machining (EDM) machine, kerosene would not be used as a purging fluid because of the combination of flammable liquid, spark, and oxygen. Thin oil with a flash point of over 200°F would be used, and it would also be necessary to make sure that the spark gap was well below the surface to eliminate the oxygen from the spark zone. In this case, flammability is listed on the machine tool, cutting conditions, MWF, and ES&H branch.

Another example of how various elements in a machining system interact is illustrated by drilling and tapping operations. As a drill or tap diameter drops below approximately 0.100 in., the percentage of the drill cross-sectional area for the flutes increases so that chips can be expelled from the hole. Since the core diameter is decreasing at a higher rate than the tool diameter, torsional strength of the drill is decreasing at a higher rate than the size. If we can increase the lubricity or aid in moving the chips, we could partially make up for the loss of torsional strength of the drill or tap by reducing the torsional load. Friction occurs in three places: at the interface between the cutting tool and the workpiece, between the chips generated and the cutting tool, and between the tool and the machined surface when the chips are being ejected from the workpiece. The more severe the machining operation (threading, boring, broaching), the greater the benefit of using MWF to reduce friction.

A different example is the machining of stainless steel. This family of steels is high in nickel and chrome content. When machined, a microscopic amount of dust is generated at the chip–workpiece interface. These microscopic particles are breathable and present a health hazard (based on the ACGIH TLV). Nickel being a listed carcinogen, many shops that machine large amounts of stainless steel use either MWF or dust collectors to minimize the dust hazard. Using MWF mitigates the nickel dust hazard and washes the very small and breathable particles down into the sump where they can be controlled. When we use cutting tools that are designed to run dry (obtain better tool life) in an enclosed machine tool, we use the MWF to wash the dust from the part, fixture, and table before unlocking the enclosure. We use a similar approach when machining alloys with other hazardous components.

As we consider the cost and benefits of using MWF, the discussion will focus on the categories listed below:

1. MWF materials
   - Concentrates and oils
   - Additives
   - Water
2. Labor
   - Distribution
   - Service
   - Disposal of fluids
   - Machine cleaning
3. Waste disposal
   - Hauler
   - City
4. Equipment
   - Distribution — pumper and refill trucks
   - Storage — tanks
   - Laboratory equipment
   - Recycling equipment
5. Testing costs
   - Manpower
   - Consumables
6. Acquisition costs
   - Purchasing
   - Receiving
7. Storage costs
   - Space
   - Other requirements
8. Effects of sump size
9. Requirements of ES&H

This discussion is based upon the operation of Honeywell FM&T (Federal Manufacturing and Technologies), LLC, which operates the National Nuclear Security Administration's Kansas City Plant. This is an OSHA Star Site that is also ISO 9001 and 14001 certified.

A number of annual and/or long-term contracts are used to obtain commodities for the facility. Any business that uses MWF is encouraged to try to contract both their MWF and additives on this basis in order to obtain the best pricing. Annual and/or long-term contracts should also be established with the waste hauler/treater. This will ensure that costs during the period of the contract will remain stable.

The estimates used in this review are based on a shop with approximately 200 machine tools. Sump sizes varied from 1 qt on a minilathe to 850 gal in a Blanchard grinder. In addition, there is one 6500-gal central system supplying MWF to six of the four-axis machining centers and one washhouse in a flexible machining system. Primary workpiece materials machined are a variety of stainless steels, aluminums, titanium, carbon steels, brass, bronze, specialty, and precious metal alloys, as well as a number of plastics. Owing to the variety of workpiece materials machined, machine tools and requirements of the machine manufacturers, workpiece size, and customer requirements, this facility uses all three types of water-based MWF, as well as three oil-based (straight oil) fluids. This is contrary to the recommendation of all MWF suppliers to standardize on one or two MWF.

II. METALWORKING FLUID — MATERIAL AND ADDITIVE COSTS

MWF are broken up into two main categories: oil based and water based. Water-based concentrates are diluted in water in a range between 2.5 and 10%. In this facility, machine tools are refilled with a 5% mix. As of this writing (June 2005) MWF concentrates used at this facility cost $12 to $15/gal, and the mix cost is between $0.60 and $0.75/gal. Straight oil MWF cost approximately $8.50 to $11/gal. The cost to charge the machine tool with a 50-gal sump would be $30 to $37.50 for a water-based cutting fluid or $425 to $550 for an oil-based cutting fluid.

Straight oil cutting fluids do not support bacterial growth and have a very long sump life. There are no evaporative losses, and carry out and splash losses are about 2% per week (1 gal/week or $11). The main contaminants are the spindle, way, hydraulic, and lubricating oils that leak into the sump and combine with the oil-based MWF and particulate from workpiece materials. Fine screens are usually adequate to remove chips from the sump. In the case of grinders, we also have to include the swarf generated from the grinding wheels. The particle loads are much higher and require a secondary filtering setup. These include magnetic particle traps to disposable filter media for nonmagnetic materials. The life for an oil-based fluid can be from 6 months to 2 years.

A water-based fluid mix requires more attention. Usually, water must be added to make up for evaporative losses. A 5 to 10% evaporative loss per day is not unusual. Variation is dependent on the relative humidity within the machining area. During the winter when humidity is low, losses are higher. It is best to add reverse osmosis (RO) or deionized or distilled (DI) water to maintain hardness levels. Typically, a 2.5% mix is delivered to the machine tool sump to make up splash, evaporative, and carry out losses. Adding a dilute mix rather than plain water helps to maintain components that are used up by normal machining operations. The weekly add would be 25 gal at $0.375/gal or $9.38. Other additives that might be needed to maintain the proper balance of biocides, buffers, or inhibitors required for operation would add about $1.00/week. A sump life of 60 days before a pump/rinse/refill of the sump is assumed. Certain machining departments obtain better life from their MWF than do others. This is dependent on the workpiece material and the sizes of the parts, age of the machine, size of the sump, etc.

A way of significantly reducing costs is to recycle the MWF mix. When a MWF mix is recycled the way oil, debris, bacteria, and mold spores are removed, the mix is rebalanced and then returned to the sump. The life of the fluid is extended for another cycle. The cost of the recycling is paid for by the reduction in MWF concentrates used and the cost of waste disposal saved.

Since 90 to 98% of MWF mix is water, it is important to know a little about municipal water costs. The Kansas City, Missouri, industrial rate for water is approximately $0.002/gal plus a water meter charge of $81.70 for a 6-in. line. The average water hardness is 100 ppm, alkalinity of 37 ppm with a 9.8 pH. As you can see, the cost of the water for the MWF mix is insignificant. Associated with the water charge is a sewer charge of $0.002/gal if the biochemical oxygen demand (BOD), suspended solids (SS), and oil and grease (O&G) are within limits. If the waste is above these limits, the charges for processing are BOD $0.17/lb, SS $0.10/lb, and O&G $0.058/lb. The plant waste stream is monitored by the city. It becomes much cheaper for our waste hauler to handle the waste than to arrange with the city to process the waste.

In order to upgrade the water supply to control hardness and purity, tap water may be processed through a reverse osmosis system. A self-contained reverse osmosis system that produces 20 gal of high purity water (resistivity greater than 5 MΩ, impurities under 50 ppb) every hour will cost approximately $18,000.

This facility has an RO water plant with piping that extends over the entire facility. RO water drop lines are located in all manufacturing departments. MWF are a small user of RO water compared with other manufacturing and laboratory operations. There is a small premium for RO water, but even if it cost 15 times the cost of tap water, the charge would be insignificant compared with the cost of the MWF concentrate. Reject water from the RO water facility is used in the cooling towers. In the case of an operation that was buying an RO water system to supply water for MWF, the cost of the capital equipment and operating cost would have to be calculated and added to the cost of the MWF mix.

III. LABOR — METALWORKING FLUIDS

Water-based fluids are handled by the waste management organization within the plant. There is one (1.0) waste management full-time employee (FTE) assigned to MWF. This 1.0 FTE is broken down as follows: 0.6 FTE to pump/flush/refill and service machine tools (including filling out the coolant card) in the factory with water-based MWF; 0.3 FTE for moving carboys, filling storage tanks, transfers to tanker trucks; 0.1 FTE for engineering and clerical support, bills of lading, and auditing shipments and suppliers. All associates who handle MWF are trained. Training will be discussed under the ES&H section.

In addition, we have 0.25 FTE lubrication technicians that are responsible for handling all the oil-based fluids in-plant. Waste management handles the disposal of oil-based waste, and this time is included above.

All machine tools that use MWF have a coolant card attached to the main power cabinet. The card has the capital equipment number and the sump size listed. The chemical material handler (CMH) records on the card the date, action taken (P, pump out sump; F, fill; A, add; Fl, flush; C, clean; D, left dry), the amount, and code number of the MWF supplied. The CMH fills out the card when he is performing action on the machine tool. This is done as a troubleshooting aid. The machinist, team manager, and/or troubleshooter have the history of the fluid in the sump. The CMH uses the capacity listed on the coolant card with the mixing charts that are on the delivery truck. The CMH also carries a refractometer and mixing charts (with refractometer readings) on the truck. In the case of synthetic coolant concentrate, 5-gal translucent pails with 1-gal markings are used to bring the concentrate to the machine tool.

The quality of the job the CMH performs will determine the life of the MWF and whether there are odor problems. When the CMH enters a department, he/she checks other machine tools in the area and tops off the sump before a request is made. The CMH handling MWF does a very thorough job when pumping old fluid out of the sump. The CMH diligently flushes the sump with fresh water and turns on the coolant pump. He/she then pumps out the rinse water. The efforts of the CMH have reduced our water-based MWF waste stream by at least 10%. The CMH knows that being proactive makes the job easier and also improves the environment in all the machining areas. It is important that the people handling MWF have this attitude. A person with a negative attitude can have a negative impact on MWF systems.

MWF technical support, central system monitoring, plant troubleshooting, and special mixing are conducted by laboratory personnel. In addition, the laboratory personnel conduct mist sampling, evaluations of new fluids, monitor stores inventory, and generate waste minimization plans, job hazard analysis (JHA), and prehazard analysis (PHA) plans. This represents 0.2 FTE in engineering time and 0.05 in technician time. For example, the laboratory prepares the soluble oil premixes (0.02 FTE) for the plant that is distributed by waste management. I serve as a coordinator between all the departments that use, distribute, collect, dispose, purchase, store, regulate, and process MWF. When people cannot get questions answered and/or help in solving problems, they take things into their own hands — sometimes with disastrous results. Personnel responsible for MWF should either take training from their MWF suppliers or take the STLE courses. Personnel with leadership responsibility should attend the STLE/SME certificate course and pass the examination.

IV. WASTE DISPOSAL AND CENTRAL SYSTEM CLEANING

A waste hauler is contracted to pick up, haul away, and process the water-based fluids at a rate of $0.52/gal. Waste oils are sold for fuel blending and have a disposal cost of $0.09/lb or $0.69/gal. Concentrate and oil drums received from MWF and oil suppliers are stamped DOT approved containers. Water-based fluid wastes are accumulated in 250-gal carboys. Loaded carboys are moved to our tank farm and accumulated in a large 10,000 gal tank. Other compatible water-based wastes from the plant are also added to this tank. Five thousand gallon shipments are arranged with our waste hauler. All oil waste shipments are sold for fuel blending, reducing waste stream cost. These shipments must be in DOT stamped containers. We use the empty concentrate and oil drums for oil waste shipment out of the plant. This reduces our costs for replacement waste drums.

Every 18 months the fluid in our 6500-gal central system is changed and the system cleaned. This is done to remove buildup and do preventative maintenance servicing of the system. As part of this project all the tanks, manifolds, flumes, pumps, sumps, and machine tools are cleaned. A 4-in. diameter pipe was installed that allows our waste hauler to pull his tanker truck to the outside of the building, connect, and pump the system dry. We avoid running 4-in. hoses from the system to the tanker truck in order to avoid shutting down the internal vehicle aisle. We also do not need a larger pumper truck to empty the system. Using a commercial tanker truck with a vacuum system, we can pump out our central system in 3 h.

When preparing to service our central system, a week before the scheduled start we add two drums of machine cleaner to the mix, slightly increasing the pH of the mix. To avoid possible staining of aluminum components, we schedule stainless steel parts to run during this period to minimize the risk of generating discrepant parts. The mix with cleaner starts to break down all built-up deposits in the system. During this period, on a daily basis we overflow the sumps and try to break up the deposits under the machine tools and the confinement moats. Finally, after the system is pumped out, it is refilled with 5% of machine cleaner and run for 24 h. All machines are loaded with programs that direct the cleaning mix over the complete working volume of the machine tool. The seven MWF nozzles are directed outwards instead of towards the centerline to wash a larger volume. We estimate that 95% of the buildup (combination of minerals, salts, oil, and grease — or as we refer to it, heavy duty bath tub ring) on the side wall is removed. This solution is then pumped out and the system is rinsed with the minimum amount of water needed to run the system. After taking the system apart, we found that the only problem areas were those that were not swept of chips.

Manpower costs for this cleanup project are greatly reduced because of the combination of facilities upgrade (drain lines), machine cleaner wash cycle, project planning, and experience. For example, we have found that the cleaning solution dissolves the buildup around the tracks for the moving false floor in the bottom of the steel settling tank. This floor is now removed for cleaning every 3 years instead of every 18 months. This interval could be lengthened depending on future inspections.

V. EQUIPMENT COSTS

This category is broken up into three sections: fluid handling equipment, recycling equipment, and laboratory equipment. Some editorial comment is included in each section to advise the reader where things can go wrong.

A. Fluid Handling Equipment Costs

Fluid handling equipment includes two electric trucks. One truck has an air-driven vacuum system and a 180-gal tank with a life of 10 years. This truck is used to pump out the sumps of the machine tools. The cost of the equipment is $10,000. The second truck has a 200-gal tank mounted on the bed and this truck is used to deliver fresh mix fluids at a book cost of $10,000. The third truck is used to deliver water. All trucks are electric and require charging and service of the wet cell batteries, motors, brakes, and tires. Also on an annual basis, all tanks require cleaning. Service to these trucks requires four man-days per year.

The primary fluid is a semisynthetic and there is a mixing station in the fluid handling area. A fluid mixer capable of delivering 20 gal/min of a 4 to 8% mix (easily adjustable) is used to fill the delivery truck. The cost of the mixing station is $800.

Waste management personnel mix the synthetic MWF at the machine tool when it is being refilled. The concentrate is thin and readily mixes with the water. As the water is poured into the sump, the CMH pours the concentrate into the water stream until the sump is half filled. Then the remainder of the concentrate is added to the sump before filling to the desired final volume with water. The coolant pump is started and the fluid is circulated for 5 min.

We also use soluble oil MWF. This fluid concentrate is premixed to the maximum concentration that can be mixed without separating into separate phases (approximately 3:1). Waste management personnel can add this premix to the water when making tank-side additions. They use the same mixing method as with the synthetic concentrates above.

Oil-based MWF are delivered to the machine tool in vendor drums. The waste oil is pumped out of the machine tool sump with an air-driven barrel vacuum that mounts on the large drum bung hole. Old, empty DOT stamped drums are used for the waste oil. Filled drums are accumulated by waste management until a truck shipment can be made with the waste hauler. An air-driven barrel vacuum can be obtained for approximately $500.

B. Testing Equipment Cost and Testing Requirements

Equipment to monitor water-based MWF during use can be as simple as a refractometer (measures concentration by light refraction), pocket pH meter (measures alkalinity), and some bottles, or as sophisticated as a complex laboratory. Start out simple and get only as complex as necessary for normal operation. Remember that most of the MWF suppliers have their own laboratories with much more sophisticated equipment and highly trained personnel that can be of aid when there is a problem. In most instances, this service is free. These vendors want you to succeed as much as your management.

A refractometer can cost from $150 to $1000 depending on whether you can read an optical scale or need a digital instrument. A refractometer measures the refraction angle for aqueous solutions (the change in direction of the light path when passing through a liquid). Refractometers are available with different scales, either refractive index or various "% Brix" scales. The instrument is calibrated by placing a sample of tap water on the prism and adjusting the reading to zero. Care needs to be taken to avoid getting free-floating oil from the top of the fluid in the sample. The refractive index of the oil is very different from the emulsified mix and will yield a false reading. Either take the sample from the MWF delivery line or use the technique mentioned in obtaining pH readings in order to obtain the sample. The disadvantage of measuring fluid concentration of the MWF mix with a refractometer is that some of the lubricating oils (way, spindle, transmission) will become emulsified in the fluid mix and give a false, high concentration reading.

A two-decimal point pocket pH meter costs about $125 and includes calibration fluids and a carrying case. All digital instruments are accurate to plus or minus one unit in the last digit, besides the inherent error of the device. A single-decimal point pH meter will be ± 0.1 pH plus its error.

When trying to maintain a fluid between 8.7 and 9.1 pH (a spread of 0.4 pH points), it is important not to give up 50% of the measurement range for instrument error. In the case of a two-decimal point pH meter, the uncertainty would be ±0.01 in the last digit plus 0.09% (1% reading error), or less than 10% of the range of interest. The increased cost for the pH meter of about $10 will be saved many times over by the costs saved from not adding additives or replacing the fluid.

The pH meter should be calibrated against pH standard solutions on the day of use. The standard solutions and the samples are maintained at room temperature. Calibration of the pH meter takes a couple of minutes, but the data obtained will be more accurate and dependable. If many measurements are taken during the day, it should be checked more often. The probe tip should be rinsed after every use. If oil is present on the top of the machine sump being checked, take a rod and stir the top of the fluid to expose the fluid under the oil. The oil will clog the pores in the pH probe tip, affecting this and subsequent measurements. While taking the measurement, continue to stir the fluid as this will force the oil away from the probe tip. These suggestions will make the readings obtained more repeatable than tripling the cost of the pH meter and using bad practice.

A number of manufacturers can also supply test kits that will allow measurement of the concentration of their fluids. Another way to check the concentration of soluble oils and semisynthetics is to perform an acid split to measure the oil content of the mix. To perform this test, all that is required are two 100-ml graduated cylinders. Collect a 250-ml sample from the machine delivery system. Use a cleaned pop bottle that has been thoroughly rinsed. Pour the sample into both 100-ml graduates. Add 4 to 5 ml of sulfuric acid to one sample. (Always add acid to the fluid; doing otherwise can cause violent reactions.) Within 24 h, the free oil in the system and the total oil in the system can be determined. A similar determination can be conducted on the fresh fluid mix delivered in your facility. The ratio of oil to percent mix can be calculated and used to determine the concentration of the machine sample. The glassware cost is about $25/graduate, which can be used over and over again. The amount of time to perform the test is minimal (5 min per sample).

A number of MWF manufacturers recommend measuring the dissolved oxygen in the water-based fluids. This provides an indication of bacteria growth in the fluid. A portable dissolved oxygen meter can be obtained at a cost of about $500. Bacteria growth can also be tracked through pH readings. The bacteria population waste product is acidic. Higher populations reduce the alkalinity of the fluid. By calibrating the pH meter and making sure the fluid is at a controlled temperature, the trend of decreasing pH becomes visible. A waterproof pH meter is a much more rugged instrument than the membrane in a dissolved oxygen meter. I prefer the pH measurement; others prefer the dissolved oxygen meter. The fluid supplier may have a test kit to allow checking of the microbicide level in the fluid to determine whether the change in mix pH is due to bacteria or to reduced reserved alkalinity.

If fluids are going to be recycled, basic measurements need to be made. The concentration and the pH of the fluid mix must be known. If the concentration or pH is incorrect, there will be problems with sump life, fluid quality, and corrosion resistance. Spoiled fluid cannot be recycled. Even if all the bacteria are killed and the concentrate rebuilt, the odor would still remain and be carried back to the machine tool. Eventually all the MWF, machine tools, and everything else would have this odor.

Finally, there are multiple manufacturers of dip slides that can be used to determine the concentration of bacteria and mold carried by the MWF. A box of ten dip slides can be purchased for $36.75 plus $5 shipping. After immersing the growth media in the fluid for a few seconds, the slide is reinserted into its case and incubated for 24 to 48 h at a temperature of 25 to 31°C. If bacteria or mold are present, they will grow into colonies that can be observed on the dip slide with the naked eye. An incubator can be used or the dip slides can simply be placed in a box under a lamp. The cost per test is approximately $4.50, and thus dip slides are used sparingly. If a bacteria problem is suspected, proceed to treat the MWF after taking the sample.

It must be remembered that the bacteria count in the MWF sump can double in 20 min (depending on bacteria type, temperature, and flow). By the time (24 to 36 h) the dip slide shows the presence of bacteria (visible colonies) the system will probably already be spoiled. Familiarity with the machine tools and MWF, tracking data on the MWF (pH, concentration), oil coverage on the sump, hints of odors, etc., should be your first indicators. Ask the machinist if the MWF color was different or if the fluid had an odor when it was first turned on. MWF trapped in the delivery system is usually the first place where the coolant spoils. If nothing is done, the entire sump will probably spoil within the next 2 days.

Taking the time necessary to conduct these basic tests on the MWF will greatly increase the sump life and reduce the fluid costs.

C. Recycling Equipment

Recycling equipment cost depends on how much fluid will be processed on a daily basis. There are now small units (costing $5000) on the market that can be connected to a specific machine tool. During the day, all the fluid in a machine tool sump (75 gal) will be filtered, a majority of the way oil separated, and mix ratio adjusted. These units use a cyclone filter to remove particulate and a skimmer to remove free oil. They can be moved to a new machine tool every day, and be quite effective for a small shop. There are other units of this type that can handle up to three machines per day, depending on sump size. Such units cost approximately $12,000.

There are larger recycling units for larger shops. Larger units must have the fluids brought to them, and the recycled fluid delivered back to the machine tool sump. These larger units either use a centrifuge or a pasteurizer to remove bacteria and mold from the fluid. They have both dirty and clean tanks, belt skimmers to remove oil, proportioners to add make-up, and additional filter media. The cost for these units can approach $50,000 depending on size. They require separate transportation equipment for the cleaned and dirty fluid. If not, the residue from the dirty fluid will contaminate the cleaned fluid.

A centrifuge does not use as much power as a pasteurizer. Both need to be periodically cleaned. To assure the performance of the fluid, the systems need to be cleaned when different fluid types are processed. If they are not cleaned properly, the residue from the previous fluid will be mixed into the new fluid. If fluid incompatibility is a concern, then the costs for recycling will escalate.

Oil-based fluids are not easily recycled. Way, spindle, hydraulic, and gear oils eventually drip into the sump and dilute the cutting oil. They mix together, and unless the oil is rerefined, they cannot be separated. As the machine ages, the way wipers wear and allow more of the way lubricant to drip into the sump. Eventually, the extreme pressure additives in the cutting oil are depleted from previous machining operations. People have stated that they have used the same cutting oil for 3 years. By asking how much way, spindle, and hydraulic oil they use on a weekly basis, it can usually be shown that they are probably machining with this lubricant oil mixture rather than with cutting oil.

VI. ACCEPTANCE TESTING

Before approving a new MWF or MWF additive for production use, we run some initial tests. They include both short-term staining and long-term corrosion tests on our normal production materials. We also expose standard materials to the concentrates for 10 days and then subject the samples to our normal cleaning material and procedures. Specimens are evaluated by x-ray photoelectron spectroscopy (XPS) and ion chromatography (IC) to make sure the new fluids can be removed from the surfaces. If the material under test stains the surfaces or cannot be removed by our production cleaning processes, it will not be accepted for use in our facility. After completing these tests, in which approximately a third to a half of the products fail, we then run a machining test. We do not approve any material that will not improve our processes.

All materials used in our manufacturing area have acceptance criteria. It is part of the way of life as a government contractor. All MWF users should do some kind of acceptance testing. If you are making a part from a copper alloy, check to make sure the cleaning detergent does not have an active sulfur compound in it.

Some examples of problems from our experiences include a production worker who saw oil on his parts. He added a mild dishwashing soap to the process with a water rinse. The dishwashing liquid had something in it to prevent hands from chapping, which caused a reaction between the copper and the dishwashing soap, resulting in stained parts. We have also had cases of residual lubricants on extruded parts reacting with the MWF. In this particular case, there was no staining on the machined surfaces.

Determining the type of acceptance testing required depends on what is being made and the customer's requirements. This cost varies, depending on the complexity of the tests. However, this is a one-time charge for each product, which will not be included in this analysis.

VII. ENVIRONMENTAL, SAFETY, AND HEALTH (ES&H) CONSIDERATION

In all cases, before obtaining a new product for evaluation, our internal procedures require that the product literature and material safety data sheet (MSDS) be on file and the material be approved by our ES&H organization. This information is reviewed by the requester, safety engineer, industrial hygienist, fire protection, environmental compliance, and waste management personnel. We determine if the new material expands our operating envelope and whether we will remain within our federal, state, and local permits. Personal protective equipment (PPE) is specified, storage codes assigned, diamond label coding verified and assigned, and procedures for handling the waste are developed. This information is listed on our computer network so that we can respond properly from the receiving dock until the material is eliminated from the system.

This seems like a great deal of expense, but it protects our facility. There are many vendors in the marketplace that say their products are safe, even if they are not. A typical example involves products that contain mineral oil. If the mineral oil is not severely hydrotreated and/or severely solvent refined, it is considered carcinogenic. All lubricating oils and greases used at this facility comply with this refining requirement, so there is no concern that lubricating oil dripping into the MWF sump is turning the MWF carcinogenic. This database also holds information about special setups that are required when machining titanium, magnesium, and other workpiece material, and which MWF are approved for use with them. Since all materials are handled in this manner, there are no additional costs for MWF.

All the CMH are first responders for spills and are trained in hazardous waste operation and emergency response (HAZWOPER). The initial training course is 40 h in duration and the update course is 8 h. CMH are recertified yearly. Each CMH handling MWF had 8 h of on-the-job training before being allowed to operate on their own. They meet with the technical backup for MWF whenever a new MWF fluid or additive is introduced or when they encounter an unusual occurrence concerning MWF on the plant floor. Activities during shutdowns are coordinated by weekly contacts.

On a yearly basis, all associates have environmental training and are required to review all the JHA procedures for their area. All machining departments are required to have at least one safety meeting per month to review job hazards. We have tried to make our factory as safe as possible with as little impact on the environment as possible. Being ISO 14001 certified and an OSHA Star site means that OSHA and ISO auditors have verified that we are accomplishing this goal. We have no additional MWF training. We have a number of MWF programs that are available to manufacturing departments to use at their safety meetings. We estimate that each associate involved with MWF should have at least 1 h of training every year. Even if machining is done without MWF, there is an obligation to disclose the hazards of the dry dust generated (alloys containing nickel, chrome, cobalt, cadmium, molybdenum, lead, etc.). Remember that when chips are blown off a part or machine, there is a good chance that the airborne dust will become an inhalation problem.

VIII. PURCHASING

It is important to work with your local vendors to set up long-term contracts for MWF and MWF additives. If a multiyear contract is signed, an annual review is required and prices can be adjusted for the following year. An annual delivery performance incentive is included. Vendors are supplied with minimum annual usage and approximate monthly usage numbers. The local vendor can then combine our usage with other customers' usage and order in truck-load quantities at specific intervals to obtain the best pricing from the manufacturers. These materials must be delivered to the plant within five working days from the time of the order. Therefore, the plant only stores a minimum amount of material in the storage area. This is our normal mode of business operation, and no additional costs occur specifically for MWF.

IX. STORAGE COSTS

It is important that MWF be stored properly. If the building is not heated during the winter, there is a possibility that the products might freeze and separate. The labels on the drums of concentrate typically have a warning about storage temperatures. From an insurance underwriter's point of view, it is important that all chemicals and oil be stored under the proper conditions. If materials are not stored properly, an inspection by the insurance carrier could result in an increase in premiums. Owing to the large number of materials used in this facility, there are no additional costs for special storage for MWF.

X. EFFECTS OF SUMP SIZE

All of the costs discussed are based on size of the sump or the amount of MWF used. It is not reasonable or cost-effective to test the fluid in a 1-qt sump. Simply dispose of the fluid, flush, and refill. Conversely, you cannot afford to lose a 6000-gal central system because of lack of monitoring. We use a 60-gal breakeven point to determine whether to monitor the MWF.

A larger system justifies careful attention. For example, we control the temperature of the MWF in our central system. Using the chilled water return from one of the in-plant air handlers, we can control the MWF temperature to within 1°F. The temperature-controlled fluid is delivered into a manifold, which surrounds the spindle housing on top of the outside spindle bearing, before introduction into the cutting zone. In this way, we can minimize spindle growth. Accuracy studies have shown that this has improved the repeatability of the spindle movements over our 16-h work day. (An estimated 50% of machine tool dimensional error is due to thermal effects.)

The MWF circulating in the bypass loop is centrifuged on a periodic basis to remove way, hydraulic, and spindle lubes that leak into the MWF, along with fine particulate, bacteria, and mold that are carried in the fluid. This helps us to increase the life and quality of the fluid circulating in the system.

XI. COST/BENEFIT SUMMARY

After reviewing the operations of this facility, it was determined that the most meaningful cost analysis would be to break down the cost for MWF for three types of systems. The first is a 6500-gal central system that runs for two shifts per day, 5 d/week. The second is an individual machine tool using water-based MWF, and the third is a screw machine using oil-based MWF.

Machine shop costs are based on labor plus overhead cost. The overhead cost is dependent on the cost and age of the machine tools in a particular shop. If the shop has new CNC machining centers, their overhead costs will be higher than shops that have Bridgeport vertical mills. Local costs for utilities, taxes, insurance, and property also affect overheads. Other costs can be included in this number depending on whether the company is dealing with private industry or a government contractor. Therefore, the labor costs can vary from $40 to $100/h. In making the estimates for this chapter, I used an arbitrary burdened shop cost of $80/h for a machining operation. In estimating the cleanup and maintenance cost, I used $40/h.

The central system has been described previously and its evaluation period lasted from January 2003 to August 2004. This analysis included all the MWF concentrates and additives, all waste disposal costs, cleaning labor, cleaning materials (cleaning solutions, wipes, pig mat, PPE), monitoring labor, generation of safety plans, disassembly and reassembly labor during cleaning, wrapping materials and labor to protect electronic cabinets, bimonthly centrifuging of the MWF to remove excess way oil and particulate, secondary filter replacement, etc. The disassembly, cleanup, reassembly, refilling, and operating the system was done over a 4-d holiday weekend, and all overtime costs are included. During the time period of this evaluation, 16,420 operations were performed in our flexible manufacturing system. Machining operation time was over 0.62 h per operation, without coordinate measuring machine and inspection time included. The MWF cost was 8.9% of the cost of the operations on the central system. This does *not* include the workpiece material (metal) cost. If this system was run dry, additional labor (estimate 1/4 h/d/machine tool = 1.5 h/d) would be needed to remove chips. If we deduct the manual chip removal cost from the percentage, the cost drops to 7.2%.

While the system is disassembled, maintenance crews have access to all conveyors, gear drives, etc., and all items are checked for wear. In addition, modifications and upgrades to machine tools, material handling robots, and air and fluid delivery systems are performed. This labor was not removed from the MWF costs. Every time this maintenance project is conducted, new materials and methods are tried in order to reduce the cost, duration, or labor requirements. The use of a new machine cleaner solution will allow us to reduce the amount of disassembly, cleaning, inspecting, and reassembling time required in certain areas every 18 months. This period will be extended to 36 months, reducing the overall costs by 0.3 to 6.9% of the total cost of operations on the central system.

A vertical machining center with a 75-gal sump using water-based MWF with a 5% mix would cost $56.25 to fill, $7.03/week for 8 weeks for makeup, disposal cost of $0.52/gal for 60 gal plus 2 h of labor or approximately $350. During the 8-week period the machine tool could easily generate 25 parts/shift or 2000 parts during the period. That means the use of MWF would only add $0.15/part to the cost of the product. On a yearly basis, the sump should be cleaned. That would add approximately 2 h of labor, plus $2.50 in materials, and $50 in additional waste costs. This yearly cleaning cycle might add less than a penny to the per part cost.

In the case of an oil-based MWF in a screw machine with the same 75-gal sump, we will assume a fluid life of 6 months. The cost of cutting oil would be $825 plus $429 for makeup and labor of $160. Waste disposal costs would be $41.40. The total cost would be approximately $1500 for the 6-month usage. A typical screw machine is capable of making 200 parts/shift, which means that during 6 months this machine could generate 26,000 parts. The MWF cost/part would be about $0.06/part. The sump should be cleaned on an annual basis, and the cleaning would add 2 h of labor, plus $2.50 in material costs, and $50 in additional waste cost. This is less than $0.005 per part.

As shown above, the MWF cost for each operation on different machine tools can vary. That is why it is necessary to look at the overall operations on a yearly basis. Approximately 40% of our purchases are for oil-based MWF, although this represents only 5% of the total volume of MWF used in the plant. Based on the overall cost of our manufacturing operation (flexible manufacturing system with a central MWF system and 200 stand-alone machine tools), the MWF costs for all machining operations are *less than 3%* of our total manufacturing operation costs. MWF costs include concentrates, central system cleaning, CMHs, salaried and hourly technical support, waste costs, and maintenance costs for MWF systems. Machining operation costs include labor, perishable tooling, and some maintenance costs. Workpiece material costs are not included in this calculation. A proportion of these costs is required to be in compliance with ES&H (OSHA, ACGIH) and insurance requirements.

Products machined in this facility are made from stainless steel, which contains nickel and chrome, and other aerospace alloys. The aluminum aerospace alloys and the perishable cutting tools we use to machine products all contain a number of alloying elements that are on the ACGIH TLV list. Our air monitoring has revealed airborne dusts when machined dry. In order to protect workers, PPE for chip removal operations would have to be changed and would become more costly.

Also, most of our machining operations contain a drilling operation. Dry machining has been proved economically feasible for turning and milling but not for drilling and tapping. We plan our machining processes so that all features on a surface can be produced in one operation. We take advantage of the new high performance perishable tools in enclosed machines, and then use the MWF to wash the machine tool to control dust. MWF are used in drilling/tapping/reaming/broaching operations. In the case of machining small features, it is necessary to use MWF when performing severe operations (drilling and tapping 0–80, 2–56, 4–40 holes in stainless steel and other materials) in order to reduce tool breakage and scrap costs. If we use MWF to do these jobs, we still must maintain the MWF and absorb the cost. If all machining operations were performed dry, we would not reduce our cost further, but increase cost for PPE and additional costs in producing small holes and internal threads.

Risk assessments have shown it would be foolish to machine flammable metals without using MWF. Not using the proper MWF would be a violation of the National Fire Protection Code. The Price–Anderson act, which defines the financial responsibilities of government contractors, would hold Honeywell responsible for all costs associated with a fire if the necessary precautions for machining flammable metals were not followed.

Finally, using MWF improves the efficiency of most machining operations. For instance, when machining multiple parts in small quantities, fixture cost must be considered. Optimum cutting tools running dry and at maximum speeds, require rigid fixtures to support the machining loads. Such costs cannot be amortized over a large enough quantity of parts and remain competitive.

XII. EDITOR'S COMMENT

It is interesting to compare this excellent cost/benefit analysis for a government-owned facility with a similar analysis from a leading manufacturer of construction and mining equipment. In the 2004 STLE Metalworking Fluids education course, Doug Hunsicker and George Egger presented dollar figures for various aspects of machining costs at one manufacturing plant, comparing costs for 1974 and 2001.[1,2] They looked at the cost of MWF, perishable tooling, capital equipment purchases, maintenance and machine operator labor costs, and management salaries. Disposal costs were not considered since this plant had its own waste treatment facility handling a large volume of waste, of which MWF were a minor component. Neither was the cost of the metal being machined considered. In 1974, MWF represented 0.99% of the above manufacturing costs. By 2001, total costs had decreased significantly, and fluid costs had been reduced by half; MWF were only 0.86% of the manufacturing costs. If metal costs had been included, the cost percentage represented by MWF would have been even lower.

It is also interesting to examine the relationship between fluid costs and perishable tooling costs for those same years. In 1974 the perishable tooling was more than 12 times the cost of the fluids, and in 2001 tooling costs were five times the cost of the fluids. Since tool life deteriorates when the MWF is discontinued for almost every operation in a typical plant, one wonders how high the tooling costs in this plant would have been if no fluids were used.

REFERENCES

1. Hunsicker, D. and Egger, G., *Optimizing Metalworking Fluid Processes*, STLE Metalworking Fluids Course, Toronto, 2004.
2. Canter, N., The possibilities and limitations of dry and near-dry machining, *Tribol. Lubr. Technol.*, 59, 31, 2003.

19 Glossary

Abrasion: Removal of material by rubbing.

ACGIH: American Conference of Governmental Industrial Hygienists.

Acid: A substance that releases protons (H^+ ions) in water, lowering the pH.

Acid number: A numerical expression of acidity based on the milligrams (weight) of potassium hydroxide required to neutralize the free fatty acid in 1 g of sample.

Acute health effects: Health effects that occur rapidly as a result of short term, one-time exposures.

Adhesion: Molecular attraction between two surfaces in contact.

Adsorption: The uptake and physical bonding of the active substance contained in the lubricant on the surface of the solid metal, a mechanism for forming the intermediate layer to reduce metallic contact between two surfaces. This is different from a reaction mechanism in which the intermediate layer is formed through the transformation of the active substances into other chemical compounds (reaction products).

Agar (Agar–Agar): A polysaccharide extract from seaweed used as a base to prepare a gelatinous media for growth and enumeration.

Air entrainment: Air that is held in suspension by a fluid and is slow to rise to the surface

AISI: American Iron and Steel Institute.

Alkali: Hydroxides of ammonia, lithium, potassium, and sodium are called alkalies; see *Base*. They raise pH when added to water.

Alkalinity: The concentration of basic or alkaline components in a mixture, determined by titration with an acid.

Allergen: Any molecule that induces the body's release of histamine, characteristic of an allergic reaction.

Allotrophism: A property of metals whereby they may exhibit multiple crystal structures at different temperatures. Under equilibrium heating and cooling conditions the change in crystal structure is completely reversible.

Alloy: A mixture of two or more elements, at least one of which is a metal, combined in a solid. The resulting alloy has chemical and mechanical properties that are different than those of the separate elements that make up the alloy.

Alum: The common name for aluminum sulfate or potassium aluminum sulfate, chemical substances frequently used in liquid waste disposal to separate out small particles of suspended matter.

Amide: Organic amides are formed by a condensation reaction of a fatty acid and an amine. Used as emulsifiers and corrosion inhibitors. The amine to fatty acid ratio may be expressed as 1:1 with little or no excess free amine, or 2:1 with some unreacted amine remaining.

Amine: A nitrogen containing organic compound, basic in nature, having the general formula NR_3 where R may be either a hydrocarbon of hydrogen.

Amphoteric: (1) Molecules containing both an acid group (COOH) and a basic group (NR_3). (2) Metals that are attacked by both acids and alkalies, i.e., aluminum and zinc.

Aniline point: Defined as the lowest temperature at which equal volumes of aniline (aminobenzene) and the test liquid are miscible. Petroleum oil with lower aniline points

(160–170°F) are more naphthenic while paraffinic oils have higher aniline points (approximately 240°F).

Anionic surfactant: A surface active agent that when dissolved in water carries a slight negative charge. These tend to be sensitive to water hardness. Examples are soaps and sulfonates.

Annealing: A conditioning treatment designed to produce a soft final structure, improve ductility, remove stress, and refine the microstructure. It usually involves austenitic transformation followed by slow controlled rate cooling inside a furnace.

Anode: The terminal of an electrolytic cell at which oxidation occurs (loss of electrons).

Anodizing: An electrochemical treatment applied to aluminum to induce the formation of a hard tenacious oxide layer. Such a layer imparts corrosion and wear resistance to the aluminum.

Anoxic: Oxygen-free conditions.

ANSI: American National Standards Institute.

Antimicrobial pesticide: A chemical designed to kill microbes; see *Microbicide*.

Aqueous: Referring to water; an aqueous solution of sugar would be sugar dissolved in water

Arbor: Spindle on which a cutter is mounted.

Archaea: Group of single-celled organisms that bear many morphological similarities to Bacteria, but are genetically and physiologically unique. Originally recovered from extreme habitats (high temperature, high salt content, high pressure), members of this domain are now recognized as being ubiquitous in the environment.

ASTM: American Society for Testing and Materials.

ATP: Adenosine triphosphate or ATP is the primary energy molecule in all cells, including bacteria and fungi.

Atopic: Inherited tendency to develop as allergy.

Austenite: The allotrophic form of iron that exists above 1670°F. Austenite can dissolve up to 2 wt% carbon and is the precursor structure for most heat treatments.

Axenic: A situation in which only one species is present (adjective).

Bacteria: Single celled, microscopic organisms that have a cell wall but no readily apparent internal structure when viewed through an optical microscope. They are widespread in our environment and are found in a variety of shapes (round, rod, or spiral). Most bacteria grow well on organic substances, although some survive wholly on inorganic material.

Bacteria, Aerobic: Bacteria requiring oxygen from the air in order to grow and replicate.

Bacteria, Anaerobic: Bacteria that cannot tolerate oxygen.

Bacteria, Facultative Anaerobic: Bacteria that can grow in either the presence or absence of oxygen.

Bacteria, Obilgate Anaerobic: Bacteria that must have an oxygen-free environment in order to survive.

Bacteria count: A measure of the number of bacteria in a fluid, determined through use of pour plates, dip-slides, direct microscopic count, or other means.

Base: A substance that reacts with protons (H^+) in water, raising the pH of aqueous solutions; see *Alkali*.

Biochemical Oxygen Demand (BOD): A measure of the amount of oxygen used by microorganisms during the process of decomposing waste. The BOD is usually measured in milligrams per liter or in parts per million (ppm) of oxygen required.

Biocide: Chemical added to a metalworking fluid formulation or mix to restrict the growth of microorganisms. This broad term includes bactericides and fungicides.

Biodegradable: The capacity to be decomposed by microorganisms.

Biofilm: A layer or matrix of microbial cells (bacterial, fungal, and/or algal) attached to a surface by the extracellular biopolymers that they produce.

Biomass: The total physical amount of living matter within a given system.

Blown oil: A vegetable or animal derived oil, chemically modified by oxidation to increase viscosity for enhanced boundary lubrication.

Boring: An operation used to enlarge a hole to an exact size using a single point tool.

Boundary lubrication: A thin layer of lubricant film that physically adheres to the surface by molecular attraction of the lubricant to the metal surface. Examples are fats, fatty acids, esters, and soap.

Brinell hardness: An indentation type of hardness test that uses a 10 mm ball indentor under a load of either 500 or 3000 kg. The diameter of the impression made is measured and the hardness number calculated as a function of load divided by projected area.

Broach: A cutting operation which combines both roughing and finishing in a single pass. The broaching tool consists of a straight shank with multiple cutting teeth, each taking the cut wider or deeper than its predecessor. Significant "drag" or rubbing is involved in broaching, and the lubricant must help overcome frictional forces.

Built-Up Edge (BUE): A piece of work material adhering to a cutting edge of the tool.

Burn: A change in the material caused by the heat of the metalworking operation; may be accompanied by discoloration.

Carbide tooling: Consists of tungsten carbide particles (or combinations of other carbides) held together by cobalt or nickel binders.

Carbon equivalent: The combined effects of the elements in cast iron that affect the formation of graphite. Usually considered to be the sum of carbon plus one third the silicon plus one third the phosphorous contents. Control of the carbon equivalent determines the solidification characteristics of cast iron.

Carbon steel: Steel containing less than 2% carbon and small (residual) quantities of other elements such as manganese, sulfur, and phosphorus.

Carbonitriding: A heat treat process that enriches the surface of a low carbon steel with both carbon and nitrogen simultaneously. Upon quenching, the case becomes hard while the low carbon core remains soft. A carbonitrided case is harder than a carburized case and is frequently applied to steels that have low hardenability and do not respond well to straight carburizing.

Carburizing: A heat treat process that enriches the carbon content at the surface of a low carbon steel. Upon quenching, the high carbon surface becomes hard while the low carbon core remains soft.

Carcinogen: A substance that is known to cause cancer or malignant tumor formation in humans or animals.

Carry-off: The amount of metalworking fluid that is taken away on metal fines or parts.

CAS number: Chemical Abstract Services Registry Number. A number that uniquely identifies a specific chemical, assigned by the Chemical Abstract Services.

Cathode: The terminal of an electrolytic cell at which reduction occurs (gain of electrons).

Cationic surfactant: A surface active agent that when dissolved in water carries a slight positive charge. Examples are quaternary ammonium compounds.

Cemented carbide: Pressed and sintered carbide particles in a binder, e.g., carbide tools.

Cementite: The compound Fe_3C. In steel and iron this is the hard, wear resistant microconstituent.

Centerless grinding: A process for grinding cylindrical parts between two abrasive wheels, a grinding wheel that cuts the metal and a regulating wheel that controls the speed and horizontal movement of the part. The part is supported from underneath by a work rest blade.

Center-type grinding: External grinding process where the cylindrical part is supported between two pointed metal shafts (centers) that are placed in depressions at each end of the part. The part rotates during the grinding process.

Centipoise: Units of viscosity measurements (0.01 g/cm/sec).

Centistoke: A unit of kinematic viscosity.

Central system: A large reservoir of metalworking fluid piped to supply multiple machine tools.

Centrifuge: Used to separate oil and/or dirt from metalworking fluids. It incorporates a spinning bowl, which receives the dirty fluid.

CERCLA: Comprehensive Environmental Response Compensation and Liability Act, administered by the EPA.

CFR: Code of Federal Regulations, U.S.

Chamfering: Machining a bevel on a workpiece.

Chelating agent: A molecule capable of attaching itself to a metal ion by two or more linkages to the same molecule. A common example is ethylenediaminetetraacetic acid (EDTA).

Chemical Oxygen Demand (COD): The amount of oxygen required for total oxidation of matter in water.

Chips: Particles that are removed when a material is cut.

Chloride: A salt resulting from the neutralization of HCl, for example, sodium chloride, NaCl.

Chronic health effects: Adverse health effects resulting from long term exposure, or persistent health effects.

Chuck: A device that holds a workpiece or tool during machining or grinding. May be mechanical or magnetic.

Climb milling: Cutting operation in which the rotation of the cutting tool is in the same direction as the workpiece movement. Consequently, the chip being formed starts out thick and becomes thinner as the tooth progresses through the workpiece.

CNOMO: Committee De Normalisation De La Machine Outiels; a consortium of French automobile manufacturers setting standards for the industry.

Coagulation: The clumping together of solids to make them settle faster. Coagulation is brought about with the use of certain chemicals such as polyelectrolytes.

Coalescence: The process of growing together or uniting, such as oil droplets agglomerating on a receptive medium.

COC: Cleveland Open Cup. Device used to measure flash and fire points.

Coining: The imprinting of a design or pattern onto a metal surface.

Combustible: A term used to classify the potential for a fluid to catch fire based upon its flash point. OSHA defines a combustible liquid as "...any liquid having a flashpoint at or above 100°F (37.8°C)...". A "combustible" material has a higher flash point than a "flammable" one; see *OSHA* 29 CFR 1910.106 (a) (18) for classes of combustible liquids.

Commensalism: A relationship in which one organism benefits without significantly affecting the other organism.

Communicable: Contagious or transmittable between persons.

Composite material: Material that contains two or more different materials separated by a distinct interface.

Computer Numerical Control (CNC): A programmable automation system that utilizes a computer to control a process with numbers, letters, and symbols.

Concentration: A measurement of the content of one or more materials in a metalworking fluid.

Conductivity: A measurement of the ability of a metalworking fluid mix to conduct an electric current, which is dependent on the amount of dissolved ionic material.

Consortium: A community composed of two or more taxa interacting to produce effects that would not be produced by any individual taxon within that community.

Corrosion: Oxidation of ferrous or nonferrous metals; includes rust, staining, pitting, and etching. The process by which a refined metal returns to its natural state.

Corrosive: A term applied to a material that causes damage to living tissue or metals.

Coupling agent: A chemical additive that aids in emulsion formation by being mutually soluble in both phases, but not an emulsifier by itself.

Crater: A depression on the cutting face of the tool due to wear.

Cratering: A surface defect in painted parts characterized by the appearance of small, pin-hole craters in the paint film.

Creep feed grinding: A grinding operation in which the grinding wheel makes deeper cuts at a slower feed rate than is typical; Q-primes are dramatically higher. This type of grinding is used to generate complex shapes or deep slots.

Crystal lattice: The spatial arrangement of atoms in a repetitive pattern.

Cutaneous: Pertaining to the skin.

Cutting fluid: A substance applied to a tool to promote more efficient machining.

Cutting speed: Tangential velocity on the surface of the tool or workpiece at the cutting interface.

Cyclone filter: A device that separates a mixture according to density through centrifugal force. Cyclones are used for cleaning metalworking fluids (separating dirt particles, chips, etc.) and have no internal moving parts.

Defoamer: Chemical additive that physically alters the surface tension of a fluid to reduce or eliminate foaming.

Deformation: The forming of ductile metal into a new shape.

Demand: The sum of chemical, physical, and biological factors contributing to depletion of a chemical in a system.

Density: The mass of a material divided by its volume.

Dermatitis: Inflammation of the skin, evidenced by a rash, itching, blisters, or crustiness.

Dermis: Layer of skin beneath the epidermis; the bulk of perceptible skin thickness.

Die: A term generally applied to the female part of a tooling set used for metalforming.

Diffusion: Movement of atoms or molecules within a material or across a mating surface.

DIN: Deutsches Institut fur Normung, a German industrial standards organization.

Dirt: A broad term for metal chips, fines, swarf, grinding grit, etc. found in a metalworking fluid mix.

Dirt volume: The percentage of solids in the metalworking fluid mix that is separated from the mix by settling or centrifuging. High dirt volumes usually indicate either inadequate filtration or filter problems. A high dirt volume can affect the performance of the metalworking fluid and lead to such problems as residue, poor finish, poor tool life, and microbial growth.

Dislocation: A defect in a crystal that interrupts the normal regular atomic array. The simplest form is an edge dislocation, which is an extra plane of atoms. The motion of dislocations under stress accounts for the plasticity exhibited by metals.

Dissolved Oxygen (DO): Elemental oxygen dissolved in water. Microbial activity will cause the DO to drop.

DNA: Deoxyribonucleic acid; a double helix ladder shaped molecule that contains a cell's genetic information.

DOT: Department of Transportation, U.S.

Down milling: Same as climb milling.

Draw bar: An instrument used to apply a thin, even coating of fluid to a surface.

Draw bead: Used in metal forming dies to control metal flow during stamping operations.

Drawing: The reshaping of a sheet metal blank into an elongated round cup or square box-like form.

Dressing: To reshape the surface or profile of a grinding wheel by removing unwanted material (either abrasive grains or imbedded metal), thus exposing sharper abrasive cutting surfaces. A diamond tool is typically used to dress a grinding wheel.

Drilling: A method for producing a hole using a cutting tool with flutes or grooves spiraling around the drill body which serve as a conduit for chip removal.

Drum filter: These permanent filter media are formed by spirally wrapping triangular shaped wire ("wedge-wire") around support rods to form a hollow cylinder or "drum". The wires are welded to the support rods so as to have uniform spacing between the wires. The drum is submerged into the fluid to be filtered, vacuum is applied to the interior of the drum, and the fluid flows through the gaps between the wires, depositing any solid contaminants on the exterior flats of the wedge-wire surface. A scraper is used to remove the accumulated dirt.

DSL: Domestic Substances List (Canada).

Ductile iron: A form of cast iron that has been inoculated with certain elements that cause the free carbon to form as nodules or spheroids. Ductile iron has ductility approaching that of steel.

E-Coat: The application of paint through an electrodeposition process.

Eczematous: An inflammatory condition of the skin exhibiting redness, blistering, scales, crusts, and/or scabbing.

Edema: Swelling of body tissues due to the presence of abnormally large amounts of fluid.

EEC: European Economic Community.

Effluent: Waste water leaving a waste treatment process, relatively cleaner than before.

Elastohydrodynamic Lubrication (EHL): A condition in which high pressures produce a large increase in lubricant viscosity, making thin films behave like a wax.

Electrodeposition: The application of electric current through a paint or lubricant medium so that it is attracted to a metal surface with opposite charge.

Emulsifier: Substances that prevent dispersed droplets coming together in emulsions by reducing interfacial tension. These typically have both oil soluble and water soluble portions to their structure.

Emulsion: A disperse system consisting of several phases which arises through the mixing of two liquids that are not soluble in each other. One liquid forms the inner (or disperse) phase, distributed in droplet form in the carrier liquid (the outer or continuous phase). Emulsifiable metalworking fluids are frequently oil-in-water emulsions, wherein oil forms the inner phase.

Endotoxins: The lipopolysaccharide (LPS) component of the Gram-negative bacterial outer cell membrane. Endotoxins are both toxic and allergenic.

Endurance limit: The level of stress at which a component will have infinite life under cyclic or fatigue loading regimen.

EPA: Environmental Protection Agency, U.S.

EPCRA: Emergency Planning and Community Right-To-Know Act (aka SARA) (U.S.).

Epidemiology: Science concerned with the study of the incidence and distribution of diseases in a population.

Epidermis: Outer layer of skin.

Erythema: Redness of the skin.

Ester: A compound that may be formed by reaction of an organic acid with an alcohol. An effective boundary lubricant.

Eukaryote: The domain of all organisms whose cells have true cell walls and a clearly defined nucleus bounded by a membrane. Fungi and all higher organisms are eukaryotes.

Exotoxins: Chemicals, mostly proteins, excreted by living bacteria and fungi.

Extraneous oil: Often called "tramp oil", is the oil or oil-like material in a fluid which is not from the metalworking fluid concentrate.

Extreme Pressure (EP) additives: A compound (usually containing chlorine, sulfur, or phosphorous) which reacts with the surface of the metal or tool to form compounds (chlorides, sulfides, or phosphates) that have a low shear strength. These compounds can help provide longer tooling or grinding wheel life in heavier duty applications.

Facing: Generating a flat surface perpendicular to the axis of rotation by machining.

FAME: A gas chromatographic (GC) methodology for analyzing fatty acid methyl esters (FAME) extracted from microbial isolates, the profile of which may be used to identify microbial species.

Fat: An animal or vegetable-derived additive comprised of glycerol esters of fatty acids. Examples are lard oil, tallow, castor oil, etc.

Fatigue: A mode of metal failure that occurs under cyclic loading below a calculated yield strength. Components that fail by fatigue usually show a "beachmark" pattern on the fracture surface representing a progressive propagation of the crack under each stress cycle.

Fatty acids: A family of organic acids, so-called because some of these chemicals occur in animal fats. Fatty acids are used in many products, including metalworking fluids, soaps, cosmetics, personal care, and household products as corrosion inhibitors, emulsifiers, and lubricants.
Feed: The rate at which the grinding wheel or cutting tool is moved along or into the workpiece.
Feed rate (drilling): The distance the tool moves per revolution.
Feed rate (milling): The maximum thickness of material removed per tooth.
Feed rate (turning): The distance the tool moves per revolution of the workpiece.
Ferrite: The allotrophic form of iron that exists below 1670°F. Ferrite is essentially pure iron.
Ferrous: Refers to iron. Steel, cast iron, and other iron-based alloys are called "ferrous metals".
FEV_1: A measure of pulmonary function defined as forced expiratory volume in 1 sec.
FIFRA: Federal Insecticide, Fungicide, and Rodenticide Act (U.S.).
Filter media: Any porous material (usually paper, cloth, or screen) that traps solids when a cutting fluid passes through.
Finish: Surface quality or appearance.
Finishing: Final machining cuts taken to obtain the desired accuracy and finish.
Fire point: The minimum temperature at which a sustained flame will burn for at least 5 sec, usually slightly higher than the flash point.
Flammable: A term used to indicate that a material will readily ignite. OSHA defines a flammable liquid as "...any liquid having a flashpoint below 100°F (37.8°C)...". A "flammable" material has a lower flash point than a "combustible" one; see *OSHA* 29 CFR 1910.106 (a) (19) for classes of flammable liquids.
Flank: Surface of a cutting tool adjacent to the freshly machined workpiece surface.
Flash point: The minimum temperature at which a brief ignition of vapors is first detected.
Fluidized bed reactor: Involves a process of passing of a liquid upward through a bed of granular media at a velocity sufficient to expand the media and hydraulically suspend the individual media particles providing a large surface area.
Fly cutter: A single-tooth milling cutter.
Foam: Gas dispersed in a liquid causing an increase in the volume of the liquid. Usually seen as bubbles on the surface of the liquid which may break quickly or be quite stable.
FOG: In waste treatment, the content of fat, oil, and grease.
Folliculitis: Inflammation of the hair follicles. A particular type of skin irritation often caused by cutting oils ("oil boils"). Usually occurs on the forearms and thighs where oil-soaked clothing comes in contact with the skin. The cutting oil and grime from the work combine to plug the pores, resulting in inflammation, open sores, and often an acne condition.
Forging: The shaping of metal through the use of impact strikes or pressure to plastically deform the material.
Fracture: Failure, breakage, or fragmentation of a specimen or workpiece.
Free oil: The amount of oil in a metalworking fluid mix that easily separates and floats on the top. Typically it is tramp oil.
Fungi: (Plural form of fungus) Aerobic microorganisms consisting of either single-celled yeast, filamentous molds, or more complex structures such as mushrooms. Fungi have a true cell wall and membrane-enclosed internal structures called organelles. Fungi are one of four kingdoms in the domain eukaryota.
G-ratio: A measure of grinding performance defined as the volume of metal removed divided by the volume of grinding wheel worn away in the process.
Galling: Surface damages on moving metal surfaces caused by localized welding of high points.
Galvanic corrosion: A common type of corrosion process in which a potential difference through an electrolyte causes surface attack at the interface of two dissimilar metals.
Grains: The individual crystallites that make up the structure of a polycrystalline body.
Gram (g): Metric unit of mass. There are 453.59 g in one pound (avoirdupois U.S. or British); 1000 in 1 kg.

Gram stain: Bacteria may be categorized by a staining technique developed by H. C. J. Gram in 1884. If their cell walls retain the dye, the organism is called "Gram positive"; if they do not, "Gram negative".

Gray iron: A form of cast iron where the free carbon is present in the form of graphite flakes. A freshly fractured surface has a dull gray appearance.

Grit size: Expresses particle size distribution as defined by ANSI. The number relates to the number of holes along a linear inch of sieve screen. Thus, as the number increases, the particles are smaller.

Gun drill: A long drill with passages for coolant, used for deep holes.

Half-Life: The time required for the concentration of a material in a fluid to decrease to half of its original concentration.

Hard turning: High speed finishing process used with steels above 60HRC hardness; speeds are two to four times faster than normal.

Hardenability: The relative ease that a martensitic transformation can be achieved in steel. One measure of hardenability is the cooling rate necessary to achieve martensitic transformation. High hardenability steels may be air hardened while low hardenability steels required water quenching.

Hardness (metals): The property of a metal to resist permanent indentation.

Hardness (water): The combined calcium and magnesium content of water. Usually expressed as parts per million (ppm) of calcium carbonate ($CaCO_3$).

Heavy metals: Metals such as chromium, mercury, copper, nickel, zinc, lead, tin, and silver. These are generally toxic in dissolved form and cause waste disposal problems.

Hematopoietic toxin: A toxin that affects the production and development of blood cells, or attacks and destroys blood cells.

HEPA filter: High Efficiency Particulate Air filters.

Hepatotoxin: A toxin that attacks liver cells.

High-speed Steel (HSS): Tool steels containing tungsten, molybdenum, vanadium, cobalt, and other elements; first used in machining at high speeds.

HLB: Hydrophilic–lipophilic balance. A numbering system describing the oil-solubility/water-solubility of emulsion components. Low values (< 10) indicate oil solubility while higher numbers (> 10) indicate water solubility.

HMIS: Hazardous Materials Identification System created by the National Paint and Coatings Association (U.S.). The system uses a horizontal bar-shaped, color coded symbol to indicate health, fire, and reactivity information about a chemical.

Hobbing: An operation in which the hobbing tool and the workpiece rotate in a precise relationship to create worm, spur, and helical gears and splines. The hobbing tool has teeth arranged around its circumference in a helical pattern.

Honey oil: A high viscosity lubricant used for difficult, high stress forming operations on heavy gauge metals.

Honing: A fine finishing process using bonded abrasive "stones" or sticks containing grit; an oscillating motion is used, creating a cross-hatch pattern.

Hormonesis: The phenomenon of apparent growth stimulation of an organism by sublethal doses of a biocide.

Hydrocylone: see *Cyclone filter*.

Hydrodynamic lubrication: Exists if the surfaces sliding over each other are separated by a coherent lubricating film of liquid.

Hydroforming: Formation of metal parts through the application of hydraulic pressure. A fluid replaces the punch in forcing the metal into the die cavity.

Hypersensitivity Pneumonitis (HP): Extrinsic allergic alveolitis; lung inflammation with asthma-like symptoms. A rare group of diseases characterized by recurrent difficult breathing, cough, and systemic signs such as muscular pain and fever. It is caused by repeated exposure and subsequent sensitization to various antigens.

IARC: International Agency for Research on Cancer.

ID grinding: An internal grinding operation performed to change the inner diameter (ID) of a cylindrical part. This is a relatively more difficult operation than surface grinding or OD grinding due to a longer arc of contact.

ILMA: Independent Lubricant Manufacturers Association.

Indexable inserts: Cutting tools of various shapes and with multiple cutting edges, usually made of carbides. When one edge becomes dull, the insert may be rotated to expose another.

Influent: Waste water entering a waste treatment process.

Inorganic chemical: A chemical compound that does not contain carbon and hydrogen bonded together. Examples are metal salts, hydroxides, strong acids and bases, and mined minerals.

Interstitial: A site within a crystal structure where a small void exists. Elements that have a small atomic diameter such as carbon and nitrogen can occupy these sites. These elements are called interstitial alloying elements.

***In vitro*:** Experiments with cells or tissues from living organisms conducted outside of the organisms, generally in a test tube or petri dish.

***In vivo*:** Experiments conducted with live animals.

Ion: Molecular species with electron excesses (anions) or electron deficiencies (cations).

IP: Institute of Petroleum, London.

Ironing: A drawing process in which the thickness of the sidewall of the part is reduced.

Irritant: A chemical that, while not corrosive, causes a reversible inflammatory effect on living tissue.

ISO: Abbreviation for International Standards Organization located in Geneva, Switzerland.

Kinematic viscosity: A function of both internal friction (viscosity) and density of the fluid.

Knurling: Rolling depressions, such as a crosshatch pattern, into a knob or handle to provide a better gripping surface.

Kurtosis: The characteristic sharpness of the peaks and valleys of a surface finish. Sharp peaks and valleys have kurtosis greater than three while broad peaks and valleys have kurtosis less than three.

Land: A straight section of a cutting tool behind the cutting edge.

Lapping: A fine finishing operation where an abrasive slurry is introduced between the workpiece and a lapping tool made of a soft material such as cast iron.

Lay: The predominant directional characteristic of the surface roughness of a surface.

LCL: The lower control limit in statistical process control (SPC).

LD_{50}: Acute or single dose of a substance producing death in 50% of the test animals within 14 days of exposure.

Lubricity: The ability of a fluid to lubricate a process.

Machinability: The capability of a material to be machined with relative ease with regard to tool life, surface finish, and power consumption.

Magnetic cutting fluid: A magnetic field is used to cause these ferromagnetic fluids to penetrate into the cut zone. Recommended for special environments such as high altitude, high vacuum, or low gravity.

Make-up: Fluid added to a system to bring it back to full volume. Make-up should be a mixture of water plus product concentrate and not water alone.

Manifest: Detailed list of contents for a shipment.

Martensite: A metastable body centered tetragonal crystal structure existing in iron and steel and resulting from nonequilibrium transformation of austenite. It is the hard strong, heat treated form of steel. Similar transformations also occur in other, nonferrous alloy systems.

Mass Median Diameter (MMD): The particulate diameter at which half of the aerosol mass is on larger droplets and half on smaller.

Metal forming: An operation designed to alter the shape of metal without producing chips.

Metalworking: A broad term used to refer to the shaping of metal by cutting, grinding, bending, stretching, or stamping.

Metalworking Fluid (MWF): A liquid used to cool and/or lubricate the process of shaping a piece of metal into a useful object. The term most often refers to a water-based fluid.

Micelle: A cluster of surfactant molecules in solution held together by Van der Waals forces.

Microbicides: Active substances used to protect water-mixed metalworking fluids against microbial attack. This broad term includes both bactericides and fungicides.

Microemulsion: An emulsion of oil in water with emulsion particle size so small that the emulsion appears translucent to transparent.

Micron: A measure of length, one millionth of a meter.

Microorganisms: Minute living things, generally including bacteria, yeast, molds, algae, protozoa, rickettsia, and viruses.

Milliliter (ml): Metric unit of liquid measure. There are 946.3 ml in one quart (U.S.) and 1000 ml in a liter.

Millimeter (mm): Metric unit of length, one thousandth of a meter.

Milling: Removing material by moving the workpiece past a rotating, multitooth cutting tool.

Mineralization: The conversion of an organic molecule to carbon dioxide.

Minimum Quantity Lubrication (MQL): see *Near-dry machining*.

Miscible: The ability of a substance to mix uniformly in another.

Mist application: Application in which the metalworking fluid is propelled in a stream of compressed air into the area where work is being performed.

Mix stability: Lack of any separation (oil, scum, or sediment) in a metalworking fluid dilution.

Model: A mathematical relationship that describes the behavior of a process.

Modulus of elasticity: A measure of stiffness of a material expressed as the ratio of stress divided by strain. A material with a high modulus will act more stiffly (or deflect less under the same load) than a material with a low modulus.

Mold: Filamentous microorganisms, composed of many cells, which may grow in metalworking fluids interfering with filtration and clogging pipelines; see *Fungi*.

Molecule: The smallest unit of matter retaining all the properties of the original substance. Composed of one or more atoms.

MRF: Metal removal fluid; fluid applied to facilitate cutting or grinding metal.

MSDS: Material safety data sheet. This is a form containing safety and regulatory information regarding any chemical.

Mutagen: A substance capable of altering the genetic material in a living cell.

Mutualism: A relationship between two or more organisms wherein all benefit.

MWF: see *Metal working fluid* above.

Mycobacteria: A Gram-positive, relatively slow growing bacterium having a rough, waxy cell wall that is high in mycolic acid content. The cell wall is relatively impermeable to various basic dyes, unless the dyes are combined with phenol. However, once stained the Mycobacteria resist decolorization with acidified organic solvents, giving rise to the description of the organism as being "acid-fast".

Mycotoxins: Materials excreted by fungi that are chemically diverse and can cause symptoms ranging from hallucinations to cancer.

Naphthenic oil: A petroleum oil with a significant content of hydrocarbons with saturated, ring-type structures. These oils are more easily emulsified but are more readily oxidized or degraded than paraffinic oils. Classification of an oil as either naphthenic or paraffinic is based upon a viscosity–gravity constant (VGC) determination.

Near-dry machining: Machining with lubricant (often a vegetable oil) applied sparingly as droplets mixed with a flow of air. Chips and parts come away almost dry to the touch. There is no excess lubricant to recover, recirculate, or recycle.

Neat oil: Hydrocarbon oil with or without additives, used undiluted.

Nephrotoxin: A toxin that destroys renal (kidney) cells.

Neurotoxin: A toxin that attacks nerve cells.

NFPA: National Fire Protection Association (U.S.). This hazard warning system uses a diamond-shaped, color-coded symbol and numbers to indicate health, fire, reactivity, and special properties information about a chemical.

NIOSH: National Institute of Occupational Safety and Health, U.S.

NIST: National Institute of Standards and Technology (U.S.).

Nitriding: A heat treat process that enriches the surface of a steel with nitrogen to form iron nitrides. The resulting case is hard as formed and does not require quenching.

NOEL: No observed effect limit. The highest dose used in a toxicity test which produces no observed adverse effects.

Noncommunicable: A disease or condition not readily transmittable between persons.

Nonferrous: Any metal that is not based on iron, such as copper and aluminum.

Nonionic surfactant: A surface active agent that has no ionic character, carries no electrical charge in solution. In metalworking fluids they typically function as cleaners or lubricants.

Normalizing: A conditioning heat treatment of iron and steel where the material is heated into the austenitic crystal structure and allowed to cool naturally in the open air.

NTP: National Toxicology Program, U.S.

OD grinding: An external grinding operation to reduce the outer diameter (OD) of a cylindrical part.

OEL: Occupational exposure limit.

Oil emulsification: The property of a metalworking fluid that determines its capacity for emulsifying or dispersing oil, typically tramp oil.

Oleophilic: Is the "oil loving" characteristic that polypropylene (and certain other materials) exhibit, making them useful for scavenging tramp oils from metalworking fluid mixes.

Organic compound: Substances containing the element carbon; most contain hydrogen, and many contain oxygen, nitrogen, or other elements as well. Simple oxides of carbon are excluded.

Orthogonal: Mutually perpendicular.

OSHA: Occupational Safety and Health Administration, U.S.

Oxic: Well oxygenated conditions.

Oxidation: A reaction that removes electrons from the substance being oxidized.

Paraffinic oil: A petroleum oil in which the majority of the molecules are saturated straight or branched chain hydrocarbons. These oils are more plentiful than naphthenic oils and are more stable, but they are also more difficult to emulsify. Classification of an oil as either naphthenic or paraffinic is based upon a viscosity–gravity constant (VGC) determination.

Passivation: The formation of a tenacious, protective oxide film on the surface of a metal.

PCR: Polymerase chain reaction; a technique for characterizing microbial communities based upon their DNA.

Pearlite: A macroconstituent in iron and steel composed of alternate layers of ferrite and cementite in a lamellar structure.

PEL: Permissible exposure limit.

pH: The negative log of the hydrogen ion concentration of an aqueous solution. Can be expressed on a scale of 1 (acid) to 14 (alkaline). Pure water has a pH of 7.

Phosphate: A salt resulting from neutralization of phosphoric acid.

Pigment: A term used to denote the solid lubricants sometimes used in metalforming lubricants. Examples include talc, mica, and graphite.

PMN: Pre-manufacture notification to EPA (U.S.).

Planktonic: Microbial cells floating freely within a fluid system.

Polar additive: A molecule with positive and negative charges isolated to different portions of the molecule. Many emulsifiers and lubricants are polar in nature. Polarity aids in emulsion formation and causes attraction to the metal surface.

Polyelectrolyte: An organic polymer containing multiple, electrically charged sites. Useful in waste treatment.

Polymer: A high molecular weight chemical compound consisting essentially of repeating structural units.

Positive filter: A type of filter using some type of filtering material (paper, cloth, wire screen, etc.) to remove particulate from a metalworking fluid.

POTW: Publicly owed treatment works (sewage).

PPM: Parts per million, a unit of concentration.

Precipitate: A reaction causing a soluble compound to form an insoluble compound and settle to the bottom of the liquid. Also, the product of this process.

Precipitation hardening: Also called age hardening, involving the precipitation of submicroscopic intermetallic compounds from a supersaturated solid solution to strengthen the alloy.

Prokaryote: The domain of single cell organisms that lack a membrane enclosed nucleus. Bacteria and archaea are prokaryotic organisms.

Punch: A term generally used for the male part of a tooling set used in metal forming.

Pyrophoric: A substance that burns spontaneously in air, at or below room temperatures, and in the absence of added heat.

Q-prime (Q′): Specific metal removal rate; the volume of metal removed per unit of time per unit of effective grinding wheel width.

Quench: Rapid cooling of heated metal for the purpose of imparting certain properties, especially hardness. Quenchants may be water, oil, fused salts, air, and molten lead.

Ra: The average surface roughness computed as the arithmetic mean of the absolute value of the distance between the baseline to the maximum peak or valley height.

Rake angle: The angle between the tool face and an imaginary line perpendicular to the freshly cut surface.

Rake face: The surface of the cutting tool that is in contact with the metal chip being removed from the workpiece during machining.

Rake, negative: The face of the tool along which the chip travels is inclined forward from the cutting edge forming a greater than 90° angle with the freshly cut surface. Such a rake is useful in machining high strength materials in order to reduce chipping of the tool.

Rake, positive: The face of the tool along which the chip travels is inclined backward from the cutting edge, forming a less than 90° angle with the freshly cut surface.

Rancidity: The condition in which a metalworking fluid develops a foul odor. Typically caused by high levels of anaerobic bacteria.

RCRA: Resource Conservation and Recovery Act, administered by the EPA.

Ream: A cutting operation used to enlarge and finish a previously formed hole. The reaming tool consists of one or more cutting elements arranged along the longitudinal axis.

Rebinder effect: Modification of the mechanical properties at or near the surface of a solid, attributable to interaction with a surfactant.

Recycling: The process which is used to clean and restore used metalworking fluid mixes for reuse.

Refractometer: An optical instrument that measures the refractive index of a metalworking fluid. Used to determine concentration.

REL: Recommended exposure limit.

Residue: That part of a metalworking fluid mix which is left after the evaporation of the water.

Reverse Osmosis (RO): A separation process similar to ultrafiltration, but using higher pressures and tighter semipermeable membranes. Application of pressure reverses the natural process of osmosis causing water to flow from the more concentrated to the more dilute solution side of the membrane.

RMS: Root mean square; a numerical expression of surface roughness.

Rockwell: An indentation type hardness test that uses a variety of indentors and loads. The hardness number is based on the depth of penetration of the indentor.

Roughing: Machining without consideration of surface finish.

Roughness: Fine irregularities on a surface measured in terms of height and spacing.

RPM: Revolutions per minute; a means of expressing rotational speed.

RTECS: Registry of toxic effects of chemical substances, published by NIOSH.

Rust: Hydrated oxides of iron.

Saponification number: A measure of the fat or ester content of a material. Expressed as milligrams of potassium hydroxide required to hydrolyze 1 g of sample.

SARA: Superfund Amendments and Reauthorization Act, U.S.

Semisynthetic fluid: A metalworking fluid concentrate with moderate to low content of mineral oil, usually 5 to 30%. Generally contains a significant amount of water, 30 to 60%. These are sometimes called "preformed emulsions".

Sensitization: In metallurgy, the precipitation of chromium carbides at the grain boundaries of stainless steel caused by heating the material in the temperature range of 1100 to 1500°F. The loss of chromium from the matrix of the steel causes a marked reduction of corrosion resistance.

Sensitizer: A chemical that causes an allergic reaction in people or animals after repeated exposure to the chemical.

Sessile: Microbial cells growing attached to a surface.

SFPM: Surface Feet Per Minute; a means of expressing travel speed in cutting or grinding operations that is more meaningful than RPM.

Shank: That portion of a tool by which it is held, such as the shank of a drill held in a chuck.

Shear plane: A narrow zone along which shearing takes place in metal cutting.

Sintered carbide: see *Cemented carbide*.

Skewness: The relative comparison of peaks and valleys of a surface; a measure of the symmetry of the profile about the mean line. If the peaks are generally higher than the valleys are deep the surface has positive skew. If the valleys are deeper than the peaks are high the surface has negative skew.

SME: Society of Manufacturing Engineers, Dearborn, MI.

Snagging: Rough grinding to remove unwanted material from casting.

Soap: An emulsifier or lubricant prepared by neutralizing a fatty acid with an alkaline material such as an amine, sodium hydroxide, or potassium hydroxide.

Solid-film lubricant: Solid lubricants, such as graphite or molybdenum disulfide.

Soluble oil: A metalworking fluid concentrate with high oil content (50 to 80%) and little or no water content. As sold it consists solely of oil, emulsifiers, and oil soluble lubricants, corrosion inhibitors, etc. When mixed with water it creates an emulsion that is milky in appearance.

Solution annealed: A precursor heat treatment applied to precipitation hardening alloys to form the supersaturated crystal structure from which the strengthening precipitates can form during aging.

Spark-out: A term used in grinding to identify the final passes of the grinding wheel over the part surface, without further feeding of the wheel. Spark-out assures that the advancement of the wheel is complete, which helps to obtain proper part geometry, size, and surface finish.

SPC: Statistical Process Control, a scientific means of monitoring process variability, continuous process improvement; a quality management tool.

Specific energy (U): In grinding, the power required to remove one unit volume of material per unit of time.

Specific gravity: The ratio of the mass of any volume of material to the mass of an equal volume of some reference material, usually water, at a standard temperature. Also called the "relative density".

Specific metal removal rate (Q'): The volume of metal removed per unit of time per unit of effective grinding wheel width.

Sperm oil: Ester-type boundary lubricant from the head cavity of the sperm whale. Since 1971 this material has not been available in the U.S. because of endangerment to the whale population.

Spindle: Rotating shaft for tool holders.

Stainless steel: An alloy of iron containing at least 11% chromium and sometimes nickel, that resists almost all forms of rusting and corrosion.

Stamping: A variety of operations in which a part is formed from a flat strip or sheet stock through the use of a forming die set.

Steel: An iron-base alloy.

STEL: Short Term Exposure Limit for a material, usually over a 15 min time frame.

STLE: Society of Tribologists and Lubrication Engineers; formerly ASLE, American Society of Lubrication Engineers.

Strain: The elongation or stretching a body exhibits in response to an application of pressure.

Stratum corneum: The outer-most layer of the epidermis; the skin's principal physical barrier against penetration by chemical substances and microorganisms.
Stress: The load applied to a body projected over the area on which it acts. Stress has the dimensional units of pressure.
Subchronic health effect: Resulting from repeated daily exposure of experimental animals to a chemical for part (approximately 10%) of their life span.
Sulfate: A salt resulting from neutralization of sulfuric acid.
Sulfonate: Typically sodium petroleum sulfonate, although calcium sulfonates are also widely used. These materials are used as emulsifiers, corrosion inhibitors, lubricants, and even demulsifiers. Originally a by-product of white oil production, sulfonates may also be prepared from synthetic alkylates. Molecular weights ranging from 380 to 540 are generally most useful in metalworking. Other applications for sulfonates are in detergents, cleaning agents, and lubricating oils.
Surface finish (or roughness): Fine irregularities measured in terms of height and spacing.
Surface tension: An inward pull or internal pressure at the surface of a liquid that tends to restrict fluid flow. Polar liquids have high surface tension, while nonpolar liquids have low surface tension. Pure water has a surface tension of 73 dynes/cm at 20°C; addition of detergents will reduce surface tension.
Surfactant: A compound that reduces the surface tension of water, or the interfacial tension between two liquids or between a liquid and a solid. A surface active agent.
SUS: Units of viscosity (Saybolt Universal Seconds).
Swaging: The squeezing or compressing of metal bar stock or tubing so as to create a taper or point.
Swarf: Metal fines and grinding wheel particles generated during grinding.
Synthetic fluid: A metalworking fluid that contains no mineral oil. Some synthetics are totally water soluble (chemical solutions) while others are emulsions of water insoluble, synthetically derived lubricants (synthetic emulsions).
TAN: Total Acid Number; see *Acid number*.
Tapping: A method used to cut or form threads inside of a predrilled hole.
TBN: Total Base Number, expressed as mg of KOH per gram of sample.
TDS: Total Dissolved Solids.
Temper: Condition of an alloy, such as annealed, cold-worked, heat-treated.
Tempering: A post quenching heat treatment that occurs when the metal is reheated to an elevated temperature but below the austentite transformation temperature. The mechanism of tempering involves the rejection of carbon from the martensite lattice.
Teratogen: A substance that causes birth defects in the fetus when the mother is exposed.
Threading: To form a screw thread on the outer diameter of a cylindrical object, such as a pipe.
Titration: A procedure for determining volumetrically the concentration of a certain substance in a solution by adding a standard solution of known volume and strength until the reaction is complete. The endpoint may be detected by various means, such as color change of an indicator solution or by an electrical measurement.
TLV: Threshold Limit Value. The limit of exposure to a material at or below which workers should experience no health problems.
TOC: Total Organic Carbon.
Tolerance: Permissible variation in the dimensions of a part.
Tool life: A measure of the length of time a tool will cut satisfactorily.
Torque: The tendency to produce rotation.
Total oil: Is the percentage of all oil or oil-like material present in the metalworking fluid mix. This value includes both product oil and extraneous oil.
Toxic: A term used to describe a chemical agent that has a harmful effect on a biological or physiological system.
Toxigenic: A bacterium or fungus that produces and excretes poisonous molecules (toxins).

Toxicology: The study of the health effects of chemicals.

Tramp oil: That oil which is present in a metalworking fluid mix and is NOT from the product concentrate. The usual sources are machine tool lubrication systems and leaks.

Trepanning: Uses a single point tool to produce a hole by cutting a circumferential groove in the metal and leaving a solid core.

Tribology: The study of interacting surfaces in relative motion, which encompasses the aspects of friction, lubrication, and wear. Derived from the Greek word, "tribos," for rubbing.

TSCA: Toxic Substances Control Act (U.S.).

TSS: Total Suspended Solids.

TTO: Total Toxic Organics.

Turning: Machining on a lathe or turning center with single-point cutting tools.

TWA: Time Weighted Average exposure to a substance during a typical workday.

UCL: The Upper Control Limit in statistical process control (SPC).

Ultrafiltration (UF): A separation technique in which a liquid is applied to a semipermeable membrane under moderate pressure. The liquid passing through the membrane (the "permeate") is cleaned and the waste stream (or "effluent") becomes more concentrated with dissolved solids. Pore sizes typically range from 0.02 to 0.07 μm, and pressures are typically 30 to 70 psig.

UN number: An identification number assigned by the United Nations to hazardous materials in transportation.

Up milling: Opposite of climb (down) milling. The cutting tool rotates against the direction in which the workpiece is fed; the chip starts out thin and gets thicker as the cut progresses.

Vanishing oil: An evaporative type of solvent-based lubricant used to stamp or draw parts that will not be washed.

Vesicles: Blisters.

Vickers: An indentation-type hardness test that uses a pyramidal shaped diamond indentor and a range of loads. The diagonals of the indentation are measured and the hardness number calculated as a function of load divided by projected area.

Viscosity: The internal resistance to flow exhibited by a fluid, a property that varies with temperature.

Viscosity index: A means of expressing the relationship between viscosity and temperature.

VOC: Volatile Organic Compounds.

Waste treatable: The ability to remove an additive from water before disposal.

Water hardness: The combined calcium and magnesium content of water. Usually expressed as parts per million (ppm) of calcium carbonate ($CaCO_3$).

Waviness: A surface feature of a workpiece involving regular, long range deviations in the plane of the surface. Waviness is usually due to tool chatter, spindle eccentricity or other unintended motions of the cutting tool or machinery.

Wear: Loss of material from a surface due to rubbing.

Wedge-wire filter: A filtration media composed of wire that has been drawn in a triangular cross-sectional profile. These wires are spot-welded at their peaks onto support rods, forming either a broad flat area for filtration, or in the shape of a cylinder or "drum." Fluid passes through the narrow spaces between the "flats" of the wires, and the dirt collects on the flats.

Wetting agent: An additive which reduces surface and interfacial tension of a fluid and facilitates spreading of the fluid over a surface.

White iron: A form of cast iron that has no free or uncombined carbon present. All of the carbon is present as carbides. A freshly fractured surface has a sparkly, white appearance.

WHMIS: Workplace Hazardous Materials Information System (Canada). A form containing safety and regulatory information regarding any chemical, similar to the MSDS in the U.S.

Wire drawing: Pulling a metal rod through one or a series of dies to cause elongation and a reduction in diameter.

Workpiece: A piece of material to be machined.

Wrought: Hot or cold worked metal alloy.

Yeast: Mostly single celled fungi. These are larger than bacteria, and either round or oval in shape.

Yield strength: The stress level required to cause a measurable permanent distortion in a material. This is usually the basis (with appropriate safety factors applied) of the maximum allowable design stress for functional components that suffer bending.

Zeta potential: An electrokinetic potential exists between oil droplets in an emulsion and the surrounding liquid (water) in which they are suspended. When an electrically charged field is applied, a charged particle will be attracted to one of the poles. The zeta potential can be measured by monitoring the movement of a particle through a microscope as it migrates in the voltage field. A measure of emulsion stability.

ZDDP: Stands for zinc dialkyldithiophosphate, an additive used to impart antiwear and/or antioxidant properties in many lubricating oils.

Index

B

C

D

E

F

G

H

I

J

K

L

M

Q

R

S

X

Y

Z